Power Electronics and Power Systems

More information about this series at http://www.springer.com/series/6403

Matteo Meneghini · Gaudenzio Meneghesso
Enrico Zanoni
Editors

Power GaN Devices

Materials, Applications and Reliability

Springer

Editors
Matteo Meneghini
Department of Information Engineering
University of Padova
Padua
Italy

Enrico Zanoni
Department of Information Engineering
University of Padova
Padua
Italy

Gaudenzio Meneghesso
Department of Information Engineering
University of Padova
Padua
Italy

ISSN 2196-3185 ISSN 2196-3193 (electronic)
Power Electronics and Power Systems
ISBN 978-3-319-43197-0 ISBN 978-3-319-43199-4 (eBook)
DOI 10.1007/978-3-319-43199-4

Library of Congress Control Number: 2016947213

Printed on acid-free paper

This Springer imprint is published by Springer Nature
The registered company is Springer International Publishing AG Switzerland

Preface

Over the last few years, gallium nitride has emerged as an excellent material for the fabrication of power semiconductor devices. The high critical field (3.3 MV/cm) allows to fabricate transistors with breakdown voltage higher than 1 kV; further advantages originate from the high mobility of the 2-dimensional electron gas (2DEG), that allows to reach very low on-resistance values (<50 mΩ for a 30 A device). Finally, the low on-resistance × gate charge product ($R_{on} \times Q_g$) permits to achieve a significant reduction of the switching losses. Converters with efficiencies higher than 99 % have already been demonstrated based on GaN, and several products are already available on the market.

This book presents an exhaustive overview of the properties and issues of GaN-based devices, including application-oriented results. To put together this book, we have adopted a bottom-up approach, by including material-related aspects, a through description of the main device properties and issues, and a discussion of the system-level aspects.

In Chap. 1 (D. Ueda, Kyoto Institute of Technology) gives introductory design considerations, starting from a general introduction on the fundamental properties of GaN and related materials. Different device structures are introduced, including standard FETs and natural superjunction devices. A comparison with standard silicon devices is also presented.

Chapter 2 (S. Keller, University of California at Santa Barbara) deals with material- and substrate-related issues. After an overview of the various substrates that can be adopted for the growth of GaN, the details of metal-organic chemical vapor deposition are described. The chapter then describes the fabrication of semi-insulating AlGaN layers, the issues related to doping, and the fabrication of heterostructures for lateral and vertical devices.

Chapter 3 (D. Marcon and S. Stoffels, IMEC) describes a GaN-on-Silicon CMOS compatible process. After a general description of GaN-on-Si epitaxy, the details of GaN-on-Si gold-free processing are described in detail. Details on AlGaN barrier recess, ohmic alloy optimization and gallium contamination issues are also given. Experimental data are widely used to support the discussion.

In Chap. 4, U. Mishra and M. Guidry address the material aspects and the various device designs that are of value in different applications of lateral GaN devices for power applications. After describing the history of AlGaN/GaN HEMTs, the operating principles of N-polar and Ga-polar devices are presented. In addition, the main issues related to the use of GaN lateral devices in power converters are discussed, along with the existing technological solutions.

Chapter 5 (S. Chowdury, UC Davis) is devoted to vertical power devices, from materials to applications. After an introductory discussion on the potential applications, vertical devices are compared to lateral devices in terms of performance and structure. The Current Aperture Vertical Electron Transistor (CAVET) is used as a representative device to discuss various concepts in this chapter. Design considerations and structure-related issues are extensively described, to provide a comprehensive overview of the topic.

In Chap. 6 (E. Matioli et al., EPFL and MIT) the operating principles of GaN-based nanowire transistors are described. Experimental data and results from the literature are compared, to provide a complete overview of the topic. Bottom-up nanowires are reviewed in the first part of the chapter, while the second part covers the use of top-down nanowires for power electronic devices. Finally, the application of nanowires for RF applications is discussed.

Chapter 7 (A.M. Armstrong and R.J. Kaplar, Sandia National Laboratories) discusses the electrical and optical methods for the characterization of deep levels in GaN. After a brief introduction, the fundamentals of DLTS and DLOS are reviewed, along with the related theory. Then, the use of spectroscopic techniques for the analysis of GaN HEMTs is reviewed.

Chapter 8 (G. Curatola, Infineon, and G. Verzellesi, Univ. Modena and Reggio Emilia) describes an approach to the modelling of power GaN HEMTs, aimed at full system optimization through concurrent simulation of device, package, and application. Results from simulations are compared with experiments for both normally-on insulated-gate GaN HEMTs and normally-off p-GaN devices in real switching applications.

Chapter 9 (I. Rossetto et al., University of Padova) describes the properties of the defects that limit the performance of GaN-based transistors. The first part of the chapter summarizes the properties of the most common defects in GaN, by describing a database of defects based on more than 80 papers on the topic. The second part of the chapter describes the results of recent experiments on the impact of common native defects (vacancies, surface states, …) and impurities (such as Fe and C) on the dynamic performance of GaN HEMTs.

Chapter 10 (P. Parikh, Transphorm) deals with cascode gallium nitride HEMTs on Silicon. After discussing the motivation and configuration of the cascode GaN HEMT, the key applications and performance advantages of cascode GaN HEMTs are extensively described. An overview of the commercially available products and the related benefits is also given in the chapter. Finally, the qualification and reliability of cascode GaN HEMTs is reviewed.

In Chap. 11, T. Ueda (Panasonic) describes the main structure and issues of gate injection transistors, and the various technologies to improve the performance of these devices. Status of the reliability including current collapse is summarized, followed by results of the application of the GIT to practical power switching circuits targeting at high efficiencies. Future technologies to improve the performances are also described.

A further strategy for normally-off operation is presented in Chap. 12 (K. Chen, Hong Kong University of Science and Technology). Fluorine implanted enhancement-mode transistors are described: a comprehensive discussion on the underlying physical mechanisms of the fluorine implantation is presented, including atomistic simulation and experimental studies. Further development of the F-implant technique and its integration with other advanced techniques such as gate recess and AlN passivation is described. Finally, the robustness of the F-implant technique is summarized.

In Chap. 13, J. Wuerfl (FBH, Berlin) describes the most relevant drift effects that limit the performance of GaN-based power transistors. After an introduction of the physical mechanisms responsible for drift effects, the most important drift phenomena observed in GaN power switching transistors and their influence on device performance are described. In addition, the chapter discusses proven technological concepts to minimize device drift.

The reliability of 650 V GaN power devices is described in Chap. 14 (P. Moens et al., ON Semiconductor). Gate Ohmic contact reliability, gate dielectric reliability and buffer stack reliability are extensively investigated. The need to do reliability investigations based on statistical data on large area power transistors (100^{+}mm gate width) instead of small test structures is emphasized. Acceleration models and statistical distribution models (Weibull) are also discussed.

Finally, Chap. 15 (F. Lee et al., Virgina Tech) reviews the system level issues related to the switching of GaN-based transistors. Specific issues related to cascoded GaN HEMTs are described. An evaluation is presented of the cascode GaN based on a buck converter in hard-switching and soft-switching modes, which shows the necessity of soft-switching for cascode GaN devices at high frequencies. Additionally, this chapter illustrates the use of GaN in a wide range of emerging applications.

The reader will be impressed by the incredible performance and potential of GaN electronics: as concluded by U.K. Mishra and M. Guidry in Chap. 4, *The market size for GaN-based electronics promises to be over $1B by 2022 and rapidly increases thereafter. This is alongside a photonics market of over $10B today with a steep growth trajectory. From an academic and research perspective it is remarkable that such a market is available with the limited science currently explored in the nitrides. With the full effort of the community and a more complete understanding of the science of the materials and devices the future will be even brighter.*

Padua, Italy

Matteo Meneghini
Gaudenzio Meneghesso
Enrico Zanoni

Contents

Chapter 1
Properties and Advantages of Gallium Nitride

Daisuke Ueda

1.1 General Background

Modern human life and industries are dependent on energy as a social infrastructure. Since much effort has been devoted to implement renewable resources, such as solar, wind, geothermal, and bio energies, a variety of electrical power conversion systems have come to be required. The renovation of traditional power systems by using wide bandgap semiconductors has become a challenging topic. Wide bandgap semiconductor will be a key to break through the tasks of blocking voltage and conversion efficiency. At this point of time, it will be meaningful to look back the evolution of Si power devices technology since they have been adjusting their structure to satisfy the application demands. I believe that the wide bandgap semiconductor may find a technological direction in their old footprint.

Early-stage semiconductor transistors, from a point-contact type to alloy one, soon targeted for the power applications. Therefore, technologies to increase current handling capability, breakdown voltage, and switching frequency have been a major concern from the beginning of semiconductor technology.

At present, power devices have been differentiated into three major application layers, which are depending on the required voltage range as shown in Fig. 1.1. Those layers, namely infrastructure, home, and personal ones, mean the application/market segments. Recently, a segment of low-voltage power application has emerged as a personal layer. This is originated to the expanding market of personal information systems represented by mobile communications. The supply voltage in this layer is approaching to 1 V, though extremely high current handling capability, e.g., order of 10 A, is required. One of those power conversion systems is known as point-of-load (POL), making growing application layer as low-voltage power one [1]. The emerging layer came from the trend of lowering the operation

D. Ueda (✉)
Kyoto Institute of Technology, Kyoto, Japan
e-mail: dueda@m.ieice.org

M. Meneghini et al. (eds.), *Power GaN Devices*,
Power Electronics and Power Systems, DOI 10.1007/978-3-319-43199-4_1

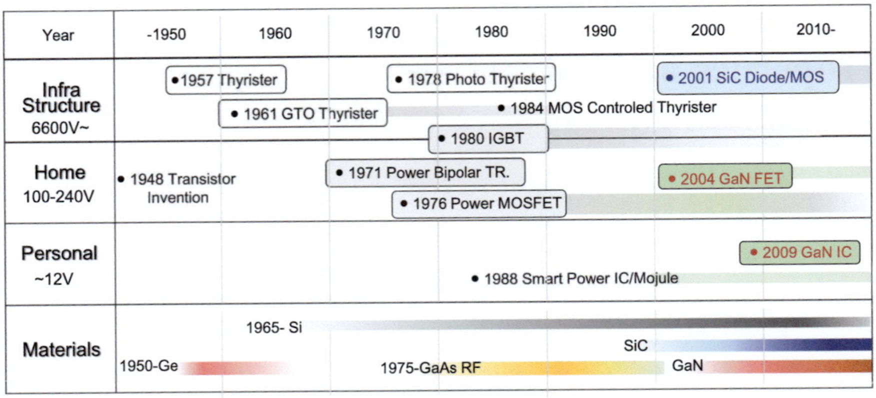

Fig. 1.1 Evolution of power semiconductor devices in infrastructure, home, and personal application layers

voltage of scaled-down CMOS. It is well known that recent system LSIs have hundreds of million transistors that consume extremely high current, though have poor breakdown voltage.

After explaining the basic material properties of III-nitride semiconductors, device design criteria depending on the application layers are introduced. Finally, future tasks of GaN-based power devices will be mentioned, considering Si evolutions.

1.2 GaN Material

GaN material's research originates from the achievements by Akasaki, Amano, and Nakamura, such as crystal growth, characterization, and fabrication process to realize blue LED [2]. The physical properties of GaN and related semiconductor materials are summarized in Table 1.1. Beside optical applications, nitride semiconductor materials have excellent electron transport properties, high breakdown voltage, and high thermal conductivity. In order to make use of these advantages, research and development started to realize groundbreaking electron devices.

In the meantime, wide spread of communication systems require to increase the power delivering capability in the microwave frequency range, where GaAs or InP devices had been playing major roles. Such situation accelerated the development of GaN-based microwave power devices, which demonstrated high output power with high gain in extremely small chip size.

It is necessary to shorten the carrier transit time τ between the spacing l for the high-frequency operation. On the contrary, applicable voltage becomes lowered due to the electric field increase between the electrodes, resulting in lowering the available power. This is a primary trade-off between breakdown voltage B_V and

Table 1.1 Material parameters of nitride semiconductors

Parameters	Symbol	Unit	Si	GaAs	AlN	InN	GaN	SiC	Diamond
Crystal structure	–	–	Diamond	Zincblend	Wurtzite	Wurtzite	Hexagonal, cubic	Hexagonal, cubic	Diamond
Density	–	g/cm^3	2.328	5.32	3.26	6.81	6.1	3.21	3.515
Mole	–	g/mol	28.086	144.64	40.9882	128.83	83.73	40.097	12.011
Atomic density	–	$atom/cm^3$	5.00E+22	4.42E+22	4.79E+22	3.18E+22	4.37E+22	4.80E+22	1.76E+23
Lattice constant	–	Å	5.43095	5.6533	3.1114/a, 4.9792/c	3.544/a, 5.718/c	Hex 3.189/a, 5.185/c, cubic 4.52	(6H)3.086/a, 15.117/c, (4H) 3.073/a, 10.053/c, (3C) 4.3596/a, 3.073/c	3.567
Melting point	–	°C	1415	1238	2200	1100	2573, @60 kbar	2830	4373, @125 kbar
Specific heat	–	J(g °C)	0.7	0.35	0.748	0.296	0.431	0.2	0.52
Linear thermal expansion coefficient	–	$°C^{-1}$	2.60E−06	6.90E−06	5.27E−6/a 4.15E−6/c	3.8E−6/a 2.9E−6/c	5.6E−6/a 3.2E−6/c	~5E−6	8.00E−07
Thermal conductivity	–	W/(cm °C)	1.5	0.46	2.85		2.1	2.3–4.9	6–20
Transition type	–	–	Indirect	Direct	Direct	Direct	Direct	Indirect	Indirect
Bandgap energy	Eg	eV	1.12	1.42	6.2	0.65	3.39(H)	3.02/6H 3.26/4H 2.403/3C	5.46–5.6
Separation energy	Γ-L, Γ-X	eV	Indirect	Γ-L0.29, Γ-X0.48s	Γ-ML 0.7 Γ-K 1.0	Γ-A > 0.7 Γ-Γ > 1.1 Γ-K > 2.7	Γ-Γ′1.9 Γ-M 2.1	Indirect	Indirect

(continued)

Table 1.1 (continued)

Parameters	Symbol	Unit	Si	GaAs	AlN	InN	GaN	SiC	Diamond
Dielectric constant	εr	–	11.7	12.9, 10.89@RF	8.5	15.3	12	10.0(6H) 9.7(4H)	5.7
Electron affinity	χ	eV	4.05	4.07	1.9	5.8	3.4	4	(NEA)
Intrinsic carrier density	ni	cm^{-3}	1.45E +10	1.79E+06	9.40E−34	9.20E+02	1.67E−10	1.16E−8(6H) 6.54E−7(4H)	1.00E−26
Effective density of states	Nc	cm^{-3}	2.80E +19	4.70E+17	4.10E+18	1.30E+18	2.24E+18	4.55E+19(6H) 1.35E+19(4H) 1.53E+19(3C)	1.00E+20
Effective density of states	N_V	cm^{-3}	1.04E +19	7.00E+18	2.84E+20	5.30E+19	1.16E+19	1.79E+19	1.00E+19
Effective mass	m^*_e	m^*el/m_0 m^*et/m_0	0.9163/l, 0.1905/t	0.067	0.4	0.1–0.05	0.2	1.5/l, 0.25/t	1.4/l, 0.36/t
Effective mass	m^*_h	m^*lh/m_0 m^*hh/m_0	0.16/l, 0.49/h	0.082/l, 0.45/h	0.6	1.65	0.6/h	0.8	0.7/l, 2.1/h
Electron mobility	μ_e	$cm^2/(V\ s)$	1500	8500	300	3200	1000	460–980	2200
Hole mobility	μ_h	$cm^2/(V\ s)$	450	400	14	220	~5	20	1800
Lattice matching	–	–	~SiGe	AlAs, InGaP	~GaN	–	~SiC, sapphire	~GaN, lN	–

cutoff frequency f_T, which is described by using critical electric field E_c and carrier saturation velocity v_{sat} in the spacing of electron path of l. The relationship is written as follows:

$$B_V = E_c \cdot l \tag{1.1}$$

Considering the transit time of electron in the path of l, cutoff frequency is described as,

$$f_T = \frac{1}{2\pi\tau} = \frac{v_{sat}}{2\pi l} \tag{1.2}$$

Then, the product of f_T, and B_V becomes constant independent from the spacing of l as

$$B_V \cdot f_T = \frac{E_c \cdot v_{sat}}{2\pi} \tag{1.3}$$

The present trade-off means that you have to reduce the breakdown voltage when you want to increase the high-frequency operation. The product is specifically determined by the material. This index is used as one of the FOM (figure of merit) of the materials, which is called Johnson's limit [3]. Wide bandgap materials have extremely high E_c since impact ionization would not take place till the hot carriers obtain the energy equivalent to the bandgap to generate electron–hole pairs. The impact ionization eventually causes the avalanche breakdown. The index of Si and GaN materials are approximately 300 and 90,000 GHz V, respectively.

Furthermore, saturation velocity of GaN materials is also estimated to be higher than conventional semiconductor materials, which is explained by the relatively large energy separation of electrons between the conduction band Γ and L valleys. The theoretically calculated separation energies are summarized in the Table 1.1. Relatively large energy separation of GaN-related materials suppresses intervalley transition under high electric field than the conventional compound semiconductors [4]. Consequently, wide bandgap devices can achieve both the high-power and high-frequency operation capabilities owing to their superior material characteristics.

Thus, GaN-based devices have desirable properties for the high-power amplification in microwave and millimeter-wave frequency ranges. Therefore, practical use of GaN-based microwave devices has started as PA (power amplifier) for base stations of mobile communication systems [5].

After demonstrating the high-frequency power devices, GaN-based devices became expected to renovate the traditional power systems to improve the conversion efficiency. Major motivation is to reduce the resistance loss by shortening the space between anode and cathode keeping even higher blocking voltage. Additionally, by using high switching capability of GaN, traditional power systems can be downsized owing to the miniaturization of passive components such as inductors, capacitors, and transformers. This means that the system cost will be lowered by using GaN-based devices. The inherent high temperature stability of wide bandgap materials is also advantageous to simplify the attached cooling system for the high-power dissipation systems.

1.3 Polarization Effect

Figure 1.2 shows the periodic table of III-V compound semiconductors, where you can see the combinations of III-N semiconductor such as GaN, AlN, and InN. III-N materials usually have crystal structure called wurtzite type, while typical GaAs and InP crystal families have structure called a zinc blende type [6]. The relationship of the bandgap and lattice constants for nitride materials are shown in Fig. 1.3.

Wurtzite-type GaN, AlN, and InN have a hexagonal crystal structure, where the axial direction of the hexagonal column is called as *c*-axis. GaN crystal has different faces perpendicular to *c*-axis as shown in Fig. 1.4. Focusing the weak breaking bonds in the figure, you can see the opposite bond direction from Ga to N atoms in the left and right figures. This means c-plane of GaN crystal has either Ga-face or N-face at the cut face, which leads the different electrical characteristics. It is reported that

Fig. 1.2 Periodic table to make compound semiconductors

	II	III	IV	V	VI
2	Be	B	C	N	O
3	Mg	Al	Si	P	S
4	Zn	Ga	Ge	As	Se
5	Cd	In	Sn	Sb	Te

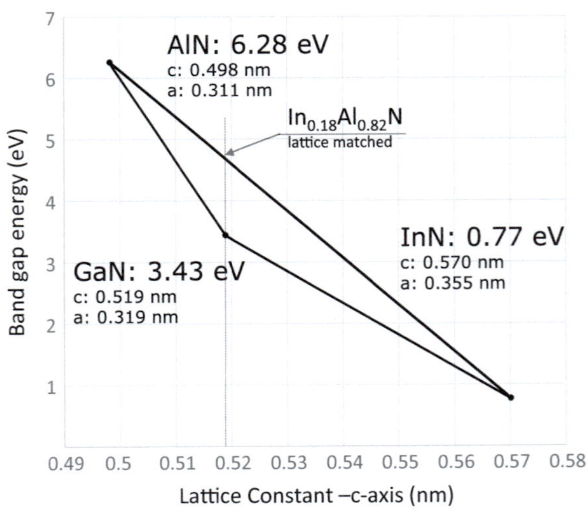

Fig. 1.3 Relationship of the bandgap energy and lattice constant for nitride-based semiconductors. Note that *c*-axis represented as lattice constant

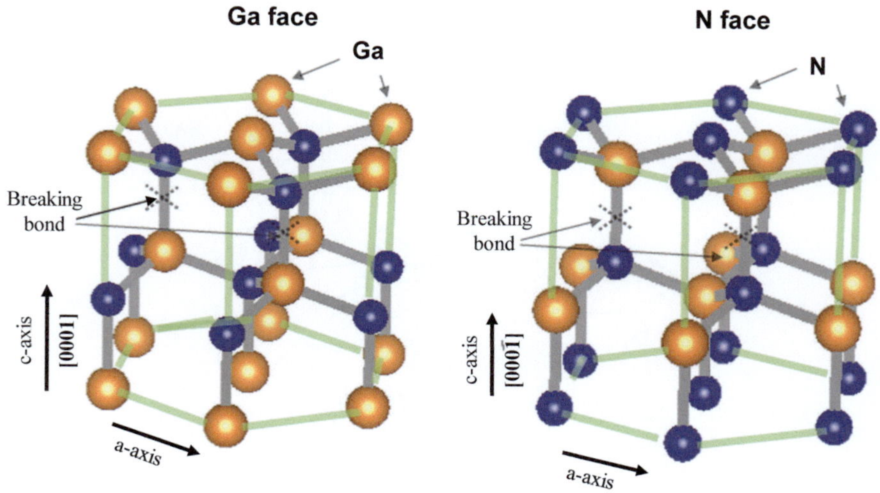

Fig. 1.4 Asymmetrical hexagonal GaN crystal structure along *c*-axis, where either Ga- or N-face appears at the surface of the crystal

Ga-face is usually obtained by MOCVD growth over the c-plane sapphire substrate, while N-face can be grown by MBE under some specific preparations [7].

Since nitrogen has higher electronegativity than gallium, Ga and N atoms have anionic (+) and cationic (−) characteristics, respectively, resulting in causing electric polarization. Though the polarization is canceled inside the material, the asymmetry at the cut face, there arises specific polarization along the *c*-axis, which is called as spontaneous polarization. The spontaneous polarization of III-N materials are reported to be 0.032, 0.029, and 0.081 C/m^2 for InN, GaN, and AlN, respectively, having the same direction from column III atom to nitrogen [8].

The mechanical stress inside the epitaxial layers with different lattice constants causes new type of polarization, which is referred to as piezoelectric polarization. Figure 1.5 explains how the piezoelectric polarization is caused in nitride semiconductors. As shown in the left figure, the resultant internal polarization vectors $P_1 + P_2 + P_3 + P_4$ become zero in a freestanding tetrahedral structure due to the crystal symmetry. However, when the crystal is deformed by the lattice mismatch, as shown in the right-hand side, the angle θ become widened when tensile stress is applied. Thus, the internal electric field becomes unbalanced to appear the piezoelectric field P_{PE} appears as $P_1 + P_2 + P_3 + P_4 = P_{PE}$. Such piezoelectric polarization was investigated in detail by Ambacher et al. [9]. Figure 1.6 shows theoretically calculated piezoelectric charge density of $Al_xGa_{(1-x)}N$, $In_xGa_{(1-x)}N$, and $Al_xIn_{(1-x)}N$ as a function of mole fraction x, assuming that all the materials are grown over GaN substrate. You can see the polarity of P_{PE} becomes either positive or negative corresponding to either compressive or tensile stress caused by the difference of lattice constant.

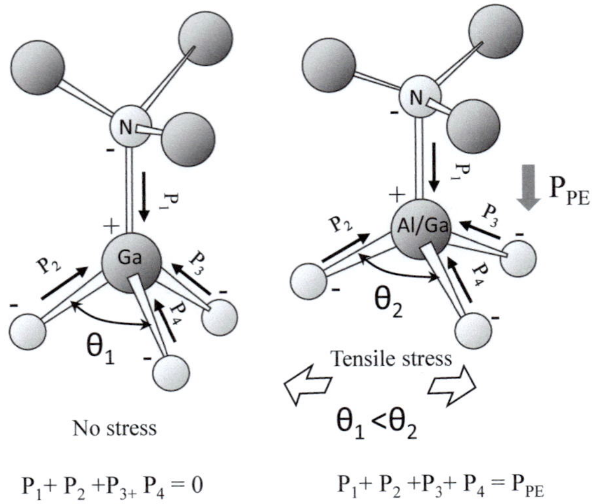

Fig. 1.5 The balanced polarization of ionic bonds in the tetrahedron shape (*left*) cause the unbalance of polarization field under the stress originated to the difference of lattice constant of underlying material. The resultant electric field is called piezo electric one, shown as E_p in the figure (*right*)

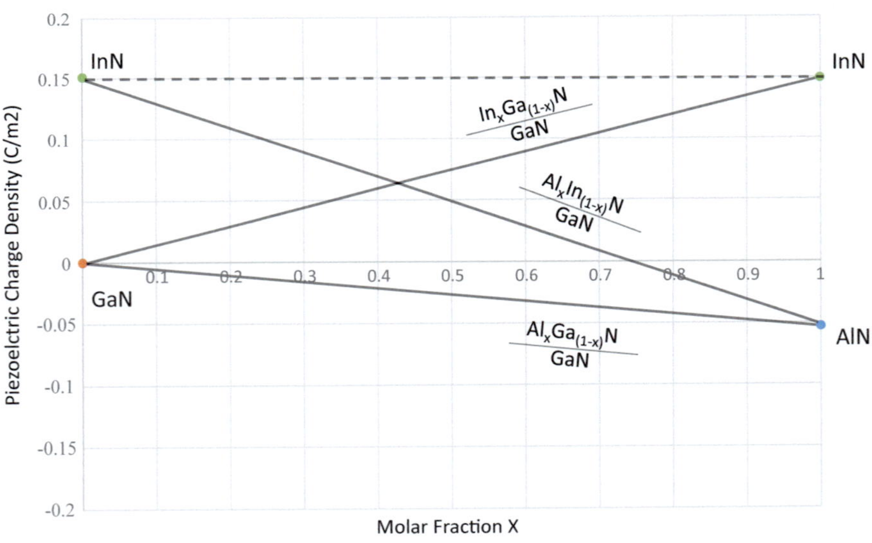

Fig. 1.6 Theoretically calculated piezoelectric polarization charge density for three ternary alloys, namely $Al_xGa_{(1-x)}N$, $In_xGa_{(1-x)}N$, and $Al_xIn_{(1-x)}N$ over GaN substrate, as a function of the mole fraction x. The calculation is based on Ambacher et al. [9]

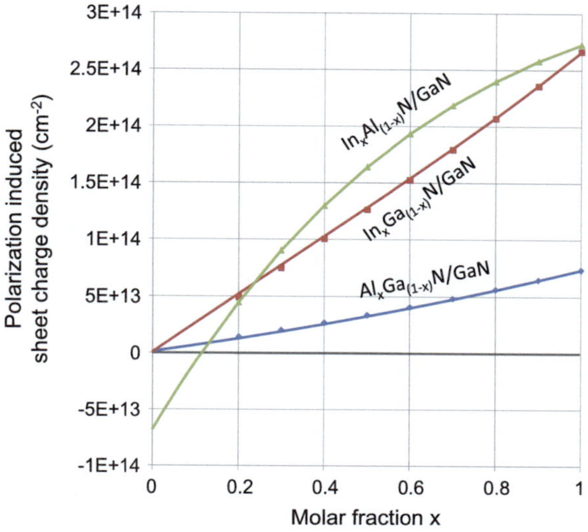

Fig. 1.7 Theoretically calculated carrier densities at the heterostructure are calculated caused by spontaneous and piezoelectric polarizations for three ternary alloys of $Al_xGa_{(1-x)}N$, $In_xGa_{(1-x)}N$, and $Al_xIn_{(1-x)}N$, as a function of the their molar fraction x grown over GaN [10]

Most notable feature of nitride semiconductors is that the free carriers are generated at the heterointerface to neutralize the fixed spontaneous and piezoelectric polarizations. Figure 1.7 shows the theoretically calculated 2DEG (2-Dimensional Electron Gas) density for three types of ternary semiconductor layer coherently grown over GaN layer [10]. You can see significantly high sheet carrier density is obtainable by InAlN/GaN combination. It is also noted that $In_{0.16}Al_{0.84}N$ can be lattice matched to GaN keeping high 2DEG density without piezoelectric effect, which is being studied at present.

Typical nitride-based HEMTs can use this 2DEG, which shows high mobility at the heterointerface. This is the most significant feature of the GaN-based FETs.

1.4 GaN-Based FET

Although AlGaN/GaN HEMTs grown over Ga-face substrate are the most widely investigated, the material system usually shows the normally on characteristic. This means the negative threshold voltage. By using the potential diagram shown in Fig. 1.8, threshold voltage V_T is described as follows

$$V_T = \phi_b - \Delta E_c - V_{AlGaN} = \phi_b - \Delta E_c - \frac{qN_s d_{AlGaN}}{\varepsilon_0 \varepsilon_{AlGaN}} \tag{1.4}$$

where ϕ_b, ΔE_c, N_s, d_{AlGaN}, ε_{AlGaN}, and q denote Schottky barrier height, conduction band offset between GaN and AlGaN, 2DEG density, AlGaN thickness, relative dielectric constant of AlGaN, and an electron charge, respectively.

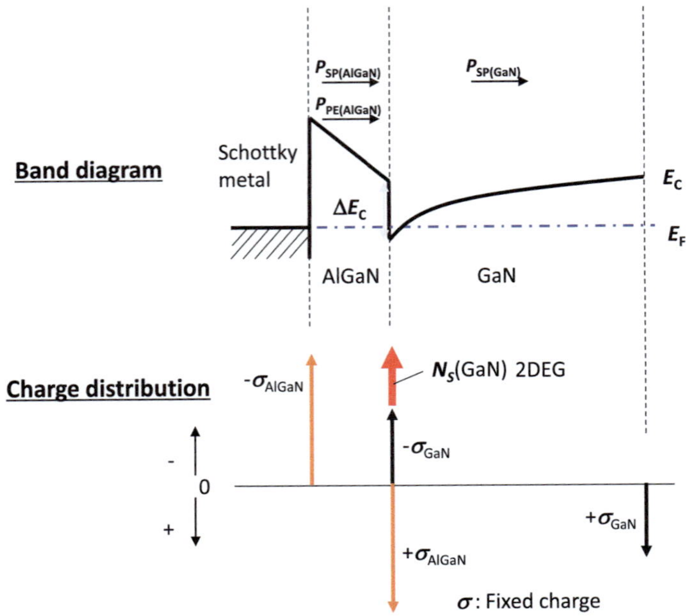

Fig. 1.8 Band diagram of Schottky junction formed over $Al_{0.25}Ga_{0.75}N$ grown on Ga-face GaN

To maintain system safety in case of gate driver failure, normally off characteristics are necessary for the switching devices. Thereby, to realize normally off characteristics has been one of the essential tasks for the practical use of GaN-based devices. As inferred from the above equation, following device design approaches are known to be effective to obtain the positive threshold voltage.

1. Increasing the work function of the gate material,
2. Lowering the 2DEG concentration by reducing Al mole fraction,
3. Thinning the AlGaN layer.

One of the methods to increase the work function of gate material is to use p-type GaN or AlGaN, which provides equivalently large work function. Some of them were reported as GIT (gate injection transistor) [11], which will be fully explained in the Chap. 11. Lowering the Al mole fraction is also effective to reduce 2DEG density, or thinning the AlGaN layer is effective to reduce the 2DEG carrier concentration [12], though either method lowers the current handling capability. Besides these approaches, to introduce fluorine into AlGaN/GaN interface [13] is effectively shifting the threshold voltage to the enhancement mode, which is also explained in the Chap. 12.

Since nonpolar or semi-polar planes have little spontaneous or piezoelectric polarization at the heterointerface, no free carriers are generated at the interface. Nonpolar and semi-polar planes are obtained in the crystal structure as shown in Fig. 1.9. It is noted that nonpolar plane has Ga and N atoms in the same plane,

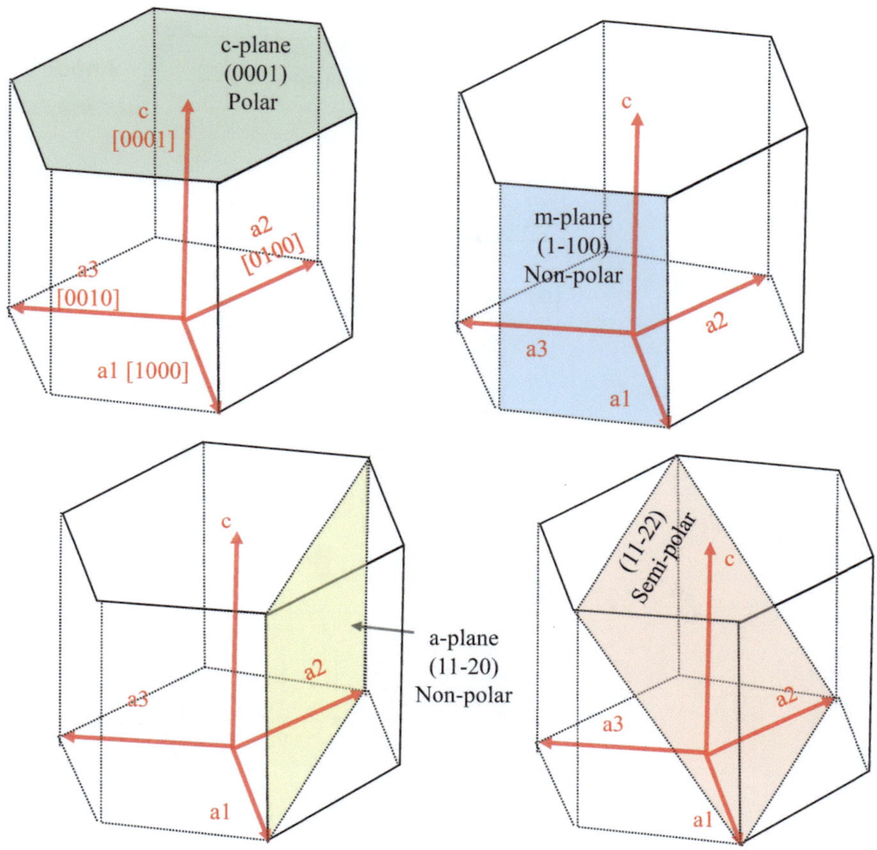

Fig. 1.9 Representative crystal planes that have little polarization effect. Note that a-plane and m-plane are called nonpolar plane, while (0012) plane is called semi-polar plane

while semi-polar one has averaged polarizations close to zero [14]. The enhanced mode GaN/AlGaN FET was successfully fabricated on those planes [15].

1.5 Natural Super Junction (NSJ) Structure

When two types of nitride layers with different polarization are alternately grown as shown in Fig. 1.10a, the same number of electrons and holes are generated at the heterointerfaces. Though the present multilayer structure is laterally conductive, the structure is presumed as multilayer capacitor with pre-stored charges at the zero bias condition. If appropriate electrical contacts are provided for electrons and holes, the pre-stored charges can be depleted as shown in Fig. 1.10b, c. The depletion voltage V_0 is described as follows.

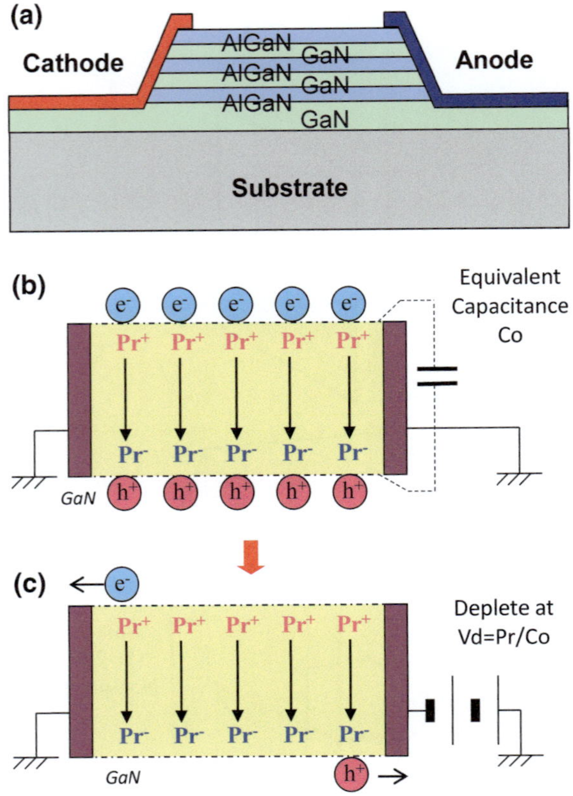

Fig. 1.10 **a** NSJ (natural super junction) structure, **b** electron and hole distribution in specific layer, **c** the depleted condition based on the model of parallel plate capacitor with pre-stored charge of N_s

$$V_0 = \frac{N_s}{C_0} = \frac{N_s C_{GaN} C_{AlGaN}}{C_{GaN} + C_{AlGaN}} \tag{1.5}$$

where N_s, C_0, C_{GaN}, and C_{AlgaN} are pre-stored charge density, capacitance between anode and cathode, GaN layer capacitance, and AlGaN layer capacitance, respectively.

When the capacitor is fully depleted, the structure is regarded as an insulator. Thereby, the attainable breakdown voltage is simply determined by the product of critical electric field and the spacing between anode and cathode. Since the depletion condition is independent from the spacing or number of layers, the obtainable on-resistance can be lowered by increasing the number of layers. NSJ made by the combination of GaN-based materials enables unlimited breakdown voltage without sacrificing the on-resistance, which overcome the trade-off between blocking voltage and on-resistance.

The fabricated NSJ structure with chip photograph is shown in Fig. 1.11a, where Ti/Al and Pt/Au were used for cathode and anode, respectively. The obtained relationship of the breakdown voltage and the spacing between electrodes are shown in Fig. 1.12 [16].

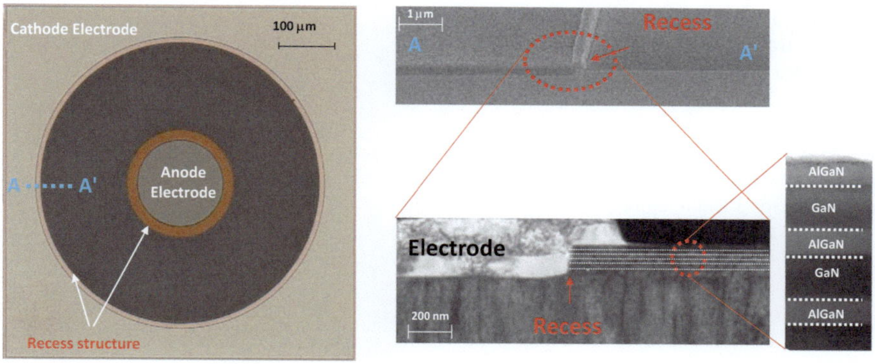

Fig. 1.11 Chip photo micrograph of fabricated NSJ diode with SEM cross section of the three layer structure. It is noted that Ti/Al and Pt/Au were used for cathode and anode, respectively

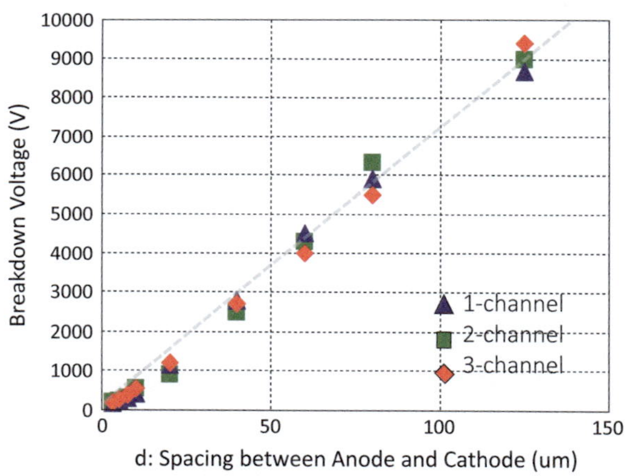

Fig. 1.12 Experimentally obtained breakdown voltage of NSJ diode as a function of the spacing between anode and cathode. Note that the breakdown voltage is not dependent on the number of layers

The present NSJ structure is similar to the well-known "Super Junction" in Si power devices [17]. The structure of Si super junction diode is shown in Fig. 1.13. You can see the same mechanism of carrier depletion can be established by using the strip-shaped PN junction, where equal number of donor and accepter charges is provided in the striped shapes. The obtainable blocking voltage is also determined by the distance d shown in the figure. It is noted that the Si super junction has the manufacturing difficulty of precise charge control of donor and accepter. If unbalanced charges remain in the junction, which becomes averaged doping concentration corresponding to the charge difference, resultant breakdown voltage is decreased. Thereby, Si super junction requires the accurate box shape of the stripes with precise doping in the manufacturing technology. On the contrary, GaN-based NSJ inherently have the same amount of electrons and holes, thereby it is called as natural super junction.

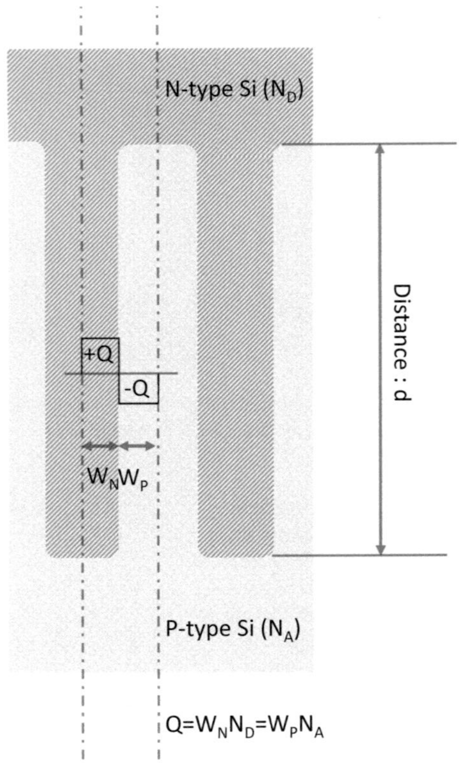

Fig. 1.13 Typical Si super junction structure under the precise charge controlled structure, where precise geometry of parallel stripe and precise doping for each layer to give equal charge of +Q and -Q for p and n regions

1.6 On-Resistance and Breakdown Voltage

GaN-based lateral devices, which use the carriers generated by the polarizations, have the breakdown voltage determined by the distance between the electrodes based on the principle of NSJ. Since the on-resistance is in proportion to the distance, it has linear relationship with breakdown voltage.

The R_{on} (on-resistance) of GaN-based FET can be simply represented by the sum of R_{ch} (channel resistance) and R_{d} (drift region resistance). Note that drift region is provided to support the drain voltage. The relationship is written as follows.

$$R_{\mathrm{on}} = R_{\mathrm{ch}} + R_{\mathrm{d}} \tag{1.6}$$

$$R_{\mathrm{ch}} = \frac{L_{\mathrm{g}}}{W_{\mathrm{g}}} \frac{1}{q\mu N_{\mathrm{s}}} \tag{1.7}$$

$$R_{\mathrm{d}} = \frac{L_{\mathrm{d}}}{W_{\mathrm{g}}} \frac{1}{q\mu N_{\mathrm{s}}} = \frac{1}{W_{\mathrm{g}}} \left(\frac{B_{\mathrm{V}}}{E_{\mathrm{c}}} \frac{1}{q\mu N_{\mathrm{s}}} \right) \tag{1.8}$$

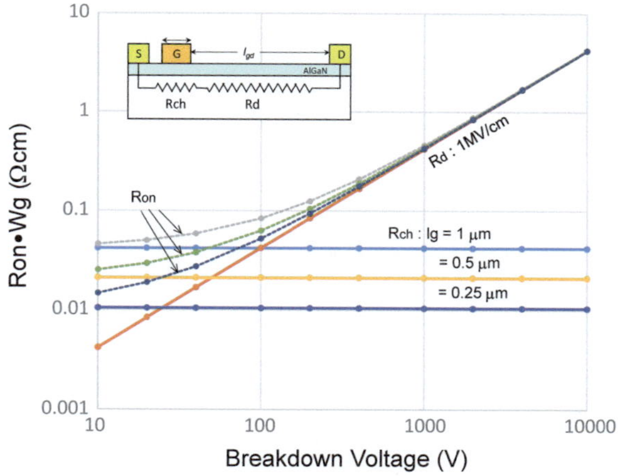

Fig. 1.14 Calculated $R_{on} \cdot W_g$ as a function of breakdown voltage for the of typical GaN/AlGaN FETs with 1.0, 0.5, and 0.25 mm gate length. It is noted that R_{on} is the sum of R_{ch} and R_d. The calculation was done under the conditions of 2DEG mobility of 1500 cm^2/Vs and critical electric field of 1 MV/cm, which is a couple of times smaller than the theoretically predicted value

where W_g: gate width, L_g: gate length, L_d: drift region length, N_s: sheet carrier density, μ: 2DEG mobility.

Figure 1.14 shows the calculated R_{on} as a function of the breakdown voltage for the GaN HEMT with unit gate width, where R_{ch}, R_d, and R_{on} are plotted for three different gate lengths. You can see the break-even point where R_{ch} is equal R_d in the figure. R_{on} asymptotically approaches to R_{ch} in the low-voltage range, while R_{on} approaches to R_d in high-voltage range as well. It is clearly seen that to reduce the gate length is useless for the device design in high-voltage range, though the design is the only way to reduce R_{on} in the low-voltage range.

It is noted that higher E_c (critical electric field) over 1 MV/cm is attained by the recent bulk GaN materials [18]. The used parameters should be updated, corresponding to the maturing of device fabrication and the crystal growth technologies, to know the break-even point along the same design criteria.

1.7 Low-Voltage Devices

Considering the device design in the personal appliances layer, where typical blocking voltage is below 100 V, lowering the R_{ch} is the most important. Table 1.2 shows a well-known scaling rule, which has been a guiding principle in silicon MOSFET technology. For power conversion systems in this voltage range, there is a scaling-down limit for miniaturized Si MOSFET because the critical electric field of Si is approximately 30 V/μm. This means that providing certain length of source-to-drain spacing is necessary in the MOSFET structure to satisfy the

Table 1.2 Typical scaling law, expected performance, and chip size

Device dimension is miniaturized by 1/k	
Blocking voltage	1/k
R_{on}	1
Capacitance	$1/k^2$
Chip area	$1/k^2$
Switching loss ($R_{on} \cdot Q_g$)	$1/k^2$
Chip cost	$1/k^2$

Table 1.3 Estimated device dimension to satisfy the demands by the application voltage, e.g., 12 and 100 V for three types of materials

Blocking voltage (V)	Required dimension		
	Si	SiC	GaN
12	400 nm	60 nm	60 nm
100	3.3 μm	0.5 μm	0.5 μm

required blocking voltage. On the contrary, wide bandgap devices still have a room of miniaturization to keep the same voltage. Table 1.3 shows the estimation of device size, expected performance, and chip size. You can see about 40 times smaller chip size is available to make the same on-resistance in the low-voltage operation. It is also noted that switching loss, which is determined by $R_{on} \cdot Q_g$, can be reduced by the factor of 40 compared to Si one. These estimation means wide bandgap power devices have a potential of superior conversion efficiency in low-voltage range with extremely small device size.

It is also effective to introduce a vertical structure in the low-voltage applications, since R_{on} reduction is attainable by increasing the gate width. Currently, Si trench power MOSFETs are widely used for the low-voltage applications. Figure 1.15 shows the evolution path of low-voltage Si power MOSFET to increase the packing density [19]. The device fabrication process moved to a self-aligned one, which are mainly used for low-voltage applications. Provided that self-aligned structure is available in wide bandgap devices, expected packing densities for lateral and vertical structures are schematically compared in Fig. 1.16. As seen in the figure, vertical structure could achieve 9 times higher packing density than lateral, which means that further reduction of R_{on} is possible using the same chip size.

However, it is not sufficient only to reduce R_{on} to improve the total conversion efficiency because of the increased parasitic capacitance giving a negative impact to the switching loss. The total energy loss is determined by the sum of DC-loss (R_{on} loss) and AC-loss (switching loss). Figure 1.17 shows typical switching waveforms of a power FET, where you can see the DC-loss during the on state and AC-loss during transient state. By integrating the gate current, we can extract the gate charge Q_g to turn on the device. You can see the slower slope of the gate voltage during

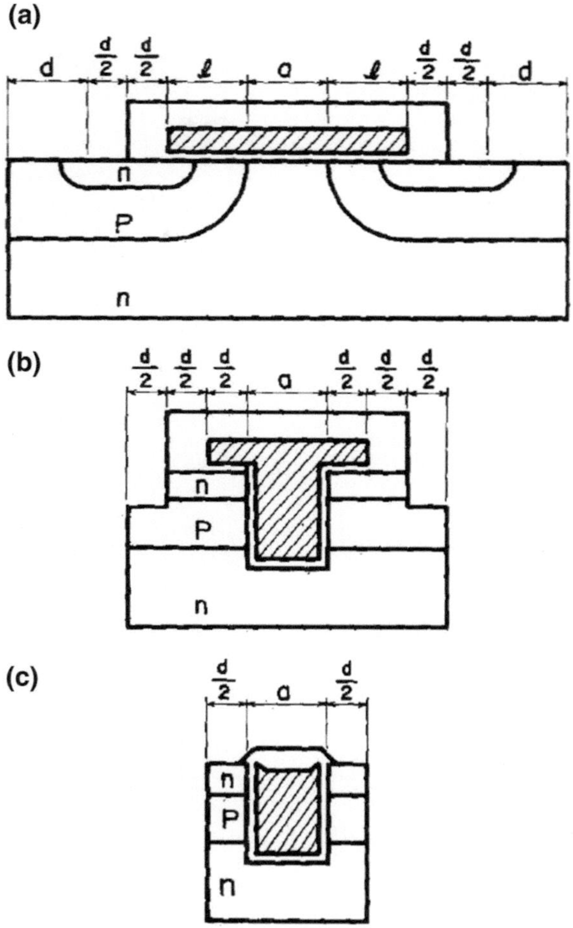

Fig. 1.15 Evolution path of R_{on} reduction of low-voltage Si power MOSFET, namely **a** DMOS, **b** early trench, and **c** fully self-aligned trench structures. Note that fully self-aligned Si trench power MOSFET **c** widely used since R_{on} is determined by packing density for low-voltage application

turn-on, which is caused by the feedback (Miller's) capacitance appearing as the decrease of drain voltage.

Based on the following facts,

$$\mathrm{Loss_{DC}} \propto R_{on} W_g, \tag{1.9}$$

and

$$\mathrm{Loss_{AC}} \propto \frac{Q_g}{W_g}, \tag{1.10}$$

the product of $R_{on} \cdot Q_g$ becomes intrinsic device parameter independent from the device size, which enables us to estimate the switching efficiency. To reduce $R_{on} \cdot Q_g$ is the most important objective for the device engineering to improve

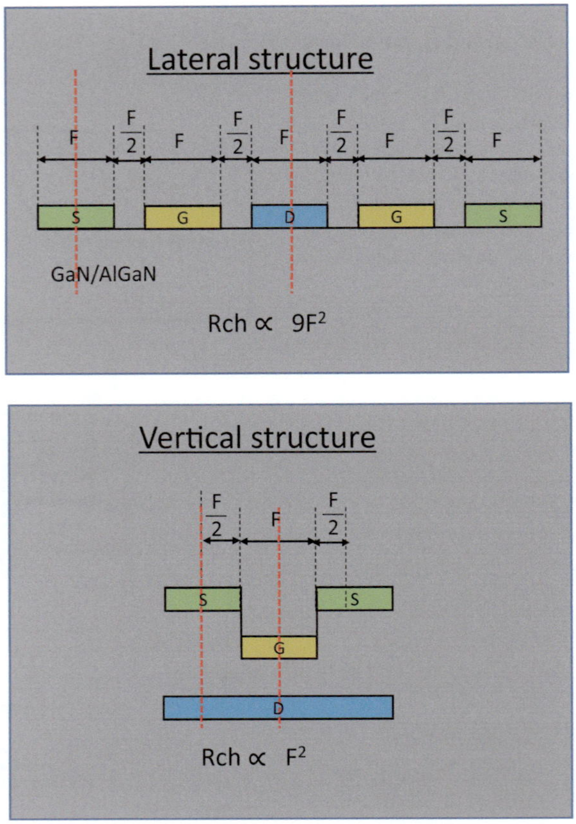

Fig. 1.16 Comparison of estimated channel resistance for lateral and vertical structure using the minimum design rule F

power conversion efficiency. Figure 1.18 shows the experimentally obtained conversion efficiency of the half bridge using GaN FETs with three different $R_{on} \cdot Q_g$. You can see the higher efficiency is obtainable for the device with smaller $R_{on} \cdot Q_g$ [20].

It is also noted that there is an optimum device size to obtain the maximum conversion efficiency for switching power systems. Figure 1.19 shows the estimated DC and AC energy loss as a function of gate width. You can see the total energy loss is minimized at the optimum gate width, where DC-loss becomes equal to AC-loss. Further, the optimum device size is also determined by the switching frequency.

Scaling-down technique is effective to the reduction of $R_{on} \cdot Q_g$ in wide bandgap devices, while that of Si has a bottom line of the scaling down. Therefore, GaN power devices have a potential to achieve superior conversion efficiency to Si one by using smaller die size.

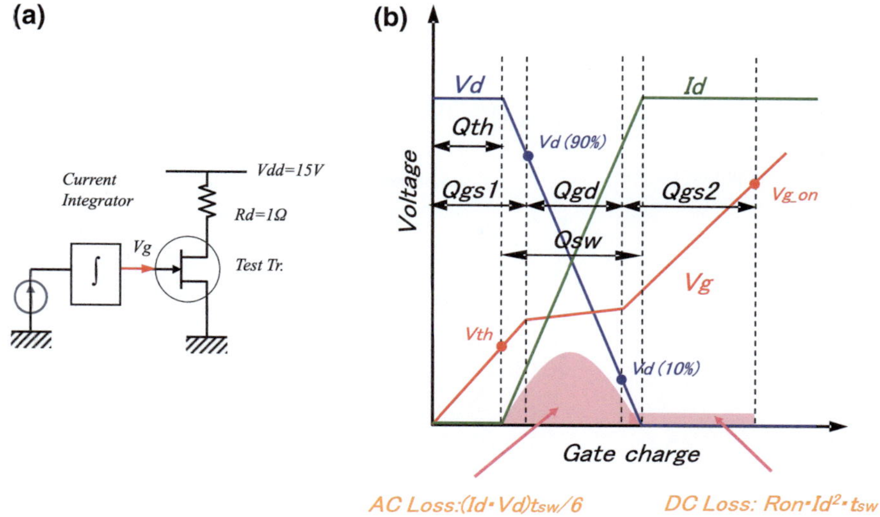

Fig. 1.17 The typical measurement system of R_{on} Q_g and observed current/voltage waveform during the switching, where Qg represents the sum of Ggs1, Qsw, and Qgs2

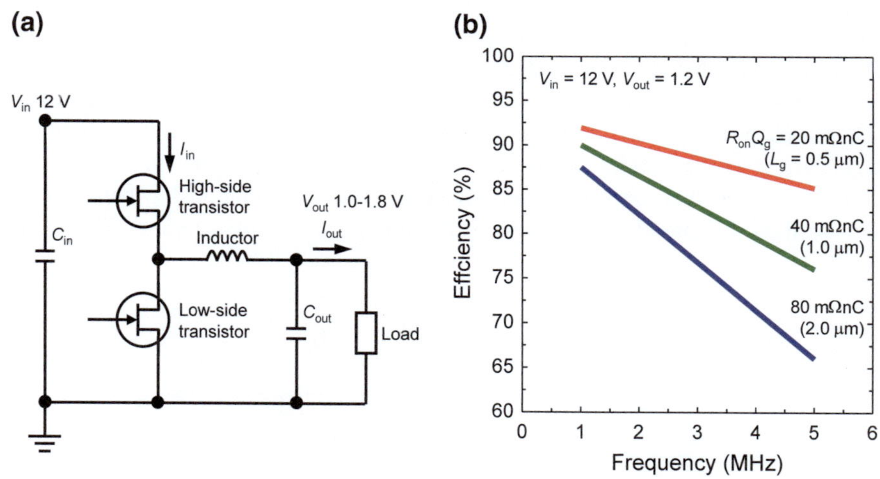

Fig. 1.18 **a** Half-bridge circuit to make DC-to-DC converter and **b** obtained conversion efficiency as a function of switching frequency for devices with three different $R_{on} \cdot Q_g$

1.8 High-Voltage Devices

AlGaN/GaN FETs have the breakdown voltage proportional to the spacing of the drift region based on the principle of NSJ structure. Extremely high breakdown voltage of more than 9000 V was fabricated by using sapphire substrate [21].

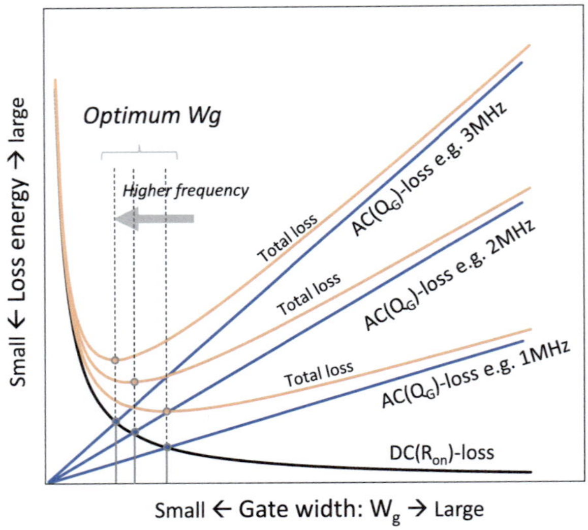

Fig. 1.19 Conceptual drawing to determine the optimum gate width of switching FET to minimize the energy loss. It is noted that the relatively smaller gate width become optimum as the increase of switching frequency

Recently, GaN/AlGaN FETs on Si substrate are gaining attention in view of reducing the material cost. However, those devices fabricated on Si substrate usually show the prematured breakdown determined by the thickness of the GaN epitaxial layers. This is because that vertical electric field is increased when the conductive Si substrate is grounded.

Since the n^+-type bulk GaN substrates become commercially available, vertical GaN-based power device is thought to be a candidate of future high-voltage and high current power devices. The schematic illustration is shown in Fig. 1.20. It should be noted that vertical power device has different relationship between R_{on} and B_V than the lateral one because of the doping of the drift region. This is the

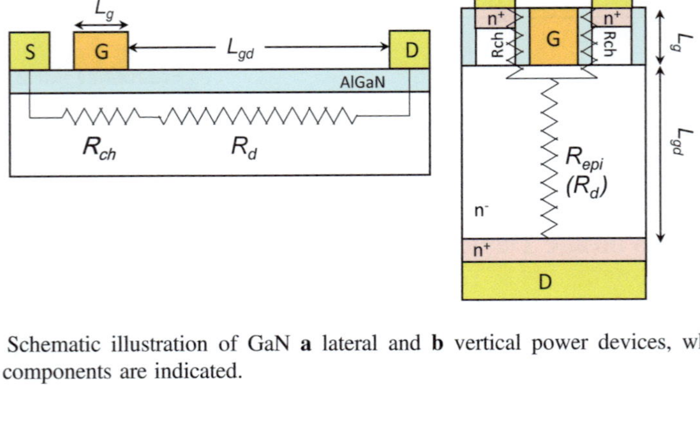

Fig. 1.20 Schematic illustration of GaN **a** lateral and **b** vertical power devices, where major resistance components are indicated.

well-known relationship in Si vertical power MOSFET, where on-resistance is proportional to the square of breakdown voltage. Provided that epitaxially grown drift region has the concentration N_d, and the depletion widening W_d is obtained under the critical electric field E_c, the relationship of one-dimensional structure is written as follows:

$$E_c = \frac{qN_d}{\varepsilon_{GaN}} W_d \tag{1.11}$$

The blocking voltage at the above condition is simply written as follows.

$$B_V = \frac{1}{2} E_c W_d \tag{1.12}$$

Since the resistance of the epitaxial drift region R_{epi} is proportional to $W_d/\mu N_d$, the resultant R_{epi} is obtained as follows.

$$R_{epi} \propto \frac{W_d}{\mu N_d} = \frac{B_V^2}{\mu \varepsilon_{GaN} E_c^3} \tag{1.13}$$

You can see R_{epi} is proportional to the square of B_V, which can be overlaid in Fig. 1.11 with twice steeper slope of drift region resistance. The dominator in Eq. (1.11) is sometimes used as Baliga's FOM (figure of merit) of the material [22], which means that the wide bandgap material can achieve extremely low R_{epi} that is inversely proportional E_c^3. It is noted that typical critical electric field E_c of GaN and Si are 300 and 30 V/μm, respectively. It is also noted that the practically obtained blocking voltage of Si power devices is sometimes in proportion to $B_V^{2.0-2.5}$, since the other factors like insufficient edge termination and/or relatively thin epitaxial layer are employed as practical device.

Although the lateral structure can easily achieve high breakdown voltage by expanding spacing of drift region, the current handling capability become lowered since the resultant gate width is reduced under the condition of same chip size. Therefore, GaN-based vertical power devices are expected to provide both high blocking voltage and high current handling capability. As explained in the previous section, to increase the packing density has little effect to reduce R_{on} in the high-voltage range because that R_{on} is determined by the resistance of epitaxially grown drift region. To reduce R_{epi} is the most important task for vertical GaN power devices.

In order to increase the current handling capability of Si power device, conductivity modulation technique has been used to dynamically increase carrier density in the drift region during the on state. Figure 1.21 shows three types of Si power devices that introduced conductivity modulation. The devices were named GTO (gate turn-off) thyristor [23], SI (static induction) thyristor [24], and IGBT

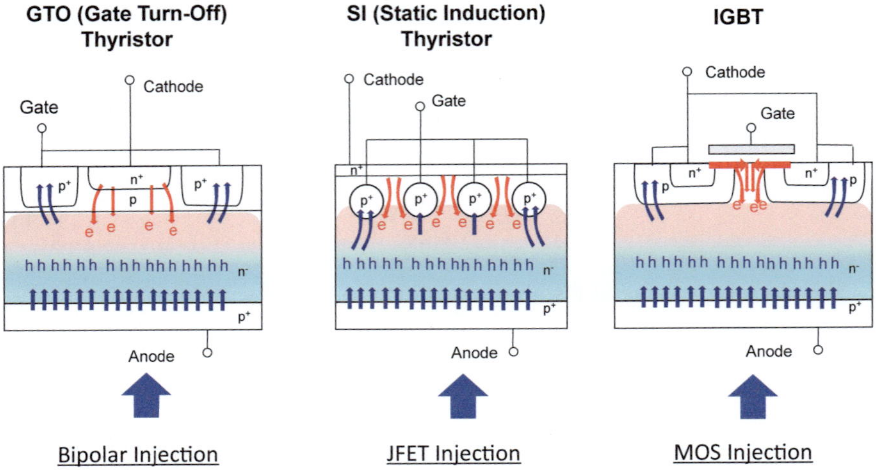

Fig. 1.21 Representative Si power devices that use conductivity modulation in the drift region. You can see the difference of electron leading structure into the drift region to occur the hole injection from the bottom p region

(insulated gate bipolar transistor) [25]. All of the three structures commonly have heavily doped p-type substrates that can inject minority carrier in the drift region, which results in reducing the drift region resistance. These structures have similar operation mechanism that employs different upper structure, which leads the electrons into the drift region, triggering the conductivity modulation. It can be seen that GTO thyristor, SI thyristor, and IGBT employ the upper device structure as BJT, JFET, and MOSFET, respectively, to lead electrons.

Since the same number of electrons and holes are generated in the drift region, drastic R_{epi} reduction can be realized in the drift region. It is noted that some of the injected holes recombine with electrons lead by the upper device structures, and the rest of injected holes flows into the upper p-type region. This means the built-in pnp bipolar operation, where bottom p-layer plays a role of emitter in the equivalent circuit. This injection mechanism contributes to increase the available current.

Although Si IGBTs have high current handling capability, they suffer from poor switching speed due to the long lifetime of minority carriers because that Si has indirect recombination process. Introducing recombination center into the drift region is necessary to decrease the life time in Si. “Killer doping” by using gold, platinum diffusion, and/or electron beam irradiation are practically employed toreduce the life time [26]. There is another advantage of the killer doping, which suppresses the undesirable latch-up of the parasitic thyristor by reducing the current gain of the bipolar operation. Though transistor named as GIT incorporates conductivity modulation, the injected carriers have very short lifetime to spread fully into the drift region because of direct recombination process of GaN-based materials.

1.9 Future Study in GaN Vertical Power Device

Recently, some technological trials are reported to reduce drift-region resistance by Conductivity modulation, where minority carrier injection structure is introduced in GaN-based devices. Mochizuki et al. proposed photon recycling in GaN pin diode that widens the space of conductivity modulation as shown in Fig. 1.22. They reported meaningful reduction of series resistance by the experimentally fabricated GaN pin diode as shown in Fig. 1.23 [27].

Incorporating bipolar mechanism into GaN-based material is also studied by Makimoto et al. [28]. Their DHBT device structure is shown in Fig. 1.24, where InGaN heavily doped p-type base region is used. They reported very high current handling capability over 100 A/cm^2 by using DHBT structure. The obtained I-V

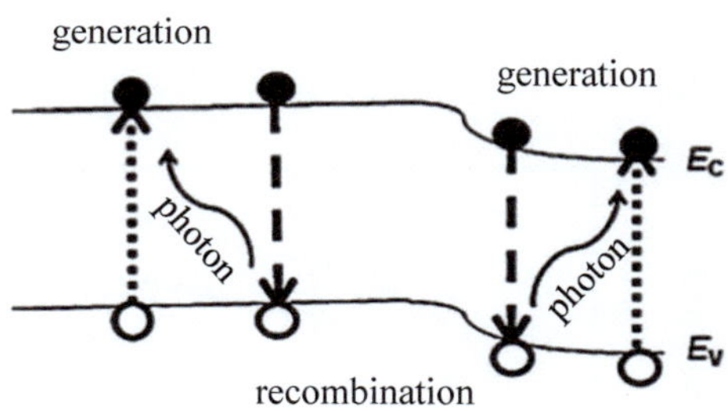

Fig. 1.22 Conceptual illustration of photon recycling that can widen the region of conductivity modulation

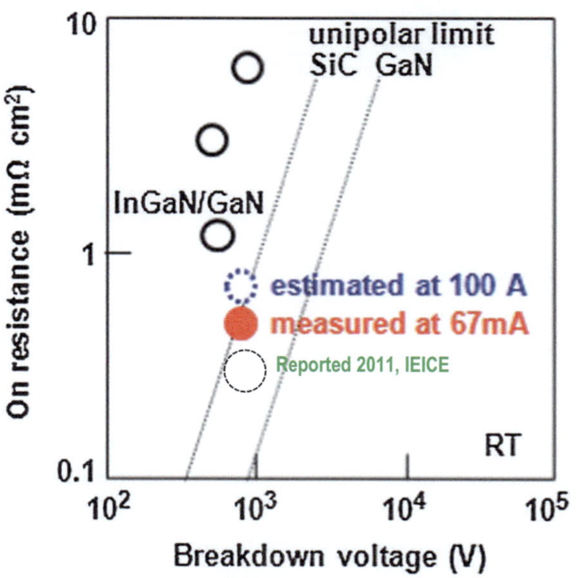

Fig. 1.23 Experimentally obtained $R_{on} \cdot A$ product as a function of breakdown voltage of GaN pin diode by Mochizuki et al., where the product could be lower than the unipolar limit [27]

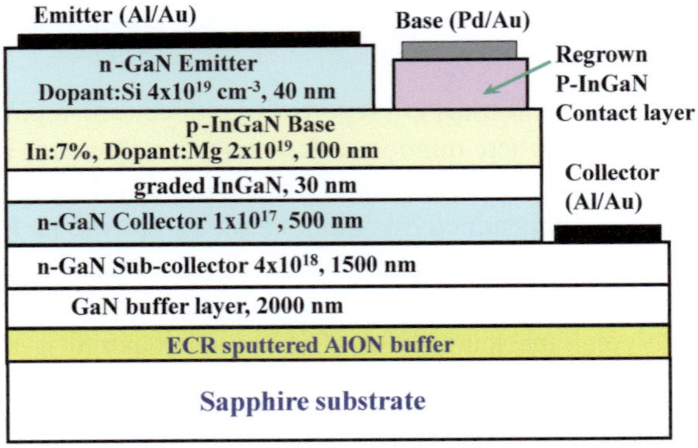

Fig. 1.24 Experimentally fabricated GaN/InGaN DHBT structure with regrown P-InGaN base contact layer

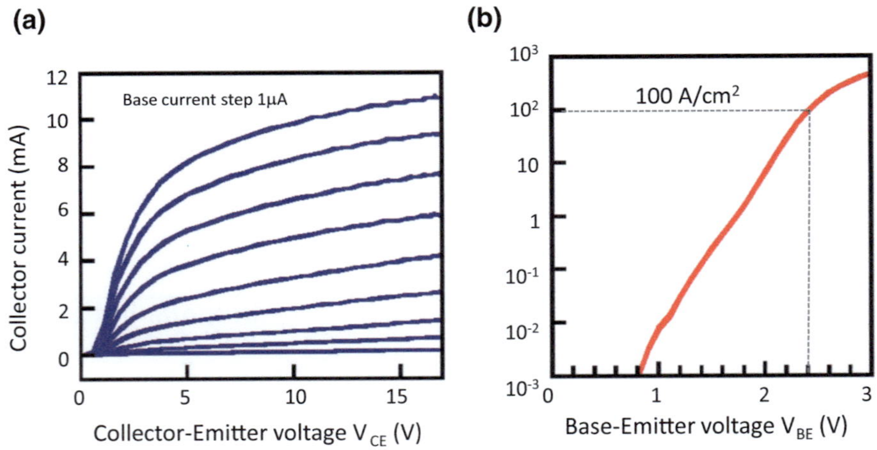

Fig. 1.25 **a** Obtained I-V characteristics of GaN/InGaN DHBT, where hfe over 1000 is seen, **b** obtained collector current density as a function of base-emitter voltage (V_{BE}), where current handling capability over 100 A/cmm^2 is seen [28]

characteristics and achieved current handling capability by HBT are shown in Fig. 1.25a, b.

The layer of infrastructure application requires both extremely high blocking voltage and high current handling capability at the same time. As explained in the above sections, to reduce the drift region resistance is the most important in the high-voltage applications. Thereby, to know the conditions of conductivity modulation, which is determined by the material parameters such as life time and diffusion length of minority carriers, will be most important.

References

1. Lee FC et al (2013) High-frequency integrated point-of-load converters: overview. IEEE Trans PE 28(9):4127–4136
2. Nakamura S, Chichibu SF (2000) Introduction to nitride semiconductor blue lasers and light emitting diodes. CRC Press
3. Johnson EO (1965) Physical limitations on frequency and power parameters of transistors. RCA Rev 163–177
4. O'Leary K, Fout BE, Shur MS, Eastman LF (2006) Steady-state and transient electron transport within the III–V nitride semiconductors, GaN, AlN, and InN: a review. J Mat Sci Mater Electron 17(2):87–126
5. Pengelly RS, Wood SM, Milligan JW, Sheppard ST, Pribble WL (2012) A review of GaN on SiC high electron-mobility power transistors and MMICs. IEEE Trans MTT 60(6):1764–1783
6. Yao T, Hong SK (2009) Oxide and nitride semiconductors. Springer, Berlin
7. Sumiya M, Fuke S (2004) Review of polarity determination and control of GaN. MRS Internet J Nitride Semicond Res 9:e1
8. Wood C, Jena D (2008) Polarization effects in semiconductors. Springer, Berlin
9. Ambacher O, Majewski J, Miskys C, Link A, Hermann M, Eickhoff M, Stutzmann M, Bernardini F, Fiorentini V, Tilak V, Schaff B, Eastman LF (2002) Pyroelectric properties of Al (In)GaN/GaN hetero- and quantum well structures. J Phys Condens Matt 14:3399–3434
10. Ambacher O, Foutz B, Smart J, Shealy J, Weimann N, Chu K, Murphy M, Sierakowski A, Schaff W, Eastman L, Dimitrov R, Mitchell A, Stutzmann M (2000) Two-dimensional electron gases induced by spontaneous and piezoelectric polarization in undoped and doped AlGaN/GaN heterostructures. J Appl Phys 87(1):334–344
11. Uemoto Y, Hikita M, Ueno H, Matsuo H, Ishida H, Yanagihara M, Ueda T, Tanaka T, Ueda D (2006) A normally-off AIGaN/GaN transistor with $R_{on}A = 2.6$ mΩcm2 and $BV_{ds} = 640$ V using conductivity modulation. IEEE IEDM Tech Digests 907–910
12. Ohmaki Y, Tanimoto M, Akamatsu S, Mukai T (2006) Enhancement-mode AlGaN/AlN/GaN high electron mobility transistor with low on-state resistance and high breakdown voltage. JJAP 45 Part 2(44):L1168–1170
13. Chen KJ, Zhou C (2011) Enhancement-mode AlGaN/GaN HEMT and MIS-HEMT technology. Phys Stat Sol (a) 208(2):434–438
14. Schwarz UT, Kneissl M (2007) Phys Stat Sol (RRL) 1:A45
15. Kuroda M, Ishidal H, Uedal T, Tanaka T (2007) Nonpolar (11–20) plane AlGaN/GaN heterojunction field effect transistors on (1–102) plane sapphire. J Appl Phys 102:093703. doi:10.1063/1.2801015
16. Ishida H, Shibata D, Yanagihara M, Uemoto Y, Matsuo H, Ueda T, Tanaka T, Ueda D (2008) Unlimited high breakdown voltage by natural super junction of polarized semiconductor. IEEE EDL 29(10):1087–1089
17. Lorenz L, Marz M, Deboy G (1998) COOLMOS—an important milestone towards a new power MOSFET generation. Proc Power Convers 151–160
18. Otoki Y, Tanaka T, Kamogawa H, Kaneda N, Mishima T, Honda U, Tokuda Y (2013) Impact of crystal-quality improvement of epitaxial wafers on RF and power switching devices by utilizing VAS-method grown GaN substrates with low-density and uniformly distributed dislocations. In: CS MANTECH conference digest, pp 109–112
19. Ueda D, Takagi H, Kano G (1987) An ultra-low on-resistance power MOSFET fabricated by using fully self-aligned process. IEEE T-ED ED-34(4):926–930
20. Umeda H, Kinoshita Y, Ujita S, Morita T, Tamura S, Ishida M, Ueda T (2014) Highly efficient low-voltage DC-DC converter at 2–5 MHz with high operating current using GaN gate injection transistors. PCIM Europe Digest 1025–1032
21. Uemoto Y, Shibata D, Yanagihara M, Ishida H, Matsuo H, Nagai S, Batta N, Li M, Ueda T, Tanaka T, Ueda D (2007) 8300 V blocking voltage AlGaN/GaN power HFET with thick poly-AIN passivation. IEEE IEDM Digest 861–864

22. Baliga BJ (1989) Power semiconductor device figure of merit for high-frequency applications. IEEE EDL 10(10):455–457
23. Wolley ED (1966) Gate turn-off in p-n-p-n devices. IEEE T-ED ED-13(7):590–597
24. Nishizawa J, Nakamura K (1977) Characteristics of new thyristors. Jpn J Appl Phys 16:541–544
25. Scharf BW, Plummer JD (1978) A MOS-controlled triac devices. In: IEEE International solid-state circuits conference, SESSION XVI FAM 16.6
26. Baliga BJ (1983) Fast-switching insulated gate transistors. IEEE EDL 4(12):452–454
27. Mochizuki K, Mishima T, Ishida Y, Hatakeyama Y, Nomoto K, Kaneda N, Tshuchiya T, Terano A, Tsuchiya T, Uchiyama H, Tanaka S, Nakamura T (2013) Determination of lateral extension of extrinsic photon recycling in p-GaN by using transmission-line-model patterns formed with GaN p–n junction epitaxial layers. JJAP 52(8S52):8JN2
28. Makimoto T, Yamauchi Y, Kumakura K (2004) High-power characteristics of GaN/InGaN double heterojunction bipolar transistors. APL 84(11):1964–1966

Chapter 2
Substrates and Materials

Stacia Keller

The choice of substrate and the material requirements for GaN-based power transistors for switching applications strongly depend on the device architecture. While to date most efforts have been directed toward the fabrication of lateral devices, vertical device layouts have recently gained interest, catalyzed by the progress in the development of larger size bulk GaN substrates. The vertical devices have the advantage that the high fields are held within the bulk of the material rather than on the surface. Large-area GaN substrates, however, are still very expensive, making a lateral device layout on a foreign substrate such as silicon, which is available in wafer sizes up to 12″, currently more attractive.

Independent of the device layout, power switching devices operate at high current densities and are required to exhibit high breakdown voltages in combination with a low on-resistance and low switching losses. Thereby, GaN-based devices take advantage of the high breakdown field, electron mobility, and saturation drift velocity of GaN, as discussed in the previous chapters of this book. At a free carrier concentration of 4×10^{16} cm^{-3}, electron mobilities as high as 1265 cm^2/Vs have been reported [1], and values of approximately 1750 cm^2/Vs were extracted from I–V curves for material with a carrier concentration of about 3×10^{15} cm^{-3}, both determined for epitaxial layers grown on bulk GaN substrates [2]. Even higher electron mobilities can be achieved in GaN-based heterostructures, with values as high as 2200 cm^2/Vs at room temperature for AlGaN/AlN/GaN heterostructures, demonstrated on foreign substrates [3].

S. Keller (✉)
Electrical and Computer Engineering Department,
University of California Santa Barbara, Santa Barbara, CA 93106, USA
e-mail: stacia@ece.ucsb.edu

M. Meneghini et al. (eds.), *Power GaN Devices*,
Power Electronics and Power Systems, DOI 10.1007/978-3-319-43199-4_2

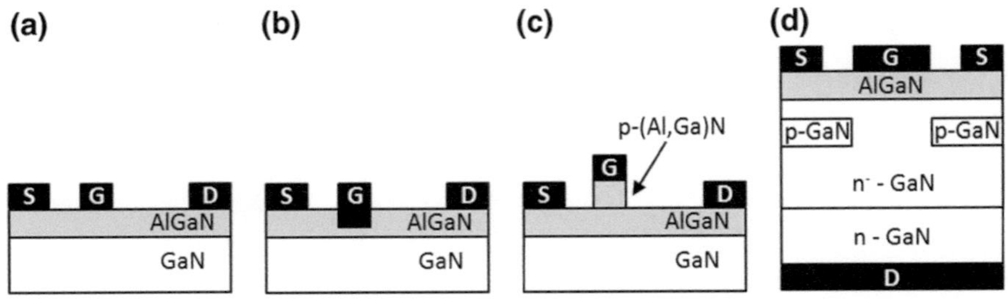

Fig. 2.1 Schematics of different AlGaN/GaN transistors: **a** D-mode structure and E-mode structures with **b** recess gate or **c** p-(Al,Ga)N cap layer, **d** vertical transistor (CAVET). *S* source, *D* drain, *G* gate

For power switching applications, enhancement-mode (E-mode) devices are preferred, either using an intrinsic E-mode transistor or utilizing a depletion-mode (D-mode) device in a cascode arrangement [4]. D- and E-mode transistor structures can be designed using combinations of GaN with (Al,Ga)N and/or p-(Al,Ga)N layers, the growth of which will be discussed in more detail in the following sections (Fig. 2.1).

While GaN-based transistors for high-frequency applications utilizing thinner GaN base layers have been fabricated by both molecular beam epitaxy (MBE) [5] and metal-organic chemical vapor deposition (MOCVD), to date transistors for power switching applications have been predominantly fabricated using the latter method, taking advantage of the higher growth rates in the MOCVD process. In addition, large-scale MOCVD reactors allow for cost-effective growth of the epitaxial structures, a development which was largely driven by GaN light-emitting diode (LED) production needs. In most applications, epitaxial layer structures based on Ga-polar GaN grown in the typical (0001) or *c*-direction are used for device fabrication.

2.1 Substrate Overview

Due to the long absence of native GaN substrates, GaN epitaxy was initially developed on the foreign substrates [6]. Historically, the first GaN-based transistors for power switching applications were demonstrated on (0001) *c*-plane sapphire and (0001) silicon carbide (6H-SiC and 4H-SiC) [7], while more recently, the majority of efforts have been pursued on silicon substrates. Typically, the (111) Si plane is used because of its trigonal symmetry, which supports the epitaxial growth of (0001) GaN [8]. The lattice mismatch between (0001) GaN to both *c*-plane sapphire and (111) silicon is very large, with 16.1 and −16.9 %, respectively (Table 2.1). The development of sophisticated growth schemes enabled the deposition of GaN films with threading dislocation densities in the order of 10^8 cm^{-2} on all three substrates. A further reduction in the threading dislocation density can be achieved through

Table 2.1 Properties of different substrates for GaN epitaxy and threading dislocation density of GaN epitaxial layers grown on top [6, 8, 18]

	Sapphire	SiC	Si	GaN
Lattice mismatch (%)	16	3.1	−17	0
Linear thermal expansion coefficient ($\times 10^{-6}$ K^{-1})	7.5	4.4	2.6	5.6
Thermal conductivity (W cm^{-1} K^{-1})	0.25	4.9	1.6	2.3
Cost	Cheap	Expensive	Cheap	Very expensive
Dislocation density of GaN films grown on substrate (optimized) (cm^{-2})	Low 10^8	Low 10^8	Low 10^8	10^4–10^6

insertion of silicon nitride interlayers [9] and implementation of epitaxial lateral overgrowth techniques [10].

Besides the lattice mismatch, differences in the thermal expansion coefficients between epitaxial layer and substrate play an important role in the epitaxial process. Whereas sapphire possesses a higher thermal expansion coefficient compared to GaN, leading to residual compressive stress in the epitaxial GaN layer, those of SiC and silicon are smaller, resulting in residual tension [11]. This problem is particularly challenging in the case of silicon, and a variety of strain management techniques have been developed to mitigate crack formation in the GaN epitaxial layers after cooling down to room temperature, which will be addressed in the substrate-specific section of this chapter.

Another critical parameter for device applications is the thermal conductivity of the substrate. As at device efficiencies <100 %, energy is converted into heat during device operation, it is essential to ensure good heat transfer to cooling elements in order to prevent overheating and device failure. Among the three foreign substrates, SiC is the one with the highest thermal conductivity (Table 2.1), making it the most attractive for transistor applications. Although sapphire has a low thermal conductivity, device heating issues can be still mitigated via flip chip bonding onto a material with high thermal conductivity [12].

Furthermore, for efficient device fabrication large-area substrates are preferred. While (111) Si substrates are available in sizes up to 12″, (0001) SiC is currently provided in diameters up to 6″. (0001) sapphire exists in large diameters, but film growth on large-area substrates has been challenging due to the strong wafer bow after cooling. In contrast to silicon, bow management on sapphire substrates would require the epitaxial layers to be under tensile stress at growth temperature, which typically results in cracking, making epitaxial solutions to this problem difficult [13].

Equally, if not more important, is the cost of the substrate itself. Among the discussed foreign substrates, the cost per wafer area is by far the lowest for silicon and the highest for SiC substrates. Semi-insulating (S.I.) SiC substrates are particularly costly. While resistive (111) silicon substrates are available, their resistivity is typically not high enough to warrant their higher price for GaN device applications. Sapphire is naturally insulating.

By far the most expensive are native GaN substrates, which are currently available in diameters up to 4″. Typical fabrication methods are hydride vapor-phase epitaxy, ammonothermal growth, and sodium flux growth [14, 15]. The threading dislocation densities in GaN substrates vary, with typical values in the 10^6–10^7 cm^{-2} range and best values in the order of 10^4 cm^{-2}. Low threading dislocation density bulk GaN substrates are particularly attractive for vertical GaN power devices, as threading dislocation can significantly contribute to leakage [16]. While the breakdown voltages of GaN films with threading dislocation densities $>10^8$ cm^{-2} are typically lower compared to those grown on low dislocation density bulk substrates [17], fairly high breakdown voltages have also been demonstrated for GaN grown on foreign substrates, as will be discussed in more detail in the following sections.

2.2 Metal-Organic Chemical Vapor Deposition

In the MOCVD process, metal-organic metal precursors such as trimethylgallium (TMGa) and trimethylaluminum (TMAl) are transported in H_2 or N_2 as carrier gas into the growth chamber, where they react on the surface of a heated substrate with the N-precursor, typically NH_3, under formation of GaN or AlN, as described by the brutto reactions.

(1) $Ga(CH_3)_3 + NH_3 \rightarrow GaN + 3\ CH_4$
(2) $Al(CH_3)_3 + NH_3 \rightarrow AlN + 3\ CH_4$

Typical (Al,Ga)N growth temperatures are in excess of 1000 °C, although specific layers, such a nucleation layers, can be deposited at lower temperatures [19, 20]. MOCVD chamber pressures can vary between about 20 and 760 Torr. To suppress prereactions between metal species and NH_3 in the gas phase, low chamber pressures are used in particular for the deposition of Al-containing layers [21, 22]. The higher tendency toward prereactions of Al species results from the stronger (Al-N) compared to (Ga-N) bond strength [23]. For the deposition of AlGaN layers, the ratio of TMGa and TMAl species needs to be adjusted accordingly to achieve the desired alloy composition [24]. The (Al,Ga)N system possesses a direct band gap from 3.41 eV (GaN) to 6.1 eV (AlN) [23].

The crystalline quality of the high-temperature (Al,GaN) films is primarily determined by their threading dislocation density. When deposited on foreign substrates, threading dislocations form due to the lattice mismatch between epitaxial layer and substrate. Thereby, different dislocation types have been observed: pure screw, mixed, and pure edge dislocations [25]. While pure screw dislocations affect the width of on-axis X-ray rocking curves, for example, taken around the (0002) diffraction peak, pure edge dislocations only contribute to the width of off-axis peaks such as $(10\bar{2}1)$ or $(20\bar{2}1)$ [26]. Thereby, the edge dislocation density becomes more apparent as the inclination angle of the asymmetric plane increases [27]. Mixed dislocations influence the width of both on- and off-axis peaks. While several approaches have been developed to link

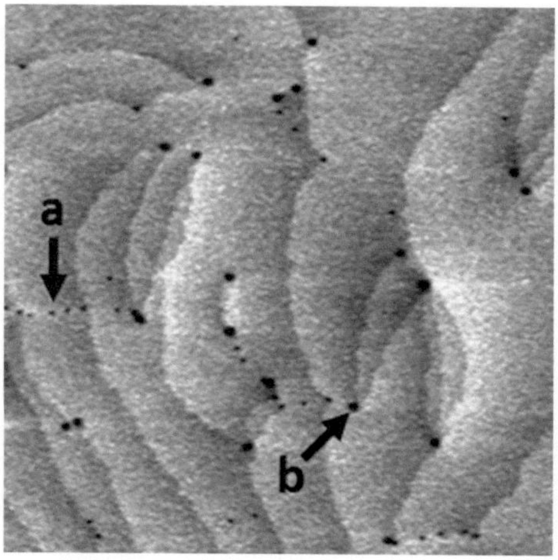

Fig. 2.2 Atomic force microscopy image of GaN film grown on sapphire. The surface steps are one GaN monolayer (0.26 nm) high. The dark spots correspond to intersections of threading dislocations with the film surface; *a* pure edge dislocation and *b* threading dislocation with screw character

the full width at half maximum (FWHM) of the X-ray diffraction peaks with the threading dislocation density in the films, an accurate determination of the threading dislocation types and their density is only possible through sample evaluation by transmission electron microscopy (TEM). For a review on the characterization of group-III nitride films by X-ray diffraction see reference [28]. Since pure screw and mixed screw–edge dislocations lead to step terminations on the crystal surface, they can also be observed by atomic force microscopy (AFM), as illustrated in Fig. 2.2. The intersections of pure edge dislocations (which do not cause a step termination) with the crystal surface can often be observed as well [29].

The conductivity of unintentionally doped (u.i.d.) (Al,Ga)N layers is controlled by several factors. (1) The concentration of residual impurities in the epitaxial layers, most commonly oxygen and carbon, which can originate from residual impurities in the precursors or carrier gasses or small leaks. Carbon is a constituent of the metal-organic precursors themselves. Oxygen occupies N sites in the GaN lattice and acts as a shallow donor [30, 31]. In $Al_xGa_{1-x}N$ films, oxygen is predicted to transition into a DX center at $x = 0.6$ [32]. Carbon preferentially occupies N lattice sites acting as a deep acceptor with an ionization energy of 0.9 eV in GaN [33, 34]. A more complex role of carbon impurities in very lightly C-doped GaN ($1–2 \times 10^{16}$ cm^{-3}) was suggested in reference [35]. (2) Native defects, with metal and nitrogen vacancies, possess the lowest formation energies in n- and p-type material, respectively. While N vacancies act as shallow donors, metal vacancies result in deep levels with acceptor character [32]. Metal vacancies are well known to form complexes with O or C impurities [36], also acting as deep traps [37]. The formation energies of the complexes are often lower compared to isolated vacancies. Both gallium vacancies and C doping can contribute to the so-called yellow luminescence in GaN [38]. (3) Threading dislocations in the epitaxial layers form acceptor states as well [39–41]. In addition, threading dislocations can be surrounded by defects [42, 43]. In high-quality material, (2) and (3) play only a subordinate role.

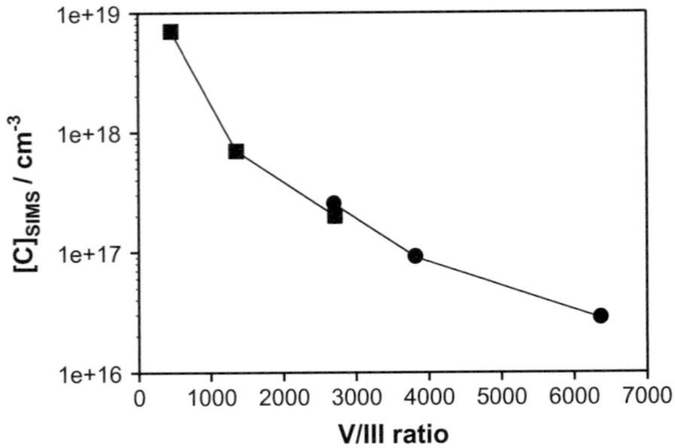

Fig. 2.3 Dependence of the carbon concentration in GaN layers on the V/III ratio during growth by varying the NH_3 flow and f_{TMGa} = constant (*squares*) and changing the TMGa flow/growth rate at a constant NH_3 flow (*circles*)

As far as residual O and C impurities are concerned, their incorporation strongly depends on the specific growth conditions [44, 45]. Since both preferentially occupy N sites in the nitride crystal, their uptake is suppressed when higher V/III ratios are applied (Fig. 2.3). The incorporation efficiency of O and C also decreases with increasing growth temperature, due to the lower thermal stability of the metal oxides compared to the corresponding nitrides, and the more efficient cleavage of the metal-C bonds, respectively [46]. Lower growth rates, allowing for more time for desorption of C-species from the surface, can also be advantageous [45]. In addition, longer residence times of the metal-organic precursor molecules in the hot zone of the reactor, for example, at higher reactor pressures, typically result in reduced C impurity incorporation [44]. Overall, the incorporation efficiency of O and C impurities increases with increasing Al content in $Al_xGa_{1-x}N$ layers, due to the stronger bonds C and O form with Al compared to Ga atoms [47]. The concentration of C and O impurities can be evaluated by the secondary ion mass spectroscopy (SIMS), where the background levels under typical measurement conditions are around 1×10^{16} cm^{-3}.

2.2.1 *Fabrication of Semi-insulating (S.I.) (Al,Ga)N Layers*

While the C incorporation shall be kept at a minimum for high purity intrinsic (Al,Ga)N layers, it can be taken advantage of for the fabrication of insulating (Al,Ga)N films, by choosing MOCVD process conditions which favor the incorporation of carbon from the methyl groups of the TMGa and TMAl precursor molecules. These include high growth rates, low V/III ratios, low pressure, and reduced temperatures during epitaxial growth. Adjustment of these parameters allows the tuning of the C concentrations over several orders of magnitude from $\leq 10^{16}$ to 10^{20} cm^{-3}, as illustrated in Fig. 2.3. The reported optimum C-doping levels are in the order of mid-10^{19} cm^{-3} [48].

In addition to process tuning, carbon can be introduced using a C precursor such as C_2H_2, as reported by Sugiyama et al. [49].

As an alternative to carbon, iron has been used as dopant for the fabrication of S.I. (Al,Ga)N films [50, 51]. Fe is incorporated into the Ga sublattice and forms multiple states within the GaN band gap. The Fe^{3+}/Fe^{2+} charge transfer level is positioned at 2.86 eV above the valence band [52]. Similar to C, Fe compensates free electrons in the crystal. Higher vertical blocking voltages, however, have been achieved using C compared to Fe as dopant [53, 54].

Besides the C- or Fe-doping level, the vertical breakdown voltage significantly depends on the thickness of the doped layer, and the higher the desired breakdown voltage, the thicker layers are required [48, 55, 56].

2.2.2 *n- and p-Type Doping*

Commonly, n-type conductivity in the (Al,Ga)N layers is established through doping with silicon via addition of silane or disilane to the gas mixture [57, 58]. Silicon incorporates on metal lattice sites and is a shallow donor in GaN. The Si donor level, however, moves deeper into the band gap with increasing x in $Al_xGa_{1-x}N$ films [32]. The electron mobility in n-type GaN films decreases with increasing carrier density, n, due to the enhanced scattering from ionized donors, from values as high as 1265 cm²/Vs at $n = 4 \times 10^{16}$ cm^{-3} to ~200 cm²/Vs at $n = 1 \times 10^{19}$ cm^{-3} [1]. In addition, the electron mobility decreases with increasing threading dislocation density (TDD) in the crystal [39]. While the value of 1265 cm²/Vs was obtained on bulk GaN substrates with a TDD of ~2×10^6 cm^{-2}, the corresponding value on GaN-on-sapphire base layers with a TDD of ~5×10^8 cm^{-2} amounted to only ~966 cm²/Vs [1]. Issues related to very low n-type doping required for drift layers in vertical devices will be addressed in Sect. 2.5.

The most common p-type dopant for (Al,Ga)N layers is magnesium which is provided using the precursor cyclopentadienyl magnesium $(C_5H_5)_2Mg$ [59]. Mg incorporates on metal sites forming an acceptor level with an activation energy of 110–160 meV in GaN [60, 61]. When GaN:Mg layers are grown under conditions typical for GaN using H_2 as carrier gas, all Mg atoms are passivated with hydrogen, which is removed from the crystal during a post-growth thermal treatment at temperatures between 600 and 850 °C in a hydrogen- and ammonia-free atmosphere [59]. Partial activation of the Mg acceptors was observed when the GaN:Mg layers were deposited in N_2 instead of H_2 as carrier gas [62]. The high activation energy of the Mg-related acceptor level (which is still lower than that of Zn, the acceptor typically used for GaAs and InP and related materials) limits the amount of holes which can be achieved in p-GaN:Mg layers to values of about 1×10^{18} cm^{-3}. Note that these doping levels require Mg concentrations of 10^{19}–10^{20} cm^{-3} in the crystal. At even higher Mg concentrations, the hole concentrations were shown to decrease, due to defects resulting from polarity inversion and/or Mg cluster formation [63]. The reported hole mobility values in GaN range between 5 and 25 cm²/Vs. Higher hole concentrations can be obtained in InGaN, as with decreasing band gap the Mg acceptor level moves closer to the valence band edge [64]. Conversely, lower hole concentrations are observed in AlGaN films as the Mg acceptor level moves deeper into the band gap with increasing Al composition [65].

Additional challenges arise when abrupt p-doping profiles are required. Mg precursor species tend to stick to the walls of tubing and growth chamber, resulting in a delayed incorporation into the epitaxial layers if not accounted for. In addition, Mg atoms tend to accumulate on the surface during growth of the Mg-doped layers, "riding" on the surface even after the Mg precursor flow is turned off, resulting in Mg incorporation into subsequent layers [66]. The carryover of Mg atoms into the following layer could be drastically reduced when the growth process was interrupted after the deposition of the GaN:Mg layer and the sample etched in an acid prior to deposition of the following n-GaN layer [66]. The Mg redistribution could also be suppressed by inserting low-temperature GaN [66] or AlN layers without any growth interruption [67, 68].

As alternative to epitaxial p-GaN layers, Mg can also be introduced into the crystal via implantation. Implant activation and stability of the Mg implantation profile during further growth and processing steps are critical when using this technique [69]. More recently, a high-temperature and high-pressure treatment of Mg-implanted GaN films has been reported to result in high-quality p-GaN layers [70]. This process is very attractive as Mg implantation can be performed locally on the wafer, eliminating complex etching and regrowth processes (see Sect. 2.5) [71]. Further investigations will show whether this process can be successfully transferred into a large-scale production environment.

In addition to the typical doping with donor or acceptor elements, in group-III nitrides grown in the typical *c*-direction, n- and p-type conductivity can be established through polarization doping [72]. In contrast to an abrupt AlGaN/GaN heterostructure, for example, where a two-dimensional electron gas (2DEG) forms at the heterointerface due to the internal electrical fields in the crystal [73], in structures where the composition of the $Al_xGa_{1-x}N$ layers is graded from $x = 0$ to a certain x value, a volume electron charge is established, which is proportional to the width and slope of the grade [74, 75]. Similarly, p-type layers form when the Al composition is graded downward [76], or the In composition in $In_yGa_{1-y}N$ layers is graded upward from $y = 0$ to a given y value [77]. Characteristics for polarization-doped layers are relatively high carrier mobilities associated with the absence of charged donors or acceptors and the nonexistence of carrier freeze out at low temperatures. Further details about polarization doping can be found in references [76] and [78].

2.2.3 AlGaN/GaN Heterostructures

The growth of $Al_xGa_{1-x}N$/GaN 2DEG structures has been extensively studied [79]. Because of the 3 % lattice mismatch between AlN and GaN, the $Al_xGa_{1-x}N$ layer thickness needs to be reduced with increasing x to avoid cracking. Relaxation of the

AlGaN layer can be suppressed by growing a very thin GaN cap layer on top [80]. Similar, though lesser, effects were observed when finishing the growth with a Si_3N_4 layer [81]. Due to the internal electric fields present in heterostructures based on (0001) GaN [72], no additional doping is required to form a 2DEG at the GaN/AlGaN interface, and the 2DEG sheet carrier density, n_s, can be tuned via composition and thickness of the $Al_xGa_{1-x}N$ layer, d ($n_s \sim x, d$) [73]. To further enhance the electron mobility, often a thin AlN interlayer is inserted between the GaN and AlGaN layers to mitigate alloy scattering [82], leading to an increase in the electron mobility from ~1400 to ~2100 cm^2/Vs at a sheet carrier density around 9×10^{12} cm^{-2}, for example. Note that recent atom probe tomography investigations revealed the presence of residual gallium in MOCVD grown AlN interlayers, which, however, did not affect the electrical properties of heterostructures with sheet carrier densities around 1×10^{13} cm^{-2} typically used in transistors [83]. While the electron mobility is little affected by the presence of threading dislocations at dislocation densities in the 10^8 cm^{-2} range or below, higher dislocation densities result in a reduction of the electron mobility [84].

2.3 Traps and Dispersion

Dispersion is a well-known phenomenon in GaN-based transistors, which can be caused by trap states in the bulk epitaxial layers [85] as well as by surface traps [86]. Thereby, the traps affect not only the switching speed, but also the breakdown voltage of the devices [7, 87]. The surface traps result from the spontaneous and piezoelectric polarization in group-III nitride films grown in the typical c-direction [72] and form to ensure charge neutrality in the structure. Their influence can be mitigated through surface passivation [88] or the implementation of epitaxial cap layers [89, 90]. The presence of trap states in the epitaxial layers can be associated with impurities, dopants, intrinsic defects, and dislocations, as discussed earlier. Their influence can be minimized by careful optimization of the epitaxial process and sufficient spatial separation between S.I. buffer and active device layers [91]. For a detailed discussion of trap-related issues see Chap. 7 of this book.

2.4 Fabrication of Epitaxial Structures for Lateral Power Switching Devices

Lateral power transistors have been demonstrated on sapphire, silicon carbide, and silicon substrates. In this layout, devices have to withstand the high electric fields along the surface and the breakdown voltage increases with increasing gate to drain spacing. For this reason, lateral devices are typically relatively large, making silicon

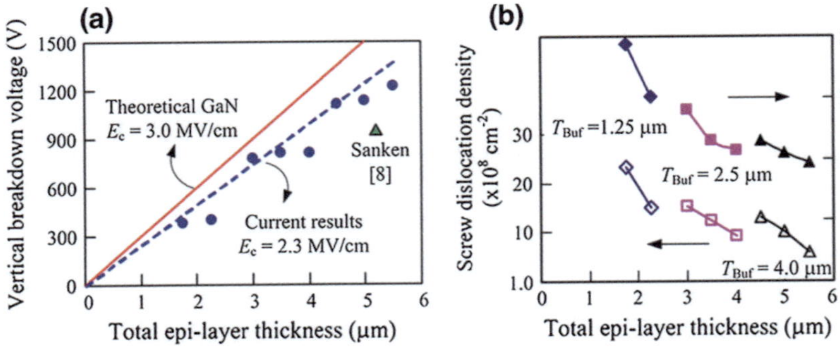

Fig. 2.4 **a** Vertical breakdown voltage as a function of the total epilayer thickness. **b** Screw (*hollow symbols*) and edge (*solid symbols*) dislocation densities derived from XRD measurements for (Al,Ga)N epitaxial layers grown on Si with various total thicknesses. Reprinted with permission from Ref. [56]. Copyright 2011, IEEE

the most attractive substrate among the three, as silicon substrates are cheap and available in diameters up to 12″.

The vertical breakdown voltage of the current blocking base layers used for these devices is determined by their resistivity and thickness [48, 55, 56]. As discussed in Sect. 2.2.1, the resistivity of the epitaxial layers depends on the C and/or Fe doping and the presence of residual impurities and native defects, as well as threading dislocations. The impact of the latter is twofold. Threading dislocations can introduce additional defects with acceptor character (Sect. 2.2) and, however, can also act as leakage path [16]. Screw-type dislocations are particularly harmful, whereas edge and mixed dislocations are less problematic [92, 93]. Hence, higher breakdown voltages are typically observed on low threading dislocation density bulk GaN substrates compared to heteroepitaxially grown layers [17]. While the specifics of the blocking layer design and growth process depend on the type of substrate, the epitaxy process is typically designed in such a way that the threading dislocation density decreases with the increasing epitaxial layer thickness, enabling the deposition of GaN layers with dislocation densities in the order of 10^8 cm^{-2} on all three substrates. The reduction in threading dislocation density with increasing (Al,Ga)N current blocking layer thickness contributes to the increase in breakdown voltage with film thickness (Fig. 2.4) [55, 56].

To minimize the dispersion, on top of the (Al,Ga)N:C,Fe base layers typically an u.i.d. GaN layer is deposited prior to GaN channel and (Al,Ga)N gating layers. Thereby, channel and gating layer designs are interchangeable between substrates (Sect. 2.4.4).

Note that the device breakdown voltage can also be enhanced using non-epitaxial techniques, for example, by the implementation of via-holes [94, 95].

2.4.1 *Current-Blocking Layer Deposition on Silicon Substrates*

GaN transistors have been successfully demonstrated on 6″ and 8″ (111) Si substrates. Often, conductive Si substrates are used as highly resistive Si substrates are unavailable. Besides the large lattice mismatch of −16.9 %, challenges of the GaN growth on silicon are gallium silicide formation and the large thermal mismatch between silicon and GaN, leading to cracking of the GaN film during wafer cool down if not accounted for [8]. To prevent gallium silicide formation, often called meltback etching, the growth on Si is typically initiated with an AlN layer. The growth conditions and properties of the initial AlN layers were shown to have a strong influence on the final GaN film quality [96]. The atomic arrangement at the AlN/Si(111) interface was investigated in detail in reference [97]. To mitigate film cracking during cool down, the epitaxial layer stack is then grown in such a way that the final stack is under compression at growth temperature. Several strain management procedures have been developed, such as grading the composition from AlN to GaN [98, 99], insertion of an AlN-GaN superlattice [100, 101], deposition of AlN interlayers [102], or nucleation with AlN followed by the growth of AlGaN layers where the composition is reduced in steps (Fig. 2.5) [103]. All these approaches require careful process control to minimize the final curvature of the wafer. In addition, defect formation in the silicon substrate, such as slip, has to be prevented [104]. State-of-the-art in situ control systems allow monitoring the wafer curvature throughout the epitaxial growth process, as illustrated in Fig. 2.6 [104–106]. Thereby, strain and curvature evolution are strongly affected by threading dislocations and their annihilation in the course of the growth process. Dislocation bending, which enhances threading dislocation annihilation, has been observed in particular at the interfaces between layers of different composition [103] or in the presence of small pits which can form at the intersection of threading dislocations with the sample surface [107]. While enabling the growth of low threading dislocation density top layers, the dislocation annihilation process leads to tension [108], counteracting the compressive strain introduced by the step-graded AlGaN layers or AlN interlayers.

For additional dislocation filtering, Si_3N_4 interlayers can be implemented [109]. In this process, patches of Si_3N_4 are deposited onto the sample surface, which are overgrown in the following (Al,Ga)N growth steps [9]. The Si_3N_4 patches stop threading dislocation propagation into sequential layers. To fulfill their insulation purpose, the bottom (Al,Ga)N layers are typically doped with carbon, as discussed earlier.

Although up to 14-μm-thick (Al,Ga)N layers have been successfully demonstrated on Si substrates [110], the strain management challenges significantly increase with layer thickness, as do the required growth time and resulting wafer cost. About 3–5-μm-thick base layers have been used for devices operating at 600 V [55, 56, 111]. As expected from their larger band gap, AlGaN base layers were shown to support higher voltages compared to GaN [112, 113]. The thickness and properties of the initial AlN layers influenced the vertical breakdown as well [101, 111, 113].

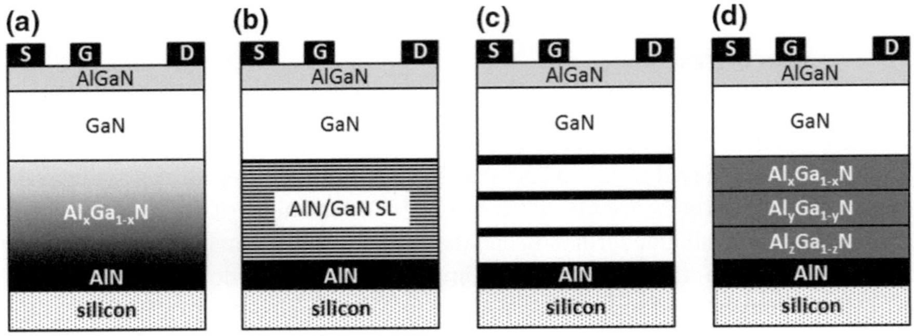

Fig. 2.5 Schematics of transistors with different base layers for strain management on Si substrates: **a** graded AlGaN layer, **b** AlN/GaN superlattice (SL), **c** AlN interlayers, and **d** step-graded $Al_xGa_{1-x}N$ layers where $x<y<z$

2.4.2 Current-Blocking Layer Deposition on Silicon Carbide Substrates

The second most widely used substrate for power transistors is SiC. Thereby, both 4H- and 6H (0001)-SiC substrates can be used, both available as conductive or highly resistive substrates. While S.I. substrates are preferred, comparable properties have been reported for devices with thick S.I. GaN base layers (≥6 μm) fabricated on n-type and S.I. SiC substrates [48]. The lattice mismatch between GaN and 4H- or 6H-SiC is about 3 % (Table 2.1), and the mismatch to AlN amounts to less than 1 %. Due to the reduced lattice mismatch and the better wetting behavior, the growth on SiC is typically initiated with AlN as well [114]. The reported AlN layer thickness values vary between several tens and hundreds of nanometers. Thereby, smooth AlN layers can be achieved after deposition of less than 50 nm of AlN [115]. For most applications, next a GaN:C layer is deposited directly onto the initial AlN layer. Typically, the GaN renucleates forming islands which coalesce in the course of the GaN deposition process [116]. U.i.d. GaN and the (Al,Ga)N gating layers are deposited similar to the process described on Si substrates. Due to the smaller thermal mismatch between GaN and SiC, relatively thick GaN layers can be deposited on SiC without cracking. As on silicon substrates, the vertical breakdown voltage of the devices increases with the thickness of the resistive buffer layer [48].

2.4.3 Current Blocking Layer Deposition on Sapphire Substrates

C-plane sapphire is the most widely used substrate for GaN-based LEDs and historically the first substrate on which thin, high-quality GaN layers were demonstrated, enabled by the introduction of low-temperature AlN or GaN nucleation layers

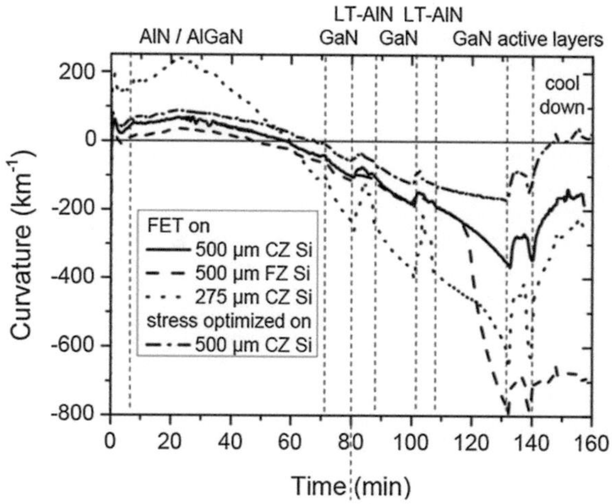

Fig. 2.6 In situ curvature measurements of a transistor structure grown on different Si substrates. Reprinted with permission from Ref. [106]. Copyright 2014 Wiley-VCH Verlag Berlin GmbH, Fed. Rep. of Germany

[19, 20]. In the typical 2-step growth process, after the deposition of an approximately 20-nm-thick nucleation layer at 550–700 °C, the wafer temperature is increased to around 1000 °C for the growth of the main GaN layer. In the initial stage, Volmer–Weber GaN islands form on the nucleation layer, which increase in diameter and height before coalescing into a planar film [117]. The GaN layer quality is strongly affected by the size and the density of the high-temperature islands, and the lower their density, the lower the threading dislocation density in the film [27].

When growing on sapphire substrates, typically an enhanced oxygen incorporation is observed in the early stages of the GaN growth. The oxygen source is primarily the sapphire substrate itself. For cleaning purposes, the substrate is typically baked in H_2 at temperatures in excess of 1000 °C prior to the deposition of the nitride layers, resulting in the formation of volatile Al and O species. Since the oxygen incorporation efficiency on the semi-polar sidewalls of the high-temperature GaN islands is orders of magnitude higher compared to planar Ga-polar GaN [118], remaining oxygen species are readily incorporated into the non-planar film. The oxygen concentration decreases to the typical low background levels (~1×10^{16} cm^{-3}, corresponding to the SIMS detection limit) as soon as the high-temperature islands have coalesced [119] and can drop further with the increasing GaN thickness [120]. Simultaneously, the threading dislocation density decreases (Fig. 2.7) [120]. Counterdoping with Fe has become a standard method to compensate for the increased oxygen incorporation in the early stages of GaN growth, enabling the fabrication of high-quality semi-insulating GaN films on sapphire substrates [50, 51]. As sapphire substrates are cheap and readily available, initial device development is often pursued on sapphire substrates and transferred to the more expensive SiC or the more challenging-to-grow-on silicon substrates at a later stage. As described before, the GaN layers can be rendered semi-insulating by doping with carbon as well (Sect. 2.2.1).

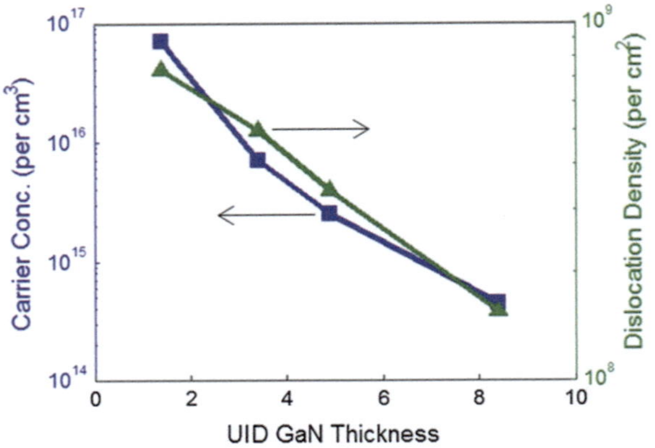

Fig. 2.7 Carrier concentration measured by Hg C–V (*squares*) and dislocation density (*triangles*) versus u.i.d. GaN thickness for p–n diodes grown on sapphire substrates. Reprinted with permission from Ref. [120]. Copyright 2016 Wiley-VCH Verlag Berlin GmbH, Fed. Rep. of Germany

2.4.4 Gating Layer Growth

The growth of AlGaN/GaN heterostructures was already discussed in Sect. 2.2.3. For D-mode devices, the gating layer is typically composed of a 25–30-nm-thick $Al_xGa_{1-x}N$ layer with $0.2 < x < 0.3$. An about 1-nm-thick AlN interlayer can be inserted between GaN channel and AlGaN layers to enhance the electron mobility.

For the fabrication of (0001) GaN enhancement-mode devices, the (Al,Ga)N layers are either grown very thin [121, 122], or recessed under the gate using etching techniques [123]. Alternately, fluorine plasma treatments [124] and epitaxial p-(Al,Ga)N cap layers have been used to eliminate the charge under the gate (Fig. 2.1) [125].

Currently, the development of lateral power switching devices is most aggressively pursued on silicon substrates; however, excellent results have also been reported on SiC substrates. Transistors targeted for 600 V operation have been demonstrated on both substrates, and laboratory results with breakdown voltages >1 kV have been reported by several groups [55, 56, 111]. Since no standardized procedure for breakdown measurements has been determined yet, a direct comparison between data from different sources is often difficult. The performance of specific devices will be discussed in detail in the device-related chapters of this book. Note that promising results were also reported on sapphire, making it an attractive substrate in particular for devices operating between 300 and 600 V, when substrate heating is less problematic [126–128].

2.5 Vertical Devices

In vertical devices, the strong electric fields are held in the bulk of the nitride crystal rather than along the surface, giving this device architecture the potential of withstanding very high electric fields. Since the electrically active device area is equal to the geometric chip area, the chip size can be markedly reduced compared to the lateral devices. As the voltage in the off-state is held in the vertical dimension, they greatly benefit from the use of very low threading dislocation density bulk GaN substrates. Fairly high breakdown voltages, however, have also been achieved for pseudo-vertical devices fabricated on foreign substrates.

The first GaN-based vertical transistor was introduced in 2002 and called current aperture vertical electron transistor (CAVET) [129, 130]. Analogous to a silicon double-diffused metal-oxide semiconductor (DMOS) structure, the CAVET consisted of a source region separated from a drain region by an insulating layer containing a narrow aperture filled with conducting material (Fig. 2.1d). The source region was comprised of a 2DEG formed at the AlGaN/GaN heterointerface, and the drain region consisted of n-type GaN. Since the virtual drain is located underneath the gate, charge does not accumulate at the gate edge (as in a lateral device) and no large fields exist near the gate. The high-field region is buried in the bulk below the gate metal, giving the CAVET the potential to support very large source–drain voltages as surface-related breakdown is eliminated. In contrast to the Si DMOS, the CAVET was, however, a normally on transistor. Since then, a variety of alternate normally off vertical device designs have been explored, such as metal–insulator–semiconductor field-effect transistors (MISFETs) and junction field-effect transistors (JFETs) (Fig. 2.8) [131, 132], and also polarization-engineered structures following the silicon superjunction transistor concept [133–135]. P–n and Schottky diodes have been investigated as well.

To simultaneously enable a low on resistance, fast switching, and high breakdown voltage, the drift region in these devices needs to be n-type doped at a level around 1×10^{16} cm^{-3} or lower and has to be sufficiently thick (Fig. 2.9). Thereby, the devices take the advantage of the high electron mobility measured for GaN layers grown on bulk GaN substrates with 1265 cm^2/Vs at $n = 3.7 \times 10^{16}$ cm^{-3} and 1750 cm^2/Vs extracted from I to V curves for layers with a free electron concentration of 3×10^{15} cm^{-3} [1, 2].

Controlled n-type doping in the regime $n \leq 1 \times 10^{16}$ cm^{-3} with a low compensation ratio to maximize the electron mobility represents new challenges for the GaN crystal growth process as well as for analytical methods to measure such low concentrations.

As discussed before, the free electron concentration in u.i.d. materials corresponds to the difference between residual donors and acceptors in the crystal (Sect. 2.2). The background concentration of C and O impurities in SIMS measurements performed under typical conditions is around 1×10^{16} cm^{-3}. While lower net carrier concentrations can be determined in electrical measurements, the nature of the residual donors and acceptors remains unknown. In order to allow stable and reproducible doping levels, relying on residual impurity doping is not desirable. Instead, the doping should be actively controlled by introducing a n-type dopant such as silicon while suppressing the concentration of the unintentional dopants C and O ideally to a level at least one order of magnitude below the targeted n-type doping. The impact of the growth conditions and the related residual impurity incorporation on the controllability of n-type doping is illustrated in Fig. 2.10. The high electron mobility reported for the layers with $n = 3 \times 10^{15}$ cm^{-3} in reference 2 alludes that

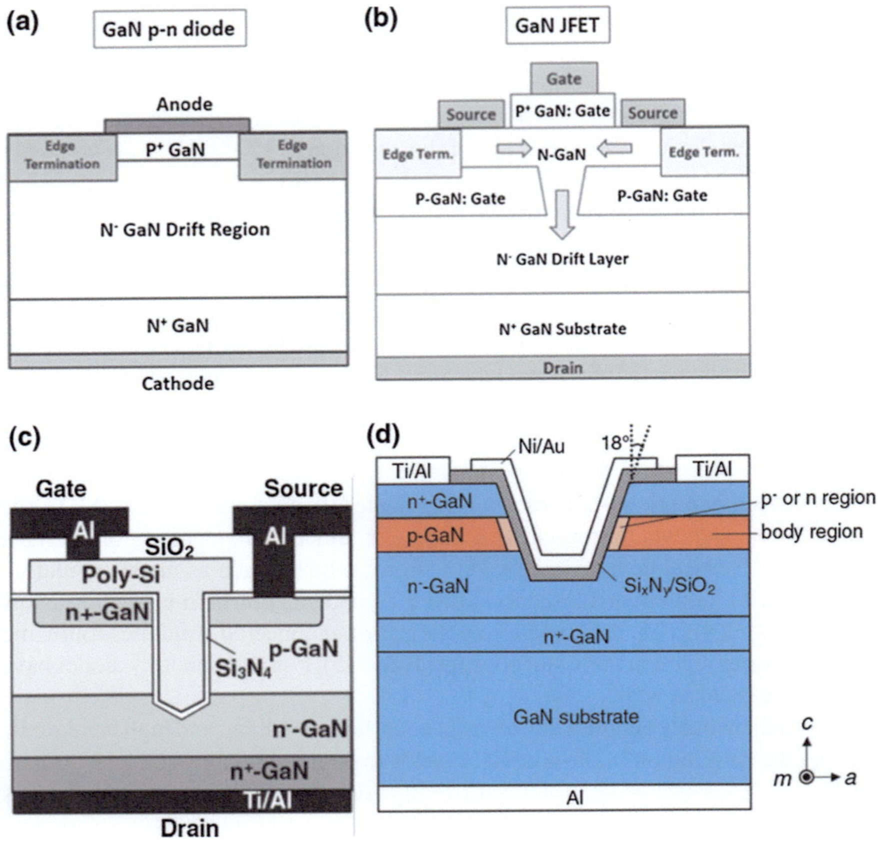

Fig. 2.8 Schematics of vertical GaN devices: **a** p–n diode, **b** JFET, and **c** and **d** MISFETs. **a**, **b** Reproduced with permission from Ref. [132], Copyright 2015, IOP Publishing. All rights reserved. **c** Reprinted with permission from Ref. [4], Copyright 2014, the Japan Society of Applied Physics. **d** Reprinted with permission from Ref. [131], Copyright 2008, the Japan Society of Applied Physics

the compensation ratio in the GaN layers was low, suggesting that the C and O levels in the films may have been indeed lower than 1×10^{15} cm^{-3}. Advances in the analytical techniques will greatly contribute to the understanding of these phenomena.

While nominally appearing straight forward, homoepitaxy of GaN has not been without problems. Dependent on substrate properties such as surface finish and unintentional miscut, the properties of thick GaN epitaxial layers can vary, as illustrated in Fig. 2.11. Generally, smoother films were obtained on lightly miscut substrates (1° to 2°) [136], preferentially toward the m-plane [2]. Devices fabricated from GaN films with rough surfaces exhibited significantly reduced breakdown voltages, which were attributed to peaks in the electric field created by the metal contacts formed on top [2, 137].

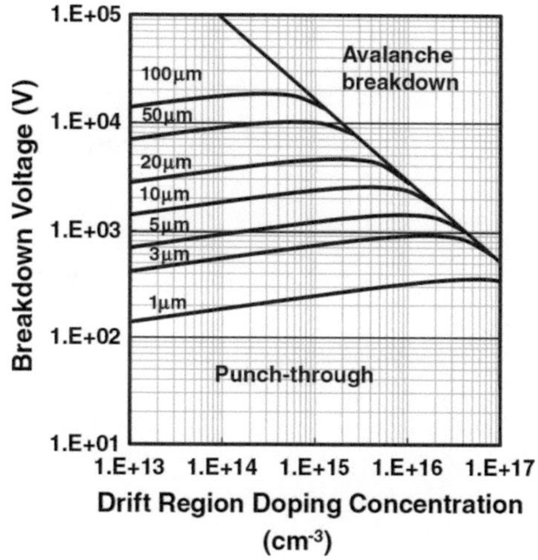

Fig. 2.9 Breakdown voltage versus drain region doping concentration for drift layer thicknesses from 1 to 100 μm. Reprinted with permission from Ref. [4]. Copyright 2014, the Japan Society of Applied Physics

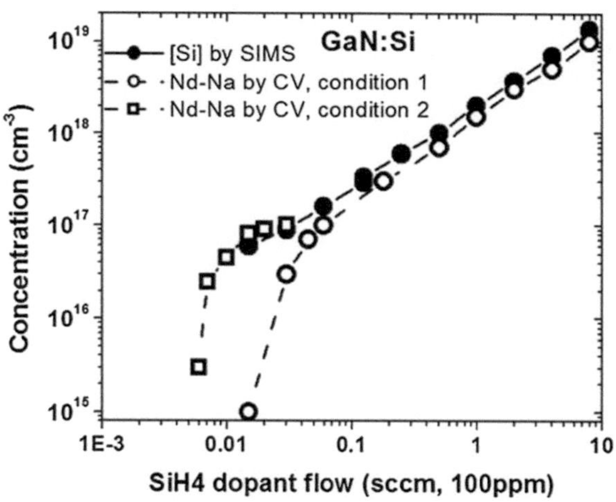

Fig. 2.10 Net doping and silicon concentration as function of silane flow. Reprinted with permission from Ref. [2]. Copyright 2015, IEEE

Under optimized conditions, breakdown voltages as high as 3.7 kV were obtained for p–n diodes with 40-μm-thick drift layers [2, 138]. The Mg doping in the 0.5-μm-thick p-GaN layers in those devices was 2×10^{19} cm^{-3}. In a different study, the breakdown voltages were found to increase with decreasing Mg doping in the p-GaN layer [139]. More recently, excellent results were reported for diodes with a hole concentration of only 7×10^{16} cm^{-3} in the p-GaN film exhibiting an avalanche breakdown voltage of >1.4 kV for devices with 8-μm-thick n-GaN layers [140].

As discussed earlier (Sect. 2.2.2), the classically doped n- and p-type layers can be replaced by the polarization-doped ones and investigations on high voltage polarization-induced vertical heterostructure p–n junction diodes have already begun (see Sect. 2.2.2) [133, 141].

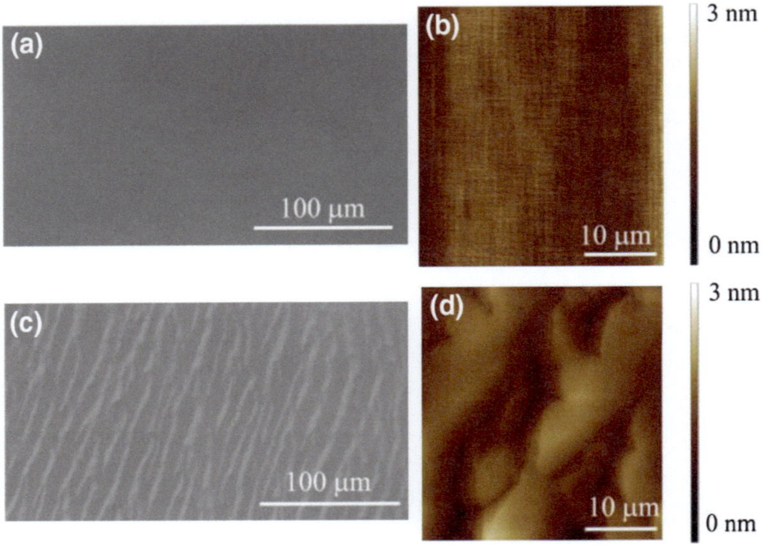

Fig. 2.11 Normaski images of **a** smooth and **c** rough surface of (Al,Ga)N films grown on bulk GaN substrates, corresponding atomic force microscopy images of **b** smooth and **d** rough surface. Reprinted with permission from Ref. [137]. Copyright 2016, Wiley-VCH Verlag Berlin GmbH, Fed. Rep. of Germany

While exhibiting reduced breakdown voltages compared to similar structures grown on low dislocation density bulk GaN substrates, relatively high breakdown voltages have also been demonstrated for pseudo-vertical p–n diodes grown on sapphire substrates [142], with a soft breakdown of 730 V for devices with 5-μm-thick drift region [120, 143]. Contrary to the previous reports [144], the impurity concentrations in GaN-on-sapphire films can be as low as those observed on bulk GaN substrates [120]. For GaN-on-Si p–n diodes with 1.5-μm-thick drift layer, a soft breakdown voltage higher than 300 V was measured, corresponding to a peak electric field of 2.9 MV/cm [145, 146]. Also, fabricated Schottky diodes exhibited a breakdown voltage of 205 V. Lower breakdown voltages in Schottky compared to p–n diodes have also been reported by others [143].

For vertical transistors with apertures (Figs. 2.1 and 2.8), additional challenges in the device fabrication arise from the need to locally block the current path in the lateral direction, requiring potentially multiple regrowth steps, often on etched surfaces. For example, in an epitaxial CAVET structure, the growth is either interrupted after p-GaN layer growth to locally remove the p-GaN layer via etching and continues the growth with the GaN aperture and the HEMT top part (Fig. 2.12a) [130, 147], or after deposition of the n-drift layer to etch the surface and regrow the p-GaN blocking layer and again the HEMT top part of the device structure (Fig. 2.12b) [71, 148]. Similar process steps are required for the fabrication of a junction field-effect transistor (Fig. 2.8).

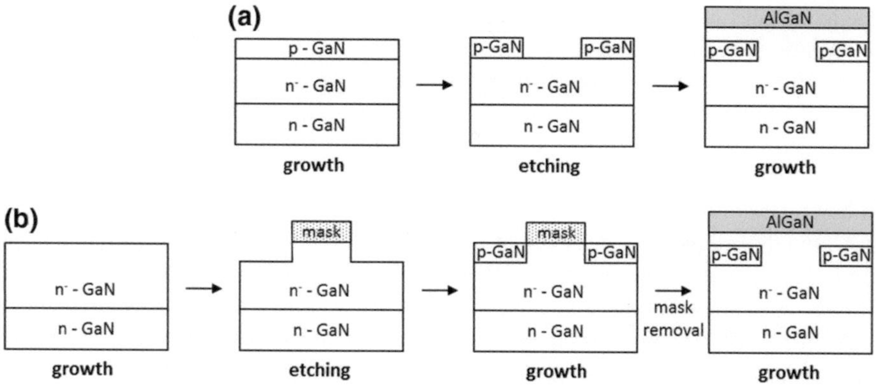

Fig. 2.12 Process flow for the fabrication of a vertical FET (CAVET) via regrowth of **a** aperture and top HEMT and **b** p-GaN current-blocking region and top HEMT

Note that in order to mitigate repassivation of p-GaN:Mg layers in subsequent MOCVD growth steps (Sect. 2.2.2) in the past also NH_3-MBE has been used to complete the CAVET layer structure [148].

Dry etching processes are well known for damaging the nitride crystal, and careful process optimization is required to minimize the harm [149]. In order to heal etch-related damage, methods such as nitrogen plasma treatment [150, 151] and annealing in ammonia/nitrogen mixtures have been used [152, 153]. The impact of the annealing conditions on the 2DEG properties of regrown AlGaN transistor layers was investigated in reference [154]. Very good results were also obtained when combining dry with wet etching steps [146, 155].

Photoelectrochemical wet etching techniques have also been employed for aperture fabrication in CAVETs [156].

2.6 Outlook

2.6.1 InAlN and AlInGaN Barrier Layers

Restrictions in the device design related to the lattice mismatch between $Al_xGa_{1-x}N$ and GaN can be overcome by replacing $Al_xGa_{1-x}N$ with InAlN or AlInGaN layers. InAlN is lattice matched to GaN at an indium composition of ~18 % and offers a high spontaneous polarization charge and band gap [157], enabling the fabrication of devices operating at very high currents [158, 159]. Details related to the MOCVD growth of InAlN can be found in references [160] and [161], on the deposition of InAlN/AlN/GaN heterostructures in reference [162], for example. Note that in InAlN heterostructures, the insertion of AlN interlayers is even more critical than for AlGaN heterostructures, as InAlN/GaN heterostructures exhibit very low electron mobilities due to scattering caused by alloy fluctuations in the InAlN layer [163], similar to the observations in InGaN [164].

In addition the exploration of transistors with AlInGaN, gating layers have begun [165]. In heterostructures with quaternary layers, the spontaneous and electrical polarization can be tuned over a wide range [78], allowing the design of depletion as well as enhancement-mode transistors [166]. Interestingly, the electron mobility in heterostructures with quaternary AlInGaN layers was found to be higher than in InAlN/GaN heterostructures even at low Ga concentrations in the quaternary films [167, 168]. In addition, devices with quaternary gating layers exhibited less gate leakage, also contributing to the superior transistor performance compared to devices with InAlN layers [169]. For a detailed discussion of the design space of polarization-engineered transistors utilizing quaternary layers see reference [133].

2.6.2 Devices Based on Non-c-plane GaN

While all device architectures discussed so far were based on the growth of typical (0001) or *c*-plane GaN, an alternate approach to E-mode devices are heterostructures on nonpolar *a*- or *m*-plane GaN. In the absence of spontaneous and piezoelectric polarization, the 2DEG in these AlGaN/GaN heterostructures only forms if electrons are supplied by n-type doping, typically using Si as dopant [170], as well known for transistors based on GaAs and InP. First, results on *m*- [171] and *a*-GaN [172]-based E-mode transistors have been demonstrated. Hampering their widespread investigation and application is the high defect density of nonpolar GaN grown on foreign substrates and the small size of the currently available native *m*- and *a*-plane GaN substrates. Further developments may overcome these limitations. Future investigations will also have to show whether there are any benefits in fabricating vertical transistors on *m*- or *a*-GaN.

Another attractive alternative are transistors based on $(000\bar{1})$ or N-polar GaN. The opposite direction of the internal electric fields in N-polar compared to Ga-polar GaN heterostructures results in several advantages [173]. Since the 2DEG now forms on top of the (Al,Ga,In)N barrier layer (Fig. 2.13), the latter resembles a natural back-barrier leading to improved carrier confinement. Furthermore, when a thin AlGaN cap layer is added on top of the GaN channel, the electric field direction in the cap layer is reversed compared to the GaN channel, opposing electron leakage from the gate [174]. Additionally, no barrier thinning occurs under pinch-off conditions. Moreover, E-mode devices can be easily devised by implementing thicker AlGaN or p-type GaN:Mg cap layers (Fig. 2.13) [175].

Resolving N-polar growth issues described in earlier reports, high-quality and high purity N-polar (Al,Ga,In)N heterostructures have been demonstrated in the past years [176], enabling the fabrication of high-performance N-polar RF and high-frequency transistors [173], and paving the way for the exploration of N-polar transistors for power switching applications.

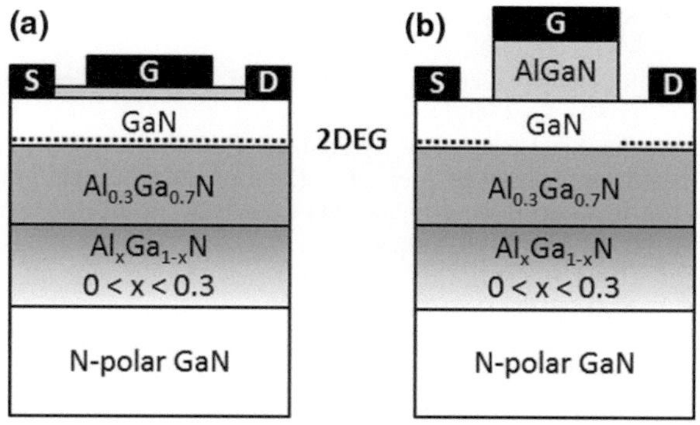

Fig. 2.13 Schematics of N-polar transistor structures: **a** D-mode structure with thin AlGaN cap layer and **b** E-mode structure with thick AlGaN layer under the gate

Further developments in the growth of lateral and vertical (Al,Ga)N heterostructures are expected in the future. While LED production processes have matured greatly in the past decades and GaN-based blue and white LEDs have become commodity products, the investigation and utilization of GaN-based transistors for power switching applications is still in the beginning. This chapter hopes to have introduced the reader to key components in the epitaxial process for these devices.

The author would like to note that the references in this chapter are by no means complete and in many cases should be viewed as examples. The work at UCSB was supported by DARPA, ONR, AFOSR and DoE through programs monitored by Drs. Harry Dietrich, Paul Macki, Gerald Witt, and Timothy Heidel. In addition, the author would like to thank Drs. Brian Swenson and Matthew Laurent for critical reading of the manuscript.

References

1. Kyle E, Kaun SW, Burke PG, Wu F, Wu YR, Speck JS (2014) J Appl Phys 115:193702
2. Kizilyalli IC, Edwards AP, Aktas O, Prunty T, Bour D (2015) IEEE Trans Electron Devices 62:414
3. Chen Z, Pei Y, Newman S, Chu R, Brown D, Chung R, Keller S, DenBaars SP, Nakamura S, Mishra UK (2009) Appl Phys Lett 94:112108
4. Kachi T (2014) Jpn J Appl Phys 53:100210
5. Kaun SW, Wong MH, Mishra UK, Speck JS (2013) Semicond Sci Technol 28:074001
6. Liu L, Edgar JH (2002) Mater Sci Eng R 37:61
7. Zhang NQ, Moran B, DenBaars SP, Mishra UK, Wang XW, Ma TP (2001) IEDM 01-589
8. Krost A, Dadgar A (2002) Mater Sci Eng B 93:77
9. Lahreche H, Vennegues P, Beaumont B, Gibart P (1999) J Cryst Growth 205:245
10. Beaumont B, Vennegues P, Gibart P (2001) Phys Stat Sol (B) 227:1
11. Choi S, Heller E, Dorsey D, Vetury R, Graham S (2013) J Appl Phys 113:093510
12. Xu JJ, Wu YF, Keller S, Parish G, Heikman S, Thibeault BJ, Mishra UK, York RA (1999) IEEE Microwave Guided Wave Lett 9:277
13. Dadgar A, Hums C, Dietz A, Schulze F, Bläsing J, Krost A (2006) Proc SPIE 6355:63550R
14. Paskova T, Evans KR (2009) IEEE J Sel Top Quantum Electron 15:1041

15. Kawamura F, Umeda H, Morishita M, Kawahara M, Yoshimura M, Mori Y, Sasaki T, Kitaoka Y (2006) Jpn J Appl Phys 45:L1136
16. Kozodoy P, Ibbeston JP, Marchand H, Fini PT, Keller S, Speck JS, DenBaars SP, Mishra UK (1998) Appl Phys Lett 73:975
17. Cao XA, Hu H, LeBoeuf SF, Cowen C, Arthur SD, Wang W (2005) Appl Phys Lett 87:053503
18. Mion C, Muth JF, Preble EA, Hanser D (2006) Appl Phys Lett 89:092123
19. Amano H, Sawaki N, Akasaki I, Toyoda Y (1986) Appl Phys Lett 48:353
20. Nakamura S (1991) Jpn J Appl Phys 30:L1705
21. Han J, Figiel JJ, Crawford MH, Banas MA, Bartram ME, Biefeld RM, Song YK, Nurmikko AV (1998) J Cryst Growth 195:291
22. Zhao DG, Zhu JJ, Jiang DS, Yang H, Liang JW, Li XY, Gong HM (2006) J Cryst Growth 289:72
23. Edgar J (ed) (1994) Properties of group-III nitrides, INSPEC
24. Koide Y, Itoh H, Khan MRH, Hiramatsu K, Sawaki N, Akasaki I (1987) J Appl Phys 61:4540
25. Wu XH, Brown LM, Kapolnek D, Keller S, Keller B, DenBaars SP, Speck JS (1996) J Appl Phys 80:3230
26. Heying B, Wu XH, Keller S, Li Y, Kapolnek D, Keller BP, DenBaars SP, Speck JS (1996) Appl Phys Lett 68:643
27. Fini PT, Wu X, Tarsa EJ, Golan Y, Srikant V, Keller S, DenBaars SP, Speck JS (1998) Jpn J Appl Phys 37:4460
28. Moram MA, Vickers ME (2009) Rep Prog Phys 72:036502
29. Kapolnek D, Wu XH, Heying B, Keller S, Keller BP, Mishra UK, DenBaars SP, Speck JS (1995) Appl Phys Lett 67:1541
30. Zywietz TK, Neugebauer J, Scheffler M (1999) Appl Phys Lett 74:1695
31. Wetzel C, Suski T, Ager JW III, Weber ER, Haller EE, Fischer S, Meyer BK, Molnar RJ, Perlin P (1997) Phys Rev Lett 78:3923
32. Gordon L, Lyons JL, Janotti A, Van de Walle CG (2014) Phys Rev B 89:085204
33. Lyons JL, Janotti A, Van de Walle CG (2010) Appl Phys Lett 97:152108
34. Lyons JL, Janotti A, Van de Walle CG (2014) Phys Rev B 89:035204
35. Tanaka T, Kaneda N, Mishima T, Kihara Y, Aoki T, Shiojima K (2015) Jpn J Appl Phys 54:041002
36. Tuomisto F, Makkonen I (2013) Rev Mod Phys 85:1583
37. Armstrong A, Arehart AA, Moran B, DenBaars SP, Mishra UK, Speck JS, Ringel SA (2004) Appl Phys Lett 84:374
38. Armitage R, Hong W, Yang Q, Feick H, Gebauer J, Weber ER, Hautakangas S, Saarinen K (2003) Appl Phys Lett 82:3457
39. Weimann NG, Eastman LF, Doppalapudi D, Ng HM, Moustakas TD (1998) J Appl Phys 83:3656
40. Look DC, Sizelove JR (1999) Phys Rev B 82:1237
41. Albrecht M, Cremades A, Krinke J, Christiansen S, Ambacher O, Piqueras J, Strunk HP, Stutzmann M (1999) Phys Stat Sol B 216:409
42. Li G, Chua SJ, Xu SJ, Wang W, Li P, Beaumont B, Gibart P (1999) Appl Phys Lett 74:2821
43. Lei H, Leipner HS, Schreiber J, Weyher JL, Wosinski T, Grzegory I (2002) J Appl Phys 92:6666
44. Koleske DD, Wickenden AE, Henry RL, Twigg ME (2002) J Cryst Growth 242:55
45. Stringfellow GB (1989) Organometallic vapor phase epitaxy. Academic Press, San Diego
46. Parish G, Keller S, DenBaars SP, Mishra UK (2000) J Electron Mater 29:15
47. Kuech, TF, Wolford DJ, Veuhoff E, Deline V, Mooney PM, Potemski R, Bradley J (1987) J Appl Phys 62:632
48. Bahat-Treidel E, Brunner F, Hilt O, Cho E, Würfl J, Tränkle G (2010) IEEE Trans Electron Devices 57:3050
49. Sugiyama T, Honda Y, Yamaguchi M, Amano H, Imade M, Mori Y (2012) International workshop on nitride semiconductors, Sapporo, Japan, 14–19 Oct 2012
50. Heikman S, Keller S, DenBaars SP, Mishra UK (2002) Appl Phys Lett 81:439

51. Heikman S, Keller S, Mates T, DenBaars SP, Mishra UK (2003) J Cryst Growth 248:513
52. Malguth E, Hoffmann A, Gehlhoff W (2006) Phys Rev B 74:165202
53. Würfl J, Hilt O, Bahat-Treidel E, Zhytnytska R, Kotara P, Krüger O, Brunner F, Weyers M (2013) Phys Stat Sol C 10:1393
54. Würfl J, Bahat-Treidel E, Brunner F, Cho M, Hilt O, Knauer A, Kotara P, Krueger O, Weyers M, Zhytnytska R (2012) ECS Trans 50:211
55. Ikeda N, Niiyama Y, Kambayashi H, Sato Y, Nomura T, Kato S, Yoshida S (2010) Proc IEEE 98:1151
56. Rowena IB, Selvaraj SL, Egawa T (2011) IEEE Electron Device Lett 32:1534
57. Koide N, Kato H, Sassa M, Yamasaki S, Manabe K, Hashimoto H, Amano H, Hiramatsu K, Akasaki I (1991) J Cryst Growth 115:639
58. Rowland LB, Doverspike K, Gaskill DK (1995) Appl Phys Lett 66:1495
59. Nakamura S, Iwasa N, Sehoh M, Mukai T (1992) Jpn J Appl Phys 31:1258
60. Kozodoy P, Xing H, DenBaars SP, Mishra UK, Saxler A, Perrin R, Elhamri S, Mitchel WC (2000) J Appl Phys 87:1832
61. Tanaka T, Watanabe A, Amano H, Kobayashi Y, Akasaki I, Yamazaki S, Koike M (1994) Appl Phys Lett 65:593
62. Keller S, Kozodoy P, Mishra UK, DenBaars SP (1999) US patent 5891790
63. Fichtenbaum NA, Schaake C, Mates TE, Cobb C, Keller S, DenBaars SP, Mishra UK (2007) Appl Phys Lett 91:172105
64. Kumakura K, Makimoto T, Kobayashi N (2003) J Appl Phys 93:3370
65. Suzuki M, Nishio J, Onomura M, Hongo C (1998) J Cryst Growth 189/190:511
66. Xing H, Green DS, Yu H, Mates T, Kozodoy P, Keller S, DenBaars SP, Mishra UK (2003) Jpn J Appl Phys 42:50
67. Tomita K, Itoh K, Ishiguro O, Kachi T, Sawaki N (2008) J Appl Phys 104:014906
68. Chowdhury S, Swenson BL, Lu J, Mishra UK (2011) Jpn J Appl Phys 50:101002
69. Chowdhury S (2010) PhD thesis, University of California, Santa Barbara
70. Feigelson BN, Anderson TJ, Abraham M, Freitas JA, Hite JK, Eddy CR, Kub FJ (2012) J Cryst Growth 350:21
71. Chowdhury S, Swenson BL, Wong MH, Mishra UK (2013) Semicond Sci Technol 28:074014
72. Bernardini F, Fiorentini V, Vanderbilt D (1997) Phys Rev B 56:R 10024
73. Ambacher O, Foutz B, Smart J, Shealy JR, Weimann NG, Chu K, Murphy M, Sierakowski AJ, Schaff WJ, Eastman LF, Dimitrov R, Mitchell A, Stutzmann M (2000) J Appl Phys 87:334
74. Jena D, Heikman S, Green D, Buttari D, Coffie R, Xing H, Keller S, DenBaars SP, Speck JS, Mishra UK, Smorchkova I (2002) Appl Phys Lett 81:4395
75. Rajan S, Xing H, DenBaars SP, Mishra UK, Jena D (2004) Appl Phys Lett 84:1591
76. Simon J, Protasenko V, Lian C, Xing H, Jena D (2010) Science 237:60
77. Enatsu Y, Gupta C, Laurent M, Keller S, Nakamura S, Mishra UK. Submitted for publication
78. Jena D, Simon J, Wang A, Cao Y, Goodman K, Verma J, Ganguly S, Li G, Karda K, Protasenko V, Lian C, Kosel T, Fay P, Xing H (2011) Phys Stat Sol A 208:1511
79. Keller S, Parish G, Fini PT, Heikman S, Chen CH, Zhang N, DenBaars SP, Mishra UK (1999) J Appl Phys 86:5850
80. Li H, Keller S, DenBaars SP, Mishra UK (2014) Jpn J Appl Phys 53:095504
81. Derluyn J, Boeykens S, Cheng K, Vandersmissen R, Das J, Ruythooren W, Degroote S, Leys MR, Germain M, Borghs G (2005) J Appl Phys 98:054501
82. Smorchkova IP, Chen L, Mates T, Shen L, Heikman S, Moran B, AKeller S, DenBaars SP, Mishra UK (2002) Appl Phys Lett 81:439
83. Mazumder B, Kaun SW, Lu J, Keller S, Mishra UK, Speck JS (2013) Appl Phys Lett 102:111603
84. Kaun SW, Burke PG, Wong MH, Kyle ECH, Mishra UK, Speck JS (2012) Appl Phys Lett 101:262102
85. Binary SC, Ikossi K, Roussos JA, Kruppa W, Park D, Dietrich HB, Koleske DD, Wickenden AE, Henry RL (2001) IEEE Trans Electron Devices 48:465
86. Vetury R, Zhang NQ, Keller S, Mishra UK (2001) IEEE Trans Electron Dev 48:560

87. Hinoki A, Kikawa J, Yamada T, Tsuchiya T, Kamiya S, Kurouchi M, Kosaka K, Araki T, Suzuki A, Nanishi Y (2008) Appl Phys Express 1:011103
88. Green BM, Chu KK, Chumbes EM, Smart JA, Shealy JR, Eastman LF (2000) IEEE Electron Device Lett 21:268
89. Shen L, Coffie R, Buttari D, Heikman S, Chakraborty A, Chini A, Keller S, DenBaars SP, Mishra UK (2004) IEEE Electron Device Lett 25:7
90. Coffie R, Buttari D, Heikman S, Chini A, Keller S, DenBaars SP, Mishra UK (2002) IEEE Electron Device Lett 23:588
91. Poblenz C, Waltereit P, Rajan S, Heikman S, Mishra UK, Speck JS (2004) J Vac Sci Technol B 22:1145
92. Simpkins BS, Yu ET, Waltereit P, Speck JS (2003) J Appl Phys 94:1448
93. Shiojima K, Suemitsu T (2003) J Vac Sci Technol B 21:698
94. Yanagihara M, Uemoto Y, Ueda T, Tanaka T, Ueda D (2009) Phys Stat Sol (a) 206:1221
95. Umeda H, Suzuki A, Anda Y, Ishida M, Ueda T, Tanaka T, Ueda D (2010) IEEE, IEDM 10-480
96. Lahreche H, Vennegues P, Totterau O, Laüt M, Lorenzini P, Leroux M, Beaumont B, Gibart P (2000) J Cryst Growth 217:13
97. Liu R, Ponce FA, Dadgar A, Krost A (2003) Appl Phys Lett 83:860
98. Marchand H, Zhao L, Zhang N, Moran B, Coffie R, Mishra UK, Speck JS, DenBaars SP (2001) J Appl Phys 89:7846
99. Raghavan S, Redwing J (2005) J Appl Phys 98:023515
100. Feltin E, Beaumont B, Laügt M, de Mierry P, Vennégués P, Lahrèche H, Leroux M, Gibart P (2001) Appl Phys Lett 79:3230
101. Arulkumaran S, Egawa T, Matsui S, Ishikawa H (2005) Appl Phys Lett 86:123503
102. Reiher A, Bläsing J, Dadgar A, Diez A, Krost A (2003) J Cryst Growth 248:563
103. Cheng K, Leys M, Dergoote S, Van Daele B, Boeykens S, Derluyn J, Germain M, Van Tendeloo G, Engelen J, Borghs G (2006) J Electron Mater 35:592
104. Clos R, Dadgar A, Krost A (2004) Phys Stat Sol A 201:R75
105. Dadgar A, Schulze F, Zettler T, Haberland K, Clos R, Strassburger G, Bläsing J, Dietz A, Krost A (2004) J Cryst Growth 272:72
106. Schulz O, Dadgar A, Henning J, Krumm O, Fritze S, Bläsing J, Witte H, Dietz A, Krost A (2014) Phys Stat Sol (c) 11:397
107. Cantu P, Wu F, Waltereit P, Keller S, Romanov AE, DenBaars SP, Speck JS (2005) J Appl Phys 97:103534
108. Raghavan S, Redwing J (2005) J Appl Phys 98:023514
109. Dadgar A, Poschenrieder M, Bläsing J, Fehse K, Dietz A, Krost A (2002) Appl Phys Lett 80:3670
110. Krost A. Personal communication
111. Cheng K (2015) www.compoundsemiconductor.net, p 36, Mar 2015
112. Visalli D, Van Hove M, Derluyn J, Degroote S, Leys M, Cheng K, Germain M, Borghs G (2009) Jpn J Appl Phys 48:04C101
113. Cheng K, Liang H, Van Hove M, Geens K, DeJaeger B, Srivastava P, Kang X, Favia P, Bender H, Decoutere S, Dekoster J, del Agua Borniquel JI, Jun SW, Chung H (2012) Appl Phys Express 5:011002
114. Weeks TW, Bremser MD, Ailey KS, Carlson E, Perry WG, Davis RF (1995) Appl Phys Lett 67:401
115. Moe CG, Wu Y, Keller S, Speck JS, DenBaars SP, Emerson D (2006) Phys Stat Sol (a) 203:1708
116. Moran B, Wu F, Romanov AE, Mishra UK, DenBaars SP, Speck JS (2004) J Cryst Growth 273:38
117. Wu XH, Fini P, Keller S, Tarsa EJ, Heying B, Mishra UK, DenBaars SP, Speck JS (1996) Jpn J Appl Phys 35:L1648
118. Cruz S, Keller S, Mates T, Mishra UK, DenBaars SP (2009) J Cryst Growth 311:3817

119. Popovici G, Kim W, Botchkarev A, Tang H, Morkoc H, Solomon J (1997) Appl Phys Lett 71:3385
120. Gupta C, Enatsu Y, Gupta G, Keller S, Mishra UK (2016) Phys Stat Sol (a) 213:878
121. Khan MA, Chen Q, Sun CJ, Yang JW, Blasingame M, Shur MS, Park H (1996) Appl Phys Lett 68:514
122. Ohmaki Y, Tanimoto M, Akamatsu S, Mukai T (2006) Jpn J Appl Phys 45:L1168
123. Lanford WB, Tanaka T, Otoki Y, Adesida I (2005) Electron Lett 41:449
124. Cai Y, Zhou Y, Chen KJ, Lau KM (2005) IEEE Electron Device Lett 26:435
125. Hu X, Simin G, Yang J, Khan MA, Gaska R, Shur MS (2000) Electron Lett 36:753
126. Saito W, Kuraguchi M, Takada Y, Tsuda K, Omura I, Ogura T (2004) IEEE Trans Electron Devices 51:1913
127. Matocha K, Chow TP, Gutmann RJ (2005) IEEE Trans Electron Devices 52:6
128. Shi J, Eastman LF, Xin X, Pophristic M (2009) Appl Phys Lett 95:042103
129. Ben-Yaacov I, Sek YK, Heikman S, DenBaars SP, Mishra UK (2002) In: Device research conference, Santa Barbara, USA (Cat. No.02TH8606), p 31–32
130. Ben-Yaacov I, Sek YK, Heikman S, DenBaars SP, Mishra UK (2004) J Appl Phys 95:2073
131. Otake H, Chikamatsu K, Yamaguchi A, Fujishima T, Ohta H (2008) Appl Phys Express 1:011105
132. Kizilyalli IC, Aktas O (2015) Semicond Sci Technol 30:124001
133. Ueda T, Murata T, Nakazawa S, Ishida H, Uemoto Y, Inoue K, Tanaka T, Ueda D (2010) Phys Stat Sol (b) 247:1735
134. Song B, Zhu M, Hu Z, Nomoto K, Jena D, Xing HG (2015) In: Proceedings of IEEE 27th international symposium on power semiconductor devices & ICs (ISPSD), Hong Kong, China, p 273, May 2015
135. Li Z, Chow TP (2013) IEEE Electron Device Lett 60:3230
136. Xu X, Vaudo RP, Flynn J, Dion J, Brandes GR (2005) Phys Stat Sol (a) 202:727
137. Tanabe S, Watanabe N, Uchida M, Matsuzaki H (2016) Phys Stat Sol (a) 213:1236
138. Kizilyally IC, Edwards AP, Nie H, Bour D, Prunty T, Disney D (2014) IEEE Electron Device Lett 35:247
139. Yoshhizumi Y, Hashimoto H, Tanabe T, Kiyama M (2007) J Cryst Growth 298:875
140. Hu Z, Nomoto K, Song B, Zhu M, Qi M, Pan M, Gao X, Protasenko V, Jena D, Xing HG (2015) Appl Phys Lett 107:234501
141. Qi M, Namoto K, Zhu M, Hu Z, Zhao Y, Song B, Li G, Fay P, Xing H, Jena D (2015) In: 73rd Annual device research conference, Columbus, OH, USA, 21–24 June 2015
142. Alquier D, Cayrel F, Menard O, Bazin AE, Yvon A, Collard E (2012) Jpn J Appl Phys 51:01AG08
143. Zhang AP, Dang GT, Ren F, Cho H, Lee KP, Pearton SJ, Chyi JI, Nee TE, Chuo CC (2001) IEEE Trans Electron Devices 48:407
144. Hashimoto S, Yoshizumi Y, Tanabe T, Kiyama M (2007) J Cryst Growth 298:871
145. Zhang Y, Sun M, Ppiedra D, Azize M, Zhang X, Fujishima T, Palacios T (2014) IEEE Electron Device Lett 35:618
146. Zhang Y, Sun M, Wong HY, Lin Y, Srivastava P, Hatem C, Azize M, Piedra D, Yu L, Sumitomo T, de Braga NA, Mickevicius RV, Palacios T (2015) IEEE Trans Electron Devices 62:2155
147. Kanachika M, Sugimoto M, Soejima N, Ueda H, Ishiguro O, Kodama M, Hayashi E, Itoh K, Uesugi T, Kachi T (2007) Jpn J Appl Phys 21:L503
148. Yeluri R, Lu J, Hurni CA, Browne DA, Chowdhury S, Keller S, Speck JS, Mishra UK (2015) Appl Phys Lett 106:183502
149. Haberer ED, Chen CH, Hansen M, Keller S, DenBaars SP, Mishra UK, Hu EL (2001) J Vac Sci Technol B 19:603
150. Lee J-M, Chang K-M, Kim S-W, Huh C, Lee I-H, Park S-J (2000) J Appl Phys 87:7667
151. Mouffak Z, Bensaoula A, Trombetta L (2004) J Appl Phys 95:727
152. Moon Y-T, Kim D-J, Park J-S, Oh J-T, Lee J-M, Park S-J (2004) J Vac Sci Technol B 22:489

153. Keller S, Schaake C, Fichtenbaum NA, Neufeld CJ, Wu Y, McGroddy K, David A, DenBaars SP, Weisbuch C, Speck JS, Mishra UK (2006) J Appl Phys 100:054314
154. Chan SH, Keller S, Tahhan M, Li H, Mishra UK (2016) Semicond Sci Technol 31:065008
155. Kodama M, Sugimoto M, Hayashi E, Soejima N, Ishiguro O, Kanechika M, Itoh K, Ueda H, Uesugi T, Kachi T (2008) Appl Phys Express 1:021104
156. Gao Y, Ben-Yaacov I, Mishra UK, Hu EL (2004) J Appl Phys 96:6925
157. Kuzmık J (2001) IEEE Electron Device Lett 22(11):510
158. Medjdoub F, Ducatteau D, Gaquière C, Carlin J-F, Gonschorek M, Feltin E, Py MA, Grandjean N, Kohn E (2007) Electron Lett 43:309
159. Sarazin N, Jardel O, Morvan E, Aubry R, Laurent M, Magis M, Tordjman M, Alloui M, Drisse O, Di Persio J, di Forte Poisson MA, Delage SL, Vellas N, Gaquière C, Théron D (2007) Electron Lett 43:1317
160. Sadler T, Kappers M, Oliver R (2009) J Cryst Growth 311:3380
161. Chung RB, Wu F, Shivaraman R, Keller S, DenBaars SP, Speck JS, Nakamura S (2011) J Cryst Growth 324:163
162. Gonschorek M, Carlin JF, Feltin E, Py MA, Grandjean N, Darakchieva V, Monemar B, Lorenz M, Ramm G (2008) J Appl Phys 103:093714
163. Kaun SW, Ahmadi E, Mazumder B, Wu F, Kyle ECH, Burke PG, Mishra UK, Speck JS (2014) Semicond Sci Technol 29:045011
164. Wu YR, Shivaraman R, Wang KC, Speck JS (2012) Appl Phys Lett 101:083505
165. Reuters B, Wille A, Holländer B, Sakalauskas E, Ketteniss N, Mauder C, Goldhahn R, Heuken M, Kalisch H, Vescan A (2012) J Electron Mater 41:905
166. Reuters B, Wille A, Ketteniss N, Hahn H, Holländer B, Heuken M, Kalisch H, Vescan A (2013) J Electron Mater 42:826
167. Ketteniss N, Khoshroo LR, Eichelkamp M, Heuken M, Kalisch H, Jansen RH, Vescan A (2010) Semicond Sci Technol 25:075013
168. Wang R, Li G, Verma J, Sensale-Rodriguez B, Fang T, Guo J, Hu Z, Laboutin O, Cao Y, Johnson W, Snider G, Fay P, Jena D, Xing H (2011) IEEE Electron Device Lett 32:1215
169. Makiyama K, Ozaki S, Ohki T, Okamoto N, Minoura Y, Niida Y, Kamada Y, Joshin1 K, Watanabe K, Miyamoto Y (2015) IEDM
170. Fujiwara T, Keller S, Higashiwaki M, Speck JS, DenBaars SP, Mishra UK (2009) Appl Phys Express 2:061003
171. Fujiwara T, Rajan S, Keller S, Higashiwaki M, Speck JS, DenBaars SP, Mishra UK (2009) Appl Phys Express 2:011001
172. Kuroda M, Ishida H, Ueda T, Tanaka T (2007) J Appl Phys 102:093703
173. Wong MH, Keller S, Nidhi, Dasgupta S, Denninghoff D, Kolluri S, Brown DF, Lu J, Fichtenbaum NA, Ahmadi E, Singisetti U, Chini A, Rajan S, DenBaars SP, Speck JS, Mishra U (2013) Semicond Sci Technol 28:074009
174. Wienecke S, Romanczyk B, Guidry M, Li H, Zheng X, Ahmadi E, Hestroffer K, Megalini L, Keller S, Mishra UK. Submitted for publication
175. Singisetti U, Wong MH, Mishra UK (2013) Semicond Sci Technol 28:074006
176. Keller S, Li H, Laurent M, Hu Y, Pfaff N, Lu J, Brown DF, Fichtenbaum NA, Speck JS, DenBaars SP, Mishra UK (2014) Semicond Sci Technol 29:113001

Chapter 3
GaN-on-Silicon CMOS-Compatible Process

Denis Marcon and Steve Stoffels

3.1 GaN-on-Si Epitaxy

In order to process GaN-on-Si wafers in a CMOS fab, they need to meet the basic criteria for wafer bow. However, due to the large lattice and thermal mismatch between (Al)GaN and Si, growing high-quality crack-free GaN epitaxial layer on 200-mm Si substrates with low bow requires an intensive optimization of the epitaxy [1].

Indeed, the main challenge at the epitaxial level is to obtain a high and uniform epitaxial quality combined with a sufficiently low wafer bow, below 50 μm, to allow processing in a CMOS fab. The wafer bow has been successfully controlled below ±50 μm by using stress-mitigating buffer layers as well as 1.15-mm-thick Si substrates instead of the standard 0.725-mm-thick substrates.

An example of stress-mitigating buffer layers is reported in Fig. 3.1. In this case, the buffer consists of a graded AlGaN layer where the Al % is graded from 100 % (AlN nucleation layer) down to 25 % before growing the GaN channel layer (inset Fig. 3.1). Moreover, the buffer layer is engineered in order to reduce the dislocations density in the device active region: in Fig. 3.1 it is possible to notice that most of the dislocations end in the Al(Ga)N buffer layers and only few reach the surface, i.e., the active region. More details on the GaN-on-Si epitaxy can be found elsewhere [1, 2].

The reproducibility of the wafers is also another crucial aspect for high-volume manufacturing of GaN technology. This has been assessed within a lot of identical wafers showing a uniform and reproducible 2DEG sheet resistivity (Fig. 3.2a) as well as a bow below the specification of ±50 μm (Fig. 3.2b), which is the typical maximum allowed bow of wafer to be processed in the CMOS fab.

D. Marcon (✉) · S. Stoffels
IMEC, Leuven, Belgium
e-mail: marcond@imec.be

M. Meneghini et al. (eds.), *Power GaN Devices*,
Power Electronics and Power Systems, DOI 10.1007/978-3-319-43199-4_3

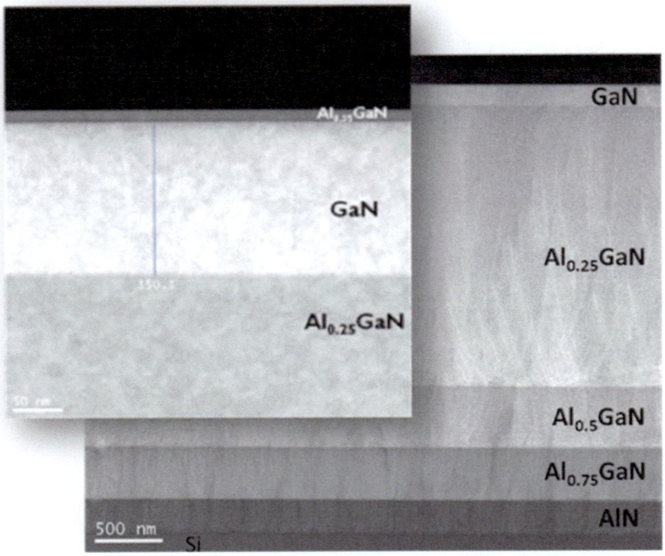

Fig. 3.1 TEM image of the full GaN-on-Si epitaxial stack. The *upper part*, i.e., active region of the stack is magnified in the *inset*

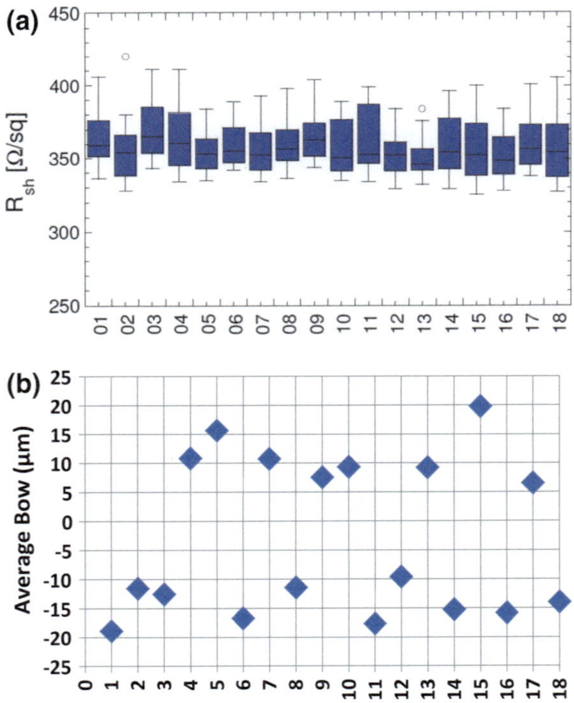

Fig. 3.2 Reproducibility within a lot of 18,200-mm GaN-on-Si wafers of the (**a**) 2DEG sheet resistance and (**b**) wafer bow

3.2 GaN-on-Si Au-Free Processing

Conventional III–V processing includes Au-containing metallization schemes patterned by lift-off. Obviously, these metallization schemes are not compatible with a CMOS fab. For this reason, an Au-free process where metals are patterned by dry-etching steps has been developed on standard CMOS tools [3].

Prior to processing, it has been assessed that the thicker and heavier GaN-on-Si wafers could be processed on most production tools without significant hardware or process modifications. Occasionally, the robot speed of the wafer transport systems had to be lowered due to the larger inertia of thicker Si substrates.

Generically, the device processing starts with the deposition of a thick high-quality Si_3N_4 layer (step 2 in Fig. 3.3). This step is needed to passivate the AlGaN surface early in the process sequence and to protect it from the next phase of the processing. Indeed, defects at the AlGaN surface have been often reported causing the so-called Rdson dispersion phenomena [4]. Alternatively, the surface of AlGaN can be passivated with Si_3N_4 grown in situ in the MOCVD chamber (Step 1 in Fig. 3.3) [5], and thus the first step described above would not be needed or only partially needed to make the Si_3N_4 layer thicker (step 2 in Fig. 3.3).

Si_3N_4 grown in situ in the MOCVD reactor on top of the AlGaN layer without any interruption of the epi-growth results in an interface with AlGaN that is crystalline and of high quality [6]. This is because in situ Si_3N_4 results in a natural

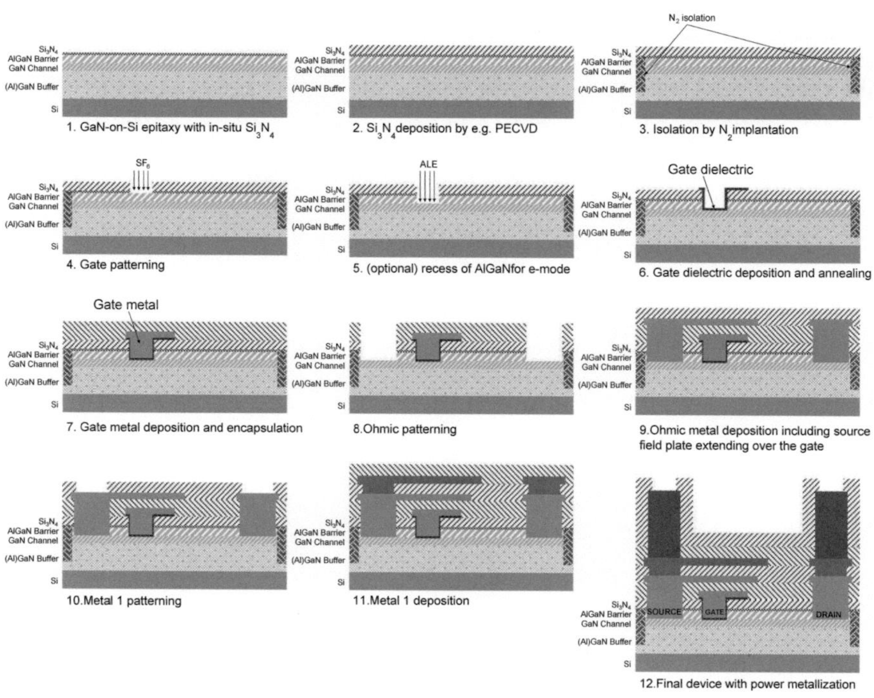

Fig. 3.3 Schematic illustration of the main process steps to obtain GaN-based power transistors

continuation of the AlGaN surface and it provides a nitrogen (N) termination of the AlGaN surface, that otherwise would result in Al or Ga dangling bonds. For this reason, in situ Si_3N_4 is preferred over ex situ layers to passivate the AlGaN surface.

The next processing step consists in a N_2 implantation for device-to-device isolation, i.e., to destroy the 2DEG outside the active device region (step 3 in Fig. 3.3).

Reactive ion etching (RIE) is used to recess the Si_3N_4 layer in the gate area (step 4 in Fig. 3.3). The main challenge here is to develop an etching procedure with high selectivity toward the AlGaN barrier, which minimizes uncontrolled damages to the very sensitive AlGaN surface [4]. An SF_6 plasma at low bias power has been shown to fulfill these requirements [3]. Moreover, after removing the Si_3N_4 layer in the gate region, it is also possible to recess the AlGaN barrier to obtain normally off (e-mode) transistors (step 5 in Fig. 3.3) [6]. The AlGaN recessing needs to be performed by a highly controllable process. For this reason, an atomic layer etching (ALE) process based on cycle of oxidation and BCl_3-based etching has been developed (Fig. 3.4). This process results in precise and reproducible etching of the AlGaN barrier of 1.1 nm/cycle (Fig. 3.4).

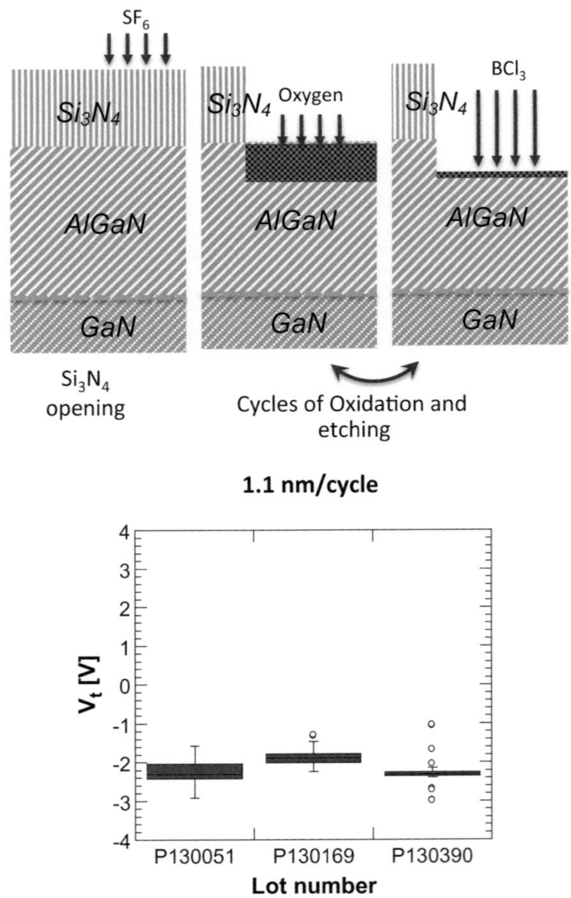

Fig. 3.4 *Upper figure* Schematic process flow for recessing the AlGaN barrier by means of atomic layer etching. *Lower figure* The threshold voltage measured on transistors over different lots processed with identical ALE recess. The *V*th variation is most probably caused by epitaxy

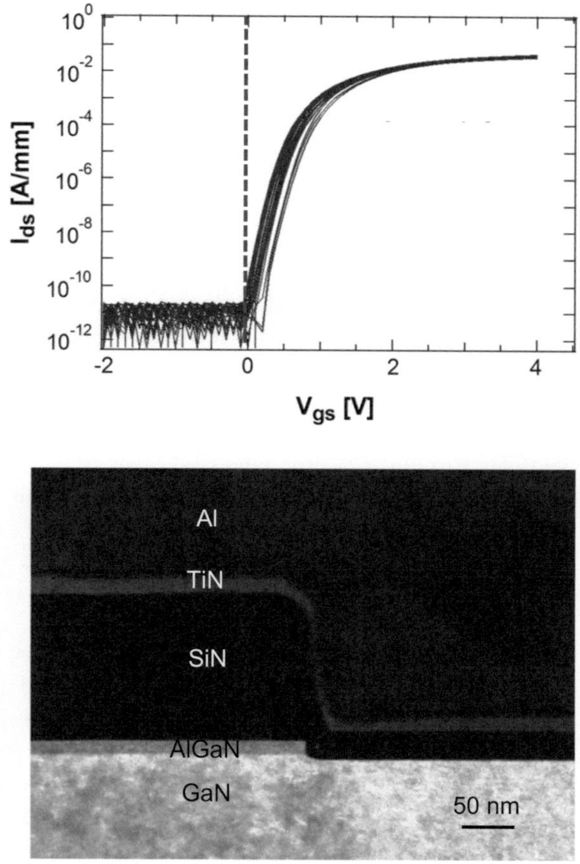

Fig. 3.5 *Upper figure* I_{DS}–V_{GS} transfer characteristics distribution. *Lower figure* TEM of the gate region of fully recessed MISHEMTs processed on a 200-mm GaN-on-Si wafer

The AlGaN barrier can be partially or fully recessed till the GaN channel. For manufacturability, full recess etching of the AlGaN barrier is preferred. Indeed, in this manner it is possible to obtain a robust process with a large process window overcoming issues such as spread in the threshold voltage caused by an non-uniform remaining AlGaN barrier layer over the wafer. Indeed, in the case of partial AlGaN recess, the uniformity of the AlGaN barrier needs to be below 1 nm to obtain a small spread of the threshold voltage over the wafer as few nm's differences in the AlGaN barrier, from transistor to transistor, can result in significant different *V*th. After a full recess of the AlGaN barrier, a very tight *V*th distribution was obtained, as is illustrated in Fig. 3.5.

The gate etching step is then followed by a surface cleaning prior to gate dielectric deposition. This step is fundamental to guarantee low hysteresis in the I_{DS}–V_{GS} transfer-characteristic. The cleaning step is then followed by the gate dielectric deposition with a subsequent anneal in forming gas (step 6 in Fig. 3.3).

The choice of the dielectric and its annealing is fundamentally important since it strongly influences the final device performance in terms of maximum breakdown voltage and trapping phenomena [7]. The gate electrode consists of a TiN/Ti/Al/Ti/TiN stack, with TiN as the work function metal. This metal stack is patterned by RIE with a $Cl_2/BCl_3/N_2$ plasma (step 7 in Fig. 3.3). The gate metal is then encapsulated in a plasma-enhanced chemical vapor deposition (PECVD) Si_3N_4 layer.

The ohmic contacts were obtained after patterning the Si_3N_4 layers. In order to obtain low ohmic contact values, this etch is subsequently extended to recess the AlGaN (step 8 in Fig. 3.3). This step is followed by the deposition of a Ti/Al-based ohmic metal stack, which is alloyed at 500 °C (step 9 in Fig. 3.3). By appropriate optimization of the ohmic module, a contact resistance below 0.5 Ω mm can be obtained as described in detail in the next paragraph. As in the case of the gate metal, also the ohmic metals are then encapsulated in a PECVD Si_3N_4 layer.

The device processing continues with the patterning and deposition of so-called metal 1 layer, which consists of Ti/Al/Ti/TiN (step 10 and 11 in Fig. 3.3). Source field plates to lower electric field peaks can be defined both in ohmic metal and in metal 1 layer. The device processing terminates with thick Al and Cu interconnect metallization layers, which are patterned in the Si_3N_4 layers and covered by a thick PECVD Si_3N_4 scratch protect layer (step 12 in Fig. 3.3).

A photograph of the fully processed 200-mm GaN-on-Si wafer is reported in Fig. 3.6, and a detail of a die is reported in Fig. 3.7. In Fig. 3.8 a cross-sectional SEM image of the gate-to-source area of a transistor is shown.

It has to be highlighted that the process described above is related to the specific MISHEMT d-mode or e-mode architectures. However, e-mode devices can also be obtained by growing a p-type layer on top of the AlGaN barrier, deposition and patterning of a gate metal and then selective recessing the p-GaN layer over the

Fig. 3.6 Photograph of a 200-mm processed GaN-on-Si wafer

Fig. 3.7 Photograph of a die with power transistors processed on 200-mm GaN-on-Si

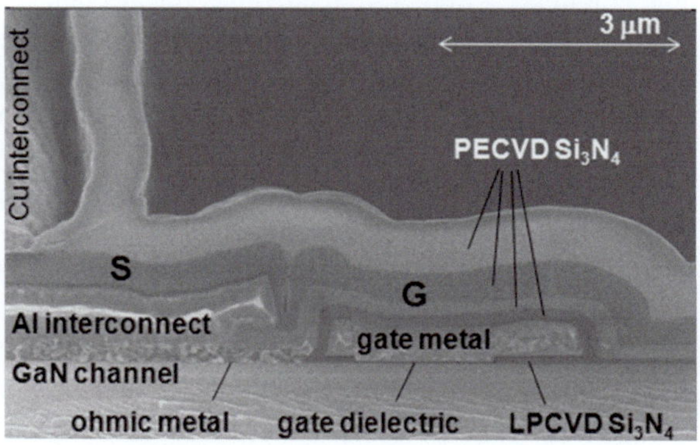

Fig. 3.8 Cross-sectional SEM picture of the gate-to-source area of a power device processed on a 200-mm GaN-on-Si wafer

AlGaN barrier [6]. In this case, the most critical steps consist in growing the p-GaN layer and controlling the recess of the p-GaN layer over the AlGaN barrier in the access areas. It is out of the scope of this chapter to illustrate in detail this alternative process for e-mode devices. More information can be found in [6].

Within the 200-mm GaN-on-Si platform, both e-mode and d-mode power transistors as well as power diodes can be obtained [6]. An output characteristic of an e-mode device is reported in Fig. 3.9.

3.3 Au-Free Ohmic Contact

Obtaining Au-free ohmic contact is a key aspect to enable processing of highly performing devices in a CMOS fab, and therefore, this topic deserves a dedicated section. The fabrication of low-resistive ohmic contacts for source/drain of

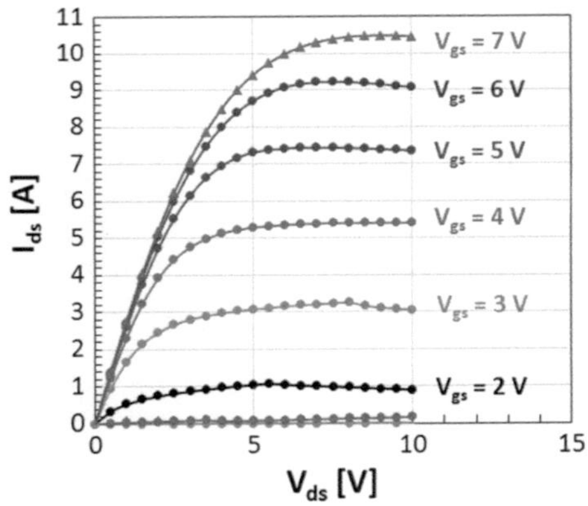

Fig. 3.9 Output I_{DS}–V_{DS} characteristic of an e-mode power bar processed on 200-mm GaN-on-Si in a CMOS fab

AlGaN/GaN devices is not obvious. Historically, the most successful metallization schemes have been using Au-containing metal stacks alloyed at relatively high temperatures (≥ 800 °C), achieving typical contact resistance (R_C) values below 1 Ω mm. However, as largely discussed above, there is the need for implementing a standard CMOS-compatible metallization scheme for GaN and for this type of metallization scheme, the use of Au needs to be avoided due to its status as a Si contaminant. On top of this requirement, a lower alloying temperature, restricted to ≤ 600 °C, is also preferable as it allows for maximum flexibility in processing and enables to have a gate-first architecture.

In recent years, several Au-free contact schemes have been proposed. The most frequently cited metallization schemes are Ti/Al-based, such as Ti/Al/W [8] and Ti/Al/Ni/Ta/Cu/Ta [9]. In these cases, R_C below 1.0 Ω mm has only been demonstrated at relatively high annealing temperatures (≥ 800 °C). Low R_C and low annealing temperature for Au-free schemes have also been published, but typically the metal schemes are not directly compatible with standard CMOS platforms. For example, using a Ta/Al metal stack, a R_C value of 0.06 Ω mm has been reported [10]. Low R_C values were also obtained with the introduction of Si doping in the GaN and/or AlGaN layer. This, however, can cause problems for the breakdown voltage of power devices, and Si implantations in GaN-based layers typically require annealing at very high temperatures (>1000 °C), not compatible with the process flow.

An interesting CMOS-compatible metal stack is Ti/Al. There are several factors which can influence the formation of a good ohmic contact to a lateral AlGaN/GaN HEMT device for these Ti/Al-based metal stacks, and several of these issues will be discussed in this section. These factors include the recess of the AlGaN barrier, the alloy temperature, the Ti/Al thickness ratio and the inclusion of silicon inside the metal stack. Each of these aspects will be highlighted in this section.

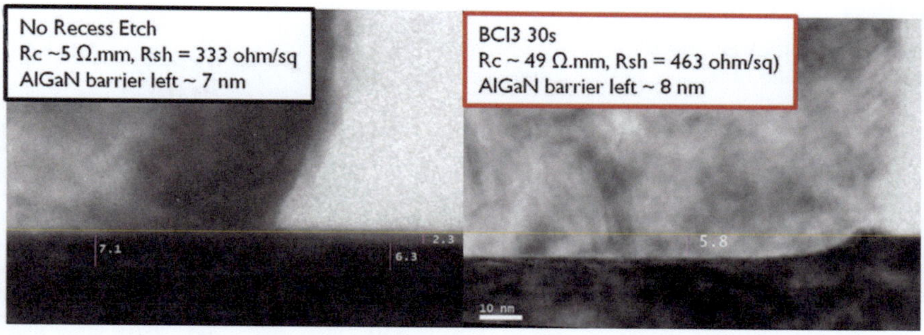

Fig. 3.10 TEM on ohmic regions with high R_C

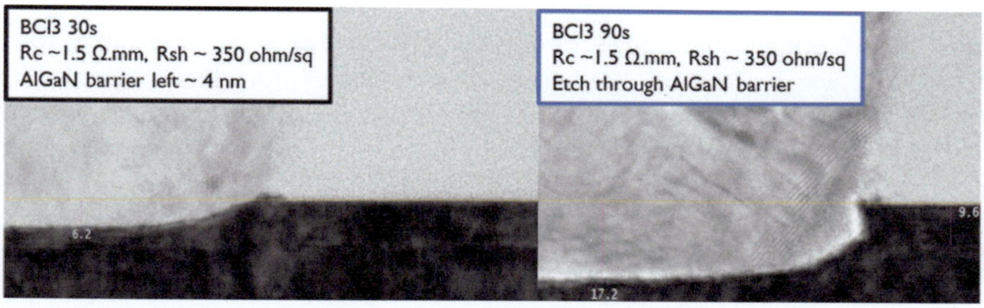

Fig. 3.11 TEM on ohmic regions with R_C of $\sim 1.5\ \Omega$ mm

3.3.1 AlGaN Barrier Recess

Without AlGaN barrier recess, high and irreproducible R_C (between 5 and 50 Ω mm) values are obtained. The AlGaN barrier forms a potential barrier between the ohmic metal and the 2DEG below, which is difficult to be lowered by simply performing an alloying step. Recessing the barrier can reduce the potential barrier, but will also at the same time reduce the 2DEG concentration below the contact. On the other hand, if the barrier becomes thin enough or is fully recessed, a more reliable contact can be achieved; as for this case, a low potential barrier is present for the formation of a lateral contact toward the 2DEG. This fact is illustrated in Figs. 3.10 and 3.11 where it is observed that the R_C increases for a partial recess ($\sim$8-nm barrier left), while a low R_C is achieved when the remaining barrier is $\sim$4 nm or lower.

3.3.2 Ohmic Alloy Optimization

As mentioned in previous paragraphs, the optimal alloy temperature is very important and a requirement imposed when performing a gate-first process would

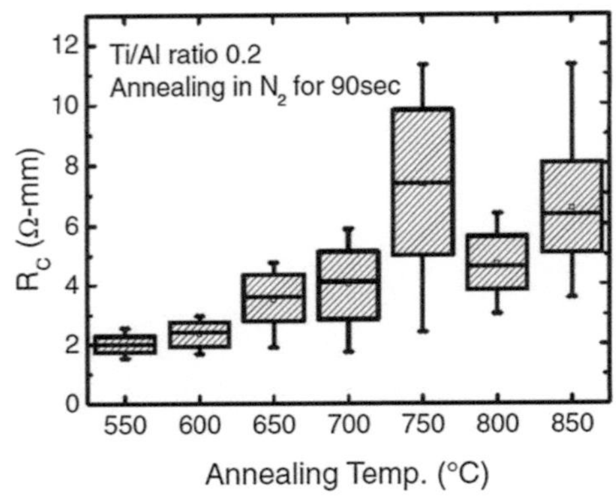

Fig. 3.12 R_C versus alloy temperature for a metal stack with a Ti/Al ratio of 0.2

be to have a temperature budget ≤ 600 °C. The optimal alloy temperature for a Ti/Al stack was verified on 200-mm GaN-on-Si wafers, for ohmic regions with AlGaN barrier recess and Ti/Al/Ti/TiN ohmic metal stack. The ohmic alloys were performed in a Rapid Thermal Anneal (RTA) system under N_2. The temperature has been varied between 550 and 850 °C, in steps of 50 °C. The lowest anneal temperature of 550 °C yields the lowest R_C values with the tightest distribution (Fig. 3.12). This optimum temperature is significantly lower than the relatively high ohmic alloy temperature (≥ 800 °C) commonly reported in the literature [8, 9].

3.3.3 Ti/Al Ratio

The Ti/Al ratio is another important factor if one wants to further lower the contact resistance. Ti is known to act as a getter and can reduce the oxidation of the semiconductor surface, which could help in reducing R_C. Moreover, it has been shown by [11] that during alloying of Ti on GaN, a N extraction occurs from GaN toward Ti to promote the formation of TiN. However, this reaction is aggressive and it can lead to void formation after the contact alloy. Al, on the other hand, has a self-limiting reaction with respect to GaN, but does not act as an oxide getter like Ti does. Also adding Al reduced the reaction rate of Ti with the GaN surface, avoiding the void formation. Therefore, there is an optimum ratio between Al/Ti when a dual-layer metal stack is used. According to this, a low Ti/Al ratio should be preferable. This trend has been confirmed in Fig. 3.13, where the Ti/Al ratio was considered as a further optimization parameter, varying the thickness of the bottom Ti and Al layers from the Ti/Al/Ti/TiN ohmic metal. The AlGaN barrier recess and alloy process conditions were fixed at their optimal values. The condition with the lowest Ti/Al thickness ratio exhibited an R_C value of 0.65 ± 0.07 Ω mm.

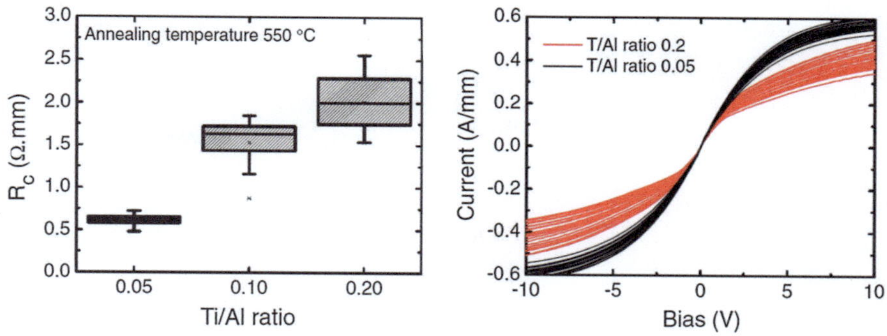

Fig. 3.13 Contact resistance, *left*, versus Ti/Al ratio, *right*, maximum TLM current for metal stacks with two different Ti/Al ratios

3.3.4 Si Layer at Bottom of Ohmic Metal Stack

Another optimization which can be done is to introduce Si in the metal stack. Si can form a eutectic melt, creating an alloy with a lower melting temperature and thus promoting the formation of ohmic contacts at lower temperatures. R_C values of non-Si- and Si-containing contacts have been directly compared in a dedicated experiment (Fig. 3.14). R_C values of non-Si-containing contacts are 0.63 ± 0.11 Ω mm (2 wafers). R_C values of Si-containing contacts are significantly lower: 0.30 ± 0.04 Ω mm (6 wafers). The saturation current at 10 V between 2 ohmic contacts spaced 12 μm (which is typically limited mainly by the 2DEG resistance) is not affected by the introduction of Si, Fig. 3.15. The Si thickness splits (2, 5, 10

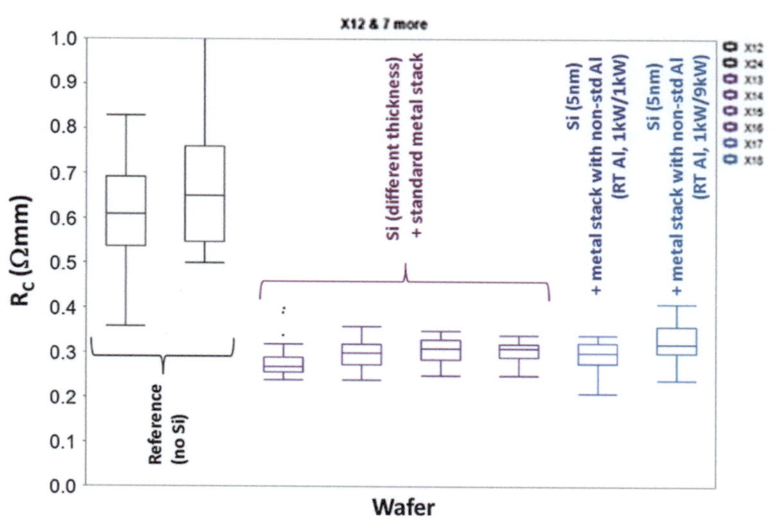

Fig. 3.14 R_C values of non-Si- and Si-containing contacts from a dedicated experiment

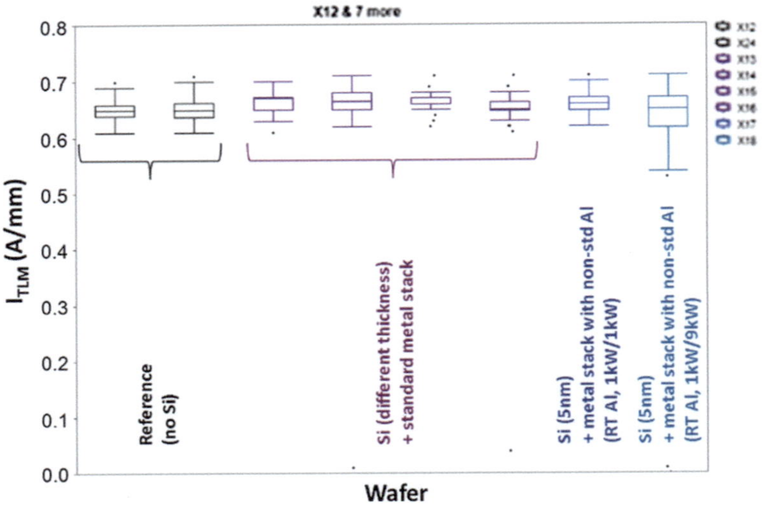

Fig. 3.15 Saturation current of non-Si- and Si-containing contacts at 10 V between 2 ohmic contacts spaced 12 μm

and 20 nm) do not significantly affect the R_C values, demonstrating a large process window toward this parameter. With the goal to maximize the process window, the introduction of a Si layer below the Ti layer helps significantly in relaxing the requirements.

3.4 Gallium Contamination Issues

Since Ga is a p-type dopant for Si, one of the major concerns of processing GaN wafers in a CMOS fab is Ga contamination. For this reason, this topic deserves a dedicated session.

A first source of Ga contamination is the presence of trace amounts of Ga on the chuck of the MOCVD epi tool. Wafers in contact with the chuck pick up part of this contamination. This is confirmed by TXRF (Total Reflection X-ray Fluorescence) analysis performed on bare Si wafers that were in contact with the chuck (Fig. 3.16). Similarly, after growth of the (Al)GaN layers in the MOCVD epi tool, the backside of GaN-on-Si wafers is Ga-contaminated. It has to be noted that there is no (Al)GaN layer growth on the backside of GaN-on-Si wafers, as far as can be detected by XSEM analysis (Fig. 3.17). It is observed that the (Al)GaN film thickness reduces along the bevel of the wafers and then suddenly drops, typically approximately at the wafer apex, to almost zero (inset Fig. 3.17).

During initial processing test loops on a strictly defined set of tools, it was observed that the Ga contamination, picked up by the backside of the GaN-on-Si wafers, was indeed spreading from these wafers to the transport systems and process chambers of the tools (Fig. 3.19). In most of the cases, the Ga contamination

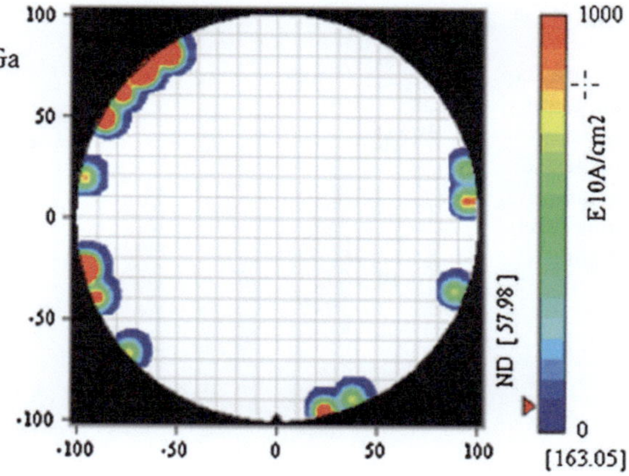

Fig. 3.16 TXRF analysis of a Si wafer surface after contact with the chuck of the MOCVD epi tool

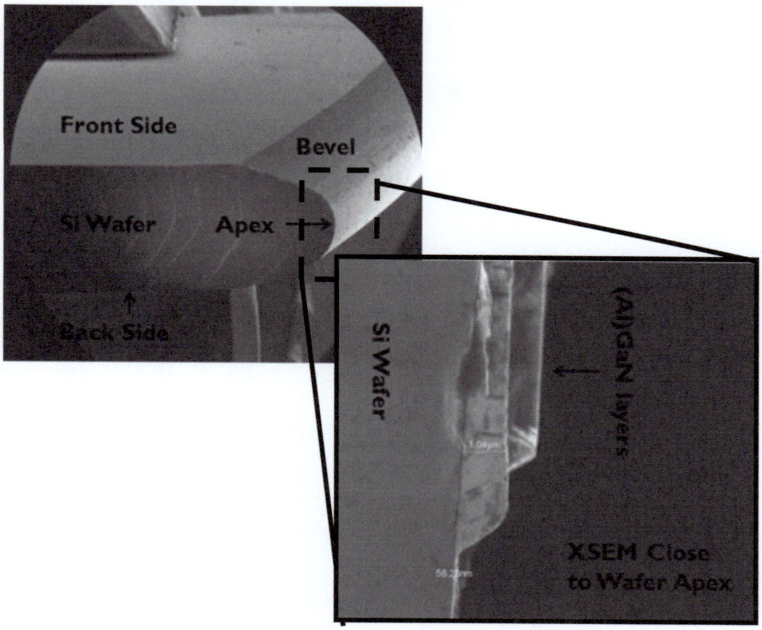

Fig. 3.17 XSEM analysis of the bevel and backside of a GaN-on-Si substrate after wafer growth. The *inset* shows a magnification of the wafer apex

level exceeded the maximum tolerated level of 10^{11} at/cm^2 on bare Si wafers cycled in these tools after GaN-on-Si wafer processing (Fig. 3.18).

To avoid Ga contamination, a HF/H_2O_2-based cleaning procedure to be applied to the backside of GaN-on-Si wafers has been developed. The Ga contamination level of a Si wafer (Fig. 3.19a) is reduced to a level close to the detection limit of TXRF after the proposed cleaning step (Fig. 3.19b) proving its effectiveness.

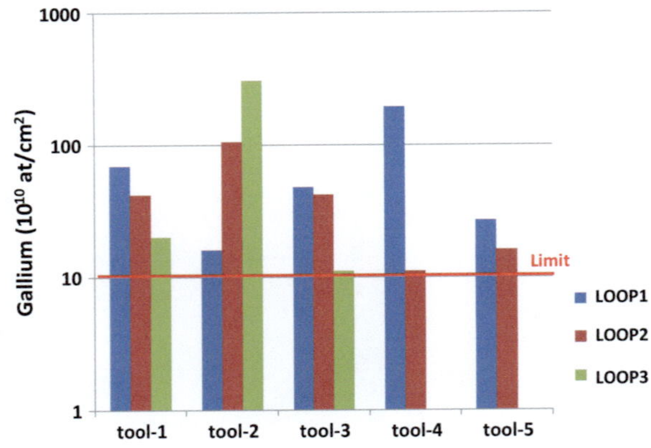

Fig. 3.18 Monitoring of Ga contamination on 5 processing tools (no etch tool was included) during three loops of GaN-on-Si wafers processing without substrate cleaning. The maximum allowed contamination level is indicated in the figure

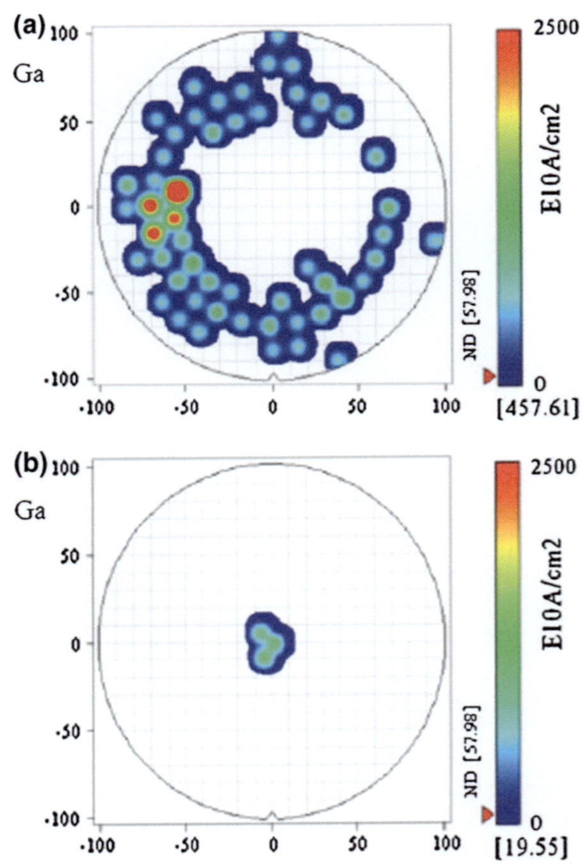

Fig. 3.19 **a** TXRF analysis of a Ga-contaminated Si wafer before and **b** after HF-based cleaning procedure

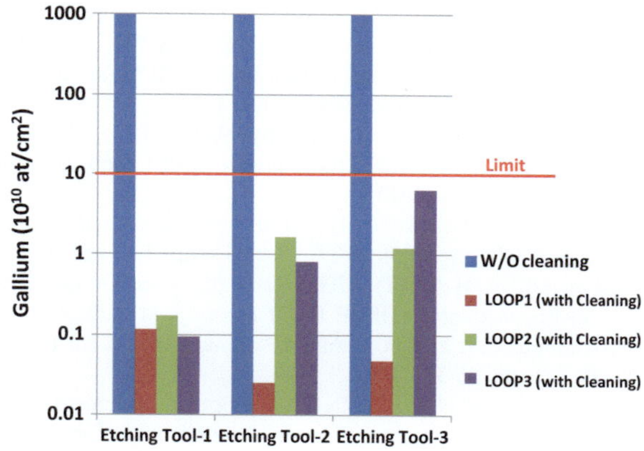

Fig. 3.20 Monitoring of Ga contamination on 3 etching tools during three loops of GaN-on-Si wafers processing. The contamination of each tool without special cleaning procedure is also reported. The maximum allowed contamination level is indicated in the figure

The second source of Ga contamination is related to etching steps. During the initial processing test loops, each etching tool got strongly contaminated when Ga-based layers were etched (Fig. 3.20). Since conventional F-containing cleaning recipes can form nonvolatile GaFx species (i.e., GaFx is not volatile below 800 °C), a new Cl_2-based clean that forms volatile $GaCl_3$ at a much lower temperatures of ~200 °C has been developed. This cleaning procedure effectively maintains the Ga contamination level well below the maximum allowed level (Fig. 3.20) for the whole period monitored.

3.5 Conclusion

In conclusion, in this chapter the feasibility of processing high-quality 200-mm GaN-on-Si wafers in a CMOS fab has been demonstrated. CMOS-compatible processing steps have been illustrated and discussed. Particular attentions have been given to the formation of Au-free ohmic contacts.

Moreover, it has been shown that Ga contamination issues can be avoided by means of cleaning procedures for the substrates and for the etching tools. This paves the way for high-volume production in CMOS fab of the next generation highly efficient GaN-based power devices cost-competitive with conventional Si power MOSFETs or IGBTs.

References

1. Cheng K et al (2012) AlGaN/GaN/AlGaN double heterostructures grown on 200 mm silicon (111) substrates with high electron mobility. Appl Phys Express 5:011002
2. Marcon D et al (2015) Proceeding of IEDM

3. De Jaeger B et al (2012) Au-free CMOS-compatible AlGaN/GaN HEMT processing on 200 mm Si substrates. In: Proceeding of ISPSD, pp 49–52
4. Mishra UK et al (2002) AlGaN/GaN HEMTs: an overview of device operations and applications. Proc IEEE 90(6):1022–1031
5. Derluyn J et al (2005) J Appl Phys 98:054501
6. Marcon D et al (2015) Proceedings of SPIE 9363, Gallium nitride materials and devices X, 936311 (13 Mar 2015)
7. Van Hove M et al (2012) CMOS process-compatible high-power low-leakage AlGaN/GaN MISHEMT on silicon. IEEE Electr Device Lett 33(05):667–669
8. Lee HS et al (2011) AlGaN/GaN high-electron-mobility transistors fabricated through a Au-free technology. IEEE Electr Device Lett 32:623
9. Alomari M et al (2009) Au free ohmic contacts for high temperature InAlN/GaN HEMT's. ECS Trans 25:33
10. Malmros A et al (2011) Electrical properties, microstructure, and thermal stability of Ta-based ohmic contacts annealed at low temperature for GaN HEMTs. Semicond Sci Technol 26:075006
11. Van Daele B, Van Tendeloo G, Ruythooren W, Derluyn J, Leys MR, Germain M (2005) The role of Al on ohmic contact formation on n-type GaN and AlGaNGaN. Appl Phys Lett 87 (6):11–13. doi:10.1063/1.2008361

Chapter 4
Lateral GaN Devices for Power Applications (from kHz to GHz)

Umesh K. Mishra and Matthew Guidry

4.1 Introduction

GaN and related materials have exploded onto the semiconductor landscape because of the broad range of applications that they address. The most visible is that it has enabled the solid-state lighting revolution with GaN-based LEDs now replacing incandescent and fluorescent light bulbs because of both their low cost and return on investment due to electricity savings afforded by their enhanced efficiency. GaN has also enabled full color displays from cell phones to stadiums, automotive and airplane mood cabin lighting to head lamps based on LEDs and most recently lasers. These are just a few of the applications that have created a >$10B and rapidly expanding market. In the wake of the photonics applications, GaN electronic applications are now emerging in their own right to serve a multi-billion market from cell phone infrastructure to RADAR and communications to power conversion applications. The dominant device that is used for electronic applications is the AlGaN/GaN HEMT. This chapter will address the materials aspects (briefly) and the various device designs that are of value in different applications.

4.2 History of AlGaN/GaN HEMTs

The AlGaN/GaN HEMT was first demonstrated in 1993 by Asif Khan et al. [1]. The structure contained 14 % Al and was grown on a sapphire substrate. The first microwave small signal performance was achieved by Khan in 1994 with an f_T of

U.K. Mishra (✉) · M. Guidry
University of California, Santa Barbara, CA, USA
e-mail: mishra@ece.ucsb.edu

M. Meneghini et al. (eds.), *Power GaN Devices*,
Power Electronics and Power Systems, DOI 10.1007/978-3-319-43199-4_4

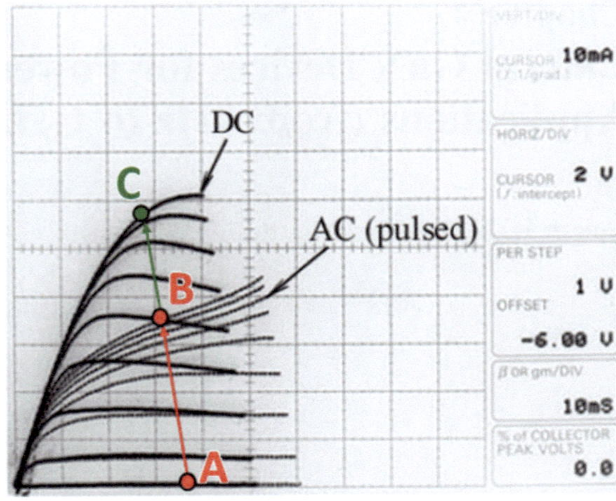

Fig. 4.1 A device with a dispersive *IV* curve []. The pulsed measurement is taken from point A to B, and it is seen that the maximum current and knee voltage are both worse than what is expected from the DC *IV* curve at point C

11 GHz and f_{max} of 35 GHz [2]. These two fundamental achievements demonstrated the viability of the AlGaN/GaN HEMT for microwave applications. The next big hurdle was the demonstration of microwave power. This was achieved in 1995 by Wu et al. [3] at UC Santa Barbara. This was a critical result as the power density of 1.1 W/mm achieved demonstrated for the first time the benefit of GaN over the competing technologies of GaAs and InP. While demonstrating the potential of GaN, this result also illustrated a major problem, that of AC/DC dispersion or current collapse. Figure 4.1 compares the DC *I–V* curve of an AlGaN/GaN HEMT on sapphire with the AC curve obtained by switching the HEMT from an off state; point A on the curve, to point B. Two issues are evident: the maximum current available from the device is reduced (current collapse), and the knee voltage at which the device current saturates increases (knee walk-out). The primary reason is that in the off state (point A), the device is under reverse bias and electrons are injected into available electron states either at the surface or the bulk layers of the device. These effects reduce both the RF output power density and efficiency of the HEMT.

Figure 4.2 demonstrates the process for surface charging; the charging of traps in the buffer is similar. As the electrons are injected and trapped into the surface states under reverse bias, the positive charge at the AlGaN/GaN interface is imaged not only on negative charge in the gate but also onto the trapped charges in the surface (and the bulk). When the device is switched on, only the mobile charges are free to respond nearly instantaneously (pS), whereas the trapped charges remain trapped for times related to the emission times of the traps (μs). As a consequence induced negative charges in the channel 2DEG in the on-state at microwave switching frequencies are reduced, since charge neutrality is satisfied in part by the immobile fixed charges. This reduces the RF maximum current as the maximum current is controlled by the number of mobile electrons in the channel under/near the source edge of the gate. In addition, the extended region of reduced charge beyond the gate manifests itself as increased resistance in the channel, which in turn

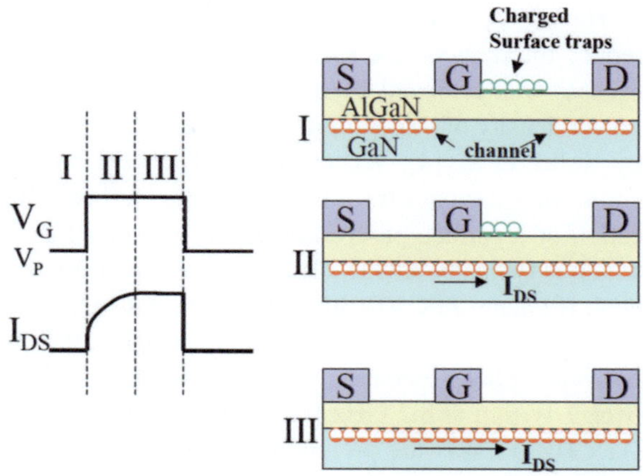

Fig. 4.2 Illustration of surface charge trapping occurring while device is pinched off which limits current flow in the channel for a period of time until the traps can respond. Figure by Zhang [103]

leads to the observed larger knee voltage. This issue of AC/DC dispersion or current collapse had to be addressed to make the AlGaN/GaN HEMT achieve its promise in RF and power conversion applications.

4.3 Addressing Dispersion

To suppress dispersion, it is necessary to (i) reduce the electric field that injects the electrons into the traps, (ii) reduce the number of traps available to inject charge into and (iii) increase the resistance of the pathways between the source of electrons and the traps.

The first step is to ensure that the GaN buffer and the AlGaN material are of the highest quality. This is addressed in another chapter in the book and so will be addressed only briefly here. On sapphire the low temperature (LT) nucleation layer conditions have to be optimized and the subsequent growth conditions tailored to maximize the grain size and reduce the dislocation density. Currently, dislocation density of the order of $5 \times 10^8 - 10^9$ cm^{-2} is routinely achieved which is sufficient for high-performance reliable AlGaN/GaN HEMTs. However, impurities such as C or Fe are added to the buffer to make the buffers more resistive. The undesired consequence of this is the addition of electron traps that can lead to dispersion. It is therefore critical to control the density and the physical separation of the traps from the channel. On SiC, the preferred substrate for RF applications, the nucleation layer is high temperature AlN on which a LT GaN layer is grown to initiate the GaN channel (similar to the case of a sapphire substrate). Similar dislocation density is achieved in this case as compared to high-quality GaN on Sapphire. GaN on Si, preferred for power switching applications, growth is initiated by using an AlN nucleation layer as well, followed by a buffer layer that manages the thermal and lattice strain encountered in the growth of GaN on Si. These buffer layers are based on graded AlGaN layers, AlN/GaN superlattice layers or their variants.

Having optimized the buffer to minimize dispersion, it is now important to reduce the traps on the surface of the HEMT, which is typically AlGaN. The first success was achieved by adding Si doping to the AlGaN layer and a maximum power density of 4.6 W/mm was achieved by this technique on a sapphire substrate by Keller et al. [4], Wu et al. [5]. The next breakthrough was achieved by disconnecting the path from the gate to the surface states. This is done by the application of a surface passivation layer, typically SiN, onto the HEMT surface layer which is typically AlGaN. This was first demonstrated in 2001 separately by Green et al. [6] using PECVD SiN and Wu et al. using sputtered SiN [7]. This innovation also eliminated the explicit need for doping the AlGaN layer. Applying a SiN passivating layer increased the power density of HEMTs grown on SiC substrates to over 9 W/mm at 8.2 GHz [8], further strengthening the argument that the AlGaN/GaN HEMT was the preferred choice for microwave high-power communication and amplification compared to the incumbent GaAs.

The last remaining technological hurdle that could spell the death of the technology was the reliability of the HEMT. There were two reasons for worry. An argument was put forth that the high density of dislocations would lead to device failure. The issues with micro-pipe density in SiC and basal plane faults in bipolar SiC were raised as examples. The 2H device structure of GaN and that the c-plane is not a glide plane for threading dislocations worked in favor of the technology. Also, very importantly, the current flow and the blocking voltage was primarily held perpendicular to the dislocation direction eliminating them from the critical failure path.

The second reason for worry was the high electric field at the drain-edge of the gate that occurred due to the high voltages applied to the devices. This could lead to both rupture of the AlGaN layer [9] and/or the surrounding dielectric which would be unacceptable. Indeed this worry turned out to be reality. The solution was the application of field plates to mitigate the electric field at the drain-edge of the gate, first demonstrated by Ando et al. in 2003 [10] with gate metal overlapping the passivation layer and second demonstrated by Chini et al. in 2004 [9] with a separate field plate metal deposition which also lowers the gate resistance. The result was an astounding jump in the available RF power density of the device to 12 W/mm even on a sapphire substrate. Optimization of the field plates geometries and combining gate-connected and source connected field plates has led to the remarkable progress that is depicted in Figs. 4.3 and 4.4, eventually reaching 41.4 W/mm [11–14].

The materials property of GaN (the relative insensitivity to dislocations) combined with electric field management via field plates, allowed the development of highly reliable GaN RF devices with exceptional electrical performance. Based on the outstanding results, the DoD under a Defense Production Act Title III program sponsored multiple microwave GaN foundries to successfully raise their manufacturing readiness level (MRL) to level 8 or higher in 2013–2014 [15–22], firmly establishing GaN as a reliable and manufacturable technology for microwave

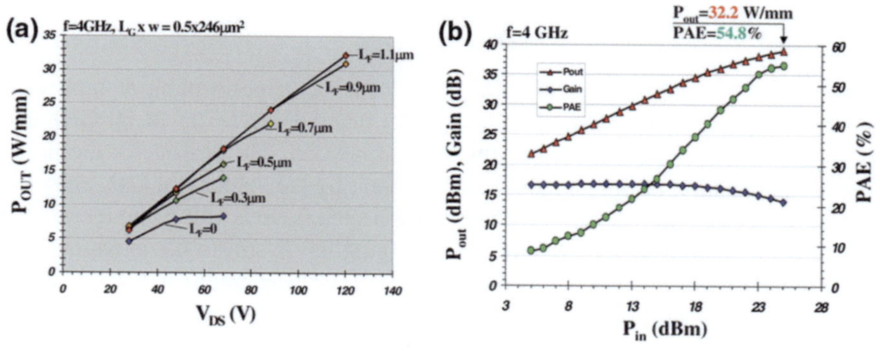

Fig. 4.3 **a** Output power versus drain bias for different field plates size L_F shows the effectiveness of field plates in reducing dispersion at high drain bias, **b** power sweep for largest field plate shows record breaking 32.2 W/mm power density [11]

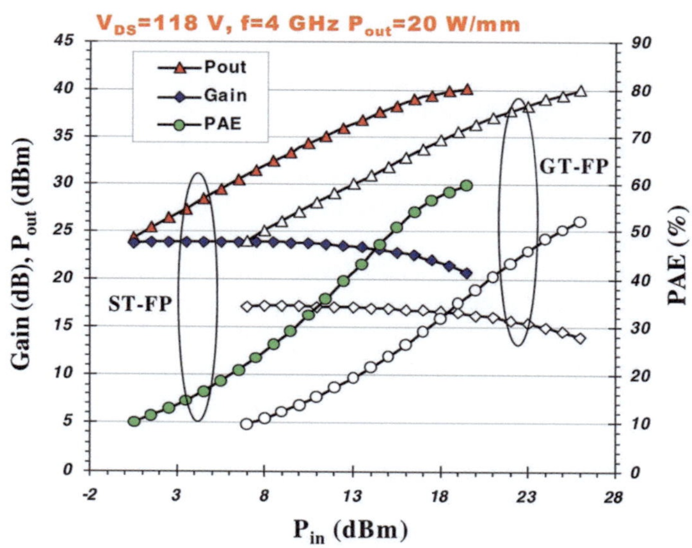

Fig. 4.4 Power sweep of source terminated field plate (ST-FP) and gate terminated (GT-FP) shows the tradeoff between the two field plate designs, with higher gain from ST-FP due to reduced gate drain capacitance with this approach [12, 13]

applications. These devices have been put into use by numerous companies, including SEDI [23] for cellular base stations, and Raytheon for defense applications including radar and electronic warfare. In 2016 Raytheon announced the release of an upgraded version of the Patriot anti-missile RADAR system using GaN [24] and the commercial deployment of GaN-based cellular base stations for 4G LTE is accelerating.

4.4 Gallium Nitride for mm-Wave Applications

There was a question of whether GaN was capable of performing at mm-wave applications. The incumbents remained GaAs based HEMTs up to 60 GHz and InP-based HEMTs and HBTs for 60 GHz and beyond. The reason for the doubt was the anticipation of hot phonon scattering limiting the velocity in GaN channels with high electron density [25, 26]. This complicated issue is still being researched [27], and the answer in a HEMT device configuration is still to be determined. In the absence of definitive proof to the contrary, work was conducted at UC Santa Barbara under ONR sponsorship and showed clearly that AlGaN/GaN HEMTs were fully capable of performing at mm-wave frequencies. The devices developed at UCSB met the three necessary criteria for mm-wave power performance: high speed, high breakdown, and low dispersion. Several important GaN mm-wave milestones were achieved in 2005: Palacios et al. reported [28] a record Ka-band power density of 10.5 W/mm at 40 GHz with an associated PAE of 33 %, where PAE was limited primarily by low linear gain of 6 dB. Moon et al. at HRL Laboratories reported [29] on devices with record GaN Ka-band efficiency 45 % at 5.7 W/mm output power, with higher efficiency resulting from higher gain. And Wu et al. at Cree reported [30] record total mm-wave GaN device power of 8 W at 30 GHz with 31 % PAE on a large-area device with 1.5 mm total gate width. At that point, the efficiency of GaN devices in Ka-band was approaching that of contemporary GaAs device and MMIC results [31, 32] but at an order of magnitude greater power density.

Research continued to extend GaN device performance to the W-band (75–110 GHz). Applications in this frequency range include wideband high data rate communication in several bands [33] and high-resolution and compact aperture imaging and radar. Exceptional power densities with good power-added efficiency (PAE) have been obtained and are summarized in Fig. 4.5. This was achieved by following rules for mm-wave devices established from GaAs and InP including ultra-short gate length, vertically scaled channel design, low channel and contact resistances, and buffer and backbarrier design to enhance output resistance. GaN W-band power performance was first reported by Micovic et al. at HRL Labs in 2006 [34] with 2.1 W/mm output power density and an associated 14 % PAE. Figure 4.6 shows the most advanced device cross-section and epitaxial layer design used by HRL to achieve record GaN device speed [35]. The gate length was 20 nm, and a 3.5 nm AlN barrier with 2.5 nm GaN cap gate layer was employed to maintain a good aspect ratio. An AlGaN buffer layer provided a barrier for electron injection into the buffer by virtue of the negative polarization charge at the AlGaN buffer/GaN interface. This provided a high output resistance which when combined with the extreme lateral and vertical scaling of the device resulted in outstanding small signal performance with an f_T of 454 GHz and an f_{max} of 444 GHz simultaneously, a GaN device record. A similar device design with a 40 nm gate length demonstrated a record GaN W-band amplifier-level PAE at 83 GHz of 27 % at 1.7 W/mm associated output stage power density in 2014 [36].

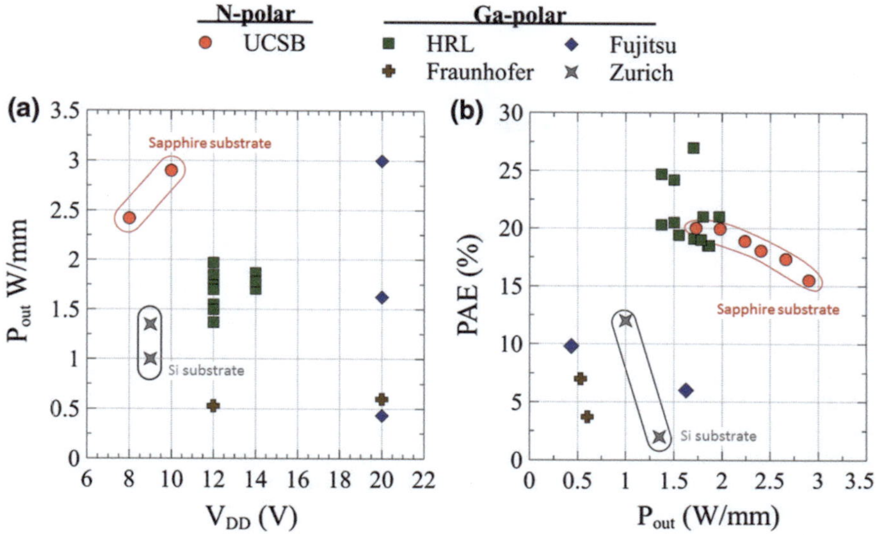

Fig. 4.5 Summary of reported W-band (75–110 GHz) power device results: **a** output power densities versus drain bias, **b** output power density versus associated power-added efficiency (*PAE*). Substrate materials are thinned SiC except UCSB which is sapphire, and ETH Zurich which is silicon. Results from UCSB, ETH Zurich, and Fujitsu are load pull, results from HRL Labs and Fraunhofer are matched devices or multi-stage amplifiers [34, 36, 39, 93, 104–115]

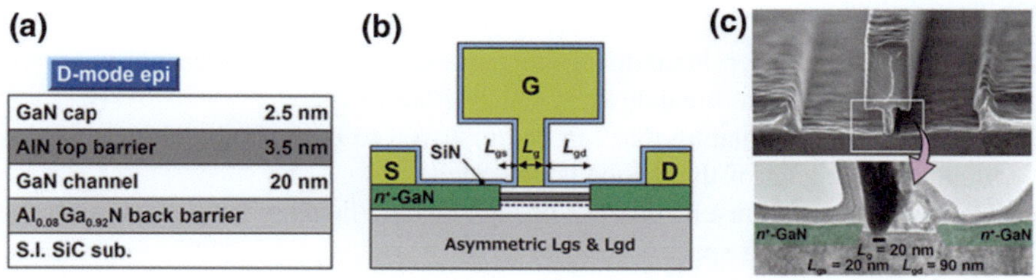

Fig. 4.6 **a** Epitaxy design, **b** device schematic cross-section and **c** TEM image of HRL Gen-IV device with 20 nm gate length, capable of record GaN f_t and f_{max} [35]

W-band GaN MMIC output power saw steady improvement from an initial value of 320 mW in 2006 to reach 2.5 W in CW operation and 3.6 W in pulsed operation by 2015, plotted in Fig. 4.7b. Device gain and thus efficiency also improved from less than 15 % to over 25 %. The MMIC power scaling was accomplished by increasing the total output stage gate periphery and implementing larger scale on-chip power combining. The highest power MMIC reported to date utilized a total of 3.75 mm of gate periphery including driver stages [37] and is indicative of increasing manufacturing maturity. It significantly surpasses 800 mW GaAs pHEMT MMIC output power [38], which used 4.8 mm of gate periphery, and shows the capability of GaN for high-power W-band applications. However, since the first

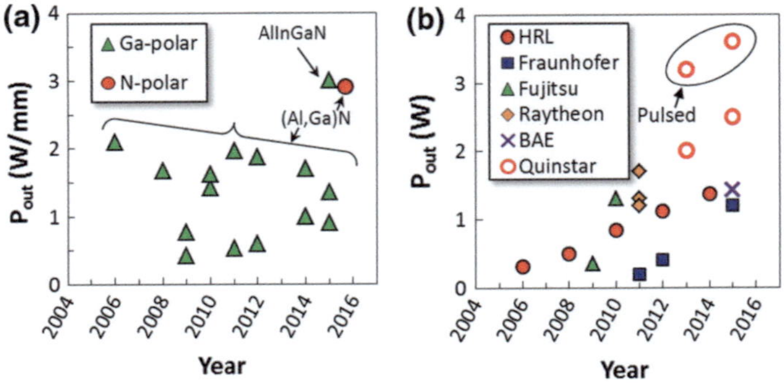

Fig. 4.7 Yearly progression in reported W-band (75–110 GHz) GaN power performance, including both MMIC and single device results [34, 36, 37, 39, 104–115]: **a** maximum GaN output power density for Ga-polar versus N-polar material. Material technology is labeled, (Al, Ga)N referring to both AlGaN/GaN and AlN/GaN devices. **b** Maximum single-chip output power levels, some measurements (labeled) were pulsed at a low duty cycle to reduce self-heating effects giving increased the output power

report in 2006, Ga-polar GaN devices have struggled to increase W-band device-level power density beyond 2 W/mm, with reported power density versus year plotted in Fig. 4.7a. Fujitsu in 2015 reported a Ga-polar record 3 W/mm power density [39], but it required use of a quaternary AlInGaN barrier layer.

The barriers to further improvement mm-wave performance in Ga-polar HEMTs are listed below

1. The trade-off between breakdown voltage and charge in the channel is severe, limiting the available breakdown voltage of the device.
2. The extent of field plating that can be employed to mitigate the electric field is limited as the gain of the device is degraded.
3. The dispersion that results because of high electric fields in the device limits the useful *I–V* operating space.
4. All the above limit the output power, efficiency, and reliability of mm-wave Ga-polar GaN HEMTs.

4.5 Historical Perspective of N-Polar GaN Development

The revolution that GaN has created in solid-state lighting (LEDs), microwave and millimeter-wave electronics, and power conversion have all used GaN grown in the c-axis direction, the direction which leads to strong polarization parallel to the growth direction. This polarization can be exploited to form the high-mobility two-dimensional electron gas (2DEG) channels in undoped GaN HEMTs [40–42]. As the material is polar, there are two possible orientations along this axis with polarization fields in opposite directions: Gallium-polar (Ga-polar) which nominally has a surface terminated with Gallium atoms and N-polar with a nominally nitrogen-terminated surface. All commercial activity to date has utilized GaN grown in the Ga-polar crystal orientation, as this was the first orientation to be grown with high structural quality and low impurities. Improving the quality of the material was one of the most critical aspects enabling the GaN revolution. N-polar can now be grown with the same quality as Ga-polar thanks to improved methods summarized by Stacia Keller in 2014 [43], but the initial poor results explain why N-polar devices were not developed earlier. This section will discuss the history of development of N-polar GaN devices and recent results which show its potential for improvements over Ga-polar GaN, especially for millimeter-wave power applications.

The 2DEG is always formed at the side of the AlGaN/GaN interface which has net positive polarization which naturally leads to two different HEMT structures, illustrated in Fig. 4.8. The basic N-polar HEMT structure can be seen to have several possible advantages. First, with the 2DEG induced from the AlGaN barrier beneath the channel, contact resistance can be less than Ga-polar where removal of the wide bandgap AlGaN barrier also removes the 2DEG under the contact. Second, with a barrier directly underneath the channel, the N-polar orientation has a more strongly confined 2DEG with the AlGaN layer serving as backbarrier. Third, as the vertical gate to 2DEG thickness is reduced to accommodate shorter gate lengths for higher speed operation, the charge-inducing barrier material thickness can be kept the same, while in Ga-polar the charge-inducing layer also determines the separation between the 2DEG and the gate.

Growth of both Ga-polar and N-polar-oriented GaN films was explored in early investigations of GaN material growth and electronic devices in the 1980s and 1990s, and the history of these investigations is covered more completely in another chapter of this book and in reviews by Wong [44] and Keller [43]. Early investigations into N-polar GaN resulted in much lower-quality material than Ga-polar GaN of the same era. Impurity concentrations were higher [45], and morphology was much worse by MBE [46] and by MOCVD [47, 48], the latter of which suffered from characteristic hexagonal hillocks. Device fabrication was also more difficult as the N-polar surface can be etched in basic solutions [47] (including photoresist developer) and so the surface must be protected during fabrication [49].

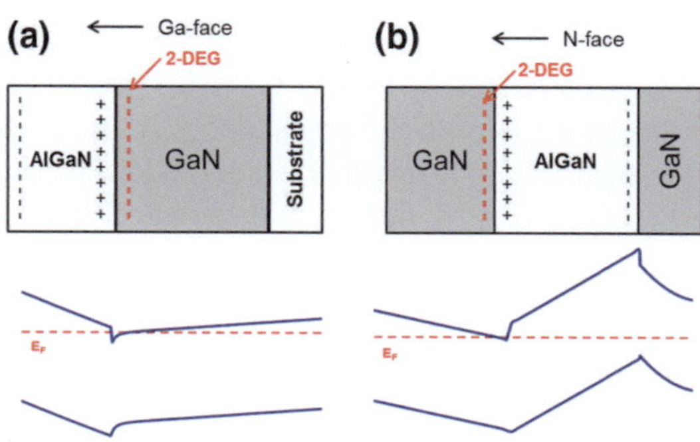

Fig. 4.8 Schematic of **a** Ga-polar HEMT and **b** N-polar HEMT structure. Net polarization charges as well as the resulting 2DEG location are labeled. Figure courtesy of S. Wienecke

The first 2DEGs in N-polar HEMT epi structures, grown by plasma induced MBE, were demonstrated at Cornell in 1999 [45, 46, 50, 51]. The 2DEG and its location on the expected top side of the GaN/AlGaN interface based on polarization were shown by capacitance-voltage charge profiling. However, while Ga-polar transistor action was shown in those reports, the N-polar crystal morphology was extremely poor and no N-polar transistor operation was shown. As high-performance devices were already being fabricated using Ga-polar AlGaN/GaN structures as described earlier in this chapter, the majority opinion [46, 47, 52] was that N-polar GaN/AlGaN HEMTs did not merit significant investment.

Nevertheless, as Ga-polar devices matured and the tradeoffs inherent in optimization made further improvements more difficult to realize, some focus at UCSB shifted to N-polar GaN. In the period of 2000–2005, some improved techniques for N-polar GaN growth by MBE [53, 54] and MOCVD [55, 56] had been reported. In 2005 S. Rajan at UCSB had developed [57] improved N-polar GaN material grown by plasma-assisted MBE on SiC by using a newly developed two-step buffer to produce films with both low dislocation density and smooth surface morphology. That two-step buffer was used to demonstrate the first high-quality N-polar 2DEG in a HEMT structure, with 1×10^{13} cm^{-2} electron density and 1020 cm^2/V s mobility. In 2007 and 2008, S. Keller and D. Brown reported studies of growth on nitrided sapphire (miscut a few degrees toward the A plane) [58] and C-face SiC (miscut toward the M plane) [59] substrates to enable smooth N-polar GaN film growth by MOCVD. These developments now provided a source of high-quality material for device investigations. With further optimization of conditions and improved reactor purity, impurity incorporation in N-polar GaN films grown by MOCVD would be reduced to the 2×10^{16} level [43], comparable to Ga-polar. With improved growth technology and epitaxy design, the room temperature mobility in N-polar can now exceed 2000 cm^2/V s [60, 61], competitive with the best Ga-polar results.

A fabrication process was developed at UCSB including a Ge sacrificial layer to protect the N-polar GaN surface from etching, which led to the first demonstration of an N-polar GaN HEMT by Rajan in 2007 [49]. Even with passivation significant dispersion was observed in the first devices, apparently caused by very deep donor-like states, or hole traps, at an energy level close to the valence band at the AlGaN/GaN interface underneath the backbarrier [62]. The effect of this type of trap has been observed in numerous Ga-polar and N-polar electronic and optoelectronic structures at interfaces with net negative polarization charge [49, 62–66], but it does not affect the typical Ga-polar AlGaN/GaN HEMT because the Fermi level does not approach the trap's energy level, so the trap is electrically inactive. The physical origin of this trap is still under investigation, but it has been shown experimentally that by doping and grading the AlGaN backbarrier to move the valence band and associated trap level further below the Fermi level the dispersion and other undesirable effects caused by this trap can be mitigated, and along with conventional surface passivation by PECVD SiN low-dispersion N-polar GaN HEMTs can be fabricated.

With the basic physics and design for N-polar HEMTs now understood, microwave performance and fabrication techniques improved in the following years. M. H. Wong reported in 2009–2011 on MBE-grown HEMTs [67, 68] which extended N-polar large-signal performance to up to 10 GHz, and microwave power density to 6.7 W/mm, which was limited by the breakdown voltage. S. Kolluri further increased large-signal performance [69]. First, while AlGaN caps had been used in prior N-polar devices to increase breakdown, Kolluri showed that a very thin but high Al composition AlGaN cap (2 nm, 60 %), with its reversed polarization which inhibits gate leakage, could be used to obtain very high breakdown voltage in excess of 150 V. Second, the gate was recessed into the PECVD SiN passivation, with integrated slanted field plates which further improved both breakdown voltage and dispersion control. The device is illustrated in Fig. 4.9a. These improvements resulted in a device with 20.7 W/mm power density with associated 60 % PAE at a 70 V bias in 2012. These results are comparable to Ga-polar devices with similar field plate schemes [70], and N-polar power results leading up to this point along with Ga-polar results are summarized in Fig. 4.9b.

The next step for N-polar GaN was to dramatically increase device speed to enable mm-wave operation, largely by reducing the gate length. As the gate length (L_G) is reduced the vertical gate to channel distance (a) must also be reduced to maintain a high enough aspect ratio (L_G/a). This is to maintain the gate's electrostatic control of the channel in order to realize the low output conductance needed for high power gain (f_{max}) and to maintain current gain (f_t) as gate length is deeply scaled (Fig. 4.10), [71–74]. The N-polar HEMT structure is distinct in that the 2DEG-inducing backbarrier layer thickness is independent of the aspect ratio, while in Ga-polar the 2DEG-inducing layer thickness and aspect ratio are directly traded

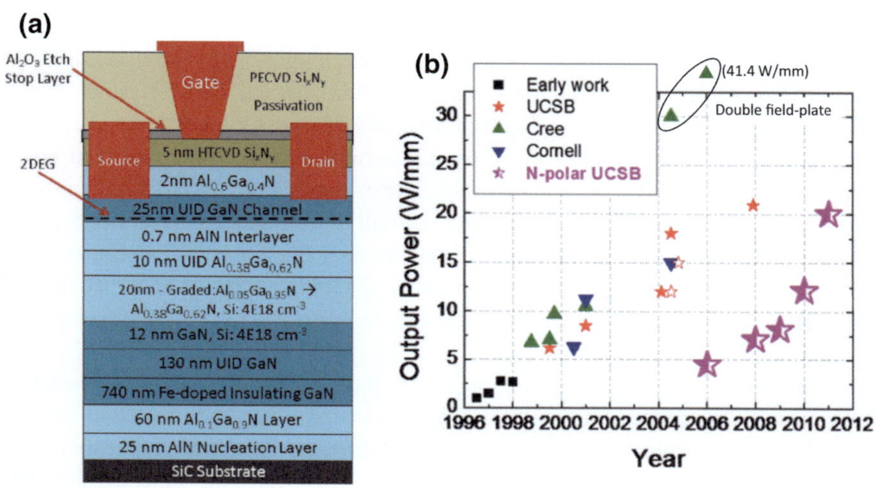

Fig. 4.9 **a** Cross-section (not to scale) of N-polar MISHEMT which demonstrated 20.7 W/mm output power density (figure by Kolluri [69]). **b** Maximum RF output power density by year for Ga-polar and N-polar GaN devices (figure after S. Keller)

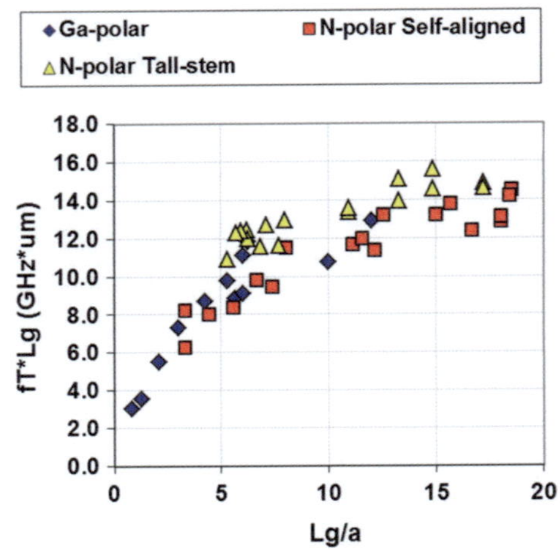

Fig. 4.10 Product of f_t L_G versus aspect ratio (L_G/a) through 2012, shows the trend that higher aspect ratio can provide a higher f_t for a given gate length. Figure by Dan Denninghoff, 2012 [74]

off against each other. In Ga-polar, to obtain high charge density with high aspect ratio in the (Al, Ga)N material system, very high aluminum composition must be used; in some cases pure AlN will be used [35], which is a highly strained material. In N-polar, the charge-inducing backbarrier also serves to further confine the charge closer to the gate, which according to physical modeling will reduce short channel effects relative to Ga-polar devices without a backbarrier [72, 73]. Therefore, as operation frequency for GaN devices increased into the W-band (75–110 GHz), the vertical scaling advantages of N-polar-oriented HEMTs are expected to become more apparent.

At UCSB, the GaN channel thickness for devices was first scaled to 10 nm in 2009 using AlGaN backbarrier materials, which along with ultra-low parasitic resistance and capacitance provided state-of-the-art $f_T \cdot L_G$ product of 16.8 GHz-μm, with 132 GHz f_t for a 120 nm gate length [75]. These results were obtained by a combination of the high aspect ratio, record low-resistance ohmic contacts of 23 Ω-μm which used with graded InGaN used to generate a 3D electron gas, capped with a thin In N layer [76], and a ohmic regrowth self-aligned to the gate for highly scaled source and drain access regions. In N-polar the channel thickness and electron density are only partially decoupled as a tradeoff exists between 2DEG density and mobility versus channel thickness [77] because of the surface pinning position. While the channel thickness with AlGaN backbarrier technology can be scaled below 10 nm, it was not pursued for device development at the time.

An alternate path to thinner channels was pursued using $In_xAl_{1-x}N$ materials. With an Indium composition (x) of 17 %, this material is lattice matched to GaN [78, 79], so thick layers can be grown without concern for cracking. Furthermore, its polarization charge and conduction band offset to GaN are both greater than what is realizable with thick AlGaN layers [80], which enables higher charge density with thinner channels. N-polar-oriented InAlN film growth and initial devices were first reported at UCSB by MOCVD in 2010 by Brown [81, 82], and by MBE in 2011 by Dasgupta [83]. State-of-the art DC device performance was reported by Nidhi in 2012 [84] including current density of 2.8 A/mm, on-resistance of 0.29 Ω-mm, and transconductance of 1.1 S/mm, with these metrics enabled by the high 2DEG density and thin channel. Those self-aligned devices had f_t up to 155 GHz, but still had low f_{max} due to the high-resistance gate used for the self-aligned process. The MBE material had significant lateral compositional fluctuations which resulted in lower mobility. Further improvements in device performance first required both material improvements and a more advanced T-gate fabrication process. Improved MBE materials were reported in 2014 by Ahmadi [85] with optimization of growth conditions to reduce fluctuations, and use of MOCVD-grown buffers for lower dislocation density, providing record N-polar InAlN-based HEMT mobility of 1850 cm^2/V s at n_s of 1.1×10^{13} cm^2 in 2015 [60]. For MOCVD-grown material, improvements were realized by J. Lu with development of a two-step backbarrier comprised of an InAlN film on top of a graded AlGaN layer [86]. Extreme vertical channel scaling was demonstrated down to a channel of 3.3 nm thickness which had a charge density of 1.8×10^{13} and a sheet resistance of 329 Ω/square [87]. A summary of N-polar InAlN-based HEMT channel scaling is shown in Fig. 4.11. D. Denninghoff developed a tall-stem T-gate process to minimize both gate resistance and parasitic capacitance, with laterally scaled gate and access region dimensions. Using that fabrication process with MOCVD-grown epi that had a 5.4 nm GaN channel thickness, Denninghoff demonstrated state-of-the-art device DC and RF performance with current density of 4 A/mm at $V_{GS} = 1.5$ V and 2.35 A/mm at $V_{GS} = 0$ V, 204 GHz f_t, and 405 GHz f_{max} [88] (Fig. 4.12).

These activities firmly demonstrated the capability of N-polar GaN/InAlN HEMTs for extreme vertical scaling and ultra-high-speed operation. However, N-polar InAlN backbarrier HEMTs still possessed some unresolved drawbacks compared to AlGaN backbarriers. 2DEG mobilities were lower, because of greater impurity incorporation in InAlN films, lateral composition fluctuations, and alloy scattering. To increase mobility and improve sheet resistance in the MOCVD-grown

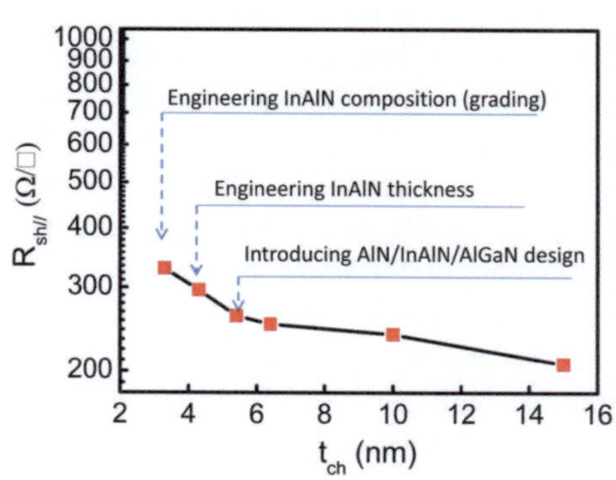

Fig. 4.11 Trend of sheet resistance vs. channel thickness for InAlN-based backbarrier HEMTs, and design features used to reduce sheet resistance for thinner channels, culminating in 329 Ω sheet resistance for a 3.3 nm channel [87]. Figure by Lu [119]

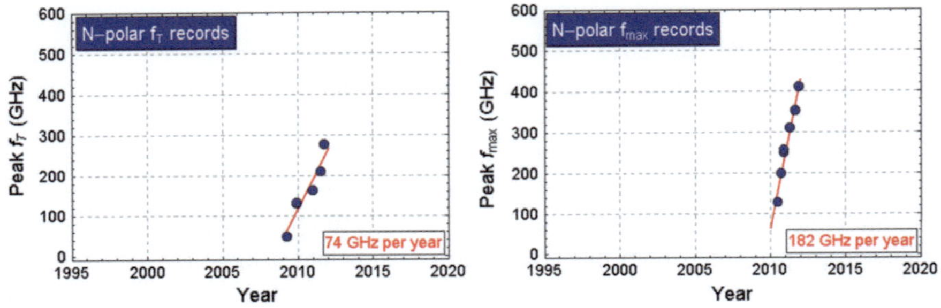

Fig. 4.12 Trend of record f_t and f_{max} versus year for N-polar HEMTs. Figure by Denninghoff [74]

epi, it was necessary to design a 2DEG density around 2×10^{13} cm^{-2} to screen the 2DEG. Such a high charge density enabled extremely high current density as Denninghoff and Nidhi showed, but it also reduced breakdown voltages. The design of the backbarrier is more complex because of the presence of multiple interfaces in multilayer backbarriers, and the difficulty in grading between AlGaN and InAlN. A quaternary alloy of AlInGaN has also shown promise [39, 89] for realizing some of the advantages of InAlN barrier materials (lower strain, larger barrier height, higher polarization charge) with improved material quality and homogeneity and is the subject of on-going development.

N-polar HEMT research at UCSB shifted back to the AlGaN backbarrier as this material is sufficient for excellent performance at W-band, and the material design is simpler and growth technology is more mature. Impurity incorporation in N-polar GaN films had been reduced to around 2E16 cm^{-3} [43], providing higher mobilities of around 2000 cm^2/V s for thick 20 nm channels [unpublished] so the tradeoff of mobility versus channel thickness for AlGaN devices was more relaxed than in earlier studies [77]. The focus shifted to optimizing the overall device design and fabrication to meet all of the metrics needed for mm-wave power performance, including high speed, breakdown voltage, and low dispersion. These metrics can be at odds with each other. Classical dispersion management methods of field plates and thick PECVD SiN passivation layers can carry significant gain penalties at mm-wave frequencies due to added capacitance. Therefore, an alternate method of dispersion control was pursued: a thick GaN cap which is deeply recessed under the gate, but which passivates the access regions.

The GaN cap serves a similar purpose as PECVD SiN passivation which is to move surface electron traps further away from the channel so that they do not influence the device's dynamic performance. SiN passivation films still have some interface traps to the underlying device, as well as bulk traps. When a GaN cap is grown in situ on top of the underlying device, with a pristine interface, there are negligible interface traps or bulk traps which are present in dielectric passivation films. This was first explored for Ga-polar HEMT designs beginning in 2004 by Shen [90] which proved that excellent large-signal performance could be obtained without any ex situ passivation, using a recessed GaN [91] or AlGaN [92] cap. However, the direction of polarization fields in the Ga-polar orientation leads to challenging tradeoffs. GaN caps depleted the channel, requiring doping in the direct vicinity of the gate, with a tradeoff between access region conductivity, drain current, and dispersion control against the breakdown voltage [91]. AlGaN caps were also explored as these exhibit reduced channel charge depletion, but the material strain and process parameter space (lack of a natural etch stop makes repeatable fabrication extremely difficult) is extremely limited. Further development of 0.5 μm gate length deep-recess GaN technology for X-band applications returned to a GaN cap to provide a precision etch stop at the AlGaN barrier, which was then further recessed using a much shorter and thus more accurate timed etch [93].

While a GaN cap in a Ga-polar HEMT depletes 2DEG charge, in N-polar because of the reversed polarization fields it enhances the 2DEG density in the access regions around the gate [94, 95]. The high charge on the source side of the gate helps supply very high drain current densities and moving the surface further from the 2DEG increases mobility. These in combination lead to very low access resistances. The initial investigation of a N-polar deep recess mm-wave device design in 2015 has already provided impressive 94 GHz large signal results with 2.9 W/mm power density at 15.5 % PAE measured at 94 GHz, and a peak PAE of 20 % with associated 1.73 W/mm [96], measured by passive load pull technique [97]. The result of the high current density is shown most clearly in Fig. 4.5 where N-polar deep recess devices have far greater 94 GHz power density at a given drain bias voltage compared to any Ga-polar device with either (Al, Ga)N or (Al, In, Ga) N barrier material. The high RF current density has also been verified more directly with C-band RF-IV measurements of a similar device [61], (Fig. 4.13), showing

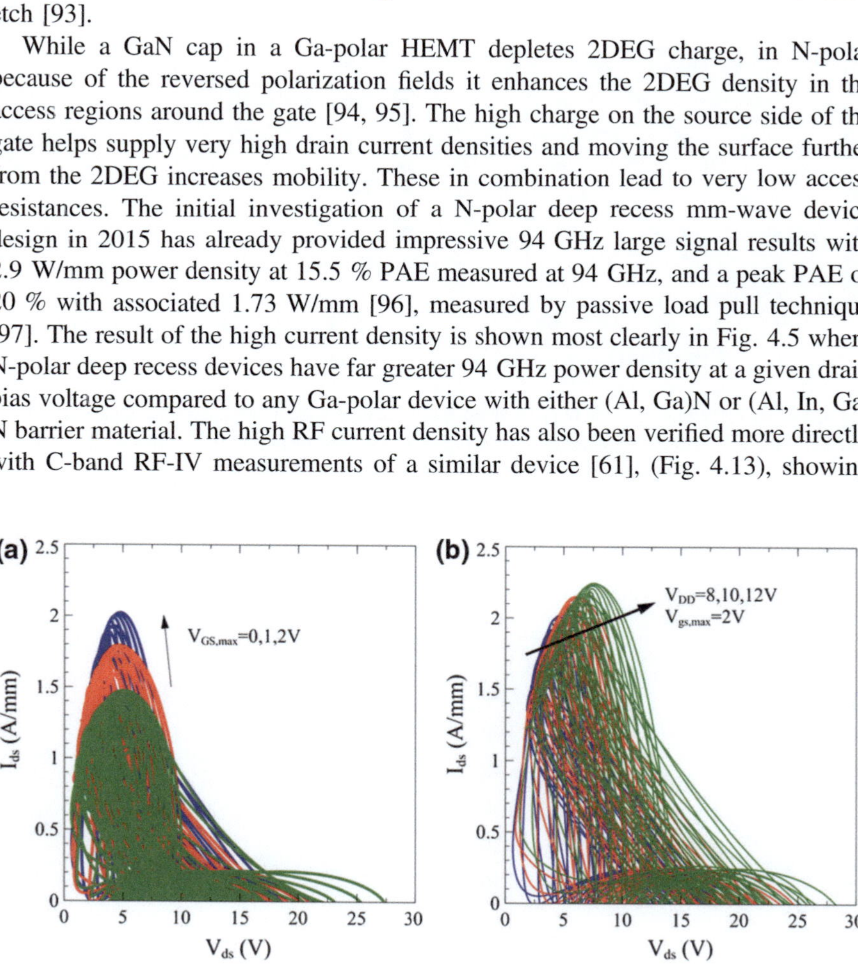

Fig. 4.13 Families of dynamic loadlines measured over multiple load impedances in active harmonic vector load pull at 6 GHz, plotted for **a** V_{DD} = 8 V bias with P_{IN} corresponding to a maximum V_{GS} of 0, 1, or 2 V during the RF cycle, and **b** V_{DD} = 8, 10, and 12 V bias at P_{IN} corresponding to a peak V_{GS} of 2 V during the RF cycle [61]. The high RF current capability of N-polar deep recess devices. Measurement provided by Maury Microwave

over 2A/mm RF peak current. The power density, while already compelling, is at present limited in large part by the use of a sapphire substrate which has low thermal conductivity, thus preventing use of higher drain bias voltages. Substantial improvements are anticipated with use of a SiC substrate. Load pull measurements of sapphire substrate devices conducted at 10 GHz, where higher efficiency reduces self-heating, show power density scaling well up to 14 V bias and 4.5 W/mm [61], (Fig. 4.15). The 94 GHz efficiency is currently limited by the gain, but modeling on similar devices indicate that substantial improvements can be made with improved

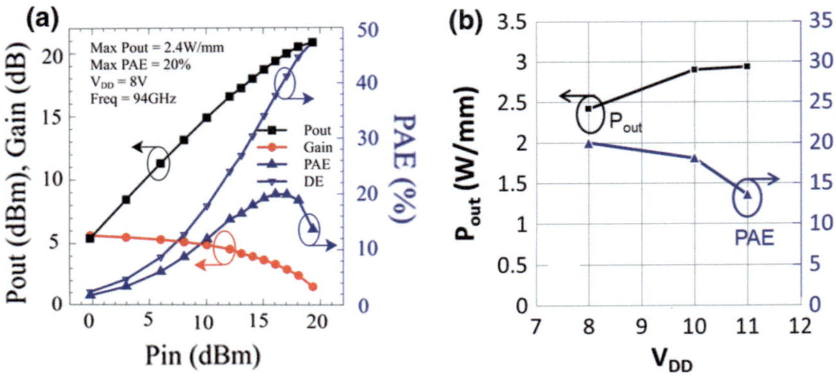

Fig. 4.14 94 GHz load pull of N-polar deep recess device with a designed gate length of 75 nm and 110 nm GaN cap passivation in access regions. **a** Power sweep at 8 V bias and **b** maximum power and PAE versus drain bias voltage, falloff in performance is likely related to self-heating [96]

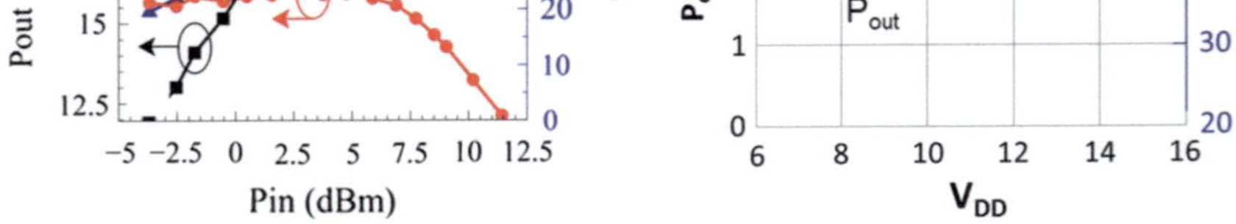

Fig. 4.15 10 GHz load pull of N-polar deep recess device with a designed gate length of 150 nm and 47 nm GaN cap passivation in access regions. **a** Power sweep at 14 V bias and **b** maximum power and associated PAE versus drain bias voltage show excellent scaling in power density and PAE versus drain bias [61]. Measurement provided by Maury Microwave

pad layout and backend processing, in addition to substantial improvements that are still possible at the device level [98] (Fig. 4.14).

Ga-polar GaN HEMTs, as discussed in the prior section, have shown impressive power density and enough process maturity to demonstrate useful MMIC continuous power levels in excess of 2 W, and are expected to supplant GaAs and InP MMICs in high-power E-band and W-band applications. The design advantages of N-polar GaN MISHEMTs, in particular with use of a deeply recessed GaN cap, appear capable of even higher performance at W-band frequencies and may provide access to much higher power levels at comparable or better efficiencies than Ga-polar GaN as the technology matures.

4.6 GaN Applied to Power Electronics

GaN electronic devices were first pursued to address defense and commercial needs for increasing RF and microwave transmitter power levels for wireless applications including radar and communication. The combination of high critical breakdown field and high electron velocity and mobility in the GaN material system provided a path to improve high-power microwave amplification.

The same physical properties that make GaN so effective at high efficiency and high-power microwave applications also lend themselves to highly efficient and compact power conversion applications. With a GaN technology baseline established from investments in microwave technology, researchers and companies began developing low-loss power switch devices. Similar to the RF devices, it was necessary to improve the material, design, and fabrication process to minimize degradation of dynamic on-resistance and current due to electron trapping, but this has been achieved over a full 600 V range by 2009 (Fig. 4.16). For fail-safe operation, power switches have to be normally off. Currently, enhancement mode operation has been achieved in two ways. One is a two-chip solution using a cascode connection of a depletion-mode device with a low voltage Si MOSFET

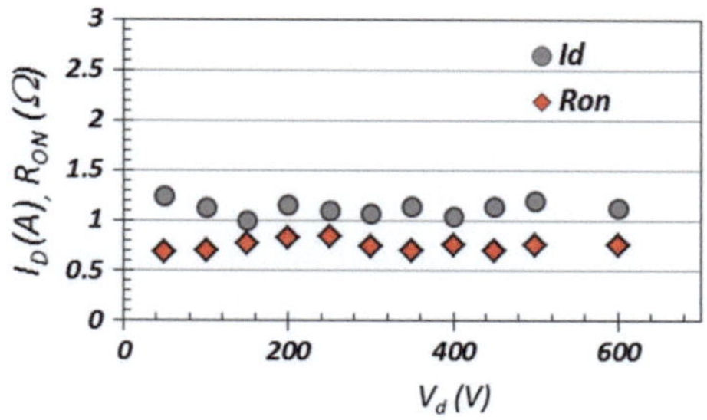

Fig. 4.16 Dynamic current and on-resistance measured directly in a DC–DC converter, showing low dispersion over the full 600 V operating range. Figure courtesy of Transphorm

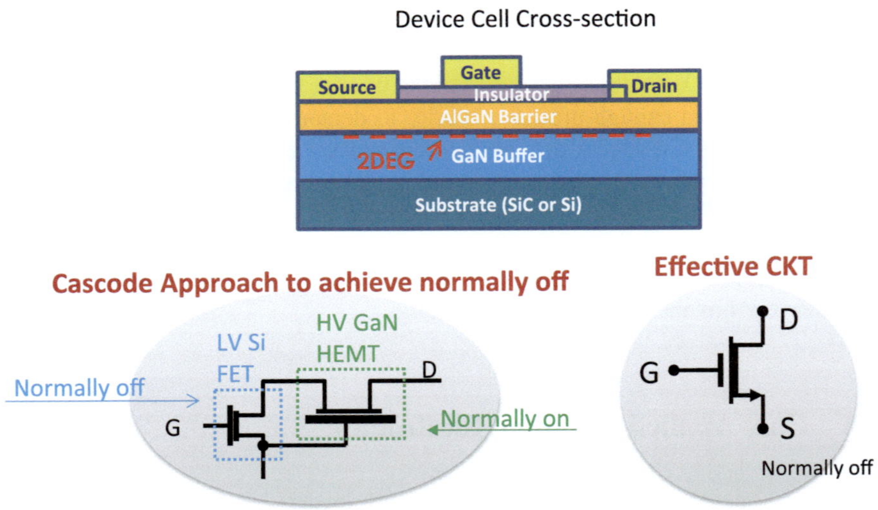

Fig. 4.17 Cascode method for normally off operation. Disadvantage: two chips required

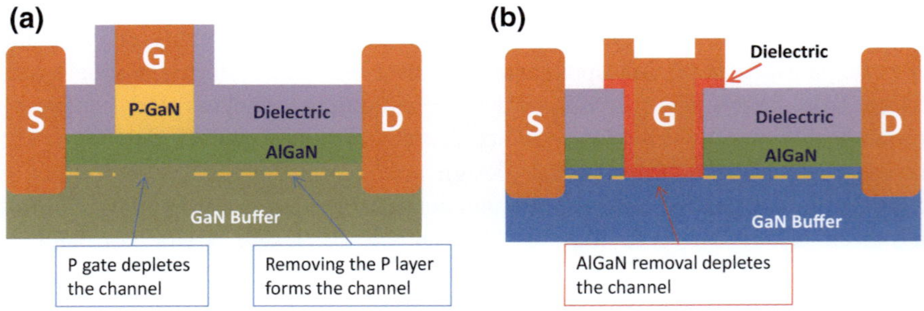

Fig. 4.18 Methods for obtaining enhancement mode devices: **a** junction gated HEMTs (Panasonic, GaN systems). Disadvantage: low threshold voltages. **b** Channel etch-through MOSFET. Disadvantage: device and dielectric not qualified

shown in Fig. 4.17 and the second is a single-chip solution by using a p-type gate to deplete the charge under the channel under zero bias in the GaN HEMT as shown in Fig. 4.18a.

600 V rated devices have been released into the market using both approaches, the first by Transphorm [99] and the second by Panasonic [100] (using an AlGaN p-type layer called the Gate Injection Transistor or GIT) and GaN Systems [101]. 200 V rated devices have been released by EPC [102]. The Transphorm cascode product is qualified using stringent JEDEC standards, and the lifetime of the

devices has been extracted using accelerated electric field and temperature stress Fig. 4.19. A lifetime of over 10 million hours has been predicted for the 600 V rated products which makes the technology on par in reliability with Silicon. The p-gated E-mode devices have demonstrated attractive performance, but there is no device lifetime data currently available. Typical performance of a 310 mΩ, 600 V device from Transphorm is shown in Fig. 4.20 and compared with a 385 mΩ CoolMOS device.

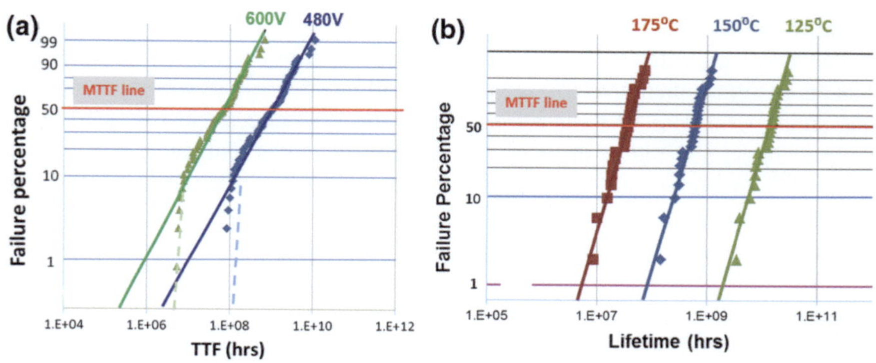

Fig. 4.19 Transphorm 600 V GaN accelerated lifetime test results: **a** voltage accelerated reliability characterization. Data collected over 900–1200 V and plotted at 600 and 480 V, using inverse power law model. MTTF 8×10^7 h at 600 V. **b** Temperature accelerated reliability characterization. Data collected at T_J from 318 to 362 °C which showed an activation energy of 1.84 eV and no indication of early failure modes. MTTF of 3×10^7 h at 175 °C

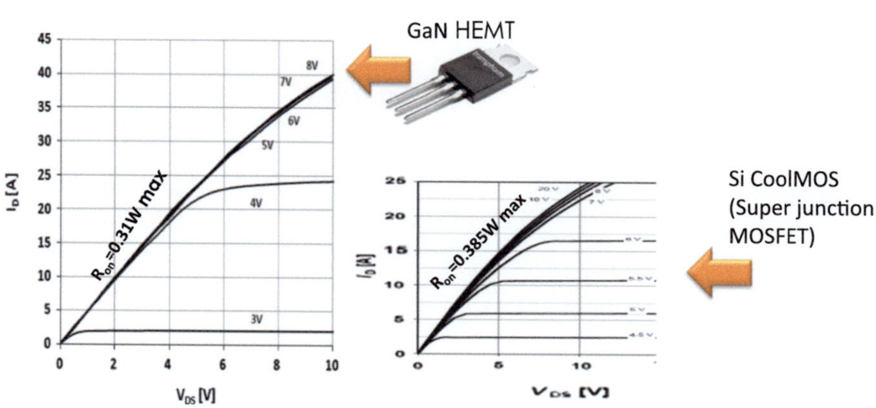

Fig. 4.20 Cascode device *IV* curves compared with Si CoolMOS

Novel architectures are possible with GaN power transistors because of the inherent Bi-directional nature of the current flow in HEMTs such as the Diode-free™ half-bridge configuration which is shown in Fig. 4.21a–d. This eliminates both the cost and power loss associated with the freewheeling diode necessary in Si-based half bridges using either MOSFETs or IGBTS (Fig. 4.21e). Furthermore, the lateral nature of the device allows either the source (low side of the half-bridge) or the drain to be connected to the case of the package employed (high side of the half-bridge) allowing the switching node to not have the capacitance of the package case associated with it (Fig. 4.22). This ensures low-EMI operation even under high switching speeds.

The ability to switch at high frequency at high efficiency and low EMI allows the systems to shrink their size and reduce the BoM costs of passive components and heat sinks. Examples of these are shown in the PFC circuits in Fig. 4.23 and a compact integrated motor drive with a filter to deliver a sinusoidal motor drive (TruSin™) in Fig. 4.24. The savings in BoM costs offsets the increased cost of a new technology such as GaN transistors and allows higher performance operation at a lower system cost as depicted in Fig. 4.28.

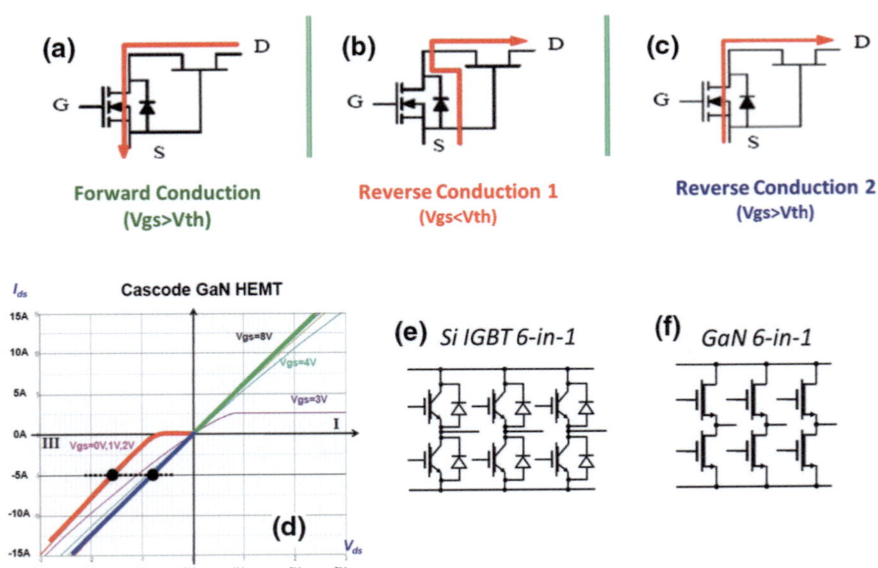

Fig. 4.21 **a–c** The forward and reverse conduction modes of the Transphorm cascode and **d** their current versus voltage behavior, illustrating the bidirectional nature of these devices. A three phase inverter with **e** silicon IGBT devices and diodes, and **f** bidirectional GaN devices which do not require external high-voltage diodes

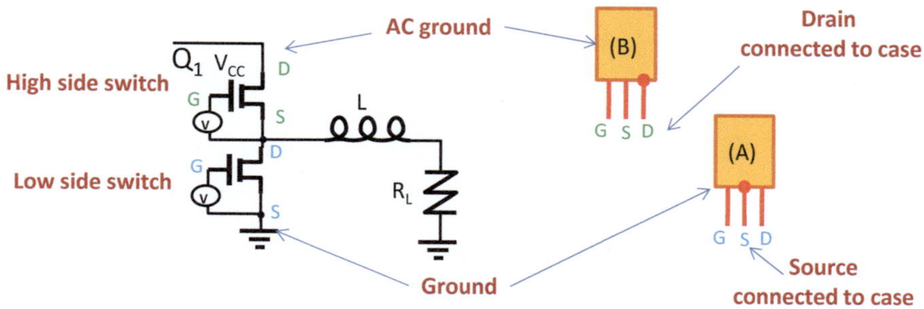

Fig. 4.22 Illustration of low-EMI packaging solution enabled by GaN

Fig. 4.23 Comparison of comparable PFC circuits: **a** silicon-based, requiring a two stage EMI filter and large inductors, and **b** GaN-based with smaller inductor size, one less EMI filter stage, enabled by the lower-loss transistors and diodes

In addition, novel power factor correction (PFC) topologies such as the totem Pole implementation shown in Fig. 4.25 is practically made by GaN transistors because of the bi-directional nature of current flow and the low capacitance necessary for this hard-switched topology. Over 99 % efficiency at 385 W is possible using this bridge-less implementation (Fig. 4.26) and is essential for achieving titanium class of power conversion performance. Finally, the GaN transistor is capable of both blocking voltage and carrying current in both directions in a

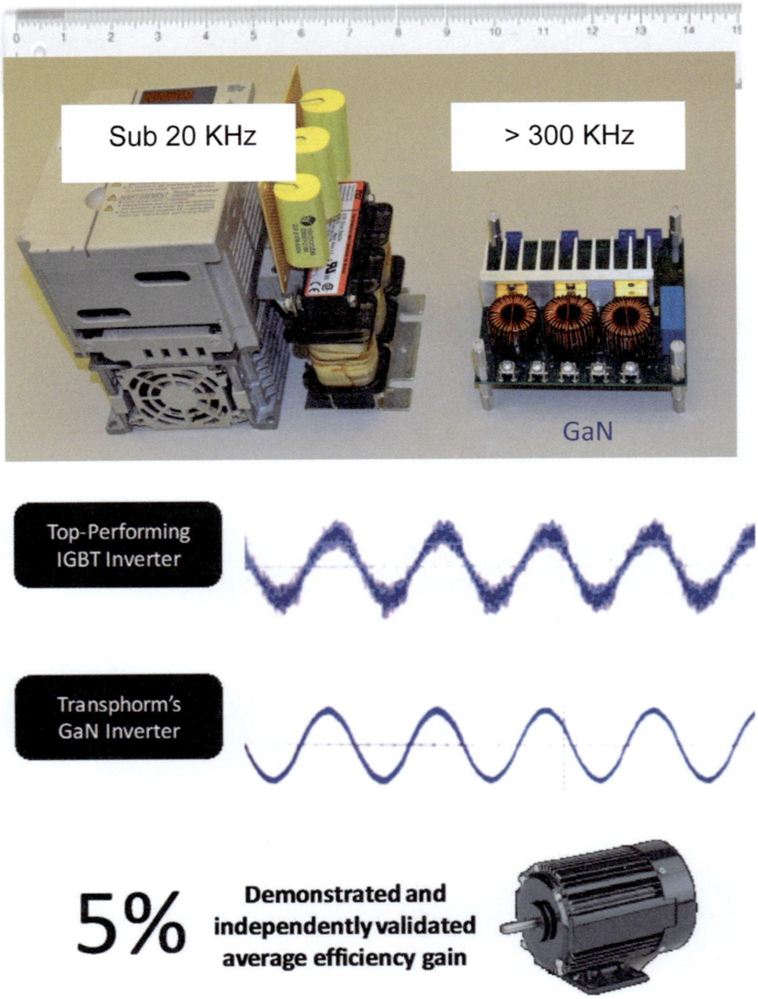

Fig. 4.24 Motor drive with a sinusoidal signal. At the left is the filter required to condition the sub 20 kHz PWM signal employed in a Si drive versus the compact filter required to condition the 300 kHz signal used for a GaN-based motor drive. An increase in the efficiency of the motor results because of the lack of charging/discharging losses caused by the pulses in an unfiltered conventional IGBT inverter

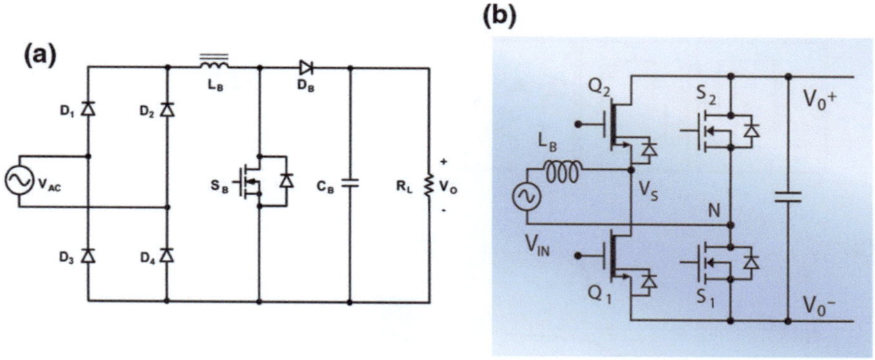

Fig. 4.25 **a** Traditional PFC with line rectifying bridge, and **b** totem pole PFC with no diode drop, Q_1 and Q_2 are fast PWM switches, and S_1 and S_2 are 60 Hz slow switches. The low reverse recovery charge (Q_{rr}) and bidirectional capability of GaN enable this bridgeless PFC

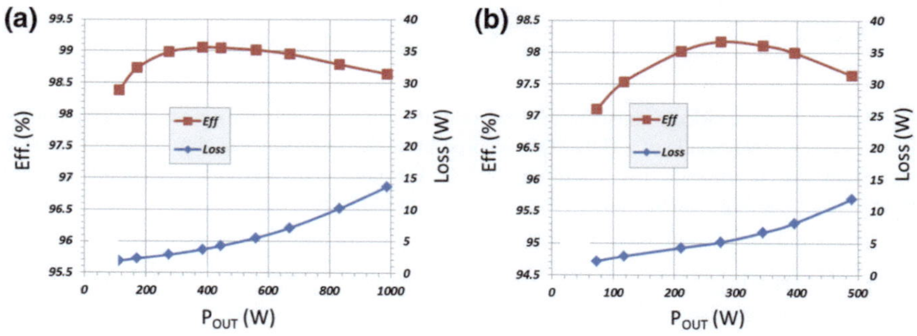

Fig. 4.26 Efficiency of bridgeless GaN totem pole PFC versus output power, at 50 kHz switching frequency and 390 V output voltage for **a** 230 V input and **b** 115 V input

four-quadrant switch (FQS). Typical implementations with Si devices require 4 or 5 separate devices Fig. 4.27a. With GaN, this is accomplished in a single device by two gates sharing a common drift region, with the choice of the source drain and gate electrodes defining which device blocks the voltage as desired, shown in Fig. 4.27. The value proposition of GaN is high system performance at reduced system cost, because of smaller heatsink, smaller passives, reduced part count, and small form factor (Fig. 4.28).

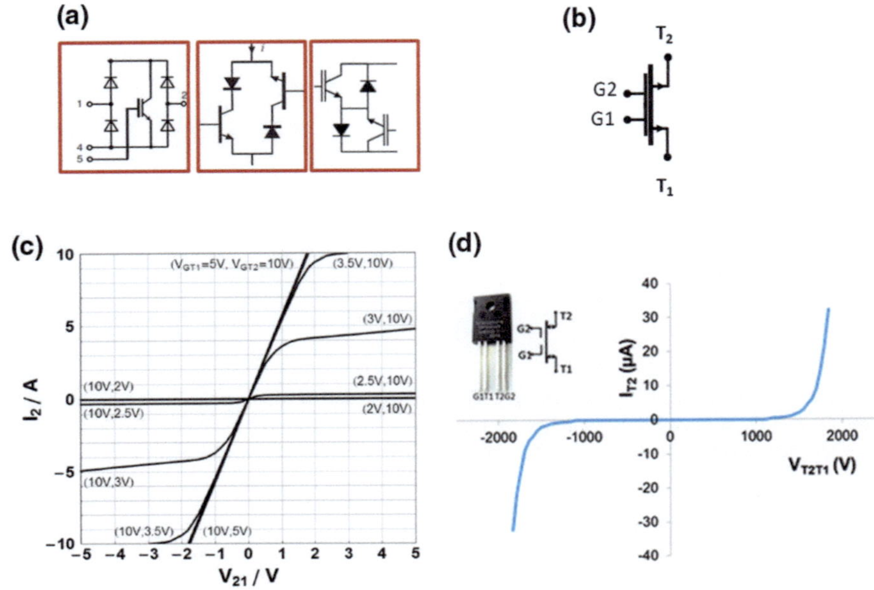

Fig. 4.27 **a** Typical silicon-transistor implementations of a four-quadrant switch (FQS), requiring 4 or 5 devices, **b** schematic of a GaN FQS, incorporating the same functionality in a single device, **c** on-state and **d** off-state *I–V* characteristics of the GaN FQS showing capability for low on-resistance and high current and voltage blocking in both directions

Fig. 4.28 GaN Power Electronics delivers higher system performance at a lower system costs because of the reduced BoM costs enabled by shrinking of passive components at higher frequencies and smaller heat sinks because of higher efficiencies

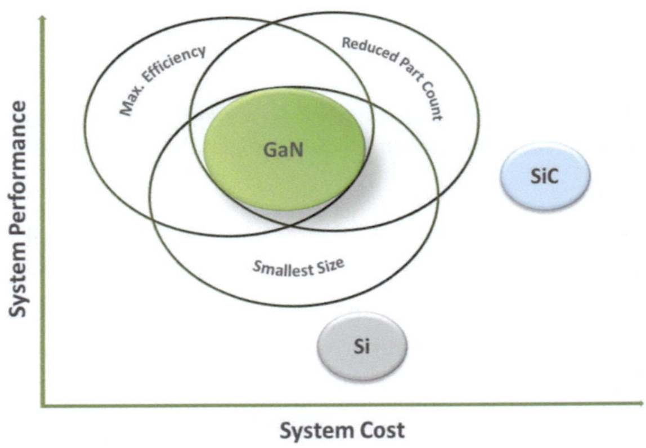

4.7 Conclusions

Gallium nitride is emerging as the next major semiconductor to supply solid-state solutions along with silicon for the next few decades. The reason is the breadth of gallium nitride-based solutions from lasers to LEDs to RF sources and power conversion applications. It is imperative for a new technology to have a breadth of applications so as to amortize the cost of the technology to provide enhanced functionality at an affordable cost. Market penetration will be proportional to the decrease in the cost of the GaN-based solution and pathways including but not limited to GaN on Si show exceptional promise. The market size for GaN-based electronics promises to be over $1B by 2022 and rapidly increases thereafter. This is alongside a photonics market of over $10B today with a steep growth trajectory. From an academic and research perspective, it is remarkable that such a market is available with the limited science currently explored in the nitrides. With the full effort of the community and a more complete understanding of the science of the materials and devices, the future will be even brighter.

Acknowledgments The sustained funding from the ONR (Max Yoder, John Zolper, Harry Dietrich and Paul Maki) is gratefully acknowledged, along with DARPA and the AFOSR.

References

1. Asif Khan M, Bhattarai A, Kuznia JN, Olson DT (1993) High electron mobility transistor based on a GaN $Al_xGa_{1-x}N$ heterojunction. Appl Phys Lett 63(9):1214–1215
2. Asif Khan M, Kuznia JN, Olson DT, Schaff WJ, Burm JW, Shur MS (1994) Microwave performance of a 0.25 μm gate AlGaN/GaN heterostructure field effect transistor. Appl Phys Lett 65(9):1121–1123
3. Wu YF, Keller BP, Keller S, Kapolnek D, Denbaars SP, Mishra UK (1996) Measured microwave power performance of AlGaN/GaN MODFET. IEEE Electron Device Lett 17 (9):455–457
4. Keller S, Wu YF, Parish G, Ziang N, Xu JJ, Keller BP, DenBaars SP, Mishra UK (2001) Gallium nitride based high power heterojunction field effect transistors: Process development and present status at UCSB. IEEE Trans Electron Devices 48(3):552–559
5. Wu YF, Keller BP, Keller S, Xu JJ, Thibeault BJ, Denbaars SP, Mishra UK (1999) GaN-based FETs for microwave power amplification. Ieice Trans Electron E82C(11): 1895–1905
6. Green BM, Chu KK, Chumbes EM, Smart JA, Shealy JR, Eastman LF (2000) Effect of surface passivation on the microwave characteristics of undoped AlGaN/GaN HEMT's. IEEE Electron Device Lett 21(6):268–270
7. Wu YF, Zhang N, Xu J, McCarthy L (2003) Group III nitride based FETs and HEMTs with reduced trapping and method for producing the same. US Patent 6 586 781
8. Wu Y, Kapolnek D, Ibbetson J (1999) High Al-content AlGaN/GaN HEMTs on SiC substrates with very high power performance. Electron Devices 20(4):925–927
9. Chini A, Buttari D, Coffie R, Heikman S, Keller S, Mishra UK (2004) 12 W/mm power density AlGaN/GaN HEMTs on sapphire substrate. Electron Lett 40(1):73
10. Ando Y, Okamoto Y, Miyamoto H, Nakayama T, Inoue T, Kuzuhara M (2003) 10-W/mm AlGaN-GaN HFET with a field modulating plate. IEEE Electron Device Lett 24(5):289–291
11. Wu Y-F, Saxler A, Moore M, Smith RP, Sheppard S, Chavarkar PM, Wisleder T, Mishra UK, Parikh P (2004) 30-W/mm GaN HEMTs by field plate optimization. IEEE Electron Device Lett 25(3):117–119
12. Wu Y-F, Moore M, Wisleder T, Chavarkar PM, Mishra UK, Parikh P (2004) High-gain microwave GaN HEMTs with source-terminated field-plates. IEDM Tech Dig IEEE Int Electron Devices Meet 25(3):1078–1079

13. Mishra UK, Shen L, Kazior TE, Wu Y-F (2008) GaN-based RF power devices and amplifiers. Proc IEEE 96(2):287–305
14. Wu Y-F, Moore M, Saxler A, Wisleder T, Parikh P (2006) 40-W/mm double field-plated GaN HEMTs. In: 2006 64th device research conference, pp 2005–2006
15. Lee C, Salzman K, Coffie R, Li V, Drandova G, Nagle T, Morgan D, Horng P, Hillyard S, Ruan J (2013) GaN on SiC HEMT process. In: Proceedings of the CS MANTECH conference, pp 91–94
16. Himes G, Maunder D, Kopp B (2013) Recent defense production act title III investments in compound semiconductor manufacturing readiness. In: Proceedings CS MANTECH conference, pp 83–86
17. Fury R, Sheppard ST, Barner JB, Pribble B, Fisher J, Gajewski DA, Radulescu F, Hagleitner H, Namishia D, Ring Z, Gao J, Lee S, Fetzer B, Mcfarland R, Milligan J, Palmour J (2013) GaN-on-SiC MMIC production for S-band and EW-band applications. In: 2013 international conference on compound semiconductor manufacturing technology CS MANTECH 2013, pp 95–98
18. Smolko J, Whelan CS, Macdonald C, Krause J, Mikesell B, Benedek M (2013) Raytheon title III gallium nitride (GaN) production program. In: 2013 International conference on compound semiconductor manufacturing technology CS MANTECH 2013, pp 87–90
19. US Government Honors Raytheon for Completing Title III GaN Production Improvement Program (2013) Semiconductor today [Online]. Available http://www.semiconductor/today.com/news_items/2013/JUN/RAYTHEON_130613.html. Accessed 05 Apr 2016
20. AFRL's Defense Production Act Title III Program for X-band GaN MMICs Completed A (2014) Semiconductor today [Online]. Available http://www.semiconductor/today.com/news_items/2014/JAN/AFRL_020114.shtml. Accessed 05 Apr 2016
21. Cree earns US DoD MRL8 Designation After Completing Title III GaN-on-SiC Production Capacity Program (2014) Semiconductor today
22. Frye B (2014) TriQuint becomes first manufacturer to achieve MRL 9 for GaN (Press Release) [Online]. Available http://www.triquint.com/newsroom/news/2014/triquint-first-manufacturer-to-achieve-mrl-9-for-gan. Accessed 05 Apr 2016
23. http://www.sedi.co.jp/
24. Judson J (2016) GaN-based patriot prototype preps for public debut. Defense news [Online]. Available http://www.defensenews.com/story/defense-news/2016/02/27/gan-based-patriot-prototype-preps-public-debut/80999806/
25. Ridley BK, Schaff WJ, Eastman LF (2004) Hot-phonon-induced velocity saturation in GaN. J Appl Phys 96(3):1499–1502
26. Matulionis A (2006) Feature article: hot phonons in GaN channels for HEMTs. Phys Status Solidi Appl Mater Sci 203(10):2313–2325
27. Khurgin J, Ding YJ, Jena D (2007) Hot phonon effect on electron velocity saturation in GaN: a second look. Appl Phys Lett 91(25):2–4
28. Palacios T, Chakraborty A, Rajan S, Poblenz C, Keller S, DenBaars SP, Speck JS, Mishra UK (2005) High-power AlGaN/GaN HEMTs for Ka-band applications. IEEE Electron Device Lett 26(11):781–783
29. Moon JS, Wu S, Wong D, Milosavljevic I, Conway A, Hashimoto P, Hu M, Antcliffe M, Micovic M (2005) Gate-recessed AlGaN-GaN HEMTs for high-performance millimeter-wave applications. IEEE Electron Device Lett 26(6):348–350
30. Wu Y-F, Moore M, Saxler A, Wisleder T, Mishra UK, Parikh P (2005) 8-watt GaN HEMTs at millimeter-wave frequencies. In: IEEE international electron devices meeting. IEDM technical digest, vol 00, no c, pp 583–585
31. Smith PM, Dugas D, Chu K, Nichols K, Duh KG, Fisher J, Xu D, Gunter L, Vera A, Lender R, Meharry D (2003) Progress in GaAs metamorphic HEMT technology for microwave applications. In: 25th annual technical digest 2003. IEEE Gallium Arsenide integrated circuit (GaAs IC) symposium, pp 21–24

32. Chen S, Nayak S, Kao M-Y, Delaney J (2004) A Ka/Q-band 2 Watt MMIC power amplifier using dual recess 0.15 μm PHEMT process. In: 2004 IEEE MTT-S international microwave symposium digest (IEEE Cat. No. 04CH37535), pp 1669–1672
33. Federal Communications Commission (2003) In the matter of allocations and service rules for the 71-76 GHz, 81-86 GHz and 92-95 GHz Bands: Loea communications corporation petition for rulemaking. FCC 03-248, 4 Nov 2003
34. Micovic M, Kurdoghlian A, Hashimoto P, Hu M, Antcliffe M, Willadsen PJ, Wong WS, Bowen R, Milosavljevic I, Schmitz A, Wetzel M, Chow DH (2006) GaN HFET for W-band power applications. In: International electron devices meeting, pp 1–3
35. Shinohara K, Regan DC, Tang Y, Corrion AL, Brown DF, Wong JC, Robinson JF, Fung HH, Schmitz A, Oh TC, Kim SJ, Chen PS, Nagele RG, Margomenos AD, Micovic M (2013) Scaling of GaN HEMTs and schottky diodes for submillimeter-wave MMIC applications. IEEE Trans Electron Devices 60(10):2982–2996
36. Margomenos A, Kurdoghlian A, Micovic M, Shinohara K, Brown DF, Corrion AL, Moyer HP, Burnham S, Regan DC, Grabar RM, McGuire C, Wetzel MD, Bowen R, Chen PS, Tai HY, Schmitz A, Fung H, Fung A, Chow DH (2014) GaN technology for E, W and G-band applications. In: 2014 IEEE compound semiconductor integrated circuit symposium (CSICS), pp 1–4
37. Schellenberg JM (2015) A 2-W W-band GaN traveling-wave amplifier with 25-GHz bandwidth. IEEE Trans Microw Theory Tech 63(9):2833–2840
38. Camargo E, Schellenberg J, Bui L, Estella N (2014) Power GaAs MMICs for E-band communications applications. In: 2014 IEEE MTT-S international microwave symposium (IMS2014), pp 1–4
39. Makiyama K, Ozaki S, Ohki T, Okamoto N, Minoura Y, Niida Y, Kamada Y, Joshin K, Watanabe K, Miyamoto Y (2015) Collapse-free high power InAlGaN/GaN-HEMT with 3 W/mm at 96 GHz. In: 2015 IEEE international electron devices meeting, pp 213–216.
40. Yu ET, Sullivan GJ, Asbeck PM, Wang CD, Qiao D, Lau SS (1997) Measurement of piezoelectrically induced charge in GaN/AlGaN heterostructure field-effect transistors. Appl Phys Lett 71(19):2794
41. Bykhovski AD, Gaska R, Shur MS (1998) Piezoelectric doping and elastic strain relaxation in AlGaN-GaN heterostructure field effect transistors. Appl Phys Lett 73(24):3577–3579
42. Smorchkova I, Elsass C, Ibbetson J, Vetury R, Heying B, Fini P, Haus E, DenBaars S, Speck J, Mishra U (1999) Polarization-induced charge and electron mobility in AlGaN/GaN heterostructures grown by plasma-assisted molecular-beam epitaxy. J Appl Phys 86(8): 4520–4526
43. Keller S, Li H, Laurent M, Hu Y, Pfaff N, Lu J, Brown DF, Fichtenbaum NA, Speck JS, Denbaars SP, Mishra UK (2014) Recent progress in metal-organic chemical vapor deposition of (000-1) N-polar group-III nitrides. Semicond Sci Technol 29:113001
44. Wong MH, Keller S, Dasgupta NS, Denninghoff DJ, Kolluri S, Brown DF, Lu J, Fichtenbaum NA, Ahmadi E, Singisetti U, Chini A, Rajan S, DenBaars SP, Speck JS, Mishra UK (2013) N-polar GaN epitaxy and high electron mobility transistors. Semicond Sci Technol 28(7):074009
45. Murphy MJ, Chu K, Wu H, Yeo W, Schaff WJ, Ambacher O, Smart J, Shealy JR, Eastman LF, Eustis TJ (1999) Molecular beam epitaxial growth of normal and inverted two-dimensional electron gases in AlGaN/GaN based heterostructures. J Vac Sci Technol B Microelectron Nanometer Struct 17(3):1252–1254
46. Dimitrov R, Murphy M, Smart J, Schaff W, Shealy JR, Eastman LF, Ambacher O, Stutzmann M (2000) Two-dimensional electron gases in Ga-face and N-face AlGaN/GaN heterostructures grown by plasma-induced molecular beam epitaxy and metalorganic chemical vapor deposition on sapphire. J Appl Phys 87(7):3375
47. Hellman ES (1998) The polarity of GaN: a Critical Review. MRS Internet J Nitride Semicond Res 3(1998):e11
48. Sasaki T, Matsuoka T (1988) Substrate-polarity dependence of metal-organic vapor-phase epitaxy-grown GaN on SiC. J Appl Phys 64(9):4531–4535

49. Rajan S, Chini A, Wong MH, Speck JS, Mishra UK (2007) N-polar GaNAlGaNGaN high electron mobility transistors. J Appl Phys 102(4):044501
50. Ambacher O, Smart J, Shealy JR, Weimann NG, Chu K, Murphy M, Schaff WJ, Eastman LF, Dimitrov R, Wittmer L, Stutzmann M, Rieger W, Hilsenbeck J (1999) Two-dimensional electron gases induced by spontaneous and piezoelectric polarization charges in N- and Ga-face AlGaN/GaN heterostructures. J Appl Phys 85(6):3222
51. Ambacher O, Foutz B, Smart J, Shealy JR, Weimann NG, Chu K, Murphy M, Sierakowski AJ, Schaff WJ, Eastman LF, Dimitrov R, Mitchell A, Stutzmann M (2000) Two dimensional electron gases induced by spontaneous and piezoelectric polarization in undoped and doped AlGaN/GaN heterostructures. J Appl Phys 87(1):334
52. Morkoç H, Di Carlo A, Cingolani R (2002) GaN-based modulation doped FETs and UV detectors. Solid State Electron 46(2):157–202
53. Kusakabe K, Kishino K, Kikuchi A, Yamada T, Sugihara D, Nakamura S (2001) Reduction of threading dislocations in migration enhanced epitaxy grown GaN with N-polarity by use of AlN multiple interlayer. J Cryst Growth 230(3–4):387–391
54. Monroy E, Sarigiannidou E, Fossard F, Gogneau N, Bellet-Amalric E, Rouvire JL, Monnoye S, Mank H, Daudin B (2004) Growth kinetics of N-face polarity GaN by plasma-assisted molecular-beam epitaxy. Appl Phys Lett 84(18):3684–3686
55. Zauner ARA, Weyher JL, Plomp M, Kirilyuk V, Grzegory I, Van Enckevort WJP, Schermer JJ, Hageman PR, Larsen PK (2000) Homo-epitaxial GaN growth on exact and misoriented single crystals: suppression of hillock formation. J Cryst Growth 210(4): 435–443
56. Zauner ARA, Aret E, Van Enckevort WJP, Weyher JL, Porowski S, Schermer JJ (2002) Homo-epitaxial growth on the N-face of GaN single crystals: the influence of the misorientation on the surface morphology. J Cryst Growth 240(1–2):14–21
57. Rajan S, Wong M, Fu Y, Wu F, Speck JS, Mishra UK (2005) Growth and electrical characterization of N-face AlGaN/GaN heterostructures. Japan J Appl Phys Lett 44, no 46–49
58. Keller S, Fichtenbaum NA, Wu F, Brown D, Rosales A, Denbaars SP, Speck JS, Mishra UK (2007) Influence of the substrate misorientation on the properties of N-polar GaN films grown by metal organic chemical vapor deposition. J Appl Phys 102(8):0–6
59. Brown DF, Keller S, Wu F, Speck JS, DenBaars SP, Mishra UK (2008) Growth and characterization of N-polar GaN films on SiC by metal organic chemical vapor deposition. J Appl Phys 104(2):024301
60. Ahmadi E, Wu F, Li H, Kaun SW, Tahhan M, Hestroffer K, Keller S, Speck JS, Mishra UK (2015) N-face GaN/AlN/GaN/InAlN and GaN/AlN/AlGaN/GaN/InAlN high-electron-mobility transistor structures grown by plasma-assisted molecular beam epitaxy on vicinal substrates. Semicond Sci Technol 30(5):055012
61. Wienecke S, Romanczyk B, Guidry M, Li H, Zheng X, Ahmadi E, Hestroffer K, Keller S, Mishra U (2015) N-polar deep recess HEMTs for W-band power applications. In: Presented at 42nd international symposium on compound semiconductors
62. Chini A, Fu Y, Rajan S, Speck J, Mishra UK (2005) An experimental method to identify bulk and surface traps in GaN HEMTs. In: 32nd international symposium on compound semiconductors (ISCS), vol 1, pp 1–2
63. Wienecke S, Guidry M, Li H, Ahmadi E, Hestroffer K, Zheng X, Keller S, Mishra UK (2014) Optimization of back-barrier doping in graded AlGaN N-face MISHEMTs. In: (poster) 2014 IEEE lester eastman conference, New York, 5–7 Aug 2014
64. Grundman M (2007) Polarization-induced tunnel junctions in III-nitrides for optoelectronic applications. PhD Thesis, University of California
65. Wong MH, Singisetti U, Lu J, Speck JS, Mishra UK (2011) Anomalous output conductance in N-polar GaN-based MIS-HEMTs. In: 69th device research conference, pp 211–212
66. Schaake CA, Brown DF, Swenson BL, Keller S, Speck JS, Mishra UK (2013) A donor-like trap at the InGaN/GaN interface with net negative polarization and its possible consequence on internal quantum efficiency. Semicond Sci Technol 28(10):105021

67. Wong MH, Pei Y, Brown DF, Keller S, Speck JS, Mishra UK (2009) High-performance N-face GaN microwave MIS-HEMTs with >70 % power-added efficiency. IEEE Electron Device Lett 30(8):802–804
68. Wong MH, Brown DF, Schuette ML, Kim H, Balasubramanian V, Lu W, Speck JS, Mishra UK (2011) X-band power performance of N-face GaN MIS-HEMTs. Electron Lett 47(3):214
69. Kolluri S, Member S, Keller S, Denbaars SP, Mishra UK (2012) Microwave Power Performance N-Polar GaN MISHEMTs Grown by MOCVD on SiC Substrates Using an Al_2O_3 etch-stop technology. IEEE Electron Device Lett 33(1):44–46
70. Pei Y, Chu R, Fichtenbaum NA, Chen Z, Brown D, Shen L, Keller S, DenBaars SP, Mishra UK (2007) Recessed slant gate AlGaN/GaN high electron mobility transistors with 20.9 W/mm at 10 GHz. Jpn J Appl Phys 46(45):L1087–L1089
71. Guerra D, Akis R, Ferry DK, Goodnick SM, Saraniti M, Marino FA (2010) Cellular Monte Carlo study of RF short-channel effects, effective gate length, and aspect ratio in GaN and InGaAs HEMTs. In: 2010 14th international workshop computational electronics, IWCE 2010, vol 1, pp 105–108
72. Guerra D, Saraniti M, Faralli N, Ferry DK, Goodnick SM, Marino FA (2010) Comparison of N- and Ga-face GaN HEMTs through cellular Monte Carlo simulations. IEEE Trans Electron Devices 57(12):3348–3354
73. Park PS, Rajan S (2011) Simulation of Short-Channel Effects in N- and Ga-Polar AlGaN/GaN HEMTs. IEEE Trans Electron Devices 58(3):704–708
74. Denninghoff D, Lu J, Ahmadi E, Keller S, Mishra U (2012) N-polar GaN/InAlN/AlGaN MIS-HEMTs with highly scaled GaN channels. In: International symposium on compound semiconductors (ISCS)
75. Dasgupta NS, Brown DF, Keller S, Speck JS, Mishra UK (2009) N-polar GaN-based highly scaled self-aligned MIS-HEMTs with state-of-the-art fT-LG product of 16.8 GHz-μm. In: 2009 IEEE international electron devices meeting (IEDM), vol 805, pp 1–3
76. Dasgupta NS, Brown DF, Wu F, Keller S, Speck JS, Mishra UK (2010) Ultralow nonalloyed Ohmic contact resistance to self aligned N-polar GaN high electron mobility transistors by In (Ga)N regrowth. Appl Phys Lett 96(14):3–6
77. Singisetti U, Hoi Wong M, Mishra UK (2012) Interface roughness scattering in ultra-thin N-polar GaN quantum well channels. Appl Phys Lett 101(1):1–5
78. Carlin JF, Ilegems M (2003) High-quality AllnN for high index contrast Bragg mirrors lattice matched to GaN. Appl Phys Lett 83(4):668–670
79. Lorenz K, Franco N, Alves E, Watson IM, Martin RW, O'Donnell KP (2006) Anomalous ion channeling in AlInN/GaN bilayers: determination of the strain state. Phys Rev Lett 97(8):1–4
80. Kuzmik J (2001) Power electronics on InAlN/(In)GaN: prospect for a record performance. IEEE Electron Device Lett 22(11):510–512
81. Brown DF, Keller S, Mates TE, Speck JS, Denbaars SP, Mishra UK (2010) Growth and characterization of In-polar and N-polar InAlN by metal organic chemical vapor deposition. J Appl Phys 107(3):1–8
82. Brown DF, Nidhi SR, Wu F, Keller S, DenBaars SP, Mishra UK (2010) N-Polar InAlN/AlN/GaN MIS-HEMTs. IEEE Electron Device Lett 31(8):800–802
83. Dasgupta S, Choi S, Wu F, Speck JS, Mishra UK (2011) Growth, structural, and electrical characterizations of N-polar InAlN by plasma-assisted molecular beam epitaxy. Appl Phys Express 4(4):045502
84. Dasgupta NS, Lu J, Speck JS, Mishra UK (2012) Self-Aligned N-polar GaN/InAlN MIS-HEMTs with record extrinsic transconductance of 1105 mS/mm. IEEE Electron Device Lett 33(6):794–796
85. Ahmadi E, Shivaraman R, Wu F, Wienecke S, Kaun SW, Keller S, Speck JS, Mishra UK (2014) Elimination of columnar microstructure in N-face InAlN, lattice-matched to GaN, grown by plasma-assisted molecular beam epitaxy in the N-rich regime. Appl Phys Lett 104(7):2014–2017

86. Lu J, Denninghoff D, Yeluri R, Lal S, Gupta G, Laurent M, Keller S, Denbaars SP, Mishra UK (2013) Very high channel conductivity in ultra-thin channel N-polar GaN/(AlN, InAlN, AlGaN) high electron mobility hetero-junctions grown by metalorganic chemical vapor deposition. Appl Phys Lett 102(23):16–21
87. Lu J, Zheng X, Guidry M, Denninghoff D, Ahmadi E, Lal S, Keller S, Denbaars SP, Mishra UK (2014) Engineering the (In, Al, Ga)N back-barrier to achieve high channel-conductivity for extremely scaled channel-thicknesses in N-polar GaN high-electron-mobility-transistors. Appl Phys Lett 104(9):092107
88. Denninghoff D, Lu J, Ahmadi E, Keller S, Mishra UK (2013) N-polar GaN/InAlN/AlGaN MIS-HEMTs with 1.89 S/mm extrinsic transconductance, 4 A/mm drain current, 204 GHz f T and 405 GHz f max, vol 33, no 7
89. Laurent MA, Gupta G, Suntrup DJ, DenBaars SP, Mishra UK (2016) Barrier height inhomogeneity and its impact on (Al, In, Ga)N Schottky diodes. J Appl Phys 119(6):064501
90. Shen L, Coffie R, Buttari D, Heikman S, Chakraborty A, Chini A, Keller S, DenBaars SP, Mishra UK (2004) High-power polarization-engineered GaN/AlGaN/GaN HEMTs without surface passivation. IEEE Electron Device Lett 25(1):7–9
91. Shen L, Pei Y, McCarthy L, Poblenz C, Corrion A, Fichtenbaum N, Keller S, Denbaars SP, Speck JS, Mishra UK (2007) Deep-recessed GaN HEMTs using selective etch technology exhibiting high microwave performance without surface passivation. IEEE MTT-S international microwave symposium digest, pp 623–626
92. Shen L, Palacios T, Poblenz C, Corrion A, Chakraborty A, Fichtenbaum N, Keller S, Denbaars SP, Speck JS, Mishra UK (2006) Unpassivated high power deeply recessed GaN HEMTs with fluorine-plasma surface treatment. IEEE Electron Device Lett 27(4):214–216
93. Chu R, Shen L, Fichtenbaum N, Brown D, Chen Z, Keller S, DenBaars SP, Mishra UK (2008) V-Gate GaN HEMTs for X-Band power applications. IEEE Electron Device Lett 29(9):974–976
94. Dasgupta NS, Keller S, Speck JS, Mishra UK (2011) N-polar GaN/AlN MIS-HEMT with fMAX of 204 GHz for Ka-Band applications. IEEE Electron Device Lett 32(12):1683–1685
95. Kolluri S, Brown DF, Wong MH, Dasgupta S, Keller S, Denbaars SP, Mishra UK (2011) RF performance of deep-recessed N-polar GaN MIS-HEMTs using a selective etch technology without Ex Situ surface passivation. IEEE Electron Device Lett 32(2):134–136
96. Wienecke S, Romanczyk B, Guidry M, Li H, Zheng X, Ahmadi E, Hestroffer K, Megalini L, Keller S, Mishra U (2016) N-polar deep recess MISHEMTs with record 2.9 W/mm at 94 GHz. IEEE Electron Device Lett 37(6):713–716. doi:10.1109/LED.2016.2556717
97. Guidry M, Wienecke S, Romanczyk B, Zheng X, Li H, Ahmadi E, Hestroffer K, Keller S, Mishra UK (2016) W-band passive load pull system for on-wafer characterization of high power density N-polar GaN devices based on output match and drive power requirements vs. gate width. In: 87th ARFTG Conference, San Fransisco, CA pp 1–4
98. Guidry M, Wienecke S, Romanczyk B, Li H, Zheng X, Ahmadi E, Hestroffer K, Keller S, Mishra UK (2016) Small-signal model extraction of mm-wave N-polar GaN MISHEMT exhibiting record performance: analysis of gain and validation by 94 GHz loadpull. In: International microwave symposium 2016 (IMS2016), San Fransisco, CA, (accepted)
99. http://www.transphormusa.com/
100. http://www.semicon.panasonic.co.jp/en/products/powerics/ganpower/
101. http://www.gansystems.com/
102. http://epc-co.com/epc
103. Zhang N (2002) High voltage GaN HEMTs with low on-resistance for switching applications", PhD dissertation. ECE Dept, Univ. of California, Santa Barbara
104. Quay R, Tessmann A, Kiefer R, Maroldt S, Haupt C, Nowotny U, Weber R, Massler H, Schwantuschke D, Seelmann-Eggebert M, Leuther A, Mikulla M, Ambacher O (2011) Dual-gate GaN MMICs for MM-wave operation. IEEE Microw Wirel Compon Lett 21(2):95–97
105. van Heijningen M, Rodenburg M, van Vliet FE, Massler H, Tessmann A, Brueckner P, Mueller S, Schwantuschke D, Quay R, Narhi T, IEEE (2012) W-band power amplifier

MMIC with 400 mW output power in 0.1 μm AlGaN/GaN technology. In: 2012 7th European microwave integrated circuits conference, pp 135–138
106. Marti D, Tirelli S, Teppati V, Lugani L, Carlin JF, Malinverni M, Grandjean N, Bolognesi CR (2015) 94-GHz large-signal operation of AlInN/GaN high-electron-mobility transistors on silicon with regrown ohmic contacts. IEEE Electron Device Lett 36(1):17–19
107. van Heijningen M, Rodenburg M, van Vliet FE, Massler H, Tessmann A, Brueckner P, Mueller S, Schwantuschke D, Quay R, Narhi T, IEEE (2012) W-band power amplifier MMIC with 400 mW output power in 0.1 μm AlGaN/GaN technology. In: 2012 7th European microwave integrated circuits conference, pp 135–138
108. Makiyama K, Ozaki S, Okamoto N, Ohki T, Niida Y, Kamada Y, Joshin K, Watanabe K (2014) GaN-HEMT technology for high power millimeter-wave amplifier. In: Lester Eastman conference, pp 5–8
109. Masuda S, Ohki T, Makiyama K, Kanamura M, Okamoto N, Shigematsu H, Imanishi K, Kikkawa T, Joshin K, Hara N (2009) GaN MMIC amplifiers for W-band transceivers. In: Proceedings 39th European microwave conference EuMC, pp 1796–1799
110. Joshin K, Makiyama K, Ozaki S, Ohki T, Okamoto N, Niida Y, Sato M, Masuda S, Watanabe K (2014) Millimeter-wave GaN HEMT for power amplifier applications. IEICE Trans Electron E97.C(10):923–929
111. Makiyama K, Ohki T, Kanamura M, Joshin K, Imanishi K, Hara N, Kikkawa T (2009) High-power GaN-HEMT with high three-terminal breakdown voltage for W-band applications. Phys Status Solidi 6(S2):S1012–S1015
112. Micovic M, Kurdoghlian A, Moyer HP, Hashimoto P, Hu M, Antcliffe M, Willadsen PJ, Wong WS, Bowen R, Milosavljevic I, Yoon Y, Schmitz A, Wetzel M, McGuire C, Hughes B, Chow DH (2008) GaN MMIC PAs for E-band (71 GHz–95 GHz) radio. In: 2008 IEEE compound semiconductor integrated circuits symposium, pp 1–4
113. Micovic M, Kurdoghlian A, Shinohara K, Burnham S, Milosavljevic I, Hu M, Corrion A, Fung A, Lin R, Samoska L, Kangaslahti P, Lambrigtsen B, Goldsmith P, Wong WS, Schmitz A, Hashimoto P, Willadsen PJ, Chow DH (2010) W-band GaN MMIC with 842 mW output power at 88 GHz. In: 2010 IEEE MTT-S international microwave symposium, pp 237–239
114. Brown DF, Williams A, Shinohara K, Kurdoghlian A, Milosavljevic I, Hashimoto P, Grabar R, Burnham S, Butler C, Willadsen P, Micovic M (2011) W-band power performance of AlGaN/GaN DHFETs with regrown n + GaN ohmic contacts by MBE. In: 2011 international electron devices meeting, pp 19.3.1–19.3.4
115. Micovic M, Kurdoghlian A, Margomenos A, Brown DF, Shinohara K, Burnham S, Milosavljevic I, Bowen R, Williams AJ, Hashimoto P, Grabar R, C. Butler, A. Schmitz, P. J. Willadsen, and D. H. Chow, "92-96 GHz GaN power amplifiers," in *2012 IEEE/MTT-S International Microwave Symposium Digest*, 2012, pp. 1–3
116. Brown A, Brown K, Chen J, Hwang KC, Kolias N, Scott R (2011) W-band GaN power amplifier MMICs. In: 2011 IEEE MTT-S international microwave symposium, pp 1–4
117. Schellenberg J, Kim B, Phan T (2013) W-band, broadband 2 W GaN MMIC. In: 2013 IEEE MTT-S international microwave symposium digest (MTT), pp 1–4
118. Xu D, Chu KK, Diaz JA, Ashman M, Komiak JJ, Pleasant LM, Creamer C, Nichols K, Duh KHG, Smith PM, Chao PC, Dong L, Ye PD (2015) 0.1 μm atomic layer deposition Al_2O_3 passivated InAlN/GaN high electron-mobility transistors for E-band power amplifiers. IEEE Electron Device Lett 36(5):442–444
119. Lu J (2013) Design and epitaxial growth of ultra-scaled N-polar GaN/(In, Al, Ga) N HEMTs by metal organic chemical deposition and device characterization. PhD Dissertation, University of California, Santa Barbara

Chapter 5
Vertical Gallium Nitride Technology

Materials, Devices and Applications

Srabanti Chowdhury

5.1 Introduction

Popularized by the need for solid-state power electronics, vertical GaN devices are becoming an essential solution for next-generation power converters. Conversion of power is necessary in almost all electrical application that we witness in the modern world ranging from charging our cell phones to satellite power systems. While the complexity and the stages of the power conversion system behind these applications can be widely different, the basic building blocks are similar. Solid-state power converters consist of building blocks such as boost converter (to step-up voltage), buck converter (to step-down voltage), inverters (to convert DC to an AC), phase converter (from one phase to 3-phase), power factor controller (to reduce the reactive power from the output) and rectifiers (AC to DC). Solid-state electronic devices, called power electronic "switches," in the form of diodes and transistors are an integral part of these power converters. A switch often called the *heart* of a converter plays a central role in achieving high-efficiency power conversion. Figure 5.1 shows a 3-phase inverter circuit where the role of the switching transistor is highlighted. Silicon (Si) has been the backbone of semiconductor industry shaping up power electronics together with computational electronics. In the form of metal–oxide–semiconductor field-effect transistor (MOSFET) and its various adaptations particularly CoolMOSFET [1, 2], insulated-gate bipolar transistor (IGBT) [1], bipolar junction transistor (BJT) and diodes, Si has served the power electronics industry extremely well, creating new platforms for innovation at circuits and systems level. However, to support the rising diversity and power requirements of these applications in both industrial and domestic sectors, solutions beyond Si are necessary. Benefits obtained at the system level, by driving circuits at higher frequencies or at higher temperatures, stimulated the need for wide bandgap

S. Chowdhury (✉)
University of California, Davis, CA, USA
e-mail: chowdhury@ucdavis.edu

M. Meneghini et al. (eds.), *Power GaN Devices*,
Power Electronics and Power Systems, DOI 10.1007/978-3-319-43199-4_5

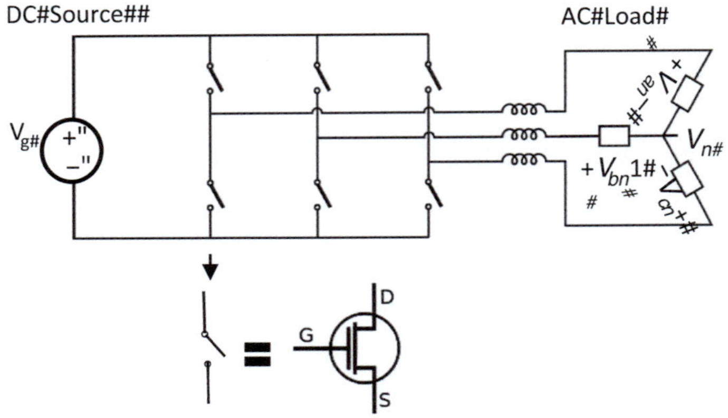

Fig. 5.1 A 3-phase inverter circuit showing conversion of DC to 3-phase AC signal. The switches represented in the figure schematically are transistors in a solid-state power converter

semiconductors, which offers the required material properties to support such operations. SiC and GaN have been the two leading technologies for the next-generation power converters pushing the roadmap toward higher power density, with improved efficiency and greater functionality.

Just like with any other energy conversion, power conversion incurs losses at all stages. Recent innovations made with AlGaN/GaN HEMTs show that in the medium-power conversion domain, the energy savings accomplished at the device (switch) level can be translated to savings at the system level, in terms of both energy and cost. Similar or more outcomes can be anticipated from vertical device technology in power converters. Our increasing dependency on data servers, mobile computing, communications and automation at every level raises the importance of loss-less power conversion. Figure 5.2 shows the entire range of power requirements for targeted application. GaN vertical devices will be in the forefront of next-generation power electronics demanded by these applications.

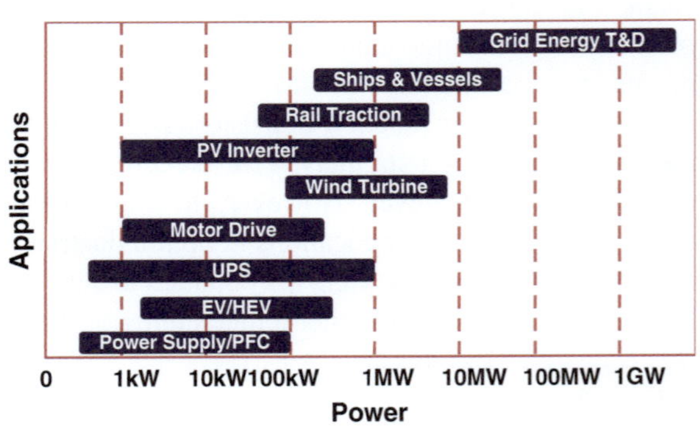

Fig. 5.2 Power requirements by various applications. GaN vertical devices will play a major role in power conversions above 30 kW (chart is based on Yole Development report [1])

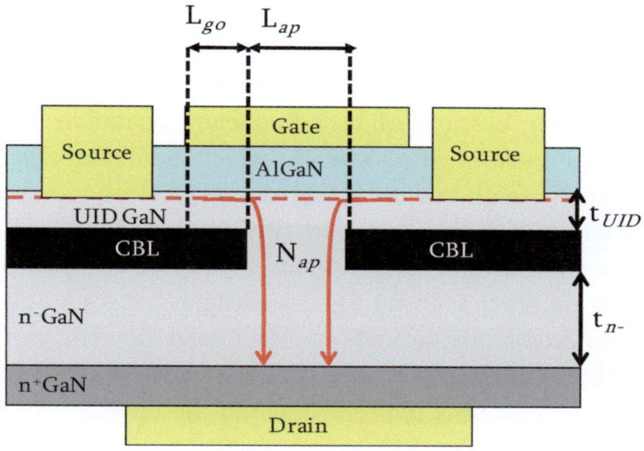

Fig. 5.3 A schematic diagram of a CAVET with its key parameters marked. The five parameters labeled, viz. L_{ap}, L_{go}, N_{ap}, t_{UID} and $t_{\mathrm{n-}}$, are of utmost importance and were modeled to baseline the design

The underlying approach toward a "low-loss" system therefore starts with the making of a "zero-loss" switch. The key to "low-loss" switching relies on the attainable properties of the semiconductor material to enable a virtually low-resistance channel when the switch is on. While there are several different semiconductors with promising material property, the leading advantage that GaN (along with other members of its family, like AlGaN) provides, besides it high critical electric field (3 MV/cm), is a channel with very high mobility, at least 10 times higher than that the maximum achievable MOSFET channel mobility. By appropriate design with AlGaN or other III-nitrides, one can enhance the mobility in selective region. A classic example utilizing such advantage is seen in a current aperture vertical electron transistor (CAVET), which will be used as the representative to discuss various concepts in this chapter.

Figure 5.3 shows the schematic of a CAVET. While the ingenuity in the design of a CAVET lies in the merging of the lateral channel to the vertical drift region, it is similar to most other vertical devices and relies on an aperture and a drift region. Common to all vertical devices are current blocking layers (CBLs) to block current flowing into any parasitic leakage paths and a drift region to block the off-state voltage.

5.2 Device Topology

5.2.1 Vertical Devices Versus Lateral Devices

Most of the effort in gallium nitride power electronics, so far, has been dedicated to the lateral geometry in the embodiment of a HEMT, rightly so, because the major benefit (of low specific R_{on}) that comes from AlGaN/GaN system lies in the 2DEG, which is horizontal in nature [3–8]. However, lateral devices have always given way to vertical counterparts in the high power limit. DMOS, IGBT, HEXFET and CoolMOS all have vertical topologies to reduce chip area for a given current rating. Increasing demand of high-frequency operation at high power density makes it crucial looking into devices, which hold the voltage in the vertical direction. Contrary to Si and SiC, GaN-based devices present the distinctive advantage of providing a high charge density channel along with high mobility, owing to the formation of polarization-induced 2D electron gas (2DEG) between Al_xGa_yN and GaN. In order to lower the on-resistance (R_{ON}), CAVET confines the electron flow first horizontally through the 2DEG and then vertically through a thick-drift region. CBL, an integral part of CAVET as well as other vertical devices, is placed such that the electrons (current flows in the opposite direction) are guided through an aperture into the drift region similar to a DMOS. Under the off-state operation, majority of the voltage is held by depleting the drift region causing the peak of the electric field to occur inside the aperture (at the p–n junction, if CBL is a p-type layer).

Similar to a Si DMOSFET in many respects, a CAVET differs from the latter in one major way. The CAVET as shown in Fig. 5.3 is a normally-on device unlike a DMOSFET, which is normally-off, and the conduction starts only at a positive threshold voltage inducing an inversion channel layer. The biggest advantage in a vertical device like CAVET and other variants of CAVET comes from the fact that in these devices the high electric field region is buried under the gate into the bulk material. In a lateral HEMT, the drain edge of the gate has high field region, which peaks to very high value and may cause surface arcing, dielectric breakdown, channel breakdown and encapsulation breakdown, leading to premature device failure. Furthermore, this high field favors charging of surface states, leading to dispersion in a HEMT. Field-plate (FP) technology with surface passivation helps reduce dispersion, and field shaping leads to higher breakdown voltage. CAVET, where the high field region is buried in the bulk material, is expected to be free from surface state-related dispersion. Without involving FPs, a CAVET should be capable of sustaining high voltages corresponding to the critical field of GaN. The CAVET utilizes the whole area under the source and the drain to hold voltage, unlike in HEMT where the region from the source to the gate does not hold any significant voltage. Owing to the buried electric field, a CAVET should offer higher volts per micron. Figure 5.4 sketches the argument how a CAVET (or any of its variant like vertical JFET or MOSFET) offers electric field management without extensive FP requirements. The CBL plays a vital role in mitigating the electric field generated under the off-state. For higher-voltage designs, appropriate edge and field terminations would be necessary to distribute the electric field uniformly in the drift region CBL and the aperture. The vertical structure allows distribution of the field more uniformly than HEMTs with FP, which is a key to maximize the "volts per micron."

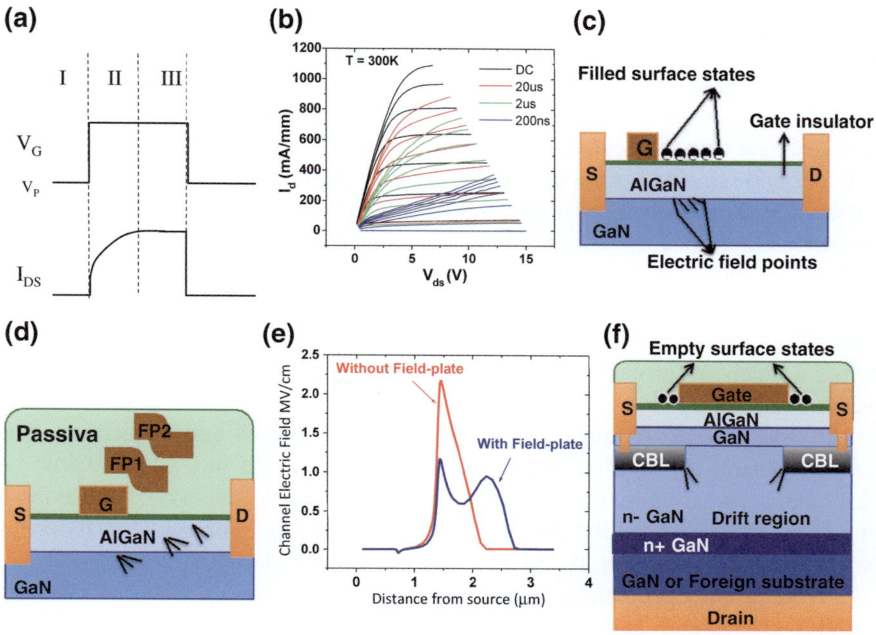

Fig. 5.4 **a** *I–V* characteristics of dispersive device. Dispersion can lead to switching losses. **b** Lateral devices without field plate causes high electric field at the drain edge of the gate, leading to occupied surface states causing dispersion. **c** Field plates are used to mitigate and mange peak electric field to minimize the dispersion and enhance breakdown voltages. **d** Two field plates flatten the single peak which occurs at the drain edge of the gate without any FP. **e** vertical transistors have the peak electric field region buried inside the bulk material. This leaves the surface states unoccupied causing no dispersion

5.3 Evolution of a CAVET

Ben Yaacov et al. [9] designed first-generation CAVETs for RF power performance, which was successfully demonstrated between 2000 and 2004 where the structures were grown on sapphire and with the drift region grown no thicker than 0.5 μm. The CBL was achieved by doping the GaN layer with Mg. Any CAVET fabrication requires a regrowth of the AlGaN/GaN layers, which creates the 2DEG. In this design, where Mg-doped GaN formed the CBL, the aperture region needed to be regrown as well along with the channel and the AlGaN cap. Since the design was directed toward realizing RF power, the gate and the aperture were kept as small as possible for a starting point to minimize capacitances. The devices successfully showed a CAVET operation validating the concept and design, but suffered from high leakage. A major contribution to the leakage came due to the technique of regrown apertures. The next design reported by Gao et al. [10] employed photoelectrochemical (PEC) etching techniques to etch InGaN aperture layer to form air gaps as CBLs. This method did not require any regrowth, and the whole structure was grown with an InGaN aperture layer, which was then etched away using the bandgap selectivity to form the CBL. The devices fabricated by this technique suffered from a turn-on voltage drop as the polarization charges between the InGaN and the GaN create a barrier to the electron flow. With a lot of design modifications using Si delta doping, the barrier could be reduced but a voltage drop of 0.7 V still remained. Although it was a significant improvement and the method circumvented the problem of regrowth, the air gap as a current blocking layer is not appropriate at higher voltages since air is a poor dielectric.

Later in 2004, CBLs created by ion implantation were incorporated into CAVETs, which delivered much improved results. The technique involved regrowth of the AlGaN/GaN layers, but not the aperture. Since the aperture was not regrown, many of the earlier problems that existed in the first-generation CAVETs were alleviated.

In 2006, CAVETs were revisited foreseeing a different end application. The ultimate benefit of a DMOSFET-like structure represented in CAVET would be best realized in power switching applications. Targeting a goal toward achieving a low-loss power switch, the first research work on high-voltage CAVETs was pursued by Chowdhury et al. [11, 12] between 2006 and 2010.

Toyota Motor Corporation, Japan, focused their attention on GaN CAVET and CAVET-like devices. In their report of a vertical insulated-gate AlGaN/GaN HFET, Kanechika et al. reported vertical insulated-gate HFETs [13] on bulk GaN substrates in 2007 with a Ron of 2.6 mΩ cm^2. The device structure, similar to a CAVET, was achieved with Mg-doped GaN as the CBL enclosing an n-type GaN aperture. Toyota Motor Corporation, Japan, funded and supported the development of CAVET for power electronics at UCSB, which resulted in the development of CAVETs and CAVET-like vertical devices targeting 60 kW and higher power switching applications.

Some other key developments that steered the vertical GaN device research are listed in chronological order in Table 5.1.

In 2007, Otake et al. [14] also reported on bulk GaN their successful demonstration of "vertical GaN-Based trench gate metal–oxide–semiconductor FET" with a threshold voltage of 3.7 V and a Ron of 9.3 mΩ cm^2.

In 2014, Nei et al. from Avogy Inc. reported vertical GaN transistors exhibiting larger than 2.3 A saturation current, breakdown voltages of 1.5 kV, area differential specific Ron of 2.2 mΩ cm^2 [15].

In 2015, Toyota Gosei reported MOSFET-based vertical GaN devices with a MOS gate structure blocking over 1.2 kV and a Ron of 2 mΩ cm^2 [16]. Table 5.1 captures the development of CAVET and other vertical devices over the past decades.

The suitable CAVET design for holding high voltages while offering low R_{ON} was developed through extensive simulation where every component of the device was analyzed. Some of the key components required for the working of a CAVET have been discussed in the following section.

Table 5.1 A chart showing the key achievements (reported in public journal) on vertical GaN device

Year	Title	Group	V_{br} (V)	R_{on} (mΩ cm^2)	V_{th} (V)
Feb. 2004	AlGaN/GaN current aperture vertical electron transistors with regrown channels [9]	Ben Yaacov/UCSB	65	–	−6
May 2007	A vertical insulated-gate AlGaN/GaN heterojunction field-effect transistor [13]	Kanechika, Toyota Motor Corporation	–	2.6	−16
Jan. 2008	Vertical GaN-based trench gate metal–oxide–semiconductor field-effect transistors on GaN bulk substrates [14]	Otake, ROHM Co.	–	9.3	+3.7
June 2008	Enhancement and depletion mode AlGaN/GaN CAVET with Mg-ion-implanted GaN as current blocking layer [11]	Chowdhury, UCSB	Low	2 (D-mode), 4 (E-mode)	−2, +0.6
June 2010	Dispersion-free AlGaN/GaN CAVET with low Ron achieved with plasma MBE regrown channel with Mg-ion-implanted current blocking layer [12]	Chowdhury, UCSB	200	2.5	−7
Oct. 2010	Vertical heterojunction field-effect transistors utilizing regrown AlGaN/GaN two-dimensional electron gas channels on GaN substrates [17]	Okada, Sumitomo Electric Industries	672	7.6	−1.1
Sept. 2013	1000 V Vertical JFET using bulk GaN [18]	Q. Diduck, Avogy Inc	1000	4.8	+1
Jan. 2014	Vertical GaN-based trench metal–oxide–semiconductor field-effect transistors on a free-standing GaN substrate with blocking voltage of 1.6 kV [19]	Oka, TOYODA GOSEI Co	775 (no FP) 1605 (FP)	12.1	+7
June 2014	Low on-resistance and high current GaN vertical electron transistors with buried p-GaN layers [20]	Yelluri, UCSB	–	0.37	−9
Sept. 2014	1.5-kV and 2.2-mΩ-cm^2 vertical GaN transistors on bulk GaN substrates [15]	Nei, Avogy	1500	2.2	0.5
April 2015	1.8 mΩ-cm^2 vertical GaN-based trench metal–oxide–semiconductor field-effect transistors on a free-standing GaN substrate for 1.2-kV-class operation [16]	Oka, TOYODA GOSEI Co	1250	1.8	3.5

5.4 Design of a CAVET

5.4.1 A Discussion of the Key Components Required for the Successful Functioning of the Device

Before going into the discussion of the design space of a CAVET, a sketch of its application space will help set up the stage for the reader to appreciate the design constraint in a vertical device.

To realize the next-generation power electronic switches, the following performance will be required.

1. *Normally-off operation*

Most efforts on GaN-based devices have been directed toward normally-on ones, and only a few **reliable** normally-off ones have been reported to date. In normally-off devices, the drain current does not flow at the gate voltage of 0 V. Si-IGBTs used in the present power converters are normally-off devices. Likewise, a normally-off device is required for automobiles (Electric and Hybrid Electric vehicles) in order to simplify the inverter circuit and make effective use of design techniques and mounting technologies for the inverters. For GaN to make an entry into the conventional power conversion architecture, normally-off operation is critical. AlGaN/GaN HEMTs are inherently normally-on devices and thereby necessitate techniques at the circuit or at the device level to make it normally-off to suit PE applications. Some of the methods practiced in HEMTS, also applicable to vertical GaN devices, to achieve normally-off operation at the device level are:

(i) Junction-gated devices to pinch-off the channel under the high work function gate, whereas maintaining the channel in the access regions that do not have the p-region has been implemented by EPC and Panasonic, the latter as the GIT or Gate Injection Transistor [21].

(ii) A multi-channel transistor with two AlGaN/GaN interfaces has been developed by Toshiba, Fujitsu, NTT Transphorm [22]. Here, a conductive access region is defined by a 2DEG formed at the upper AlGaN/GaN interface (as in a normal HEMT). A second channel below is accessed via gate recess etching and MIS gate formation within the recess. The second channel is formed between a thin AlGaN layer and GaN such that there is no charge in the channel at zero bias and charge is induced under forward bias (normally-off).

(iii) Normally-off operation by introducing F under the gate structure to shift the threshold voltage positive under the gate while maintaining a conductive channel in the access regions has been developed by HKUST [23].

(iv) Finally, the Channel Etch Through (CET) structure MOSFET where the 2DEG is etched through and an MOS gate deposited has been developed by RPI [24].

All these structures, listed above, are designed to ensure single-chip solution. However, in MIS-gated devices the gate insulator on GaN and AlGaN is not reliable as yet, and the reproducibility of the interface properties and stability under operation needs to be significantly improved. The junction-gated devices have low threshold voltage and limited gate-bias drive range because of the turn-on of the p–n junction gate.

At the circuit level, the simplest embodiment so far of normally-off GaN transistors incorporates a cascade approach where a normally-off low-voltage Si FET is connected to a normally-on high-voltage GaN HEMT in series while the gate of the GaN HEMT is connected to the source of the Si FET as shown in Fig. 5.5 [8]. This hybrid configuration delivers a normally-off solution compatible with existing Si drivers, as well as the freedom to optimize the GaN HEMTs without the complication of special gate drive circuits. Although attractive for the simplicity of design, the single-chip solution will be necessary to reduce losses due to bond wire inductances and PCB traces.

2. *High breakdown voltage*

Higher power density is the overarching need for all power conversion applications that shape the need of high power density switches. An example best shown with Toyota HVs [25] paints a comprehensive picture, which applies to many other applications.

Figure 5.6 shows the relationship between the motor power of Toyota's HVs and other EVs and power source voltages of these systems. In a HV system, the battery voltage is once raised to a power source voltage by a voltage booster and then supplied to the motor through the inverter. The raised voltage can take values from 202 V of the battery up to a maximum of 500 V. The new HVs need high motor power with high power source voltage as shown in Fig. 5.5. The breakdown voltage of devices used in these inverters is about 1.2 kV. The breakdown voltage of devices used in the inverters will probably become higher in the future due to protection against surge voltage and higher efficiency due to wiring losses, etc.

3. *Low on-resistance and high current density*

In order to increase the energy conversion efficiency of any converter systems, it is necessary to decrease the on-resistance of the devices. Moreover, to enable miniaturization of converters/inverters one has to increase the current capacity up to several hundred A/cm^2.

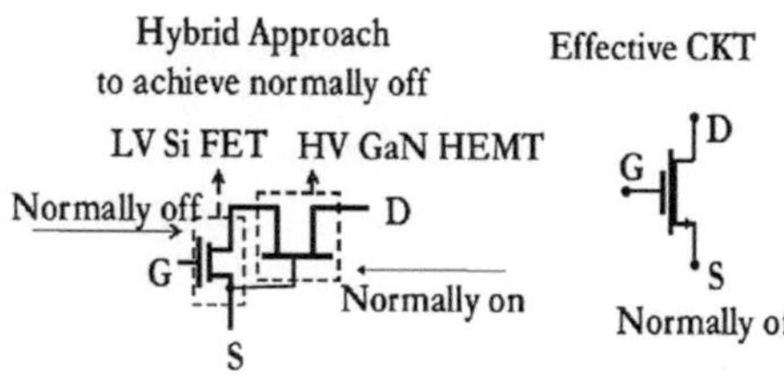

Fig. 5.5 Cascade normally-off GaN HEMT allows for standard silicon-based drive circuits while delivering the high-voltage performance of GaN devices

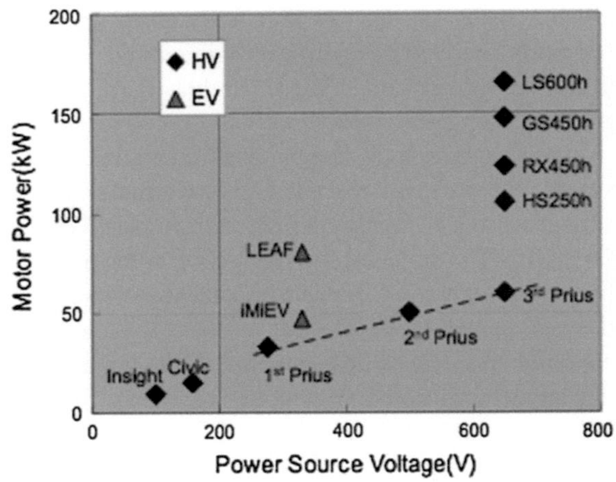

Fig. 5.6 Power source voltage versus motor power in Toyota HVs. Picture adopted from Kanechika et al. [25]

4. *High-temperature operation*

Power Si devices are not used over 150 °C, because power loss increases due to an increase in leakage current in the off-state under high-temperature environment, thereby decreasing their reliability. GaN devices are expected to work over 200 °C and possibly higher, because of its wider bandgap. An example best shared with the present hybrid vehicles having two cooling systems, one is for the engine and the other is for the inverter, and the temperature of coolant water for the inverter is lower than that for the engine. If GaN devices work over 200 °C, the cooling system of hybrid vehicles will be simplified and its cost will be reduced. The same example can be extended toward other applications like PV inverters, where heat sink forms the majority of the volume. Higher-temperature operation will reduce cooling requirement and cooling cost, effectively shrinking the size and overall system cost.

GaN vertical devices have the potential to meet the entire requirement stated in 1–4. Moreover, packaging technologies developed for Si IGBT can be utilized effectively with minimum changes in modules.

5.5 The Key Components of a CAVET

CAVET differs from conventional vertical devices like vertical JFET or MOSFET in its fusion of a HEMT-like channel to a thick-drift region. A CAVET can be formed with simply n-GaN channel instead of AlGaN/GaN, often referred as MES-CAVET. In this section, first we will discuss the design space with respect to a CAVET and bring other device platforms into the discussion as relevant to the design parameter.

In a CAVET (see Fig. 5.3), the intrinsic current flow occurs in two dimensions; electrons first flow horizontally through the 2DEG and then move vertically through the aperture region. This is quite different from the HEMT or the bipolar transistor, where current flow is confined to one dimension. It is therefore critical to develop an accurate model in order to identify which parameters primarily determine the device characteristics.

Electrons (current flows opposite to the flow of electrons) first travel horizontally through the 2DEG, until it reaches the gate. The gate only modulates the electrons in the 2DEG, so the pinch-off occurs in the horizontal direction inside the 2DEG underneath the gate, just like in a standard FET.

Electrons, which pass the pinch-off point in the channel, continue to travel horizontally at their saturated velocity V_{sat} until they arrive at the aperture, travel downward through the aperture and are collected at the drain. It is critical that the conductivity of the material inside the aperture as well as in the drain region be much larger than that of the 2DEG so that the entire voltage drop between the source and drain occurs in the 2DEG. This condition ensures that the total current passing through the device is entirely determined by the conductivity of the 2DEG. If such condition is not met, then a significant amount of the applied source–drain voltage is supported across the aperture. In this case, until V_{DS} is very large, the 2DEG does not pinch-off and the current does not reach its saturation value. This is analogous to quasi-saturation in a bipolar transistor, which can occur at large injection currents when the ohmic drop $I_c \cdot R_c$ across the collector drift region becomes comparable to the total base–collector voltage V_{CB}. In addition, the conductivity of the 2DEG must be much higher than that of the adjacent bulk GaN directly below the 2DEG to ensure current flow through the 2DEG rather than through the bulk GaN.

Extensive simulation-based studies were conducted during the course of development of a CAVET to optimize the device dimensions for making it suitable for high power application. CAVET has a complex geometry, and hence, the design rules are not simple. The important parameters as identified in Fig. 5.3 are detailed out in this section.

N_{ap} and L_{ap} (Doping and length of the aperture)

It is essential to have the aperture region more conductive than the channel region to avoid any downstream choke of the current in the on-state of the transistor. In order to ensure this, the doping of the aperture region (N_{ap}) is chosen such that the conductance of the aperture region ($L_{ap} \cdot q \cdot \mu \cdot N_{ap}$. W_g/t_{CBL}) is higher than the channel resistance (where q is the electron charge, L_{ap} is the length of the aperture, t_{CBL} is the thickness of the CBL region, W_g is the gate width of the device and μ is the mobility of electrons in the drift region).

Increasing the aperture doping is not very desirable because that increases the chance of the breakdown occurring in the aperture. Therefore, lowering R_{on} primarily means increasing the $L_{ap.}$ To give an idea of the impact of aperture doping on I–V characteristics, two CAVETs were modeled, one with a higher aperture resistance than the other. The aperture resistance was changed using a combination of L_{ap} and $N_{ap.}$

From the simulation (Silvaco ATLAS) result shown in Fig. 5.7a, b, it can be seen that the device with higher aperture resistance does not show saturation of the current. Or in other words since a major part of the applied V_{DS} is now absorbed in the aperture, much higher voltage needs to be applied to gain saturation. Device shown in 5.7b with lower aperture resistance easily saturates.

The results indicate that for lower aperture doping which is necessary for high breakdown voltage design, longer apertures (wider opening) are needed. The simulation was done with a 4-μm aperture and as high as 8×10^{17} cm^{-3} doping was needed in the aperture to saturate the current without the quasi-saturation region. This suggests that the aperture needs to be wider than 4 μm for lower N_{ap}. The simulations were done for qualitative understanding of the device. Later experimental results showed the validity of the model. The doping in the aperture was maintained as low as the drift region to avoid premature breakdown in the

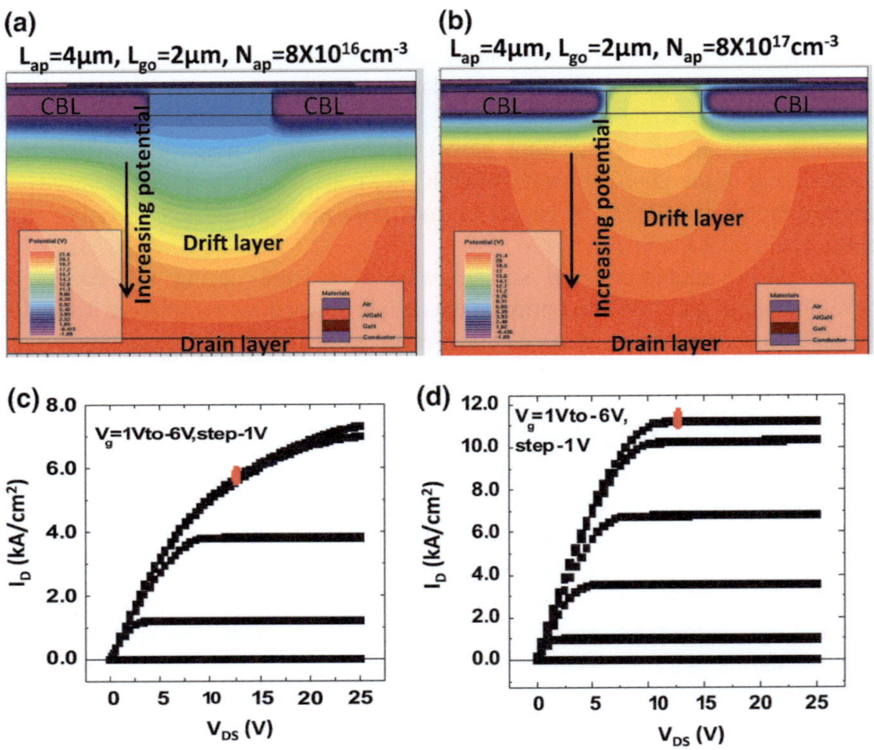

Fig. 5.7 **a** A CAVET simulated with Silvaco ATLAS with a resistive aperture (**b**) CAVET in (**a**) with increased conductivity of the aperture. CAVET with resistive aperture leads to slow saturation as seen from the *I–V* curve in (**c**) as the resistance of the aperture region is decreased, the current saturates as shown in (**d**). The voltage distribution is shown in the potential contour, which is shown in the simulated snap shots above. The *red dots* in the plots indicate the bias condition at which the above pictures were taken

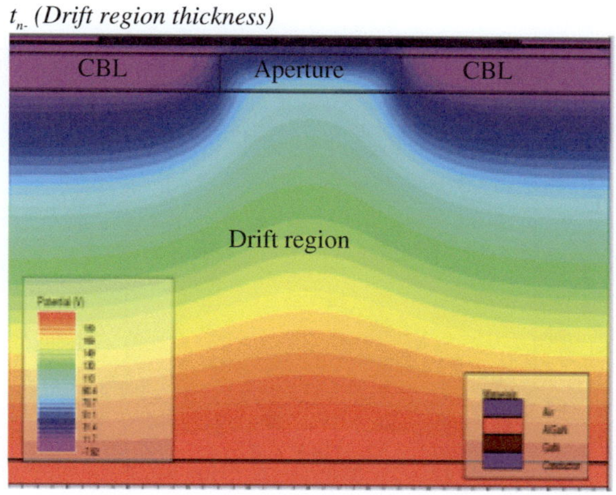

Fig. 5.8 Potential contour in a CAVET under the off-state operation showing the majority of the voltage dropping across the drift region

aperture region, which necessitated an aperture region longer than 4 μm and was experimentally verified (Fig. 5.8).

The CAVET is designed to support most of the applied voltage in the drift or lightly doped (n-) layer. This is the very principle based on which the vertical devices work. While the channel architecture can vary to form different types of vertical transistors, viz. JFET or MOSFET, the expected properties of the drift region and therefore its design are common to all. Majority of the voltage drop under the off-state operation occurs in the drift region. Therefore, the drift region has to be lightly doped to deplete under the off-state drain bias. However, increasing the thickness and lowering the doping of the region increase the R_{on} under the on-state. The effect is more pronounced for higher-voltage classes of vertical devices, where drift region is made thicker to sustain the high-voltage switching. From the potential contours, one can see that most of the applied voltage is supported by the drift region. As the applied voltage is increased, the depletion region extends and the potential contours are pushed into the aperture. The peak field region is buried in the bulk material and occurs at the lower edge of the CBL or the (p)–n junction.

What will play an important role in vertical transistor performances is the mobility of the drift region. The improvement in the quality of the bulk GaN material over the past decades leads us to predict that the differentiator of GaN technology with other competing technologies like SiC (which presents similar advantages due to its comparable bandgap and critical electric field) is indeed the mobility of the material [26]. Increase in mobility of the bulk GaN that forms the drift region will lower the R_{ON} without any significant penalty paid. The drift region mobility is therefore a key component and impacts the performance and the roadmap just like the critical electric field. A simple simulation done in Silvaco ATLAS to re-emphasize this statement is shown in Fig. 5.9. Dependence of R_{on} on the L_{ap} for the same doping densities and the same aperture to gate overlap

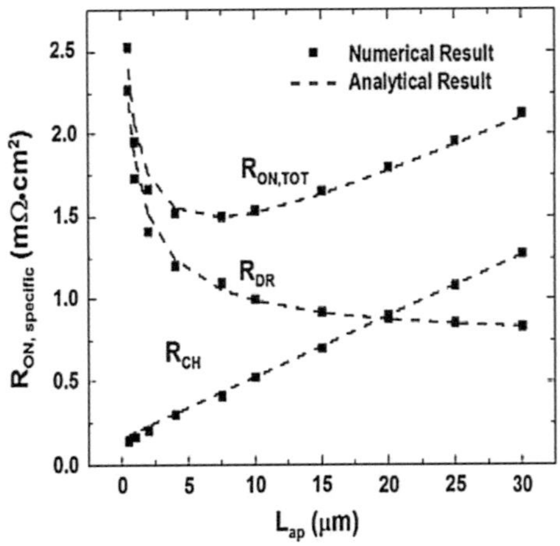

Mobility (drift region) in cm^2/Vs	V_{br}^*(V)	R_{on} mΩ.cm^1 (Lateral channel JFET)
1100	1240	1.4
900	1260	1.7
700	1280	2.2
500	1300	3.1

Fig. 5.9 R_{on} of CAVET is comprised of the channel and the drift region resistance. The plot captures both of those components as a function of L_{ap}. Total R_{on} as a function of the electron mobility in the drift region is listed. Bulk GaN mobility will play a key role in distinguishing GaN vertical devices from SiC counterparts

(L_{go}), optimized for minimal parasitic leakage, is also illustrated in Fig. 5.9. R_{on} decreases with increase in L_{ap} since the conductance in the aperture increases proportional to L_{ap}. Saturation in R_{on} is observed when the channel and drift regions offer the dominant part of the resistance [27].

The channel thickness and effective gate length (L_{go})

Besides ensuring proper gate control and maintaining a good transconductance (g_m) in the device, channel thickness and effective gate length play another important role in these devices best understood from the leakage analysis of the device. Three main leakage paths exist and are shown in Fig. 5.5.

1. *Through the CBL*

If the CBL does not provide a high enough barrier to the current, current can flow from source to drain through it. This is an undesirable path for the current and can be of large significance in the device performance.

2. *Unmodulated electrons*

Under normal functioning, all the electrons from the source should flow through the 2DEG to the gate horizontally and then vertically through the aperture to the drain.

However, if the confinement of the electrons in the channel is not sufficient, then some electrons can find an easy path from the source to the drain via the conductive aperture, bypassing the gate. Such electrons add to the leakage current and should be cut off from flowing. L_{go}, the length of the gate, that overlaps the aperture (which is analogous to a std. HEMT gate length) and t_{UID} are the two knobs that control this path.

Fig. 5.10 Three critical leakage paths in a CAVET (1) through CBL, (2) unmodulated electrons and (3) gate leakage

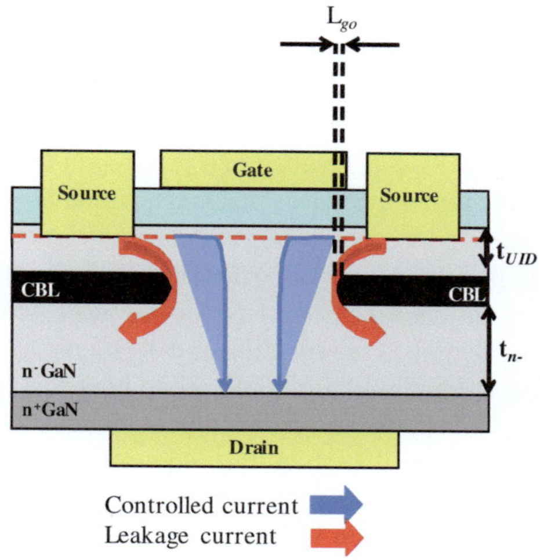

If L_{go} is small and or t_{UID} is large, the effective distance from the source to the aperture is decreased and the chance of electrons flowing to the drain, without being modulated by the gate, increases. So L_{go} has to be large enough to enhance gate control injected by the source. On the other hand, t_{UID} needs to be of minimum thickness.

However, if t_{UID} is too small, the electrons in the 2DEG would be close to the CBL. This increases the chance of the electrons getting captured (especially if the CBL has traps) causing dispersion when the gate signal is pulsed.

This sets a lower limit in t_{UID} determined by the dispersion criterion.

3. *Through the gate*

The other leakage takes place through the gate and can be addressed by suitable gate dielectric as shown in Fig. 5.10.

5.5.1 Current Blocking Layers

A discussion on doped versus implanted current blocking layer

The design of a CAVET involves designing a robust CBL, an aperture conductive enough to not choke the current but not so conductive to limit the breakdown voltage of the device and a low-doped drift region to hold the blocking voltage. The CBL is designed to provide a barrier to the electron flowing from the source to the drain through any other path except the aperture. A p-type GaN layer therefore would be a very effective current block providing a barrier of over 3 eV. Although from the band diagram as high as 3 eV barrier could be achieved by a doping (Mg) level of $10^{19}/cm^3$ (1 % active at RT), there are other fabrication-related issues that make it less attractive. To complete the device fabrication, AlGaN/GaN (25/140 nm) layers need to be regrown on top of the CBL. This implies that the p-doped layer underneath has to be activated. While buried p-layer activation is difficult to achieve, this scheme also necessitates regrowth on etched trenches. It was seen in previous studies that etching of the aperture region exposed non-c planes and regrowth over such inclined facets resulted in a depression that propagated to the surface. Other studies have found evidences that GaN, which is grown on facets other than <0001> plane, tends to incorporate large concentrations of n-type impurities. Since the peak electric field in a CAVET is located directly beneath the gate, such high n-type doping in that region causes an increase in the peak field limiting the device breakdown. Such highly doped region if under the gate metal would increase the gate leakage. Use of a good gate insulator could prove beneficial in terms of gate leakage, but premature breakdown because of increase in peak electric field, which is as a result of the unintentional high n-type impurities, still cannot be ruled out. One of the best alternative paths to circumvent these problems was identified to be the ion-implantation techniques to achieve current blocking functionality. The advantage of using an ion-implanted CBL comes from the fact that aperture layer need not be regrown. Figure 5.11 illustrates the process flow of fabricating a CAVET with ion-implanted CBL. In this approach, the aperture layer is grown along with the base structure [28]. Then, the aperture is masked and implanted to form the CBL. Since

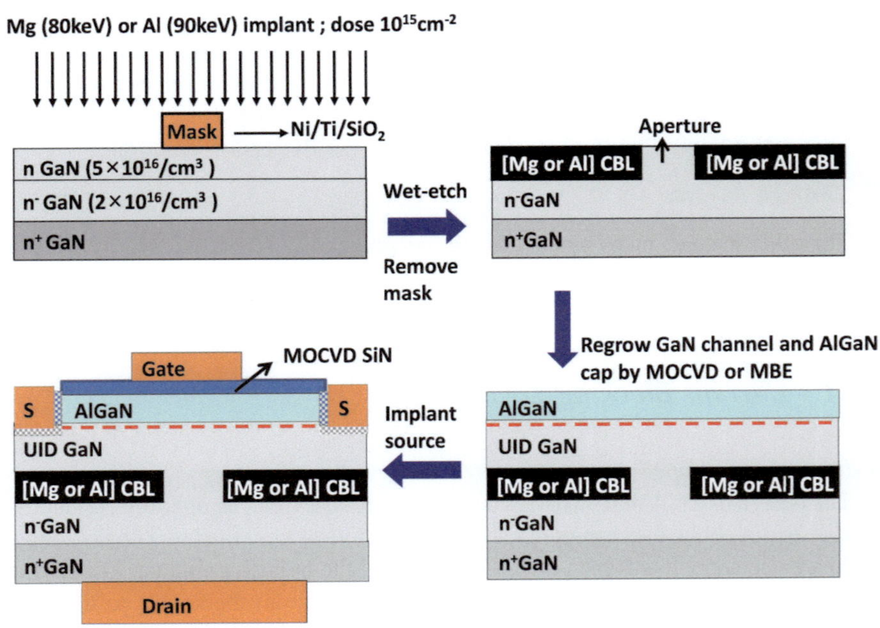

Fig. 5.11 A schematic of a CAVET process flow. Here implanted CBL scheme has been adopted. Activation of the implanted species (not shown in the flow) occurs after the implantation stage. Regrowth can be accomplished by both MOCVD and MBE techniques. However, with Mg implantation, low-temperature regrowth is necessary to prevent Mg from out diffusing into the regrown layers

this method involves no etching of the aperture layer, the regrowth of the AlGaN/GaN layers takes place on the c-plane and remains planar. While both implanted and doped p-GaN CBLs present their distinctive realm of advantages and disadvantages, one must point out that significant technological progress has been made on both fronts showing promising trends. CBL obtained by [Mg] ion-implantation technique has proved to be advantageous over a partially activated Mg-doped CBL, particularly if the implanted Mg is activated using a very high-temperature process. Although such high-temperature activation process is a common practice in SiC, it required extensive modification and development to adapt to GaN. Recent reports demonstrated excellent p-type property of Mg-implanted GaN when activated at temperatures above 1400 °C [29].

Avogy, leading the commercialization of GaN vertical devices, reports their devices based on doped CBL. The critical most part of a doped p-GaN CBL vertical FET is the realization of an excellent p–n junction, buried into the bulk GaN material. Other associated challenges related to regrowth on non-planar surface, etc. have been well tackled with extensive etch and regrowth studies. The p–n junction controls the breakdown electric field (or more appropriately the electric field at the designed blocking voltage) as well as dispersion characteristics under switching operation. The predicted electric field in GaN is around 3 MV/cm, and the operation close to the predicted limits relies on a perfect p–n junction.

On the other hand, based on the recent progress made with high-temperature activation of Mg-implanted GaN, showing excellent p(implanted Mg)–n junction diode characteristics, one can surely hope for next-generation CAVETs or CAVET-like vertical devices to rely on the implantation scheme, which alleviates the need for regrowth on etched trenches.

5.5.2 Performance and Cost

Sustainability of any technology depends on the size of market and GaN devices projects a promising trend to capture a substantial PE market size based on their vertical and lateral device technology. It is difficult to obtain a fair comparison of the two technologies since the more mature lateral GaN technology targets 600–1000 V market, and vertical GaN technology, which is yet to mature, finds it prospective market beyond 1 kV. With aging of both of these technologies, the answer will become clearer, but it is worthwhile to discuss factors that contribute to performance and cost at the die level.

High-voltage FOM, as was discussed in section II, clearly places GaN in the forefront of the race for ultimate power switch offering the lowest loss. Recent result reported by UNIPRESS-TopGaN-Ammono SA team [26] demonstrating electron mobility in bulk GaN over 1100 cm^2/Vs supports the argument in favor of GaN, stronger than ever. Higher mobility is crucial for reducing R_{on} and hence pushing the loss in a switch toward zero [27].

With higher blocking voltage, vertical device technology becomes more economical since it consumes less die area for a rated current. Avogy lately reported their device FOM (defined by V_{BR}^2/R_{on}) of 10^9 V^2 Ω^{-1} cm^{-2}. When the same metric was applied to a 300 V EPC lateral GaN HEMT, a FOM of 1.6×10^7 V^2 Ω^{-1} cm^{-2} was obtained. It is important to mention here that very few reports are available on lateral GaN power devices, sharing the information on the die size, and hence, it is difficult to compute the FOM realistically for comparison with its vertical equivalent. With maturity vertical GaN and stability of lateral GaN devices, high-voltage FOM will definitely become a very useful criterion to evaluate and compare their performance.

Measuring the total loss in a switch per switching cycle [$E_{on} + E_{off}$] is essential to evaluate the efficiency of a switching device.

The total switching loss per cycle for any switching device can be estimated by adding the two losses incurred during a switching cycle: crossover loss [$0.5\ I_l \cdot V\ (t_{on} + t_{off})$] and turn-on loss due to output capacitance (C_{oss}) during inductive switching [$C_{oss}\ V^2/2$], where t_{on} and t_{off} are the turn-on and turn-off delays per switching cycle. Assuming a 5 A drive current, the total loss per cycle for a 1.2 kV device operating at 50 A amounts to be 0.42–0.52 mJ.

The picture of the vertical GaN technology remains incomplete without discussing a thriving cost structure. A very important factor that will shape the real market for vertical GaN is the price of good quality GaN substrate. Power electronic device seeks for a low-cost, scalable GaN substrate technology that allows development of epitaxial layers supporting high electron mobility. An approximate cost of $0.04/A was estimated assuming $1000 per 6 in. of conductive GaN substrate yielding around 80 % of the dies.

Both performance and cost discussed in this section are predicted under realistic assumptions and hence look encouraging and achievable in the near future.

5.6 Role of Bulk GaN Substrate

A vertical device structure is typically fabricated on a homoepitaxially grown material where thick device layers could be grown with low extended defects on thick bulk GaN substrates ($\sim$200 μm or more).

Since the blocking voltage in the off-state of the device is sustained by the drift layer, to attain higher blocking voltages, the drift region is grown thicker so that almost all of the applied voltage between source and drain is absorbed in that region. Scaling up of voltages requires scaling up of the thickness of the drift region, and thus, vertical topology proves more economical in device area for high-voltage applications. The manufacturability and commercialization would greatly depend on the availability of good quality material with low extended defect density and low impurity concentration at a low cost. The advancement in solid-state lighting (LEDs and laser diodes) has enabled great improvement in the bulk GaN technology facilitating growth of wafers over 6 in. in diameter with dislocation densities as low as 10^4 cm^{-2}. The trend so far looks very promising with the cost of bulk GaN substrate dropping by significant amount. According to Lux Research, two-inch ammonothermal substrate costs will fall by more than 60 % to $730/substrate in 2020. While four-inch HVPE substrate costs will fall by 40 % to $1340/substrate in 2020, which looks very encouraging. Such progress in bulk GaN technology makes GaN vertical device extremely important area of research when the need for energy-efficient PE devices for high power application is becoming more and more obvious.

5.7 CAVETs for RF Application

Last but not the least, CAVETs are very well suited for RF device applications. While power electronics applications are demanding higher power density at frequencies of 100–300 kHz, RF applications are also seeking for increasing power density and higher power-added efficiency (PAE) for RADAR applications. It is critical that the power density and PAE of RF RADAR amplifiers be maximized to reduce chip size and thermal cooling requirements. This enables new RADAR amplifier classes such as switched mode amplifiers and reduces electrical and thermal system complexity and cost while simultaneously simplifying deployment.

Vertical GaN-based transistors uniquely harness the high breakdown field of GaN because the electric field distribution is uniform, easily designed and contained in the bulk of the device, compared to peaked surface/interface electric field distribution in lateral devices. Coupled with the reduced dispersion at high voltage, they provide the desired high power density and PAE not possible with current technology.

In 2002, first-generation RF-CAVETs demonstrated current gain cutoff frequency, $f_t > 12$ GHz [30]. With significant improvement in the material and device technology, CAVETs with its significant advantage of a high-mobility channel offers very promising solution for, but not limited to, L-, S- and C-bands.

5.8 Conclusion

The future of vertical GaN power electronics looks very promising, based on the recent progress made [31]. Lateral GaN HEMTs penetrating the market and establishing a GaN technology roadmap are enormously helpful to set up the vertical GaN technology and its market. On a complementary note, development of GaN-based vertical transistors to address the power level beyond that of the lateral GaN is crucial for sustaining the overall GaN-based power electronics market. Bulk GaN technology has made excellent progress over the last 5 years, thanks to the lighting market, which is stimulating the development of vertical GaN devices for power conversion. There are key issues demanding focused research, which include developing well-controlled low-doped drift region to sustain high voltages, robust and well-behaved p–n junction, where the p-region acts as the CBL and a roadmap toward low-cost material. Finally, vertical GaN can play a very important role as RF devices, a necessary stage to stay in the race for high power electronics application alongside of SiC.

Acknowledgments The author would like to acknowledge Prof. Umesh Mishra, Drs. Brian Swenson and Man hoi Wong and Dong Ji for their technical contributions at various levels to this work.

The author would also like to thank Toyota Motor Corporation, Japan (Dr. Tetsu Kachi, Dr. Masahiro Sugimoto and Dr. Tsutomu Uesugi), and ARPA-E (Dr. Timothy Heidel, Dr. Pawel Gradzki, Dr. Eric Carlson and Dr. Daniel Cunningham) for supporting the vertical GaN device development.

References

1. Jayant Baliga B (2008) Fundamentals of power semiconductor devices. Springer, New York
2. Kadavelugu A, Baliga V, Bhattacharya S, Das M, Agarwal A (2011) Zero voltage switching performance of 1200 V SiC MOSFET, 1200 V silicon IGBT and 900 V CoolMOS MOSFET. In: IEEE energy conversion congress and exposition (ECCE), Phoenix, AZ
3. Mishra UK, Parikh P, Wu YF (2002) AlGaN/GaN HEMTs-an overview of device operation and application. Proc IEEE 90(6):1022–1031
4. Dora Y, Chakraborty A, McCarthy L, Keller S, Denbaars SP, Mishra UK (2006) High breakdown voltage achieved on AlGaN/GaN HEMTs with integrated slant field plates. IEEE Electron Device Lett 27(9):713–715
5. Selvaraj SL, Suzue T, Egawa T (2009) Breakdown enhancement of AlGaN/GaN HEMTs on 4-in silicon by improving the GaN quality on thick buffer layers. IEEE Electron Device Lett 30 (6):587–589
6. Lu B, Palacios T (2010) High breakdown (>1500 V) AlGaN/GaN HEMTs by substrate-transfer technology. IEEE Electron Device Lett 31(9):951–953
7. Chu R, Corrion A, Chen M, Ray L, Wong D, Zehnder D, Hughes B, Boutros K (2011) 1200-V normally off GaN-on-Si field-effect transistors with low dynamic on–resistance. IEEE Electron Device Lett 32(5):632–634
8. Chowdhury S, Mishra UK (2013) Lateral and vertical transistors using the AlGaN/GaN heterostructure. IEEE Trans Electron Devices 60(10):3060–3066
9. Ben-Yaacov I, Seck Y-K, Mishra UK, Denbaars SP (2004) AlGaN/GaN current aperture vertical electron transistors with regrown channels. J Appl Phys 95(4):2073–2078
10. Gao Y, Stonas A, Ben-Yaacov I, Mishra U, Denbaars S, Hu E (2003) AlGaN/GaN current aperture vertical electron transistors fabricated by photoelectrochemical wet etching. Electron Lett 39(1):148
11. Chowdhury S, Swenson BL, Mishra UK (2008) Enhancement and depletion mode AlGaN/GaN CAVET with Mg-Ion-implanted GaN as current blocking layer. IEEE Electron Device Lett 29(6):543–545
12. Chowdhury S, Wong MH, Swenson BL, Mishra UK (2012) CAVET on bulk GaN substrates achieved with MBE-regrown AlGaN/GaN layers to suppress dispersion. IEEE Electron Device Lett 33(1):41–43
13. Kanechika M, Sugimoto M, Soeima N, Ueda H, Ishiguro O, Kodama M, Hayashi E, Itoh K, Uesugi T, Kachi T (2007) A vertical insulated gate AlGaN/GaN heterojunction field effect transistor. Jpn J Appl Phys 46(21):L503–L505

14. Otake H, Chikamatsu K, Yamaguchi A, Fujishima T, Ohta H (2008) Vertical GaN-based trench gate metal oxide semiconductor field-effect transistors on GaN bulk substrates. Appl Phys Express 1:011105-1–011105-3
15. Nie H, Diduck Q, Alvarez B, Edwards A, Kayes B, Zhang M, Bour D, Kizilyalli DI (2014) 1.5 kV and 2.2 mΩ cm^2 vertical GaN transistors on bulk-GaN substrates. IEEE Electron Device Lett 35(9):939–941
16. Oka T, Ina T, Ueno Y, Nishii J (2015) 1.8 mΩ cm^2 vertical GaN-based trench metal–oxide–semiconductor field-effect transistors on a free-standing GaN substrate for 1.2-kV-class operation. Appl Phys Express 8(5):054101
17. Okada M, Saitoh Y, Yokoyama M, Nakata K, Yaegassi S, Katayama K, Ueno M, Kiyama M, Katsuyama T, Nakamura T (2010) Novel vertical heterojunction field-effect transistors with re-grown AlGaN/GaN two-dimensional electron gas channels on GaN substrates. Appl Phys Express 3(5):054201
18. Diduck Q, Nie H, Alvarez B, Edwards A, Bour D, Aktas O, Disney D, Kizilyalli IC (2013) 1000 V vertical JFET using bulk GaN. ECS Trans 58(4):295–298
19. Oka T, Ueno Y, Ina T, Hasegawa K (2014) Vertical GaN-based trench metal oxide semiconductor field-effect transistors on a free-standing GaN substrate with blocking voltage of 1.6 kV. Appl Phys Express 7(2):021002
20. Yeluri R, Lu J, Hurni CA, Browne DA, Chowdhury S, Keller S, Speck JS, Mishra UK (2015) Design, fabrication, and performance analysis of GaN vertical electron transistors with a buried p/n junction. Appl Phys Lett 106(18):183502
21. Ueda T, Tanaka T, Ueda D (2007) Gate injection transistor (GIT)-A normally-off AlGaN/GaN power transistor using conductivity modulation. IEEE Trans Electron Devices 54(12): 3393–3399
22. Kanamura M, Ohki T, Kikkawa T, Imanishi K, Imada T, Yamada A, Hara N (2010) Enhancement-mode GaN MIS-HEMTs with n-GaN/i-AlN/n-GaN triple cap layer and high-k gate dielectrics. IEEE Electron Device Lett 31(3):189–191
23. Cai Y, Zhou Y, Chen KJ, Lau KM (2005) High-performance enhancement-mode AlGaN/GaN HEMTs using fluoride-based plasma treatment. IEEE Electron Device Lett 26(7):435–437
24. Niiyama Y, Kambayashi H, Ootomo S, Nomura ST, Yoshida S, Chow TP (2008) Over 2 A operation at 250 °C of GaN metal-oxide semiconductor field effect transistors on sapphire substrates. Jpn J Appl Phys 47(9):7128–7130
25. Kanechika M, Uesugi T, Kachi T (2010) Advanced SiC and GaN power electronics for automotive systems. In: 2010 international electron devices meeting
26. Kruszewski P, Jasinski J, Sochacki T, Bockowski M, Jachymek R, Prystawko P, Zajac M, Kucharski R, Leszczynski M (2014) Vertical schottky diodes grown on low-dislocation density bulk GaN substrate. The international workshop on nitride semiconductor
27. Ji D, Chowdhury S (2015) Design of 1.2 kV power switches with low RON using GaN-based vertical JFET. IEEE Trans Electron Devices 62(8):2571–2578
28. Chowdhury S (2010) PhD Thesis. AlGaN/GaN CAVETs for high power switching application
29. Anderson T, Kub F, Eddy C, Hite J, Feigelson B, Mastro M, Hobart K, Tadjer M (2014) Activation of Mg implanted in GaN by multicycle rapid thermal annealing. Electron Lett 50 (3):197–198
30. Ben Yaacov I (2004) PhD Thesis. AlGaN/GaN current aperture vertical electron transistor
31. Kachi T (2014) Recent progress of GaN power devices for automotive applications. Jpn J Appl Phys 53(10):100210

Chapter 6
GaN-Based Nanowire Transistors

Elison Matioli, Bin Lu, Daniel Piedra and Tomás Palacios

6.1 Introduction

The outstanding electronic properties of GaN semiconductors, such as large breakdown voltage, high critical electric field, high electron mobility and saturation velocity, high-temperature operation, make them an ideal material for power switches, converters, and RF power amplifiers. However, despite the large development of GaN lateral transistors, such as high-electron-mobility transistors (HEMTs), there are still several limitations in terms of performance, scalability of breakdown voltage and current density, reliability, and linearity. In addition, current GaN devices suffer from high leakage current, short-channel effect for short gate length devices, which reflect the poor control of the gate electrode (placed at the top surface of the device) over electrons in the channel.

Nanowires offer a potential solution to address such challenges. Nanostructuring the channel region yields enhanced control over electrons in the two-dimensional electron gas (2DEG) through a gate electrode that surrounds the channel. Gate control through the sidewall of the nanowires can largely improve the electrostatics due to the “wrap–around” gate geometry [1]. Moreover, the

E. Matioli (✉)
Institute of Electrical Engineering, Ecole Polytechnique Fédérale de Lausanne (EPFL), 1015 Lausanne, Switzerland
e-mail: elison.matioli@epfl.ch

B. Lu
Cambridge Electronics Inc., Cambridge, MA 02215, USA

D. Piedra · T. Palacios
Department of Electrical Engineering and Computer Sciences, Massachusetts Institute of Technology (MIT), Cambridge, MA 02139, USA

T. Palacios
e-mail: tomasp@mit.edu

M. Meneghini et al. (eds.), *Power GaN Devices*,
Power Electronics and Power Systems, DOI 10.1007/978-3-319-43199-4_6

nanowire geometry also presents a unique system to investigate material properties at nanoscale that could not otherwise be done in planar structures. Certain quantum mechanical and transport phenomena can be well studied in nanowire systems [2–4]. However, electronic conduction through nanowires are much more susceptible to the surface and external environment than in bulk devices [5–8]. Fermi-level pinning at the nanowire surface causes bending of the conduction band, which depletes electrons at the surface of the nanowire. This can drastically increase the resistance in small diameter wires [9].

Different fabrication and growth schemes based on top-down and bottom-up approaches give additional flexibility to process nanowire devices and integrate them into larger systems [10, 11]. Bottom-up approaches rely on epitaxial growth to spontaneously synthesize nanowires. Several techniques have been reported for GaN nanowires such as laser-assisted catalytic growth (LCG) [12], chemical vapor deposition (CVD) [13, 14], molecular beam epitaxy (MBE) [15], and anodic alumina template synthesis [16]. These techniques usually rely in one of two different mechanisms: catalyst-based and catalyst-free. A large body of research has been published in the subject and summarized in several review papers [17, 18].

Compared with "top-down" nanofabricated device structures, "bottom-up" synthesized nanowires offer well-controlled size in at least one critical device dimension—the channel width—which is at or beyond the limits of lithography. In addition, their small cross section can accommodate much larger lattice mismatches during growth without formation of dislocations, which allows growth on a variety of substrates [19–24]. However, from a device perspective, the use of bottom-up nanowires in transistors remains technologically challenging. Most applications require large current densities, for which it is necessary to have a large number of nanowire devices working in parallel. The alignment of bottom-up nanowires becomes an important challenge [13]. On the other hand, top-down approaches offer advantages in terms of flexible design and alignment [25] since the nanostructures are obtained through post-growth nanofabrication.

This chapter summarizes some of the most relevant work reported so far in the literature regarding the use of bottom-up and top-down nanowires in GaN-based transistors. In the first part, we will review the application of bottom-up GaN nanowires. In the second part, we will cover the use of top-down nanowires for power electronic devices relying on the use of nanowire gate contacts (tri-gate structure) for GaN transistors and Schottky barrier diodes (SBDs). Finally, we will show the application of nanowires for RF applications in increasing the linearity of the transconductance.

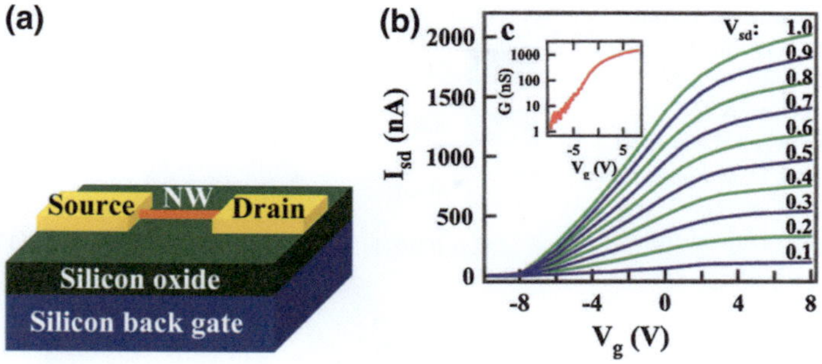

Fig. 6.1 **a** Schematic of a NW FET. **b** I–V_g data for values of $V_{sd} = 0.1$–1 V. *Inset* Conductance G versus gate voltage. Reprinted with permission from [26]—copyright 2016 American Chemical Society

6.1.1 Bottom-Up Nanowire Devices: GaN Nanowire Field-Effect Transistors

Field-effect transistors (FETs) have been demonstrated based on individual GaN nanowires (NWs), grown by bottom-up approaches. Bottom-up nanowire devices are commonly realized by transferring the nanowires from the growth substrate to a second substrate, commonly highly doped silicon covered by an oxide layer, as shown in Fig. 6.1a. Ohmic contacts are subsequently deposited at the edges of the nanowire to operate as source and drain contacts. Gate voltage can be applied either through the Si substrate or by depositing gate electrode on top of the nanowire. This process illustrates the importance of a precise alignment of the fabrication process with respect to the nanowires, which makes it very challenging.

Figure 6.1a shows an example of bottom-up GaN nanowire transistor synthesized via a laser-LCG. Gate-dependent electrical transport measurements revealed good modulation of the conductance of n-type GaN NWs by more than 3 orders of magnitude. These devices presented electron mobility as high as 650 cm^2/Vs, which is larger than the mobilities reported on NW grown by CVD methods [14, 27]. However, the mobility is largely degraded for smaller diameter wires due to surface scattering [28]. Figure 6.1b shows the IV characteristics of a single-nanowire GaN FET (10 nm of diameter) with a good switching behavior and transconductance G_m above 1 μS (which normalized by the wire diameter results in 100 mS/mm).

These results illustrate the possibility of GaN nanowires to serve as building blocks for nanodevices, especially in applications requiring single nanowires such as single electron transistors [4]. Such nanowire geometry also offers a platform for transport measurements [29], where quantum effects in low-dimensional III-nitride systems can be studied.

High-frequency FETs have also been demonstrated in GaN nanowires with sub-micrometer gate length of 500 nm [30]. The structure consisted of GaN/AlN/AlGaN/GaN nanowire metal–insulator–semiconductor field-effect transistors (MISFETs),

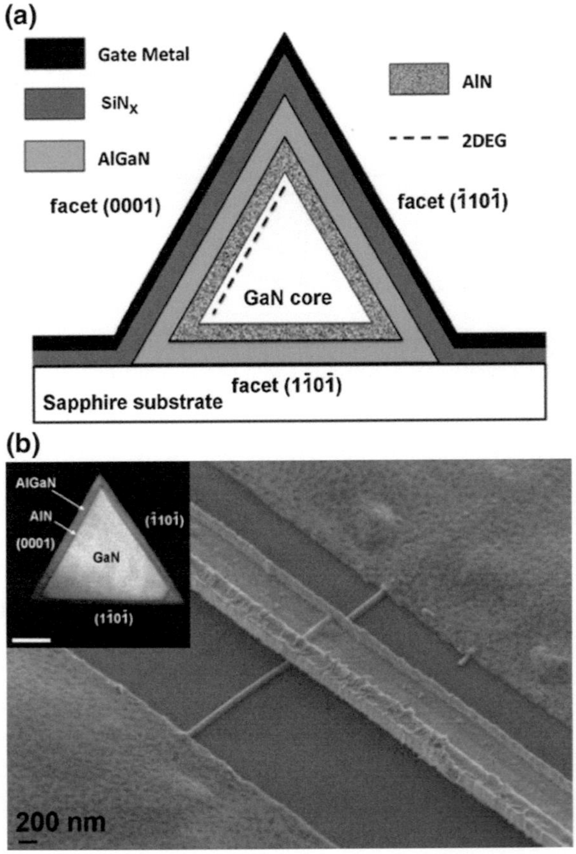

Fig. 6.2 **a** Schematic cross section of an NW covered by SiNx and the gate. **b** SEM image showing an NW connected between the source and drain. *Inset* TEM image of a GaN/AlN/AlGaN NW cross section. Figure adapted from [30]

as shown in Fig. 6.2. Devices presented intrinsic current-gain cutoff frequency (F_T) of 5 GHz as well as an intrinsic maximum available gain (F_{MAX}) cutoff frequency of 12 GHz. These results show the potential of GaN-based nanowire FETs also for microwave applications.

Despite the promising DC and RF performance of single-nanowire devices, most applications require a large set of nanowires working in parallel. The main challenge with bottom-up approaches is the difficulty in aligning several nanowires close together, especially for lateral devices where the nanowires are transferred to a different substrate. A vertical configuration also presents challenges since several of the current growth techniques result in misaligned wires with non-uniform dimensions. This makes the fabrication process of a large set of nanowires very difficult. In addition, the large surface-to-volume ratio makes the electrical performance of these devices very sensitive to the environment. Although this is undesirable for transistors, it opens interesting opportunities in sensing applications [31].

The next section discusses the top-down approaches for nanowire devices. The devices fabricated through such approach are easily scalable and can therefore be applied to large current density power transistors and diodes.

6.1.2 *Top-Down Nanowire Devices*

6.1.2.1 Tri-Gate GaN Transistors for Power Electronics Applications

Tri-gate transistors or Fin-FETs were first proposed for deeply scaled Si technology to reduce the short-channel effects and off-state leakage current [32, 33]. In 2011, Intel announced the adoption of tri-gate structure in their 22-nm microprocessor. Despite their use in low-voltage digital applications, these advanced Si device structures are of little interest to high-voltage power-switching applications, where transistors are typically much larger and p-n junctions are used to manage the electric field.

The design of GaN-based power transistors, on the other hand, is very different from high-performance Si transistors and these devices could benefit from tri-gate structures. Although the high-breakdown electric field and the superior electronic transport properties of the 2DEG in the AlGaN/GaN heterostructures are significant advantages of GaN over silicon, the lack of flexibility to easily form good p-n junctions makes it difficult to control the off-state leakage current in GaN power transistors. For example, most of the GaN power transistors are lateral devices fabricated on an AlGaN/GaN heterostructure, where the channel layer is typically 0.4–1-μm-thick undoped GaN epitaxially grown on a high-resistive III-nitride buffer layer (usually a combination of GaN, AlN, and AlGaN doped with carbon to increase the buffer resistivity [34]). Without p-n junctions in the device, the off-state leakage current between the source and drain is limited by the potential barrier induced by the gate bias in the GaN channel layer. When high voltage is applied to the drain of the device, the electron barrier in the channel is reduced and leakage current increases in the undoped GaN channel layer. This off-state leakage current becomes much more pronounced in normally off GaN transistors since negative gate voltage is not available to further pinch-off the channel. As a result, many normally—off GaN transistors suffer from either low blocking voltage or high off-state leakage current, which is one of the major challenges in GaN power-switching devices. Therefore, despite their completely different application spaces, the deeply scaled Si MOSFETs and high-voltage lateral GaN FETs bare an important similarity: Both devices suffer from the drain-induced barrier lowering (DIBL) effects, resulting in high drain-to-source leakage current. The tri-gate structure can be used to solve the off-state leakage problem in both devices.

6.2 Tri-Gate GaN Power MISFET

Top-down GaN tri-gate structure was first explored by Ohi and Hashizume [25] and by Zimmermann et al. [35]. In Ref. [25], the device was fabricated on AlGaN/GaN heterostructure and named multi-mesa-channel (MMC) HEMT. As shown in Fig. 6.3a, trenches were etched into AlGaN/GaN layers, and a Ni/Au Schottky

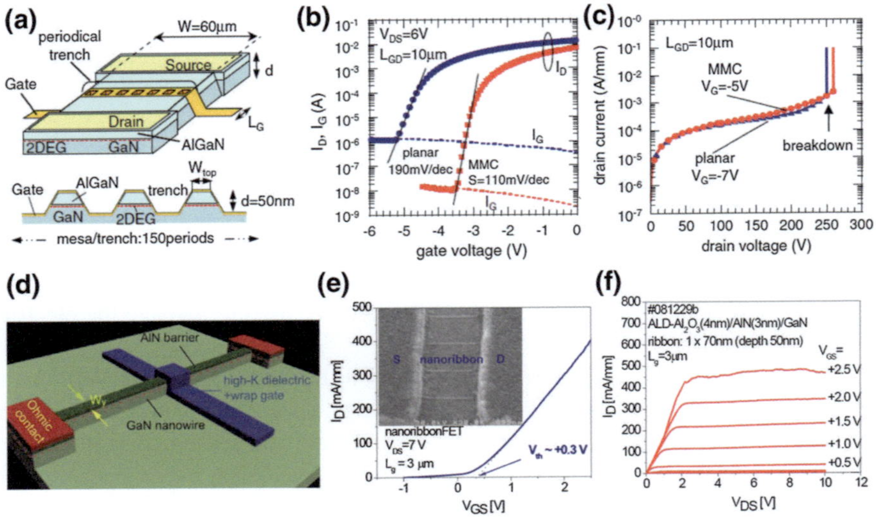

Fig. 6.3 **a–c** schematic, transfer characteristic and breakdown voltage measurement of the multi-mesa-channel (MMC) AlGaN/GaN HEMTdemonstrated by Ohi et al. The mesa top width W_{top} is 70 nm; mesa bottom width is 330 nm; gate length $L_g = 0.4$ μm and total gate width W is 60 μm; **d–f** schematic, transfer characteristic and breakdown voltage measurement of the E-mode AlN/GaN MISFET demonstrated by Zimmermann et al. The nanoribbon width is 70 nm and gate length is 3 μm (adapted from [25, 35]—copyright 2016 The Japan Society of Applied Physics)

gate contacted both the sides and top of each mesa-channel. The fabricated MMC HEMT is a normally on (depletion-mode) device. The threshold voltage of MMC HEMT is more positive compared to the planar reference device. The transfer characteristics also showed an improved subthreshold swing and lower gate leakage current than the planar HEMT as shown in Fig. 6.3b. However, the improvement of channel control from the MMC HEMT could be due to the reduced gate leakage, which can also improve the subthreshold voltage swing. In addition, breakdown measurement in Fig. 6.3c shows a similar off-state leakage current in both the MMC HEMT and the planar HEMT, which is mainly dominated by the Schottky gate leakage.

Zimmermann et al. [35] demonstrated both enhancement-mode (E-mode) and D-mode nanoribbon MISFETs on AlN/GaN heterostructure using Al_2O_3 as gate dielectric. By making the AlN/GaN nanoribbon as narrow as 70 nm, they were able to make E-mode transistors as shown in Fig. 6.3d–f. This shift in threshold voltage was attributed to the tri-gate sidewall depletion effect. A more comprehensive explanation of this effect can be found in the later part of this section.

High-voltage tri-gate normally—off GaN MISFETs were demonstrated on a standard $Al_{0.26}Ga_{0.74}N$/GaN HEMT structure by Lu et al. [36] with only 120 nm normally—off gate length. Figure 6.4 shows some major fabrication steps. The fabricated tri-gate normally—off GaN MISFETs have a tri-gate length of 660 nm.

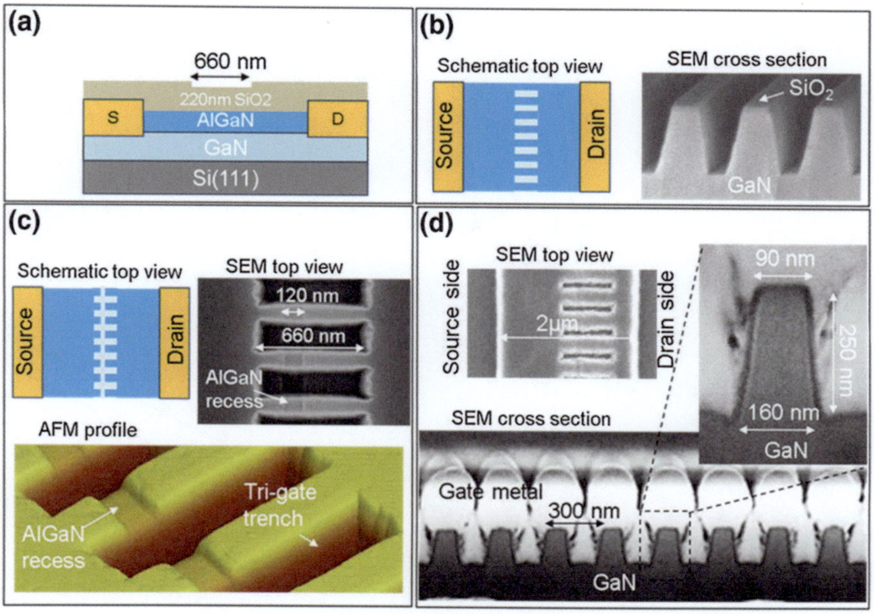

Fig. 6.4 Schematics, AFM and SEM images showing the major fabrication steps and structures of the tri-gate normally—off GaN MISFETs. Reference depletion-mode planar AlGaN/GaN HEMTs and planar recessed-gate GaN transistors FETs were fabricated together with the tri-gate normally off GaN MISFETs with the same gate dielectric and process conditions. **a** Tri-gate region pre-definition. **b** Interference lithography and etching. **c** E-beam litho and AlGaN recess. **d** Gate dielectric and metallization

Each tri-gate unit is of trapezoidal shape with top-gate width of 90 nm, a period of 300 nm and its sidewalls are 250 nm tall as shown in Fig. 6.4d. A 120-nm-long normally—off gate region was formed by recess-etching of the AlGaN barrier within the tri-gate region (Fig. 6.4c). The gate dielectric consists of 9-nm SiO_2/7-nm Al_2O_3, deposited by atomic layer deposition (ALD), with the SiO_2 contacting the GaN channel. 2-μm-long Ni/Au gate electrode covers the entire tri-gate region (Fig. 6.4d).

The tri-gate normally off GaN MISFET in Fig. 6.5a has a threshold voltage of 0.78 V extrapolated from the transfer characteristics. Even though the normally—off gate length is only 120 nm, the subthreshold slope of the tri-gate MISFET is steep and invariant as V_{ds} increases from 1 to 5 V (note that the planar depletion-mode device has a long gate length of 2 μm and as a result, its subthreshold slope also does not change with the V_{ds}). This is an indicator of good channel control from the tri-gate structure. This is more evident in Fig. 6.5b. A 160-nm-long recessed-gate device without tri-gate structure suffers from short-channel effects, such as degraded subthreshold slope and increased subthreshold current with higher V_{ds} voltage while the tri-gate GaN MISFET of 120 nm channel length has a constant subthreshold slope and stable subthreshold current (adapted from [36]).

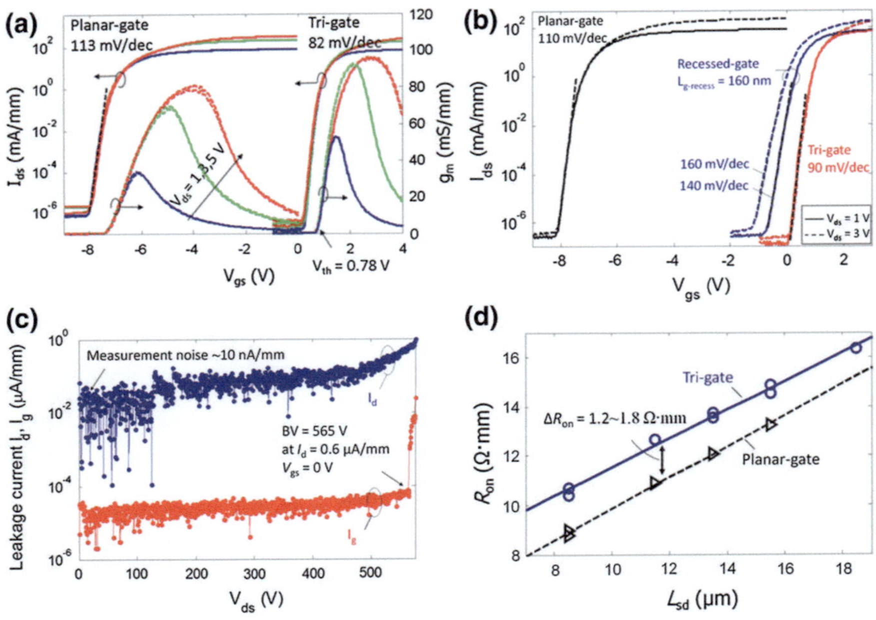

Fig. 6.5 **a** Transfer characteristics at V_{ds} = 1, 3, and 5 V with bi-direction gate sweep of a tri-gate normally off GaN MISFET and a standard planar-gate depletion-mode GaN transistor. Both devices have L_{gs} = 1.5 μm, L_g = 2 μm, L_{gd} = 10 μm and W = 100 μm. **b** I_{ds}–V_{gs} characteristics of the tri-gate normally off MISFET compared with the standard planar-gate transistor and the recessed-gate MISFET with recessed-gate length of 160 nm. All the devices have the same dimensions with L_{gd} = 8 μm and are biased at V_{ds} = 1 V (*solid*) and 3 V (*dashed*). **c** Three-terminal BV measurement of the tri-gate normally—off GaN MISFET with L_{gd} = 10 μm at V_{gs} = 0 V. **d** On-resistance R_{on} as a function of source-to-drain distance L_{sd} at V_{gs} = 7 V. The R_{on} of normally off tri-gate devices is on average 1.2–1.8 Ω mm higher than that of standard planar-gate devices. All the units are normalized by the width of the ohmic contacts (W = 100 μm) rather than the effective channel width to take into account the actual size of the device

The superior channel control from the tri-gate structure resulted in much lower off-state drain leakage current of only 0.6 μA/mm at drain bias of 565 V with gate bias of 0 V as shown in Fig. 6.5c. This low leakage current is 10–100 times better than the conventional planar normally off GaN power transistors, demonstrating the great potential of the tri-gate/Fin-FET structure in GaN power devices. Removing part of the planar channel region will likely impact the tri-gate device on-resistance. However, this impact is manageable by, for instance, limiting the tri-gate length. As shown in Fig. 6.5d, even with a 120-nm-long gate-recess region, which typically has very poor electron mobility [37], the on-resistance of the normally off tri-gate transistors is only about 1.2–1.8 Ω mm higher than that of standard planar-gate devices.

6.2.1 *Additional Considerations of Tri-gate GaN Power Transistors*

In addition to the enhanced channel electrostatic control by the tri-gate structure, there are other important factors to be considered. First, due to the 3D gate structure, there is additional sidewall-gate to channel capacitance. As shown in the simulation of a depletion-mode tri-gate AlGaN/GaN MISFET in Fig. 6.6, when the sidewall height increases, the sidewall capacitance becomes larger but eventually saturates since the bottom part of the sidewall-gates becomes too far away from the 2DEG channel. This additional sidewall-gate to 2DEG capacitance results in lateral depletion of the 2DEG from the sidewall-gates as shown in Fig. 6.7, which is one of the reasons for the observed positive threshold voltage shift of the depletion-mode tri-gate AlGaN/GaN transistors compared to the planar-gate AlGaN/GaN transistors [25, 36].

In addition, since III-nitride semiconductors are piezoelectric materials and their polarization plays an important role on the origin of the 2DEG in AlGaN/GaN or InAlN/GaN HEMTs [39], the change of mechanical strain from the 3D tri-gate structure impacts the 2DEG in the nanowire AlGaN/GaN or InAlN/GaN HEMTs. Azize et al. [10] has observed that the sheet resistance (R_{sh}) of the 2DEG channel in nanoribbon (NR) AlGaN/GaN structure fabricated by top-down dry-etching increases with the decrease of NR width as shown in Fig. 6.8a. However, the increase of R_{sh} cannot be explained by the NR sidewall surface depletion alone but rather is correlated with the in-plane tensile strain relaxation (Fig. 6.8a) as the NR width decreases. The tensile strain relaxation of the AlGaN barrier causes a reduction in the 2DEG density due to the reduced AlGaN/GaN interface piezoelectric charge. The conductivity of the AlGaN/GaN NRs can be increased (increasing in current density as shown in Fig. 6.8b) by depositing Si_xN_y stress layer. Similarly, large change in the sheet resistance of InAlN/GaN NRs was also observed when ALD Al_2O_3 was deposited over the NRs as shown in Fig. 6.8c. A reduction of R_{sh} as large as 46 % was observed once the InAlN/GaN NRs were completely buried

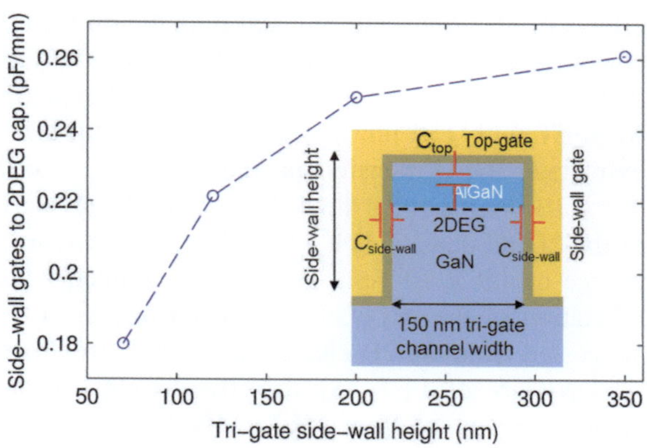

Fig. 6.6 Simulated sidewall-gates (one pair) to 2DEG capacitance at $V_g = 0$ V of each tri-gate AlGaN/GaN MIS-HEMT unit structure with 18 nm ALD SiO_2 gate dielectric and 150 nm width shown in the *inset* schematic. The capacitance is normalized to the tri-gate length direction perpendicular to the cross-section schematic *inset*. The simulation assumes an ideal side-wall interface without trap states. Adapted from [38]

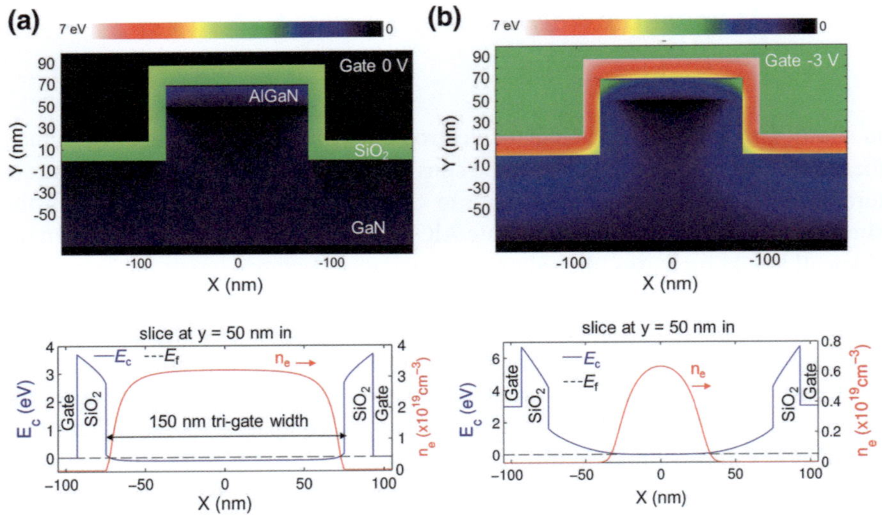

Fig. 6.7 Simulated 2D plots of the conduction band edge of the 70-nm-side-wall depletion-mode tri-gate AlGaN/GaN MISFET at gate bias of 0 (**a**) and −3 V (**b**) with 1D slices at $Y = 50$ nm positions. Adapted from [38]

in the Al_2O_3 [40]. Using convergent beam electron diffraction and finite-element simulation, the reduction of the R_{sh} is shown to be correlated with the increase of tensile stress to the InAlN/GaN NRs from the Al_2O_3 stress layer [41].

To understand the changes in NR sheet resistance, we will further investigate the change of 2DEG density and mobility, which both contribute to the NR sheet resistance. First of all, the impact to electron mobility by the effects of strain on the electron effective mass can be ignored. First-principle calculations have shown that the change in the electron effective mass due to strain is too small to contribute to the observed large change of NRs sheet resistance in the references [10, 40]. For example, the reduction of electron effective mass with 1 % tensile strain only results in less than 1 % increase in mobility [42]. On the other hand, effects of electron scattering and trapping on the effective electron mobility and change of electron density play much larger roles. Figure 6.9 shows the extracted 2DEG density and effective mobility from a depletion-mode AlGaN/GaN tri-gate MISFET with 150 nm tri-gate channel width and 70 nm sidewall height. The extracted 2DEG density in the tri-gate channel at $V_{gs} = 0$ V is about $4.8 \times 10^{12}/cm^2$ smaller than the planar AlGaN/GaN MIS-HEMT. The reduction in the 2DEG density at $V_{gs} = 0$ V is very close to the piezoelectric polarization charge of $4.9 \times 10^{12}/cm^2$ in the fully strained $Al_{0.26}Ga_{0.74}N$ layer on GaN. The relaxation of the $Al_{0.26}Ga_{0.74}N$ indeed results in the reduction of the 2DEG as suggested from the 2D electrostatic and piezoelectric simulations of the tri-gate AlGaN/GaN MISFET [38]. The effective electron mobility in the tri-gate MISFET channel is reduced from 1660 to 1260 $cm^2\ V^{-1}\ s^{-1}$, as expected from additional

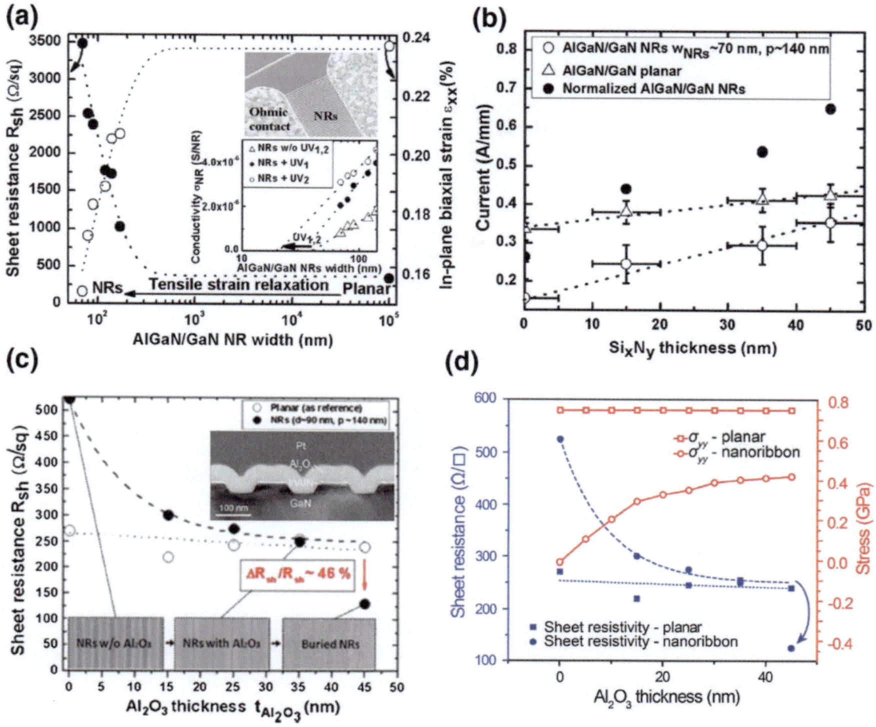

Fig. 6.8 **a** Normalized AlGaN/GaN nano-ribbon (NR) sheet resistance R_{sh} as a function of its width and in-plane biaxial strain measured by micro-Raman spectroscopy [10]. **b** Current density at 2 V bias carried by AlGaN/GaN NRs as a function of the Si_xN_y thickness deposited on the NRs [40]. **c** Normalized R_{sh} of InAlN/GaN NRs as a function of Al_2O_3 passivation thickness. Once the NRs were completely buried in Al_2O_3, their R_{sh} dropped by 46 %. **d** Finite-element simulation of the in-plane stress σ_{yy} perpendicular to the NR direction showing an increase of stress as the Al_2O_3 increases [41]

electron trapping and scattering effects from the sidewall-gates. However, the impact on channel mobility is not very severe for 150 nm tri-gate channel width.

In summary, going from the conventional planar device structure to 3D tri-gate structure, device designers have an additional degree of freedom in tailoring the transistor's performance. When designing GaN tri-gate devices, it is important not only to take into account the impact of the 3D gate structure on the gate capacitance and the electron effective mobility but also the impact on the piezoelectric charge in the devices.

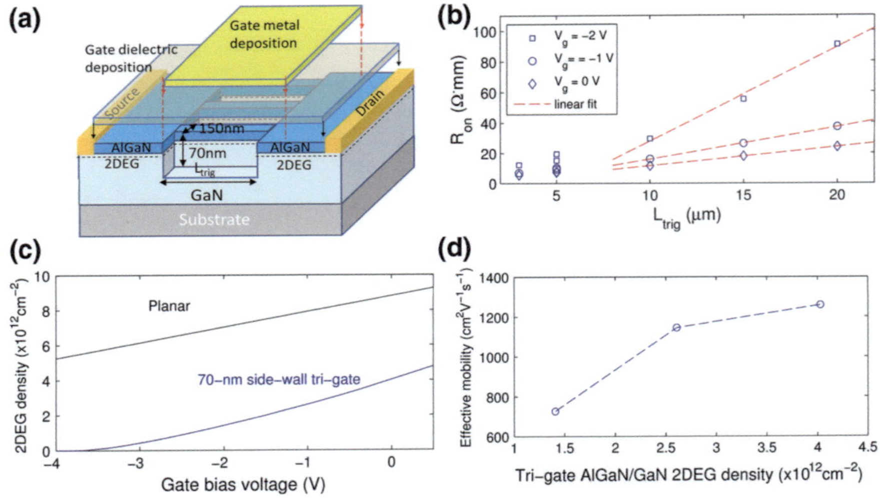

Fig. 6.9 **a** Schematic of the depletion-mode (D-mode) AlGaN/GaN tri-gate MISFET with 150 nm tri-gate channel width, 70 nm tri-gate height and ALD SiO_2 as gate dielectric; **b** on-resistance R_{on} of D-mode MISFETs as a function of tri-gate length L_{trig} was used to extract the tri-gate channel sheet resistance using $L_{trig} \geq 10$ µm. Note that the R_{on} has a non-linear relationship with $L_{trig} < 10$ µm **c** extracted 2DEG density from electrostatic simulation fitted to the quasi-static CV (QSCV) measurement; **d** calculated effective 2DEG mobility in the tri-gate channel from the equation $\mu_e = 1/qn_{2DEG}R_{sh\text{-}trig}$, where $R_{sh\text{-}trig}$ is the tri-gate channel sheet resistance and n_{2DEG} is the tri-gate channel 2DEG density. Figures are adapted from [38]

6.3 Nanowires for RF Applications: Increasing Linearity of G_m

In many RF applications, it is necessary to maintain high-frequency performance at high gate bias to allow for large input signal range. Theoretically, the transconductance, g_m, of the transistor should increase with gate voltage and level off at its maximum point [43]. The short-circuit current-gain cutoff frequency f_T should do so as well. In GaN HEMTs, however, experimental measurements show that the transconductance reaches a maximum value then decreases with increasing gate voltage (see Fig. 6.11). This effect becomes more accentuated as the device gate length is reduced. In deeply scaled devices, the drop in g_m and f_T can be greater than 30 % at high currents, thus limiting the effectiveness of the RF amplifier at higher power [44].

One explanation for the decrease in transconductance is the non-linear access resistance effect proposed by Palacios et al., DiSanto et al. and Trew et al. [44–46]. This explanation attributes the drop in g_m to the increase of the source access resistance with increasing drain current. With increasing gate bias, there are more electrons with the energy to emit optical phonons, which decreases the average electron velocity in the channel due to increased phonon scattering.

Considering that the drop in g_m may be due to increasing source access resistance, a possible method to prevent this drop is to reduce the gate-to-source distance and

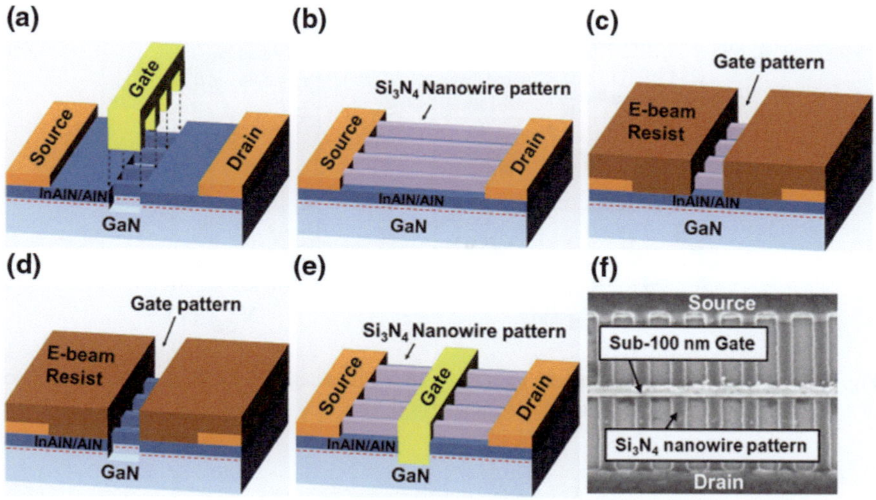

Fig. 6.10 **a** Device structure of nanowire channel HEMT, **b** Si_3N_4 nanowire pattern, **c** gate patterning and channel etching, **d** Si_3N_4 removal, **e** gate lift off, **f** SEM image after gate. Adapted from [48]

thus limit the area of the device contributing to source access resistance, which is achieved in a self-aligned device. Shinohara et al. [47] demonstrated a self-aligned device with n+ regrown contacts. As shown in this work, the transconductance roll-off with high V_{GS} is suppressed. However, as the access regions support high voltage, a low breakdown voltage could limits the range of use of self-aligned devices in high-voltage applications.

Another option to overcome the drop in g_m effect is to increase the current supplied from the source. Nanowire structures are used to increase the electrons in the source access region relative to the intrinsic channel. Nanowires are formed only under the gate electrode, but the source access region remains planar. Thus, the current supplied from the source relative to the intrinsic channel under the gate is higher. This is done with sub-100-nm gates to maintain high current and examine the phenomena under accentuated conditions.

Figure 6.10a shows the device structure and process flow of the fabricated nanowire channel GaN HEMT developed for this experiment. As shown in the figure, the source access region has higher current capability than the channel because of the larger effective width; hence, it can act as a more ideal source to the intrinsic nanowire-patterned transistor.

To evaluate the effectiveness of nanowire channel in improving the device frequency performance, the RF characteristics were measured from 100 MHz to 40 GHz. Although the nanowire channel device has a lower maximum current-gain cutoff frequency (f_T) throughout the range of input voltages than the planar device (due to the larger fringing capacitance from the side gate and access region, wider than the effective channel region), it does not suffer from as dramatic a drop in f_T at

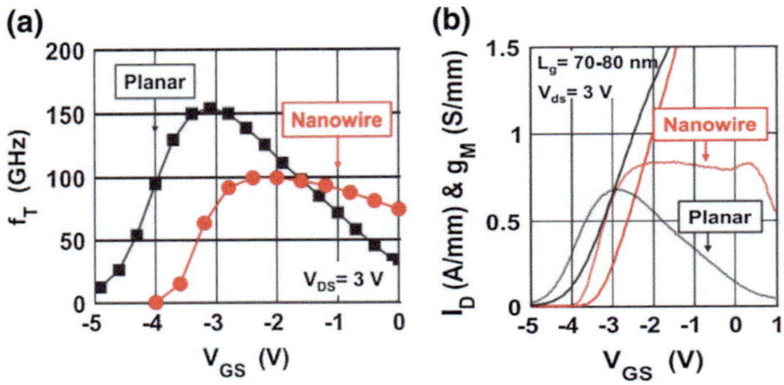

Fig. 6.11 **a** Short-circuit current gain for a range of input voltages. **b** DC transfer characteristics of a nanowire channel device and a planar device [48]

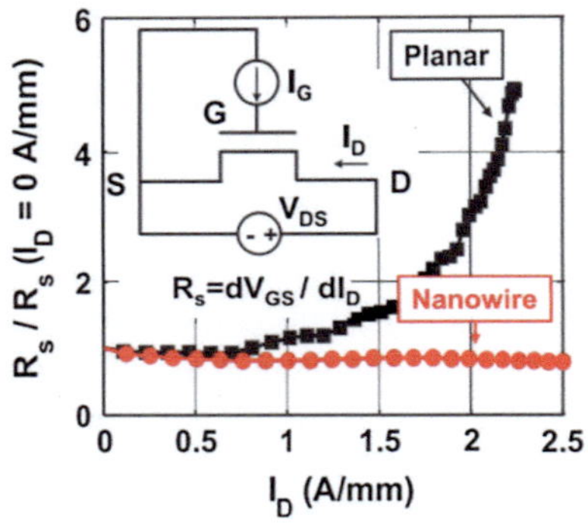

Fig. 6.12 Source resistance over a range of drain current density for a nanowire channel device and a planar device. Adapted from [50]

higher gate biases. The improved uniformity of the f_T across different input voltages can be attributed to a flatter transconductance, as shown in Fig. 6.11.

To further investigate the device behavior, the source access resistance is measured using the current injection method [49]. As shown in Fig. 6.12, the source resistance is relatively constant as a function of drain current in the nanowire device, while it increases rapidly in the planar device. These results support the explanation that the drop in g_m is due to the increased source access resistance. By decreasing the relative width of the transistor channel with respect to the source access resistance, these devices have effectively increased the relative amount of current supplied by the source to suppress the transconductance drop. Furthermore, this constant source resistance enables a higher gate overdrive, which in turn, increases the current in the device. The relatively stable RF characteristics over a wide range of input bias condition in the nanowire devices are useful to improve the linearity of these devices with respect to traditional GaN HEMTs.

6.4 Nanostructured GaN Schottky Barrier Diodes

Diodes are another important family of devices, widely used in power electronics circuits. SBDs are highly desirable due to their fast switching speed, since these are essentially majority carrier devices, and low turn-on voltage, which can lead to higher circuit efficiency.

Conventionally, GaN-based SBDs are fabricated on planar AlGaN/GaN heterostructures. When the anode contact is formed on the AlGaN barrier, at forward bias, electrons from the 2DEG in the AlGaN/GaN heterostructure need to overcome the AlGaN/GaN interface barrier and flow through the highly resistive AlGaN layer. This results in large turn-on voltage and high on-resistance [51, 52]. Recessing the barrier at the anode contact can mitigate this issue [53]. However, this also depletes the 2DEG under the anode region increasing the resistance for electrons since contact to electrons at the 2DEG occurs only at the edge of the anode region.

Another major limitation of conventional AlGaN/GaN SBDs is their large reverse leakage current, for similar reasons as previously discussed for AlGaN/GaN HEMTs. Several technologies have been proposed to reduce leakage current in these devices. They rely on passivating the AlGaN surface with SiN_x or SiO_2 [54], and on plasma treatment of the surface using O_2, N_2, C_2F_6, CF_4 [51, 54, 55, 56, 57], which resulted in a reduction of up to two orders of magnitude in leakage current, down to ~6 μA/cm^2. The use of field-plate termination in lateral SBDs improved further the leakage current down to 10^{-5} A/cm^2 (or equivalently to 100 nA/mm) at −100 V [38]. Despite the efforts to address these problems, reports on simultaneous enhancement in both forward and reverse characteristics of the SBDs are limited.

6.4.1 Nanostructured Anode for GaN SBDs

To address these issues both in forward and reverse biases, a novel SBD structure based on a 3D nanopatterned anode was proposed by Matioli et al. [9], as shown in Fig. 6.13a. Similarly to the tri-gate transistors, the patterned anode consisted of a series of periodically spaced trenches, etched far below the AlGaN barrier. However, in this case the purpose of these trenches is two-fold: to improve forward-bias characteristics and to reduce leakage current under reverse bias.

In forward bias, the sidewalls of the 3D anode structure form close-to-ideal Schottky junctions to the 2DEG in the GaN channel, bypassing the highly resistive AlGaN barrier and improving its turn-on characteristics. The function of the top barrier in this device is to induce a large density of electrons with high mobility in the 2DEG.

To reduce the leakage current in reverse bias, a section of the patterned anode between the Schottky junction and the cathode of the diode was covered with oxide to form a structured metal–oxide–semiconductor (MOS) region. This region is similar to a tri-gate MOS region, as shown in Fig. 6.13b, c. This tri-gate structure

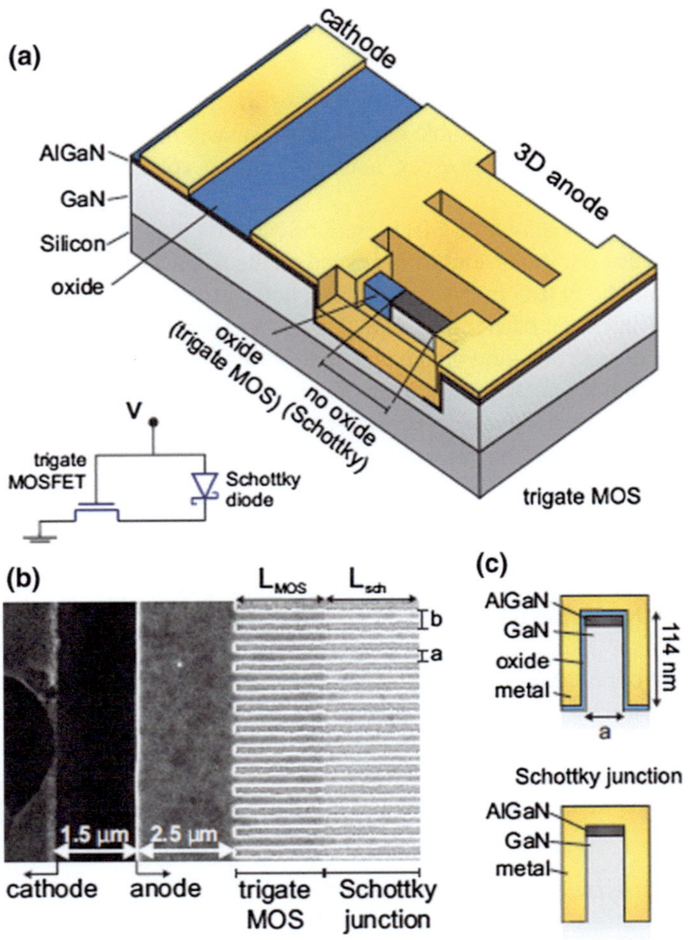

Fig. 6.13 **a** Schematics of the SBDs with 3-D anode structure. **b** Top-view SEM image of the SBDs with 3-D anode structure, where the metal layer was removed from only part of the nanostructured anode forming two different regions: trigate MOS with an oxide layer and the Schottky junction. The L_{MOS} and L_{sch} represent the length of the trigate MOS and the Schottky junction, respectively. *Inset* equivalent circuit of the SBDs with 3-D anode, corresponding to a SBD in series with a trigate MOSFET. **c** Cross-sectional representation of the trigate MOS and Schottky junction. Figures adapted from [9]

provided an electrostatic barrier for electrons under reverse bias, similarly to the trigate HEMT structure [37], but tailored to the specific operation and issues of AlGaN/GaN SBDs. The pillars offered a better geometry to deplete electrons in the channel in reverse bias compared with conventional planar field-plate structures, which drastically reduced reverse leakage current. The equivalent circuit of the SBDs with 3D

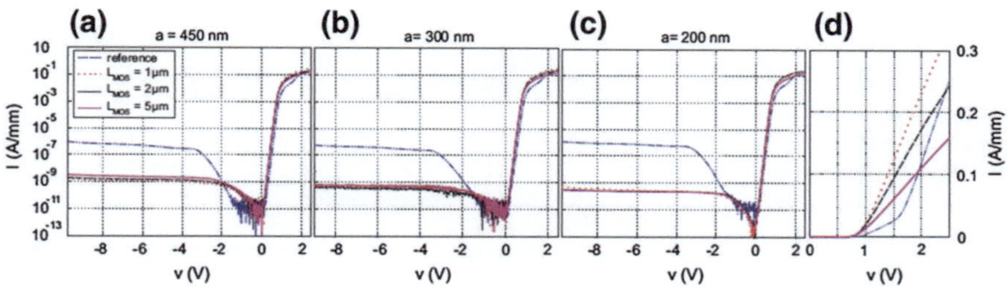

Fig. 6.14 Current–voltage (*I–V*) characteristics of SBDs with 3-D anode for different widths LMOS = 1, 2, and 5 μm, compared with the reference planar diode. The different plots correspond to pillar widths of **a** $a = 450$ nm, **b** $a = 300$ nm, **c** $a = 200$ nm. **d** Forward-bias *I–V* characteristics of the 3-D anode diodes with $a = 300$ nm and LMOS = 1, 2, and 5 μm compared with the reference planar diode. Similar behavior and turn-on voltages were observed for $a = 200$ and 450 nm. The current measured in these devices (both in the reference planar and in the 3-D anode SBDs) was normalized by the width of the diode ($W = 100$ μm). Figures adapted from [9]

anode, composed of a SBD in series with a tri-gate MOSFET, is shown in the inset of Fig. 6.13b.

As shown in Fig. 6.14a–c, a large reduction in reverse leakage current, of more than two orders of magnitude, was observed from the reference diode to the 3D anode SBD with pillar width $a = 450$ nm (Fig. 6.10a). The reverse leakage current was further reduced by shrinking the width of the pillars to 300 nm (Fig. 6.14b), down to nearly 4 orders of magnitude for $a = 200$ nm (Fig. 6.14c).

Under forward bias, the reference planar diode presented a large turn-on voltage of 1.5 V as well as two distinct knee voltages in the I–V characteristics (Fig. 6.14d). This non-ideal behavior was eliminated by the patterned anode structure, where the metal–semiconductor Schottky junction at both sidewalls of the pillars dominates the I–V characteristics, resulting in a reduction of the turn-on voltage for the 3D anode SBDs to 0.85 V. The ideality factor was also improved from 1.43 for reference diode to 1.27 for the patterned diodes, which was independent of L_{MOS}. The tri-gate MOS region was more effective in reducing leakage current for smaller pillar widths due to a better depletion of carriers by the negative potential applied to the anode contact.

Increasing the length L_{MOS} from 1 to 5 μm resulted in an increase in series resistance and a reduction of the leakage current. This shows a clear trade-off between improving forward- and reverse-bias characteristics by increasing the tri-gate length L_{MOS}. A good balance was achieved for L_{MOS} of 2 μm, which yielded a significant reduction of more than three orders of magnitude in leakage current, down to 263 pA/mm with only a mild increase of 21 % in series resistance, to 5.96 Ω mm, and an ideality factor of 1.27.

The tri-gate MOS formed over a portion L_{MOS} of the 3D anode is of fundamental importance to reduce the reverse leakage current. Otherwise, the reduced Schottky barrier at the sidewall of the 3D anode structure would just increase leakage current of the SBD.

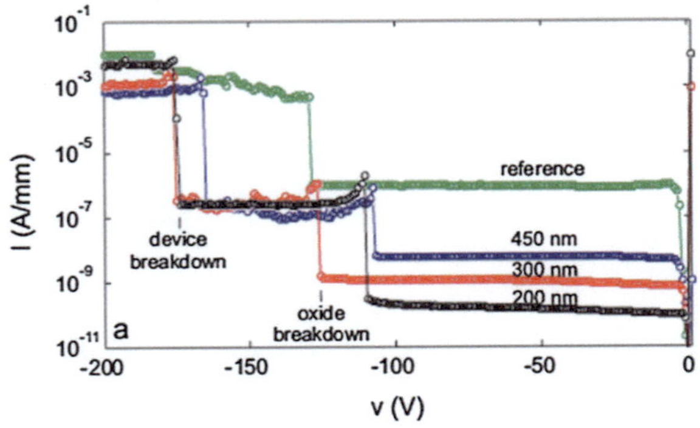

Fig. 6.15 Breakdown voltage measurements for the reference planar diode as well as the 3-D anode SBDs with $L_{MOS} = 2$ μm and $a = 450$, 300, and 200 nm. The distance between the cathode and anode electrodes was 1.5 μm in all these devices. Figures adapted from [9]

The 3D anode SBDs presented a breakdown voltage V_{BV} of up to −127 V (oxide breakdown voltage), for a distance of 1.5 μm between the cathode and anode electrodes, at a reverse leakage current lower than 10 nA/mm as shown in Fig. 6.15. The reference planar SBD showed a similar V_{BV} at a much larger leakage current. The reverse leakage current of the 3D anode SBD with $a = 200$ nm was 270 pA/mm at −110 V [9], which is much smaller than in other high-V_{BV} lateral SBDs reported in the literature, at similar reverse bias [33, 38] and, to the best of our knowledge, is among the lowest leakage current demonstrated in lateral AlGaN/GaN SBDs. These devices also presented smaller temperature dependence of the reverse leakage current compared to a planar device, with a leakage current at 200 °C of 14 nA/mm at −40 V.

In summary, ultralow leakage current SBDs could be demonstrated through with a 3D anode contact. In contrast to conventional AlGaN/GaN SBDs, this new device forms Schottky contact directly to the 2DEG at the sidewalls of the 3D anode structure to improve its turn-on characteristics. In addition, this device integrates an insulated tri-gate MOS structure to reduce its reverse-bias leakage current. By optimizing this new technology, SBDs with 3D anode structures with turn-on voltage of 0.85 V, on-resistance of 5.96 Ω mm and ideality factor of 1.27 have been demonstrated. The reverse-bias leakage current was significantly reduced by nearly four orders of magnitude, down to 260 pA/mm, with a breakdown voltage of up to 127 V for a distance of 1.5 μm between the cathode and anode electrodes. To the best of our knowledge, this is among the lowest leakage currents reported in lateral AlGaN/GaN SBDs fabricated on silicon substrate.

6.5 Conclusions

This chapter discussed the properties of nanowires to demonstrate power and RF transistors and SBDs with improved performance in both forward- and reverse-bias operations.

The first part discussed bottom-up approaches for nanowire devices. Single-GaN-nanowire devices (with 10 nm of diameter) demonstrated very good DC performances, with good switching behavior, transconductance G_m above 100 mS/mm, relatively large mobilities, as high as 650 cm^2/Vs, and excellent gate control over electrons in the channel. These devices are also promising for RF applications, presenting intrinsic f_T and f_{MAX} in tens of GHz range. Despite the promising DC and RF performance of single-nanowire devices, most applications require a large set of nanowires working in parallel to yield larger output currents. This is very challenging in bottom-up approaches, both in a growth as well as in a fabrication perspective. The large surface-to-volume ratio makes the electrical performance of these devices very sensitive to the environment, which opens great opportunities in sensing applications.

The second part of the chapter reviewed the application of tri-gate structures for GaN power electronics applications. The tri-gate structure is shown to significantly reduce DIBL effects and off-state drain leakage current. The control of off-state leakage current is extremely important for normally off GaN devices. Other important design considerations are also discussed. Due to the additional sidewall-gates, the gate capacitance in the tri-gate GaN transistors is larger than in the planar devices for per unit area of channel electron density. This increase of gate capacitance contributes to the more positive pinch-off voltage in the tri-gate depletion-mode AlGaN/GaN HEMTs. The tri-gate structure also causes the changes in the strain in the tri-gate AlGaN/GaN (or InAlN/GaN) HEMTs, which in term changes the 2DEG density in the AlGaN/GaN nanoribbons. This correlation between the mechanical strain and the 2DEG density is a unique property that can be used to modify device performance. The additional exposure surfaces in the tri-gate structure have a negative impact on the channel mobility. Fortunately, experiments showed that the effective electron mobility in the tri-gate AlGaN/GaN MISFETs with 150-nm ribbon width is still above 1000 cm^2/Vs. Later the chapter discussed the benefits adapting a top-down nanowire channel technology for RF devices. It was shown that the drop in transconductance at high gate voltages that is usually associated with planar GaN devices could be suppressed by etching a nanowire pattern into the device channel. The advantages of having a flatter g_m were further seen in the improved stability of the maximum f_T across a range of input voltages.

This chapter also discussed the implementation of nanopatterned anodes to improve the SBD turn-on voltage and reduce its reverse leakage current. In such structure, the effective Schottky junction is formed at the metal–GaN interface at the pillar sidewalls, which resulted in a reduced turn-on voltage from 1.5 to 0.85 V. A section of the patterned anode covered with oxide, forming a tri-gate MOSFET connected in series with the SBD, served as an electrostatic barrier to reduce its reverse leakage current by nearly 4 orders of magnitude, down to 270 pA/mm at −110 V, with series resistance of 5.96 Ω mm and ideality factor of 1.27.

In summary, this chapter shows that nanowire structures are very effective to decouple the device characteristics in forward and reverse bias, allowing their separate optimization, which offers a pathway for future low-leakage, fast-switching, and high-linearity GaN devices.

References

1. Lu W, Xie P, Lieber CM (2008) Nanowire transistor performance limits and applications. IEEE Trans Electron Device 55(11):2859–2876
2. Matioli E, Palacios T (2015) Room-temperature ballistic transport in III-nitride heterostructures. Nano Lett 15(2):1070–1075
3. Mastro M, Kim HY, Ahn J, Kim J, Jennifer H, Charles E Jr (2010) Quasi-ballistic hole transport in an AlGaN/GaN nanowire. ECS Trans 28(4):47–52
4. Kim J-R, Kim B-K, Lee IJ, Kim J-J, Kim J, Lyu SC, Lee CJ (2004) Temperature-dependent single-electron tunneling effect in lightly and heavily doped GaN nanowires. Phys Rev B 69:233303
5. Polenta L, Rossi M, Cavallini A, Calarco R, Marso M, Meijers R, Richter T, Stoica T, Luth H (2008) ACS Nano 2:287
6. Calarco R, Marso M, Richter T, Aykanat AI, Meijers R, Hart AVD, Stoica T, Luth H (2005) Nano Lett 5:981
7. Sanford NA, Blanchard PT, Bertness KA, Mansfield L, Schlager JB, Sanders AW, Roshko A, Burton BB, George SM (2010) J Appl Phys 107:034318
8. Simpkins BS, Mastro MA, Eddy CR Jr, Pehrsson PE (2008) J Appl Phys 103:104313
9. Matioli E, Lu B, Palacios T (2013) Ultralow leakage current AlGaN/GaN Schottky diodes with 3-D anode structure. IEEE Trans Electron Device 60:3365
10. Azize M, Palacios T (2011) Top-down fabrication of AlGaN/GaN nanoribbons. Appl Phys Lett 98:042103
11. Li Y, Xiang J, Qian F, Gradecak S, Wu Y, Yan H, Blom DA, Lieber CM (2006) Dopant-free GaN/AlN/AlGaN radial nanowire heterostructure as high electron mobility transistors. Nano Lett 6(7):1468–1473
12. Duan X, Lieber CM (2000) Laser-assisted catalytic growth of single crystal GaN nanowires. J Am Chem Soc 122:188–189
13. Peng HY, Wang N, Zhou XT, Zheng YF, Lee CS, Lee ST (2002) Control of growth orientation of GaN nanowires. Chem Phys Lett 359:241–245
14. Kim H-M, Kim DS, Park YS, Kim DY, Kang TW, Chung KS (2002) Growth of GaN nanorods by a hydride vapor phase epitaxy method. Adv Mater 14:991–993
15. Bertness KA (2011) Senior member, IEEE. In: Sanford NA, Davydov AV (eds) GaN nanowires grown by molecular beam epitaxy. IEEE J Select Top Quant Electron 17(4)
16. Cheng GS, Zhang LD, Zhu Y, Fei GT, Li L (1999) Large-scale synthesis of single crystalline gallium nitride nanowires. Appl Phys Lett 75:16
17. Songmuang R, Monroy E (2013) GaN-based single-nanowire devices, on III-Nitride semiconductors and their modern devices. In: B Gil (ed), 01/2013. Oxford University Press, USA, pp 289–364
18. Fortuna SA, Li X (2010) Metal-catalyzed semiconductor nanowires: a review on the control of growth directions. Semicond Sci Technol 25:024005
19. Glas F (2006) Phys Rev B 74:121302
20. Yoshizawa M, Kikuchi A, Mori M, Fujita N, Kishino K (1997) Japan. J Appl Phys 2(36):L459
21. Sekiguchi H, Nakazato T, Kikuchi A, Kishino K (2006) J Cryst Growth 300:259
22. Calleja E, Sanchez-Garcia MA, Sanchez FJ, Calle F, Naranjo FB, Munoz E, Molina SI, Sanchez AM, Pacheco FJ, Garcia R (1999) J Cryst Growth 201/202:296
23. Tchernycheva M et al (2007) Nanotechnology 18:385306
24. Songmuang R, Landre O, Daudin B (2007) Appl Phys Lett 91:251902
25. Ohi K, Hashizume T (2009) Drain current stability and controllability of threshold voltage and subthreshold current in a multi-mesa-channel AlGaN/GaN high electron mobility transistor. Jpn J Appl Phys 48:081002
26. Yu H, Xiangfeng D, Yi C, Lieber CM (2002) Gallium nitride nanowire nanodevices. Nano Lett 2(2):101–104

27. Stern E et al (2005) Electrical characterization of single GaN nanowires. Nanotechnology 16:2941
28. Sundaram VS, Mizel A (2004) Surface effects on nanowire transport: a numerical investigation using the Boltzmann equation. J Phys Condens Matter 16:4697
29. Songmuang R, Katsaros G, Monroy E, Spathis P, Bougerol C, Mongillo M, De Franceschi S (2010) Quantum transport in GaN/AlN double-barrier heterostructure nanowires. Nano Lett 10(9):3545–3550
30. Vandenbrouck S, Madjour K, Théron D, Dong Y, Li Y, Lieber CM, Gaquiere C (2009) 12 GHz F_{MAX} GaN/AlN/AlGaN Nanowire MISFET. IEEE Electron Device Lett 30:4
31. Chen CP et al (2009) Label-free dual sensing of DNA molecules using GaN nanowires. Anal Chem 81(1):36–42
32. Huang X, Lee WC, Kuo C, Hu C et al (1999) Sub 50-nm FinFET: PMOS. In: IEDM Technical Digest, pp 67–70
33. Doyle BS, Datta S, Doczy M, Hareland S, Jin B, Kavalieros J, Linton T, Murthy A, Rios R, Chau R (2003) High performance fully-depleted tri-gate CMOS transistors. IEEE Electron Device Lett 24:263–265
34. Kato S, Satoh Y, Sasaki H, Masayuki I, Yoshida S (2007) C-doped GaN buffer layers with high breakdown voltages for high-power operation AlGaN/GaN HFETs on 4-in Si substrates by MOVPE. J Crystal Growth 298:831–834
35. Zimmermann T, Cao Y, Guo J, Luo X, Jena D, Xing H (2009) Top-down AlN/GaN enhancement- and depletion-mode nanoribbon HEMTs. In: IEEE device research conference (DRC) Digest, pp 129–130
36. Lu B, Matioli E, Palacios T (2012) Tri-gate normally-off GaN power MISFET. IEEE Electron Device Lett 33(3):360–362
37. Lu B, Saadat OI, Palacios T (2010) High-performance integrated dual-gate AlGaN/GaN enhancement-mode transistor. IEEE Electron Device Lett 31:990–992
38. Lu B (2013) AlGaN/GaN-based power semiconductor switches. PhD Dissertation, Massachusetts Institute of Technology
39. Ambacher O, Foutz B, Smart J, Shealy JR, Weimann NG, Chu K, Murphy M, Sierakowski AJ, Schaff WJ, Eastman LF, Dimitrov R, Mitchell A, Stutzmann (2000) Two dimensional electron gases induced by spontaneous and piezoelectric polarization in undoped and doped AlGaN/GaN heterostructures. J Appl Phys 87(1), pp 334–344
40. Azize M, Hsu AL, Saadat OI, Smith MJ, Gao X, Guo S, Gradecak S, Palacios T (2011) High-electron-mobility transistors based on InAlN/GaN nanoribbons. IEEE Electron Device Lett 32(12):1680–1682
41. Jones EJ, Azize M, Smith MJ, Palacios T, Gradecak S (2012) Correlating stress generation and sheet resistance in InAlN/GaN nanoribbon high electron mobility transistors. Appl Phys Lett 101:113101
42. Dreyer CE, Janotti A, Van de Walle CG (2013) Effects of strain on the electron effective mass in GaN and AlN. Appl Phys Lett 102:142105
43. Dora Y, Chakraborty A, McCarthy L, Keller S, DenBaars SP, Mishra U (2006) High breakdown voltage achieved on AlGaN/GaN HEMTs with integrated slant field plates. IEEE Electron Device Letters 27(9):713–715
44. Palacios T, Rajan S, Chakraborty A, Heikman S, Keller S, DenBaars SP, Mishra UK (2005) Influence of the dynamic access resistance in the gm and fT linearity of AlGaN/GaN HEMTs. IEEE Trans Electron Device 52(10):2117–2123
45. DiSanto DW, Bolognesi CR (2006) At-bias extraction of access parasitic resistances in AlGaN/GaN HEMTs: impact on device linearity and channel electron velocity. IEEE Trans Electron Device 53(12):2914–2919

46. Trew RJ, Liu Y, Bilbro GL, Kuang W, Vetury R, Shealy JB (2006) Nonlinear source resistance in high-voltage microwave AlGaN/GaN HFETs. IEEE Trans Microw Theory Tech 54(5):2061–2067
47. Shinohara K, Regan D, Corrion A, Brown D, Tang Y, Wong J, Candia G, Schmitz A, Fung H, Kim S, Micovic M (2012) Self-aligned-gate GaN HEMTs with heavily-doped n^+-GaN ohmic contacts to 2DEG. In: Proceedings of the IEEE international electron devices meeting, pp 617–620
48. Lee DS, Wang H, Hsu A, Azize M, Laboutin O, Cao Y, Johnson W, Beam E, Ketterson A, Schuette M, Saunier P, Palacios T (2013) High linearity nanowire channel GaN HEMTs. In: Device Research Conference (DRC) 71st Annual, Notre Dame pp 195–196
49. Greenberg DR, del Alamo JA (1996) Nonlinear source and drain resistance in recessed-gate heterostructure field-effect transistors. IEEE Trans Electron Device 43(8):1304–1306
50. Lee DS (2014) Deeply-scaled GaN high electron mobility transistors for RF applications. Doctoral dissertation, Massachusetts Institute of Technology
51. Motayed A, Sharma A, Jones KA, Derenge MA, Iliadis AA, Mohammad SN (2004) Electrical characteristics of AlxGa1 − xN Schottky diodes prepared by a two-step surface treatment. J Appl Phys 96(6):3286–3295
52. Lee J-G, Park B-R, Cho C-H, Seo K-S, Cha H-Y (2013) Low turn-on voltage AlGaN/GaN-on-Si rectifier with gated ohmic anode. IEEE Electron Device Lett 34(2):214–216
53. Yao Y, Zhong J, Zheng Y, Yang F, Ni Y, He Z, Shen Z, Zhou G, Wang S, Zhang J, Li J, Zhou D, Zhisheng W, Zhang B, Liu Y (2015) Current transport mechanism of AlGaN/GaN Schottky barrier diode with fully recessed Schottky anode. Jpn J Appl Phys 54:011001
54. Hashizume T, Ootomo S, Oyama S, Konishi M, Hasegawa H (2001) Chemistry and electrical properties of surfaces of GaN and GaN/AlGaN heterostructures. J Vac Sci Technol, B 19(4): 1675–1681
55. Kim JH, Choi HG, Ha M-W, Song HJ, Roh CH, Lee JH, Park JH, Hahn C-K (2010) Effects of nitride-based plasma pretreatment prior to SiNx passivation in AlGaN/GaN high-electronmobility transistors on silicon substrates. Jpn J Appl Phys 49:04DF05-1–04DF05-3
56. Dimitrov R, Tilak V, Yeo W, Green B, Kim H, Smart J, Chumbes E, Shealy JR, Schaff W, Eastman LF, Miskys C, Ambacher O, Stutzmann M (2000) Influence of oxygen and methane plasma on the electrical properties of undoped AlGaN/GaN heterostructures for high power transistors. Solid-State Electron 44(8):1361–1365
57. Ha WJ, Chhajed S, Oh SJ, Hwang S, Kim JK, Lee J-H, Kim K-S (2012) Analysis of the reverse leakage current in AlGaN/GaN Schottky barrier diodes treated with fluorine plasma. Appl Phys Lett 100(13):132104-1–132104-4

Chapter 7
Deep-Level Characterization: Electrical and Optical Methods

Andrew M. Armstrong and Robert J. Kaplar

7.1 Introduction

The most direct method for measuring the influence of defects on HEMTs is to examine how the I_{ds} changes under a given bias condition due to thermally or optically stimulated transitions of electrons and holes into or out of deep levels. The primary effect of changing the occupancy of a deep-level defect on HEMT operation is the formation of a local space-charge that acts like a floating gate. Filling a defect with excess electrons creates a local negative potential that acts to partially pinch-off the channel and reduce I_{ds}. Conversely, electron emission from a defect state makes the local potential more positive and increases I_{ds}. Thus, defect states act to self-bias the HEMT and lead to instability in device operation as the occupancy of deep levels changes under dynamic operating conditions, such as switching or self-heating.

Analyzing the magnitude of variation in I_{ds} provides a straightforward means to assess the degree to which defects affect device performance, but mitigating defects requires understanding how they influence HEMT behavior. Explaining device behavior in the context of defects requires determining the electronic deep-level energy (E_t), deep-level concentration (N_t), and physical location of the corresponding defect in the HEMT. The energetic location of the defect in the bandgap impacts the temporal response of the device; e.g., shallow defects cause dynamic instabilities during switching, while deep defects influence the DC operating points of the device. The deep-level concentration determines the magnitude of variation imposed on I_{ds}. The physical location of a defect determines which aspects of device operation it

A.M. Armstrong · R.J. Kaplar (✉)
Sandia National Laboratories, Albuquerque, USA
e-mail: rjkapla@sandia.gov

A.M. Armstrong
e-mail: aarmstr@sandia.gov

M. Meneghini et al. (eds.), *Power GaN Devices*,
Power Electronics and Power Systems, DOI 10.1007/978-3-319-43199-4_7

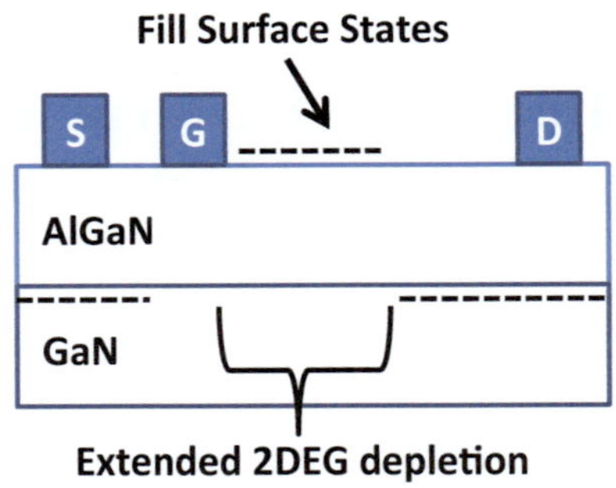

Fig. 7.1 Electron leakage from the gate electrode fills surface states. The excess negative charge on the AlGaN surface in between the gate and drain depletes the underlying 2DEG. The charged surface states are referred to as a "virtual gate"

impacts; e.g., defects under the gate impact V_{th}, while defects between the gate and drain impact R_{on}.

The major phenomena of degraded performance in high-voltage GaN HEMTs that are associated with deep levels are dispersion [1] and current collapse [2]. The term dispersion refers to depressed HEMT *I–V* characteristics with fast pulsing compared to DC conditions. This is often ascribed to the role of AlGaN surface states in creating a so-called virtual gate effect illustrated in Fig. 7.1. The virtual gate model posits that during device pinch-off, the large potential difference between the gate and drain electrodes allows electrons from the gate to tunnel into the AlGaN barrier. There, the electrons can be captured by the ionized surface donors that supplied the free carriers constituting the 2DEG. By charge conservation, a reduction of ionized donors at the bare AlGaN surface must be met with a local reduction in 2DEG density (n_s) resulting in reduced channel conductivity. If the emission rate of the surface donors is much slower than the switching frequency driving the device, I_{ds} lags the gate voltage, an effect known as gate lag. Key aspects of dispersion are that the phenomenon becomes apparent with fast pulsing of the gate and is ascribed to trapping at surface donors.

Current collapse (also termed dynamic on-resistance) describes a degradation of *I–V* characteristics after a large bias is applied between the source and drain electrodes. This effect is attributed to the charging of deep levels within the AlGaN barrier or GaN buffer by electrons that are impact-ionized from the 2DEG channel. The loss of carriers in the channel reduces the conductivity and thus the maximum achievable drain current. The drain current will not fully recover until the deep levels surrounding the channel emit their trapped electrons. Thus, the drain current lags the drain voltage, an effect known as drain lag. Note that current collapse manifests itself with large drain–source V_{ds} bias and is thought to involve deep levels in the GaN buffer. Current collapse is no longer an issue in RF GaN HEMTs, but it still persists in high-voltage GaN power HEMTs. Standing off several hundred or even several thousand volts between the source and drain often requires the intentional inclusion of deep-level defects in the GaN buffer region to suppress source-to-drain leakage. These buffer defects can also trap charge generated due to hot carriers scattering from the channel [3] or due to gate leakage [4].

Further, very deep traps in III-N materials may have very long time constants, and the filling of such traps may shift the parametric characteristics of the device, since the emission time is so long compared to the switching time that the shift is effectively DC [4, 5]. Such parametric shifts are reliability concerns analogous to the bias-temperature effects observed in silicon CMOS, since they may cause the device to drift away from the DC bias point for which the circuit that they are a part of was designed [6].

Characterizing all of these aspects of defect activity in GaN HEMTs requires a panoply of spectroscopy techniques. Measuring I_{ds} transients under different gate and drain bias conditions provides information on the lateral position of a defect, i.e., under the gate or in the access region. Lateral spatial resolution can also be achieved by measuring variation in V_{gs} or V_{ds} at fixed I_{ds}. Measuring gate-to-drain capacitance (C_{gd}) transients as a function of gate bias determines the vertical location of defects, i.e., barrier, channel, or buffer layer. The spectroscopy methods of DLTS and DLOS can be applied to either I_{ds} or C_{gd} to quantify E_t and N_t of the observed deep levels.

The remainder of this chapter describes the application of all of these techniques to GaN HEMTs. The fundamentals of deep-level transient spectroscopy (DLTS) and deep-level optical spectroscopy (DLOS) are reviewed, as they are the most often used methods to characterize defects in GaN HEMTs. The application of DLTS and DLOS to I_{ds} and C_{gd} for GaN HEMTs is then described to assess the location of defects and their impact on devices.

7.2 Fundamentals of DLTS and DLOS

This section describes the use of DLTS and DLOS to characterize the electrical and optical properties of deep-level defects. Capacitance-mode DLTS (C-DLTS) and capacitance-mode DLOS (C-DLOS) are reviewed first, followed by current-mode DLTS (I-DLTS) and current-mode DLOS (I-DLOS).

7.2.1 C-DLTS

C-DLTS is a well-known technique that is sensitive to thermally stimulated capacitance transients from deep levels as deep as ~1 eV from the conduction band edge. Thermally stimulated transitions of electrons and holes into or out of deep levels located in the depletion region of a semiconductor junction produce capacitance transients. For the simple case of an n-type Schottky diode, majority-carrier electron emission to the conduction band from a deep level is followed by the removal of the free electron from the depletion region by the built-in electric field, as shown in Fig. 7.2. This process leaves behind a defect that now has a more positive electric charge. The net increase in space-charge in the depletion region is compensated by the incursion of the free electron tail at the edge of the junction. The resultant reduction in the depletion width (*d*) can be measured as an increase in depletion capacitance. It is important to note that C-DLTS is also applicable to p–n junctions and

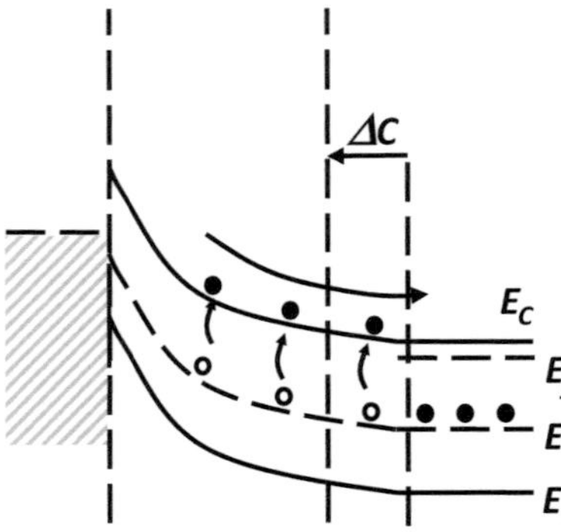

Fig. 7.2 Electron emission from defects in a depletion region. The electrons are swept out of the depletion region, leaving behind a fixed positive charge on the defects. This increase in space-charge causes the depletion region to contract, which is measured as an increase in the junction capacitance

minority-carrier processes; however, only majority-carrier processes in an n-type semiconductor will be considered here. It is further noted that from the point of view of the net charge within a region of a semiconductor device and the resulting electrical transients, it can in some cases be difficult to distinguish between emission of one carrier type and capture of the other carrier type (e.g., electron emission vs. hole capture) unless simplifying assumptions are made (e.g., assuming that processes involving holes are negligible); the device structure under consideration and the bias conditions used usually provide reasonable confidence in such assumptions. Additionally, high electric fields such as those that occur in high-voltage HEMTs may influence the relative dominance of emission versus capture [6, 7].

The electron emission rate from a deep level will influence the characteristic time constant associated with a capacitance transient. Such capacitance transients are readily measurable, so the emission rates of deep levels within a depletion region can be experimentally determined. Relating emission rates to the physical properties of a deep level allows one to characterize deep levels using depletion capacitance methods. From detailed balance considerations and neglecting degeneracy, the thermal electron emission rate of a deep level (e_{th}) can be expressed as

$$e_{th}(T) = \sigma_{th} v_{th} N_C \exp\left(\frac{E_t - E_C}{kT}\right) \tag{7.1}$$

where T is the absolute temperature, σ_{th} is the thermal carrier capture cross section, N_C is the conduction band density of states, E_C is the conduction band energy minimum, and k is the Boltzmann constant.

For the case of a fully occupied acceptor-like deep level in a depletion region at large reverse bias, the space-charge in the depletion region will evolve in time as $q[N_d - n_t(t)]$, where $n_t(t) = N_t \exp(-e_{th}t)$ is the concentration of occupied traps, N_t is the total trap concentration, and N_d is the net concentration of ionized dopants. For $N_t \ll N_d$, the capacitance can be expressed as

$$\frac{\Delta C(t)}{C_0} = \frac{N_t}{2N_d} \exp\left(-e_{th}t\right) \tag{7.2}$$

where C_0 is the final (steady-state) value of the capacitance and ΔC is the amplitude of the capacitance transient. This is the basis of depletion capacitance methods for deep-level spectroscopy.

Determining the time constant and amplitude of a capacitance transient yields information about E_t and N_t. The DLTS measurement begins with the depletion region under a quiescent reverse bias V_r, and the traps residing therein are assumed to be empty. A fill pulse bias V_f is applied for a time t_f to collapse the depletion region and bring free electrons in proximity of the empty traps; capture ensues. When the fill pulse is removed and the free carriers retract, thermal emission of the trapped electrons produces an exponential capacitance transient with time constant $\tau_{th} = e_{th}^{-1}$, which from Eq. (7.1) depends on the temperature T and both E_t and σ_t. As T increases, τ_{th} decreases and the DLTS apparatus processes the transient into a signal that peaks at temperature T_{max} when τ_{th} equals a preset value τ_{ref}. Traps with different E_{th} or σ_{th} exhibit different T_{max} for a given τ_{ref}. By using several associated values of τ_{ref} and T_{max}, one constructs an Arrhenius plot of $\ln\left(\tau_{ref} T_{max}^2\right)$ versus T_{max}^{-1}, from which E_{th} and σ_{th} are extracted via the slope and y-intercept, respectively. The deep-level concentration is calculated by evaluating Eq. (7.2) at large t. More details regarding the DLTS measurement process and instrumentation can be found elsewhere [8].

Practical limits on sample temperature as well as transient observation time typically limit DLTS sensitivity to deep levels within ~1 eV of a band edge. DLOS must be used to examine deep levels lying deep in the GaN bandgap, as discussed next.

7.2.2 C-DLOS

DLOS measures the optical characteristics of a deep level such as the optical ionization energy E_o and Franck–Condon energy d_{FC} from the spectral dependence of the optical cross section (σ^o), while N_t can be found from Eq. (7.2) similar to DLTS. The energy E_o is the energy required for an absorbed photon to promote an electron from a deep level to the conduction band for the case when the photoemission process is not assisted by phonons. Since the excitation is now optical rather than thermal, deep-level defects as deep as ~6 eV below E_c can be observed using a xenon lamp. Again, we restrict discussion of DLOS to the case of an n-type semiconductor.

DLOS is similar to DLTS except now thermal emission rates are assumed to be negligible compared to optical emission rates. Monochromatic illumination is used to excite deep-level emission. Rather than scanning T as in DLTS, for DLOS one scans the incident photon energy ($h\nu$) by using a monochromator coupled to a broadband light source. The capacitance transient for each value of $h\nu$ can then be recorded and analyzed to determine the optical emission rate (e^o) that is defined as σ_o multiplied by the incident photon flux (Φ).

To extract $\sigma_o(h\nu)$, one takes the time derivative of the photocapacitance transients $C(t)$ near $t = 0$, i.e., just at the beginning of the illumination period. Assuming that the sample temperature is low enough or that the occupied deep levels are sufficiently far (>1 eV) from the band edges, thermal processes can be neglected. It then follows that the deep levels with $E_0 > 1$ eV are fully occupied. In this case, the spectral dependence of σ^o is given by

$$\sigma^o \propto \frac{1}{\Phi}\frac{dC}{dt}\Big|_{t=0}. \tag{7.3}$$

The σ^o data are then fit to a theoretical model to determine the value of E_0. Many models exist that treat the general case of strong defect-lattice coupling in various ways. One model that is often used is that of Pässler [9], and this is the model that is used in all of the studies recounted here. More details regarding the DLOS measurement process and associated instrumentation can be found elsewhere [10].

7.2.3 Applicability of C-DLTS and C-DLOS to HEMTs

C-DLTS can be readily applied to HEMTs using the gate-to-drain Schottky diode. The only caveat is the signal-to-noise ratio of the C-DLTS apparatus. C-DLTS transients are typically less than 10 % of the total depletion capacitance. Thus, the gate electrode must have sufficiently large area to produce a ΔC that can be resolved experimentally. This is not typically a problem for power GaN HEMTs due to their large area compared to RF GaN HEMTs.

The applicability of optical spectroscopies to GaN HEMTs should be considered in terms of metal coverage on the surface of the device and the optical transmission of the substrate. Backside illumination is not ideal because the SiC substrates typical of RF HEMTs absorb in the UV and block any light from exciting defects in the AlGaN; the case is worse for Si substrates typical of power HEMTs, which absorb both visible and UV light, making both GaN and AlGaN inaccessible to backside illumination. Thus, C-DLOS application to HEMTs typically requires measurements to be performed using front-side illumination. This can be done in one of two ways. Conventional GaN HEMTs use opaque metal gates that will block the incident illumination. Nonetheless, C-DLOS is possible because the gate electrode is typically very thin. Incident light can be scattered at the surface and internally reflected several times in the AlGaN/GaN heterostructure, which affords multiple optical passes below the gate electrode. Additionally, AlGaN/GaN Schottky diodes can be fashioned from the HEMT epitaxy with semitransparent Schottky contacts on the surface. Semitransparent Ni contacts can readily be formed on AlGaN, which is very similar to the typical Ni/Au Schottky contacts used in GaN HEMTs. Thus, C-DLOS measurements performed on AlGaN/GaN heterostructures are directly applicable to HEMTs because they have identical semiconductor structure and similar metal/semiconductor interfacial electronic properties.

An important aspect of both C-DLTS and C-DLOS is that they have innate depth resolution. C-DLTS and C-DLOS only observe defects within the depletion region under the Schottky electrode. For the case of HEMTs, the extent of this depletion region under the gate is given by the usual parallel-plate capacitor approximation $C = A\varepsilon/d$, where A is the junction area and ε is the semiconductor dielectric constant (note that this neglects the quantum capacitance of the 2DEG). The depletion depth increases with more negative gate bias. Thus, the applied bias controls which region of the device under the gate will be probed by C-DLTS or C-DLOS. For example, if the bias applied to the gate is larger (more positive) than V_{th}, the 2DEG is accumulated, and the depletion region is primarily confined to the AlGaN barrier. When the gate bias is much smaller (more negative) than V_{th}, the 2DEG is pinched-off. In this case, the depletion depth is much larger than the AlGaN barrier thickness, so d is primarily constituted by the GaN spacer and buffer layers.

These arguments can be quantified. The relative contribution to the total depletion capacitance from a particular portion of the depletion region is

$$\frac{x_2^2 - x_1^2}{d^2}, \tag{7.4}$$

where $x = 0$ at the surface and the region of interest is bounded by depths $0 < x_1 < x_2 < d$ [11]. When the 2DEG is accumulated, the depletion depth is coincident with the AlGaN/GaN heterointerface where the channel resides, typically ~20 nm below the surface. Thus, the AlGaN barrier and GaN channel dominate the depletion capacitance, and C-DLTS and C-DLOS will be primarily sensitive to any defects residing in these regions. Under pinch-off, the depletion depth typically extends several microns below the 2DEG channel. The AlGaN barrier contributes less than 0.01 % to the total depletion capacitance, while the GaN spacer and buffer regions contribute the remaining 99.99 %. This means that C-DLTS and C-DLOS sensitivity to AlGaN barrier defects is reduced by 10,000× when in pinch-off compared to accumulation. This strong bias dependence of C-DLOS sensitivity can be exploited to discriminate among deep-level defects within the AlGaN barrier and GaN channel versus the GaN spacer and GaN buffer [12].

7.2.4 I-DLTS and I-DLOS

I-DLTS and I-DLOS measurements typically proceed by measuring the emission rate associated with the recovery of I_{ds} with the transistor in the on-state after application of a large positive V_{ds} bias, a large negative V_{gs} bias, or both. Spectroscopic analysis of I-DLTS is largely the same as for C-DLTS [13], and likewise, spectroscopic I-DLOS analysis is similar to C-DLOS [14]. The respective deep-level defect emission rates can be attained by substituting I_{ds} for C. Deep-level parameters E_{th} and σ_{th} or E_o and d_{FC} can then be determined in the same way as described above.

Determining the spatial location of the defects that cause I_{ds} transients is important. Knowing where defects reside in the device helps to understand how they will influence device behavior for a given operating condition. Knowledge of which layer

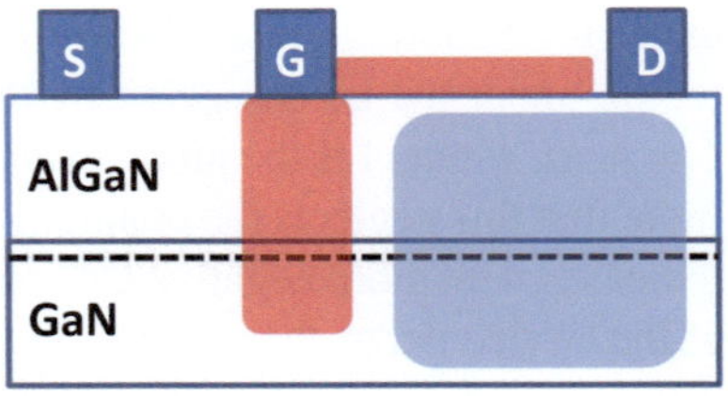

Fig. 7.3 Regions where defects are filled with electrons for various fill pulse conditions. The *red boxes* correlate to fill pulses with ($V_{gs} < V_{th}$, $V_{ds} \geq 0$ V), and the *blue box* correlates to fill pulses with ($V_{gs} > V_{th}$, large V_{ds})

a particular defect resides in enables rational strategies to mitigate or eliminate their influence through optimized crystal growth, device design, and device processing. However, unlike C-DLTS and C-DLOS, neither I-DLTS nor I-DLOS provide inherent spatial sensitivity because I_{ds} is the same everywhere throughout the device.

Nonetheless, the physical location of defects can be ascertained. One method is to use different combinations of V_{gs} and V_{ds} stress conditions to fill defects in different regions of the device, i.e., to use different fill pulse conditions [14–16]. Another method is to use different operating bias conditions that are sensitive only to defects in certain regions of the device, i.e., to use different on-state conditions [17].

Varying the HEMT fill pulse causes electron trapping in different regions of the HEMT. Application of a strongly negative $V_{gs} < V_{th}$ fill pulse can cause electrons to tunnel from the gate electrode and fill defects located under the gate region, in the surface access region between the gate and the drain, or both [15, 16]. Application of a strongly positive V_{ds} bias along with a strongly negative V_{gs} bias emphasizes access region trapping due to enhanced electron tunneling in the direction of the drain. On the other hand, application of strongly positive V_{ds} with $V_{gs} > V_{th}$ causes hot electrons to scatter out of the channel and become trapped primarily in the AlGaN barrier or GaN buffer layers between the gate and the drain [2, 14, 16]. Therefore, defect states that are evident only for the application of fill pulses with the channel pinched-off (strongly negative V_{gs}) are likely to be related to the surface in the access region, while defect states that are evident only for the application of fill pulses with the channel open ($V_{gs} \sim 0$ V) are likely to be related to the AlGaN barrier or GaN buffer. These concepts are summarized in Fig. 7.3.

Recently, I-DLTS and I-DLOS with lateral spatial resolution have been developed to differentiate deep levels under the gate from those located in the access region of HEMTs [17]. I-DLTS and I-DLOS are primarily sensitive to defects located in the access region when the device is biased to produce low mutual transconductance (g_m) and high output conductance (g_o), such as when operating in the triode regime with V_{gs} that is large relative to V_{th} and V_{ds}. In this case, I-DLTS and I-DLOS analyses are greatly simplified by operating under the condition of constant I_{ds}. If I_{ds} and V_{gs} are held fixed, the drain voltage of the intrinsic HEMT is constant. The change in gate-to-drain resistance $\Delta R_{gd}(t)$ due to defect emission in the access region is then measured as the change $\Delta V_{ds}(t)$ required to maintain constant I_{ds}. Then, R_{gd} (or V_{ds}) takes the place of I_{ds} when determining deep-level emission rates, and the areal defect density (D_t) in the access region is calculated as [17]

$$D_t = \left(\frac{n_s^2}{n_s - \frac{L'}{qW\mu(-\Delta R_{gd})}} \right) \tag{7.5}$$

where n_s is the 2DEG sheet density, W is the gate width, μ is the channel mobility, and L' is the physical length of the virtual gate extension.

Conversely, I-DLTS and I-DLOS are primarily sensitive to deep levels located under the gate when bias conditions produce a large g_m and a small g_o, such as in saturation mode [17]. In saturation, the influence of defects in the active region on I_{ds} is negligible. Changes in R_{dg} do not impact I_{ds} because the output resistance is already very large. If I_{ds} and V_{ds} are held fixed, the shift in threshold voltage $\Delta V_{th}(t)$ due to deep-level defect emission under the gate is equal to the change in gate voltage $\Delta V_{gs}(t)$ required to maintain constant drain current. Now, V_{th} (or V_{gs}) takes the place of I_{ds} when determining deep-level emission rates, and D_t under the gate is calculated as

$$D_t = \frac{2\varepsilon \Delta V_{th}}{qd} \tag{7.6}$$

where the defects are assumed to be located in the AlGaN barrier.

Maintaining constant I_{ds} by dynamic feedback control of V_{ds} or V_{gs} can require sophisticated circuitry. For this reason, it may prove more convenient to investigate defects under the gate using C-DLOS and C-DLTS rather than establishing feedback control of V_{ds} or V_{gs} for I-DLTS and I-DLOS under constant-I_{ds} conditions.

7.3 Application of DLTS and DLOS to GaN HEMTs

This section reviews multiple studies applying C-DLTS, C-DLOS, I-DLTS, and I-DLOS to GaN HEMTs. Confident assignment of deep-level defects to various regions of the device using DLTS and DLOS is demonstrated. DLTS and DLOS measurements of GaN HEMTs with different gate metals and surface passivation processes confirm that defects in different locations can be selectively probed as a function of fill pulse conditions [15, 16]. Comparison of DLOS measurements of GaN HEMTs and GaN thin films demonstrates the ability of C-DLOS to differentiate between AlGaN barrier- and GaN buffer-related defects [12, 18]. C-DLOS and I-DLOS measurements of the same GaN HEMTs also show that defects in the GaN buffer can influence V_{th} and R_{on} by trapping under the gate and trapping in the access region [18]. The lateral spatial selectivity of constant-I_{ds} mode I-DLTS and I-DLOS is also substantiated [17].

7.3.1 Using Fill Pulses to Spatially Locate Traps

A direct way to establish that different filling pulses can selectively prime defects either under the gate or in the access region of GaN HEMTs is to compare the C-DLTS of HEMTs with different gate electrodes and surface passivation layers. Such a study was conducted for GaN HEMTs and indeed validated the utility of fill pulses to distinguish defects in different regions of the device [15].

In the study, three sets of HEMTs were fabricated [15]. Set A had devices with ITO gates and silicon nitride passivation, with $V_{th} \sim -1.5$ V. Set B had devices with Ni/Au gates and silicon nitride passivation, with $V_{th} \sim 0$ V. Set C had Ni/Au gates and no surface passivation, with $V_{th} \sim -0.5$ V.

C-DLTS analysis of devices from Set A using a ($V_{gs} = -4$ V $< V_{th}$, $V_{ds} = 10$ V) fill pulse revealed a defect state with activation energy of 0.63 eV. Based on the discussion above, a deep level observed with such a fill pulse is likely to exist either under the gate or at the passivation/surface interface in the access region. Figure 7.4 shows pulsed I_{ds}–V_{gs} data for Set A using three different fill pulses. The ($V_{gs} = 0$ V, $V_{ds} = 0$ V) quiescent pulse was used as a control to produce minimal defect trapping. The ($V_{gs} = -4$ V, $V_{ds} = 0$ V) fill pulse was used to emphasize defect filling under the gate, and the ($V_{gs} = -4$ V, $V_{ds} = 10$ V) pulse was used to emphasize defect filling in the access region. The large shift in V_{th} indicates definitive trapping under the gate. However, increasing V_{ds} during the fill pulse did not change R_{gd} (i.e., dI_{ds}/dV_{gs}), suggesting that the 0.63 eV deep level is not related to access region traps. This conclusion was validated by analysis of Set B, where only the gate contact was different from the devices in Set A. Set B exhibited no significant trapping, providing conclusive evidence that the 0.63 eV deep level is due to trapping under the gate and is not related to surface states in the access region.

C-DLTS measurements of devices from Set C found a trap state with an activation energy of 0.099 eV using a ($V_{gs} = -4$ V $< V_{th}$, $V_{ds} = 10$ V) fill pulse. Again, the corresponding defect could be attributed to either a surface state or a defect under the gate based on the fill pulse used. Figure 7.5 shows pulsed I_{ds}–V_{gs} data for Set C using the same three fill pulses as used for Set A listed above. No shift in V_{th} was observed for Set C, which suggests that the 0.099 eV trap is not located under the gate. However, a large change in R_{gd} was evident that increased with increasing V_{ds} bias. This behavior points to a surface-state-related defect in the active region. To corroborate this ascription, comparison was again made with Set B, which shared the same gate processing but included surface passivation compared to the bare surface in Set C. As noted above, Set B did not suffer any significant trapping effects, which validates the conclusion that the defect giving rise to the 0.099 eV trap state is located at the surface in the access region. Analysis of the thermal dependence of I_{ds} transients resulting from the ($V_{gs} = -4$ V, $V_{ds} = 10$ V) fill pulse revealed a defect state with a small activation energy of 0.099 eV but with unusually long time constants of ~100 ms. It was found that e_{th} for the 0.099 eV defect had an exponential dependence of $1/T^3$ typical of conduction hopping along a surface rather than a $1/T$ dependence that is expected for defects in a homogeneous crystal matrix. This finding

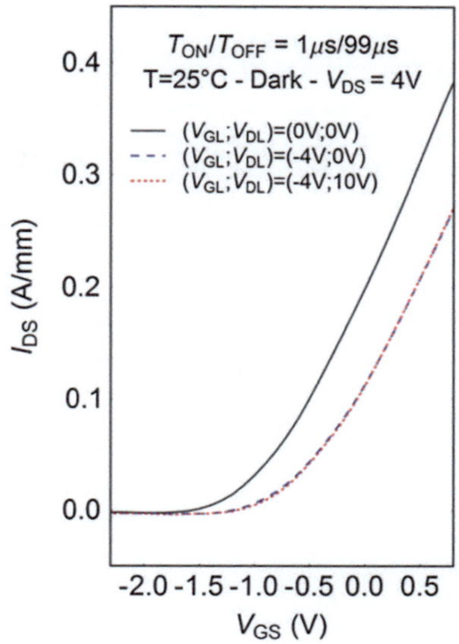

Fig. 7.4 Pulsed I_{ds}–V_{gs} data for an AlGaN/GaN HEMT with an ITO gate and silicon nitride surface passivation (Set A from text). Note that the response to the $V_{gs} = -4$ V fill pulse is invariant to V_{ds}, suggesting that the defects causing reduced I_{ds} are located under the gate and are not located in the access region. From Ref. [15]

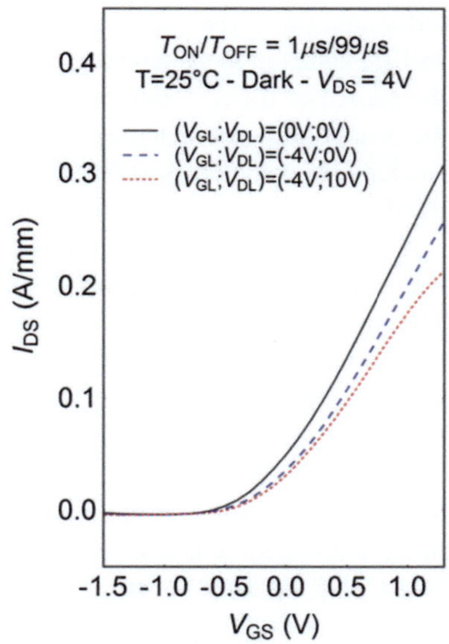

Fig. 7.5 Pulsed I_{ds}–V_{gs} data for an AlGaN/GaN HEMT with a Ni/Au gate and without surface passivation (Set C from text). Note that the response to the $V_{gs} = -4$ V fill pulse depends strongly on V_{ds}, suggesting that the defects causing reduced I_{ds} are located in the access region and not under the gate. From Ref. [15]

further supports the attribution of deep levels to surface defects when using a fill pulse with $V_{gs} < V_{th}$ and strongly positive V_{ds}.

Prior studies have established that the spatial location of defects can also be determined using on-state fill pulses [2, 3, 14, 16]. Early work in GaN HEMTs [2, 19] reported a severe increase in R_{on} when applying a large V_{ds} bias with $V_{gs} > V_{th}$ and the channel accumulated. The lack of significant gate stress using on-state fill

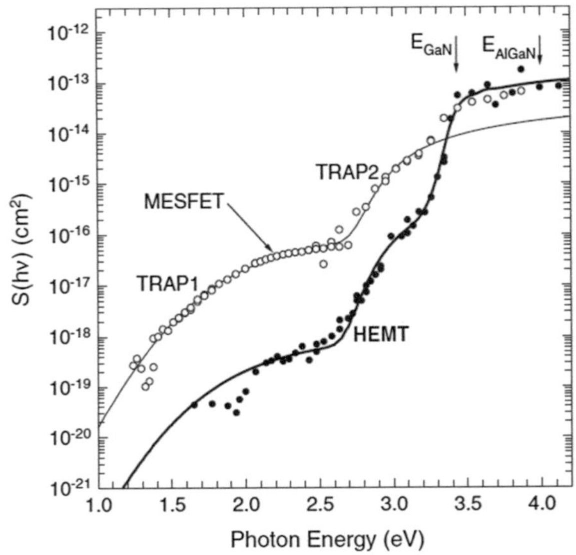

Fig. 7.6 I-DLOS spectra of a GaN HEMT and a GaN MESFET. The similar deep-level spectra indicate similar defects in both devices; these were attributed to the carbon-doped GaN buffer in each. From Ref. [14]

pulses discounts the role of gate leakage in filling defects located under the gate or surface states in the access region. In these studies, the channel was presumed to be the source of trapped electrons. The lateral location of the responsible defects was therefore likely between the gate and the drain because this is the lateral region where carriers are sufficiently accelerated to escape the channel. The vertical location of the defects was considered to be the adjacent AlGaN barrier or GaN buffer.

Comparison of DLOS of GaN HEMTs [2] and GaN MESFETs confirmed that carbon doping in the GaN buffer was indeed responsible for the large increase in R_{on}. The current collapse was the characteristic phenomenon reported for GaN MESFETs that required hours for I_{ds} recovery at room temp [20], indicating the need for optical spectroscopy to fully characterize the responsible defect states. Figure 7.6 shows the I-DLOS spectra of GaN HEMT and MESFET devices subject to current collapse [14]. I-DLOS spectra of the HEMT devices are qualitatively similar to those of the MESFET, providing strong evidence that the two primary deep levels indeed exist in the GaN layer. Carbon-related defects were suspected as the microscopic origin of the 2.85 eV level because the I-DLOS spectra were similar to previous photoluminescence excitation results reported for GaN:C [21], and calculated defect density was found to track linearly with carbon impurity concentration.

It should be noted that defects located under the gate region can also exist under the access region. Thus, the same defects can appear in both off-state and on-state fill pulse stresses and can impact both V_{th} and R_{gd}. Such defect activity has been reported using a combination of C-DLOS and I-DLOS applied to an AlGaN/GaN HEMT with a GaN:C buffer that exhibited both V_{th} and R_{gd} variations due to defect trapping [18]. Figure 7.7 shows C-DLOS spectra taken on HEMTs using the gate electrode [18], which found the same 1.8 and 2.85 eV deep

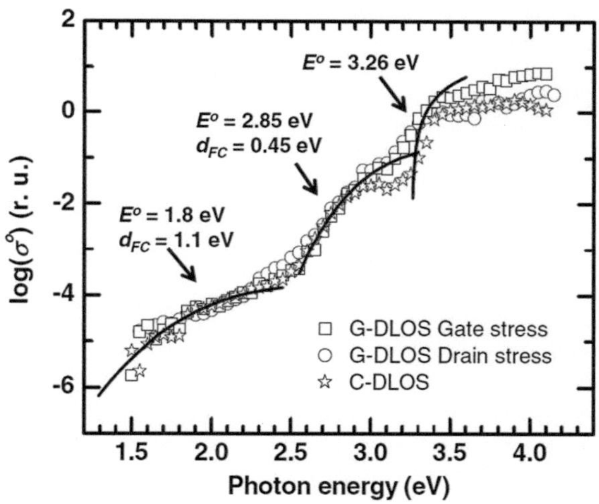

Fig. 7.7 C-DLOS and I-DLOS (marked as G-GLOS, where the G for conductivity is analogous to the I for current) spectra of a GaN HEMT. The C-DLOS spectrum is sensitive only to defects under the gate, while the I-DLOS spectra are sensitive only to defects in the access region. The similarity of the DLOS spectra shows that defects under the gate can also exist in the access region. From Ref. [18]

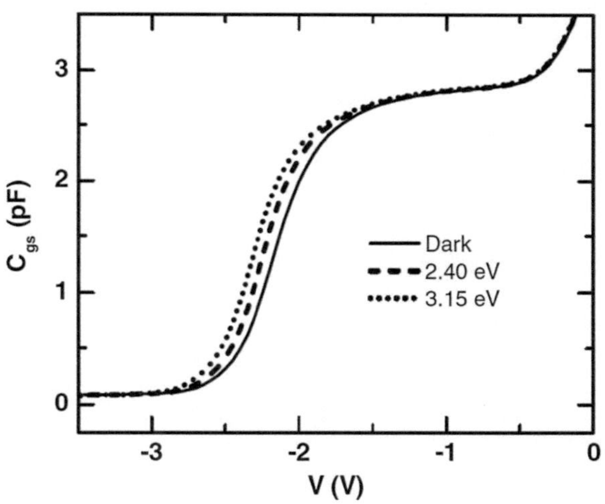

Fig. 7.8 Capacitance-voltage measurements of an AlGaN/GaN HEMT in the dark and under illumination. The lateral shifts in the curves with illumination indicate that the defects observed in Fig. 7.7 influence V_{th}. From Ref. [18]

levels reported by Klein et al. previously [2, 3, 14]. Shifts in V_{th} due to these defect levels were measured by capacitance–voltage sweeps, while sub-bandgap illumination was used to excite the deep levels optically, as shown in Fig. 7.8. This confirmed that carbon-related defects in the GaN buffer impact V_{th} as well as R_{on}. Figure 7.7 also shows I-DLOS measurements of the same device biased in the triode region to be sensitive only to trapping in the access region. Both gate stress ($V_{gs} < V_{th}$, $V_{ds} = 20$ V) and drain stress ($V_{gs} > V_{th}$, $V_{ds} = 20$ V) produced identical I-DLOS spectra that were also very similar to the C-DLOS spectrum. The conclusions from these data are that defects under the gate can also exist between the gate and the drain (in the GaN buffer in this case) and can impact R_{on}.

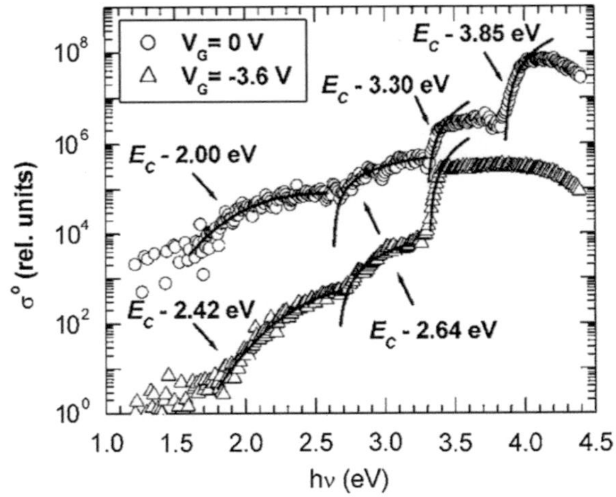

Fig. 7.9 C-DLOS spectra of an AlGaN/GaN heterostructure. The bias dependence of the spectra demonstrates the depth sensitivity of C-DLOS. The 3.85 eV deep level and the 4.0 eV AlGaN band edge confirm C-DLOS sensitivity to the AlGaN barrier at zero gate bias. Conversely, the absence of AlGaN features at reverse bias confirms exclusive sensitivity to the GaN buffer. From Ref. [12]

7.3.2 Using Measurement Bias to Spatially Locate Traps

The electrical bias used to facilitate C-DLOS or I-DLOS measurements can also determine the vertical location of defects in HEMTs. Figure 7.9 shows an example of bias-dependent C-DLOS performed on an AlGaN/GaN heterostructure that was formed from the same epitaxial structure as a fully processed HEMT. The heterostructure and HEMT contained a semi-insulating Fe-doped GaN buffer [12]. The spectral features of the C-DLOS measurements show a strong bias dependence. This is expected based on the discussion above. At 0 V, the 2DEG is accumulated, so C-DLOS is primarily sensitive to both the AlGaN barrier and the GaN channel. At −3.6 V bias, the 2DEG is depleted, so C-DLOS is primarily sensitive to the underlying GaN:Fe buffer layer. Therefore, defects that appear only in the 0 V C-DLOS spectrum can be ascribed to the AlGaN barrier, while defects only appearing in the −3.6 V spectrum can be attributed to the GaN:Fe layer, and defects appearing in both C-DLOS spectra are common to the GaN buffer and GaN channel regions.

C-DLOS sensitivity to the AlGaN barrier at 0 V is confirmed by spectral features above the bandgap energy of GaN. The C-DLOS spectrum at 0 V evidences saturation at 4 eV due to AlGaN band-edge absorption, and the 3.85 eV defect level is also necessarily related to the AlGaN barrier. This is definitive evidence that C-DLOS has requisite depth resolution to probe nanoscopic layers embedded in heterostructures and to distinguish nanoscopic layers from microns of surrounding material. The 2.00 eV defect level is also AlGaN-related due to its singular appearance at 0 V. Conversely, the 2.42 eV deep level only appears at −3.6 V, so it can be confidently ascribed to the GaN:Fe buffer layer in the HEMT. The 2.64 and 3.30 eV levels are common to both C-DLOS spectra, so the corresponding defects are located in the GaN channel and the GaN buffer layers. It is worth noting that defect levels appear at an energy near 3.3 eV in Figs. 7.6, 7.7, and 7.9. Given that these DLOS spectra

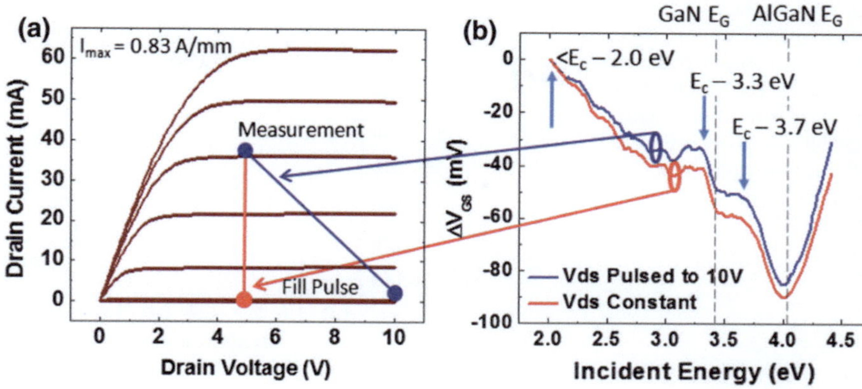

Fig. 7.10 I-DLOS of an AlGaN/GaN HEMT without surface passivation. I-DLOS was performed with the device in saturation to probe defects under the gate. The weak dependence of I-DLOS to V_{ds} for fill pulses with $V_{gs} < V_{th}$ indicates that defects in the access region do not contribute to the deep-level spectra when the device is biased in saturation. From Ref. [23]

were taken on samples grown independently, these data indicate that the ~3.3 eV defect level is quite common in GaN. Based on extensive study, the 3.3 eV defect level is most likely related to a carbon impurity [18, 22].

Layer spatial resolution has also been demonstrated using I-DLTS and I-DLOS [17, 23, 24]. Figure 7.10 shows I-DLOS measurements of an AlGaN/GaN HEMT. The measurements were performed under a saturated bias condition, and the V_{gs} (equivalently V_{th}) variation was measured in the constant-I_{ds} regime to be exclusively sensitive to defects below the gate electrode [23]. To validate exclusive gate-region sensitivity, measurements were performed on an AlGaN/GaN HEMT that was not passivated. Significant trapping by surface states in the access region is expected due to the lack of passivation. Fill pulse conditions of (V_{gs} = −8 V, V_{ds} = 5 V) or (V_{gs} = −8 V, V_{ds} = 10 V) were used to fill surface states. Only the deep-level occupancy in the access region is expected to vary for these fill pulse conditions because only V_{ds} was changed. Any I-DLOS sensitivity to deep levels arising from surface states could then be recognized because their magnitude in the I-DLOS spectra should be enhanced by the fill pulse with the larger V_{ds}. As expected, the I-DLOS spectra do not show significant increase in magnitude for the two fill pulses. Indeed, the fill pulse with the larger V_{ds} indicated a slightly reduced density of defects. The small differences in the I-DLOS spectra were attributed to variations in the source-to-drain resistance. This finding validates the exclusive sensitivity of I-DLOS to defects located under the gate when the measurement is performed in saturation mode.

I-DLOS measurements of the same AlGaN/GaN HEMT were also performed in the triode operating condition to probe surface states in the access region, as shown in Fig. 7.11. In this implementation of constant-I_{ds} I-DLOS, V_{gs} was fixed and ΔV_{ds} was measured to monitor changes in R_{gd} due to defect emission. A fill pulse of (V_{gs} = −8 V, V_{ds} = 10 V) was again used to fill surface defects. Significant ΔV_{ds} was

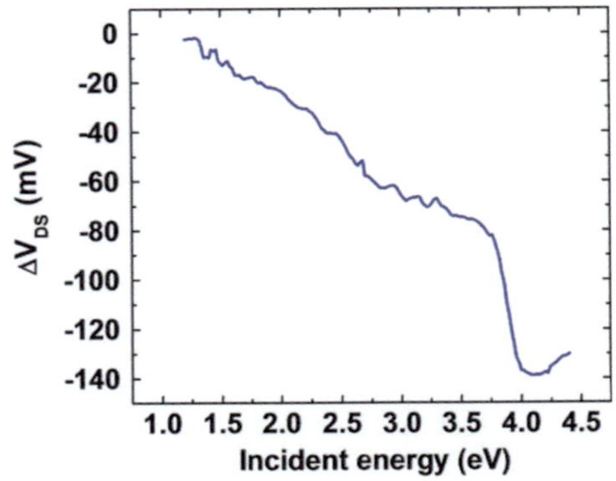

Fig. 7.11 I-DLOS measurement of an AlGaN/GaN HEMT without passivation. The I-DLOS was performed in the triode regime to enable sensitivity to defects in the access region. From Ref. [23]

measured, indicating the existence of surface states. Comparison of Figs. 7.10 and 7.11 substantiates the sensitivity of I-DLOS to access region defects when performed in the triode regime. The typical 3.3 eV defect level evident in the I-DLOS spectra in Fig. 7.10 (which measures defects only under the gate) is absent from the I-DLOS spectrum in Fig. 7.11 (which focuses on the AlGaN surface in the access region). As discussed above, the 3.3 eV defect level is commonly observed in GaN, so its absence in Fig. 7.11 corroborates the measurement's primary sensitivity to the AlGaN barrier. Thus, I-DLOS measurements performed in the triode region are demonstrated to have primary sensitivity to defects located in the access region.

7.3.3 Additional Methods to Measure Spatially Localized Traps

Drain current transient techniques may also be combined with physical characterization methods such as surface potential measurements to further refine the ability to determine where charge is trapped within the HEMT. One study of high-voltage HEMTs that correlated Kelvin force microscopy with slow drain current transients following electrical stress in the pinch-off state concluded that the thickness of the AlGaN barrier and the associated magnitude of the electric field near the gate edge largely determined where in the device charge was trapped; factors such as surface passivation and buffer doping were found to be of secondary importance in these devices [25]. Figure 7.12 shows the correlated drain current transient and surface potential measurements for a device with a thick (50-nm) AlGaN barrier, a carbon-doped buffer, and Al_2O_3-based surface passivation (Device A). Despite the surface passivation and the carbon-associated deep levels in the bulk GaN, this device shows a large change in surface potential with time, indicative of significant charge trapping during stress. In contrast, a device with a thinner (20 nm) barrier, no buffer doping, and no surface passivation (Device B) showed much less change in surface potential. This unexpected result was explained by the thicker barrier in Device A, which

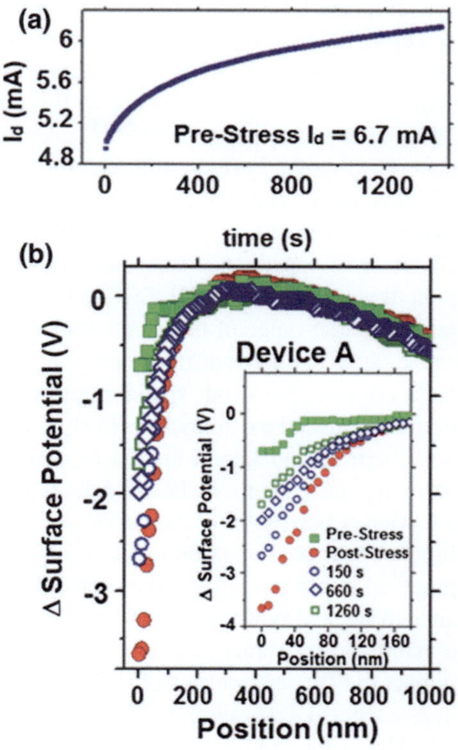

Fig. 7.12 I_{ds} transient following ($V_{gs} = -9$ V, $V_{ds} = 0$ V) stress in an AlGaN/GaN HEMT with a 50-nm AlGaN barrier, a carbon-doped GaN buffer, and Al_2O_3-based surface passivation (this device was termed "Device A" in the study). Position is measured from the drain side of the gate edge. The large change in surface potential near the gate edge is indicative of surface trapping. From Ref. [25]

resulted in a lower electric field near the gate edge and less tendency to inject electrons deep into the device.

Finally, it is noted that in order to achieve normally-off operation, many power HEMTs utilize a recessed gate to make V_{th} more positive; such devices often utilize gate dielectrics to limit gate leakage current [7]. Thus, charge trapping may occur both within the dielectric and at the dielectric–semiconductor interface for such MIS structures, and variations of the techniques described herein are applicable to the characterization of such structures. For example, constant-capacitance (CC) DLTS and DLOS have been utilized to determine interface state density in Al_2O_3/GaN MIS capacitors [26]. In this technique, the voltage across the MIS structure is adjusted such that the capacitance remains constant during the emission transient [27], analogous to constant-current DLTS/DLOS in HEMTs.

7.4 Conclusion

Optical and electrical characterization techniques such as DLTS and DLOS are powerful and effective methods that may be used to ascertain both the nature and spatial location of traps within AlGaN/GaN HEMTs for both RF and high-voltage power-switching applications. Due to its reliance on thermal emission from deep-level states, DLTS is most useful for shallow traps less than ~1 eV from the band edges. Conversely, owing to the optically stimulated emission inherent to DLOS, it is most useful for deeper states and thus finds great utility in the (Al)GaN materials system. Further, differentiation between thermal and optical energies may be ascertained, and defect-related parameters such as the Franck–Condon energy associated with lattice relaxation may be determined. While originally developed to characterize capacitance transients for simple one-dimensional structures such as Schottky and pn diodes, DLTS/DLOS of HEMTs has spurred the development of a much broader range of measurement techniques owing to the ability to utilize both capacitance and current transients for time constant characterization. Further, the ability to sample various spatial regions within the device due to the availability of both the gate and the drain to set the bias condition affords yet more flexibility. Thus, both vertical and lateral spatial discrimination of deep levels may be achieved, allowing one to identify whether defects reside within the GaN buffer/channel region, the AlGaN barrier, or at the surface of the device. While of high interest from the fundamental physics point of view, characterization of deep levels is also of great practical importance given that such defects influence many aspects of the performance of RF and power-switching HEMTs, such as dispersion, current collapse, and dynamic on-resistance; additionally, very deep levels with long time constants may impact the DC parametrics of the device. Thus, deep-level characterization techniques such as DLTS and DLOS are likely to remain vital to the robustness of III-N HEMTs in the coming years, and further refinements and enhancements of these methods are sure to be developed.

References

1. Mishra UK, Parikh P, Wu Y-F (2002) AlGaN/GaN HEMTs-an overview of device operation and applications. Proc IEEE 90:1022
2. Klein PB, Binari SC, Ikossi K, Wickenden AE, Koleske DD, Henry RL (2001) Current collapse and the role of carbon in AlGaN/GaN high electron mobility transistors grown by metalorganic vapor-phase epitaxy. Appl Phys Lett 79:3527
3. Klein PB (2002) Photoionization spectroscopy in AlGaN/GaN high electron mobility transistors. J Appl Phys 92:5498
4. DasGupta S, Sun M, Armstrong A, Kaplar RJ, Marinella MJ, Stanley JB, Atcitty S, Palacios T (2012) Slow detrapping transients due to gate and drain bias stress in high breakdown voltage AlGaN/GaN HEMTs. IEEE Trans Elec Dev 59:2115
5. Joh J, del Alamo JA (2011) A current-transient methodology for trap analysis for GaN high electron mobility transistors. IEEE Trans Elec Dev 58:132
6. Khalil SG, Ray L, Chen M, Chu R, Zhender D, Arrido AG, Munsi M, Kim S, Hughes B, Boutros K, Kaplar RJ, Dickerson J, DasGupta S, Atcitty S, Marinella MJ (2014) Trap-related parametric shifts under DC bias and switched operation life stress in power AlGaN/GaN HEMTs. Proceedings of 52nd IEEE IRPS, Paper CD.4

7. Kaplar RJ, Dickerson J, DasGupta S, Atcitty S, Marinella MJ, Khalil SG, Zehnder D, Garrido A (2014) Impact of gate stack on the stability of normally-Off AlGaN/GaN power switching HEMTs. Proceedings of 26th IEEE ISPSD, 209
8. Lang DV (1974) Deep-level transient spectroscopy: A new method to characterize traps in semiconductors. J Appl Phys 75:3023
9. Pässler R (2004) Photoionization cross-section analysis for a deep trap contributing to current collapse in GaN field-effect transistors. J Appl Phys 96:715
10. Chantre A, Vincent G, Bois D (1981) Deep-level optical spectroscopy in GaAs. Phys Rev B 23:5335
11. Blood P, Orton JW (1990) The electrical characterization of semiconductors: majority carriers and electron states. Academic Press, London (Chap. 7)
12. Armstrong A, Chakraborty A, Speck JS, DenBaars SP, Mishra UK, Ringel SA (2006) Quantitative observation and discrimination of AlGaN-and GaN-related deep levels in AlGaN/GaN heterostructures using capacitance deep level optical spectroscopy. Appl Phys Lett 89:262116
13. Mitrofanov O, Manfra M (2003) Mechanisms of gate lag in GaN/AlGaN/GaN high electron mobility transistors. Superlattices Microstruct 34:33
14. Klein PB, Binari SC (2003) Photoionization spectroscopy of deep defects responsible for current collapse in nitride-based field effect transistors. J Phys Cond Matter 15:R1641
15. Meneghesso G, Meneghini M, Bisi D, Rossetto I, Cester A, Mishra UK, Zanoni E (2013) Trapping phenomena in AlGaN/GaN HEMTs: a study based on pulsed and transient measurements. Semi Sci Tech 28:074021
16. Meneghini M, Bisi D, Stoffels S, Van Hove M, Wu T-L, Decoutere S, Meneghesso G, Zanoni E (2014) Trapping in GaN-based metal-insulator-semiconductor transistors: Role of high drain bias and hot electrons. Appl Phys Lett 104:143505
17. Arehart AR, Malonis AC, Poblenz C, Pei Y, Speck JS, Mishra UK, Ringel SA (2011) Next generation defect characterization in nitride HEMTs. Phys Status Solidi C 8:2242
18. Armstrong AM, Allerman AA, Baca AG, Sanchez CA (2013) Sensitivity of on-resistance and threshold voltage to buffer-related deep level defects in AlGaN/GaN high electron mobility transistors. Semi Sci Tech 28:074020
19. Khan MA, Shur MS, Chen QC, Kuznia JN (1994) Current/voltage characteristic collapse in AlGaN/GaN heterostructure insulated gate field effect transistors at high drain bias. Electron Lett 30:2175
20. Binari SC, Kruppa W, Dietrich HB, Kelner G, Wickenden AE, Freitas JA Jr (1997) Fabrication and characterization of GaN FETs. Solid-State Electron 41:1549
21. Ogino T, Aoki M (1980) Mechanism of yellow luminescence in GaN. Jpn J Appl Phys 19:2395
22. Armstrong A, Arehart AR, Green D, Mishra UK, Speck JS, Ringel SA (2005) Impact of deep levels on the electrical conductivity and luminescence of gallium nitride codoped with carbon and silicon. J Appl Phys 98:053704
23. Arehart AR (2011) Investigation of electrically active defects in GaN, AlGaN, and AlGaN/GaN high electron mobility transistors. PhD Dissertation, Ohio State University
24. Sasikumar A, Arehart AR, Martin-Horcajo S, Romero MF, Pei Y, Brown D, Recht F, di Forte-Poisson MA, Calle F, Tadjer MJ, Keller S, DenBaars SP, Mishra UK, Ringel SA (2013) Direct comparison of traps in InAlN/GaN and AlGaN/GaN high electron mobility transistors using constant drain current deep level transient spectroscopy. Appl Phys Lett 103:033509
25. DasGupta S, Biedermann LB, Sun M, Kaplar R, Marinella M, Zavadil KR, Atcitty S, Palacios T (2012) Role of barrier structure in current collapse of AlGaN/GaN high electron mobility transistors. Appl Phys Lett 101:243506
26. Jackson CM, Arehart AR, Cinkilic E, McSkimming B, Speck JS, Ringel SA (2013) Interface trap characterization of atomic layer deposition Al2O3/GaN metal-insulator-semiconductor capacitors using optically and thermally based deep level spectroscopies. J Appl Phys 113:204505
27. Johnson NM (1982) Measurement of semiconductor–insulator interface states by constant-capacitance deep-level transient spectroscopy. J Vac Sci Tech 21:303

Chapter 8
Modelling of GaN HEMTs: From Device-Level Simulation to Virtual Prototyping

Gilberto Curatola and Giovanni Verzellesi

8.1 Introduction

Gallium nitride (GaN) high electron mobility transistors (HEMTs) are expected to assume an increasingly important role among transistor technologies for the power electronics industry [1]. GaN technology on cheap silicon substrates has been strongly pursued by the semiconductor industries in the past decade. The properties of GaN such as wider band gap, higher breakdown voltage, larger critical electric field, and higher thermal conductivity mean that GaN-based transistors should operate at higher voltages and much higher switching frequencies and, in general, handle higher power density. This would translate into largely enhanced power efficiencies than pure silicon (Si) devices.

However, compared to very strong Si competition and rapidly changing technology and customer requirements, and due to the many technological challenges given by the co-integration of III-nitride and Si technologies, there is an increasing demand for rapid solution techniques to fasten the introduction of GaN-based products into the market.

In particular, there is the need of fast and reliable tools that allow technologists and designers to gain a detailed understanding of the GaN system and to identify and, possibly solve, the key challenges that are still open before GaN introduction into the market.

G. Curatola (✉)
Infineon Technologies AG, Villach, Austria
e-mail: Gilberto.Curatola@infineon.com

G. Verzellesi
Dipartimento di Scienze e Metodi dell'Ingegneria and En&tech,
Università di Modena e Reggio Emilia, Modena, Italy
e-mail: giovanni.verzellesi@unimore.it

M. Meneghini et al. (eds.), *Power GaN Devices*,
Power Electronics and Power Systems, DOI 10.1007/978-3-319-43199-4_8

More importantly those tools should possess two main requirements: (i) being fast and accurate in order to widen the gap with the robust, reliable and rapidly advancing Si technology; (ii) capable of helping in the decision-making process, starting from the process side, up to the customer side. The latter requirement means the important feature of being able to link the different chain elements of the GaN semiconductor industry: from technology development, to device assembly and, finally, to end customer applications.

The GaN virtual prototype (VP) approach that will be presented in this chapter possesses all the main requirements that we have just described. It allows a very accurate and reliable prediction of the performance of GaN-based field effect transistors, and it can support the GaN technology development and be used as a tool for device optimization; it allows to predict the full performance of GaN products in real customer applications and can be used by designers to predict GaN system efficiency and also as a tool for customer support. In particular, as will be described in detail in the rest of this chapter, the developed modelling approach allows to have a very detailed understanding of the complex physics of GaN HEMTs thanks to a physically based and calibrated trap modelling. This modelling approach has been carefully calibrated on several different GaN technologies and validated by means of advanced characterization techniques on devices with different electrical and geometrical properties. Moreover, our GaN VP approach allows an accurate description of the GaN technology, not only at device level but also at system level, and can carefully predict the switching performance of GaN transistors in real switch-mode power applications. In general, the GaN VP can assist both technologists and designers and could be used, in our opinion, as a general guideline tool for GaN system optimization with clear advantages in terms of decision making, budget resources, and fastening GaN products introduction into the market.

In the past three decades, there has been tremendous progress in the quality of process and device simulation for silicon-based devices. However, for GaN heterostructures, the existing tools have still some drawbacks. The wide band gap of III-nitride compounds, for example, leads to an extremely small number of free charge carriers ($n_i \propto \exp(-E_G/2)$) at room temperature; therefore, the tools have to calculate with fourfold increased number of decimals to obtain the required accuracy. This blows up the time needed for every simulation. In addition, due to the immature status of GaN technology and the lack of extensive and accurate characterization data, almost no device models have been developed and calibrated specifically for GaN-based electronics, and most common commercial TCAD tools still rely on the models which have been developed and calibrated only for silicon transistors. Another complication for GaN device simulations is represented by the intrinsic nature of a GaN power transistor, where the basic electrical features of the transistor are determined by the chemical and geometrical properties of a very thin $Al_xGa_{1-x}N$ layer, generally 10–30 nm thick. On the other side the typical pitch of a high-voltage power device is several microns. This means that a device simulator (assuming a 2D approach, for simplicity) should be able to define a mesh size of few angstroms in one direction, for a device that is several microns wide, in the other direction. This translates, when compared with standard

Si-based devices, in a much larger number of mesh points required and, therefore, in a much longer computing time needed to perform accurate GaN simulations.

Last but not least, due to the very complex nature of GaN-on-Silicon technology, we believe that a full-system approach is needed where technology, packaging, and application are all optimized at the same time. Typical case which illustrates the need of a full-system approach is given, for example, by the GaN cascode product where a high-voltage normally-on GaN transistor is co-packaged with a normally-off low-voltage (LV) Si MOSFET. In this particular case, the GaN device, the LV MOSFET, and the package must be all optimized for the specific product application. Technology optimization, packaging parasitics minimization, and application-specific requirements must be all fulfilled at the same time. This implies, first, a complete and detailed understanding of the system and, second, a fast and reliable system approach that can address the individual components of this chain and optimize them based on the final specific product requirements.

The GaN VP approach presented in this chapter allows to link accurately and efficiently the GaN technology development with the final application customer needs. In particular, the GaN VP approach is composed by four main parts:

1. TCAD device modelling;
2. Compact model development/calibration;
3. Application boards characterization and electromagnetic simulations;
4. Spice application simulations.

A schematic representation of the GaN VP is presented in Fig. 8.1 [2]. As can be noted, the GaN VP mimics one to one the different steps of manufacturing of a typical semiconductor product: from technology development, to assembling, and to final end customer application. Each step of the "virtual" approach receives as inputs, for validation and calibration, data from the corresponding "real" node. A closed-loop structure also allows to feedback inputs to each node from all other nodes of the chain and largely strengthens the efficiency of the overall system optimization.

The GaN VP approach has been tested and calibrated on both normally-on and normally-off GaN device concepts and has proven to be robust, accurate, and reliable in both cases, as it will be shown in the rest of this chapter.

8.2 Device-Level Simulation

Application of GaN HEMTs as high-voltage switching devices in power converters requires the minimization of all off-state leakage current paths in order to increase the maximum blocking voltage and boost the converter power rating. This calls for the adoption of insulated-gate structures to minimize the gate leakage current (in case of normally-on GaN HEMT), on one side, and the introduction of channel bottom confinement layers or buffer compensation techniques, to control the drain–source leakage current, on the other. In this context, carbon (C) doping is intensively investigated as an option for GaN buffer compensation for the advantages it offers

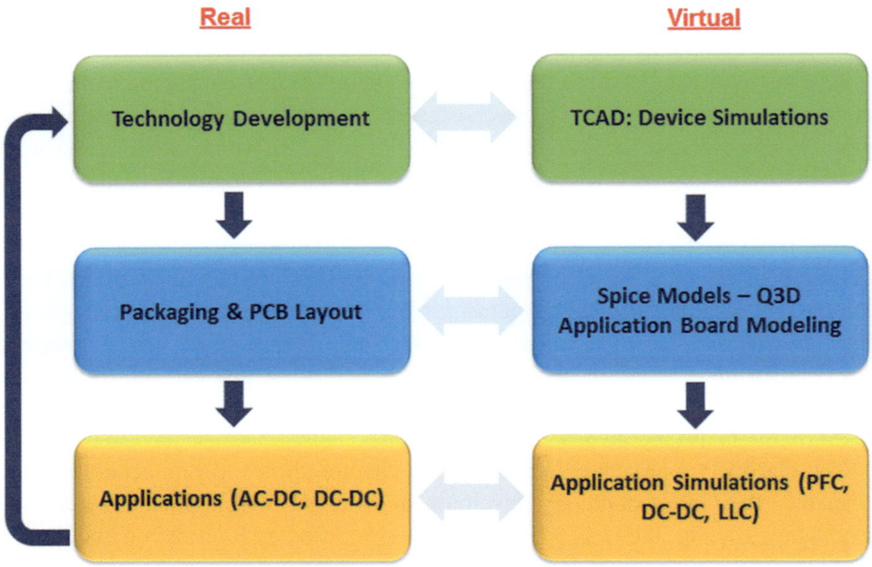

Fig. 8.1 Schematic view of the GaN virtual prototype approach (VP)

over other possible adoptable impurities (like Fe) in terms of abrupt doping switch off at controllable distances from the AlGaN/GaN interface and the consequent potential for an engineered trade-off between off-state breakdown and current-collapse effects [3–7]. Moreover, C doping poses less severe contamination issues with respect to other adoptable impurities, like Fe, when GaN technology fabrication steps are implemented in the same conventional Si line used for other power technologies. In the light of the above considerations, our efforts on the device-modelling side have mainly focused on two aspects of the electrical behavior of GaN HEMTs that are crucial for power switching applications, namely (1) the way C doping in the device buffer layer can be modelled and its impact on electrical performance [8, 9], and (2) the mechanisms underlying HEMT three-terminal off-state breakdown [10].

Analyzed devices are AlGaN/GaN HEMTs and other related test structures fabricated by means of a normally-on GaN HEMT technology characterized by the following epitaxial structure grown by MOCVD on p-type Si substrates: 4.6-μm GaN buffer intentionally doped with C (concentration in the 10^{19}–10^{20} range), 300-nm undoped GaN channel, 20-nm $Al_{0.22}Ga_{0.78}N$ barrier. Devices feature an insulated-gate and a double field-plate structure. Devices fabricated on two different epitaxial wafer types are considered, characterized by non-optimized and optimized C doping and, as a result, significant and negligible trap-related effects, respectively. We will refer to these as to non-optimized buffer technology and optimized buffer technology, respectively. The rated breakdown voltage is for both wafers 650 V (defined at 10 nA/mm and $V_{GS} = -12$ V).

Two-dimensional numerical device simulations were carried out with the commercial simulator Sentaurus Device [11]. Self-heating effects and electron quantization in the channel were neglected. Input parameters that have been kept at their simulator's default values for the different materials include: dielectric constant, bandgap and electron affinity, and electron and hole effective masses. Channel electron mobility and source/drain contact resistances have been set to 2000 $cm^2/(Vs)$ and 0.6 Ω/mm, in agreement with Hall and TLM measurements, respectively. Electron saturation velocity in the GaN channel was set to 10^7 cm/s to fit the experimental transconductance in saturation regime. Polarization charges were accounted for by means of the simulator strain-charge model [11, 12].

C-doping compensation effect has been explained in the literature as the result of two possible, alternative mechanisms, depending on growth technique and conditions, namely (i) autocompensation through interplay of C_N–C_{Ga} states incorporated during growth with comparable concentrations [13, 14], (ii) compensation by dominant C_N and/or other C-related acceptor states [14, 15]. The Fermi level is pinned at around midgap in the first case, thus making the C-doped buffer to behave as an almost ideal semi-insulating layer, while it is pulled into the bottom half of the bandgap in the second one, turning the buffer into a weakly p-type region.

We considered three different C-doping models and systematically tested them against measurements from GaN HEMTs.

(A) Autocompensation by C_N–C_{Ga} states with energies $E_{CGa} = 0.11$ eV (donors) and $E_{CN} = 3.28$ eV (acceptors) and equal concentrations $N_{CN} = N_{CGa}$ [13, 14] [hereafter referred to as model (A)];
(B) Compensation by a dominant C-related acceptor state A_C with energy $E_{AC} = 2.5$ eV and concentration N_{AC} [15] [hereafter referred to as model (B)];
(C) Partial autocompensation through A_C–C_{Ga} states with $N_{CGa} > N_{AC}$ [hereafter referred to as model (C)].

All above energies (E_{CN}, E_{CGa}, E_{AC}) are referred to the conduction band edge (E_C). A sketch of trap levels according to the three C-doping models is shown in Fig. 8.2.

In addition to above C-related traps, acceptor-like traps (A_{UID}) were distributed within all GaN layers, with energy $E_{AUID} = 0.37$ eV set in agreement with temperature-dependent drain current transient experiments carried out in devices

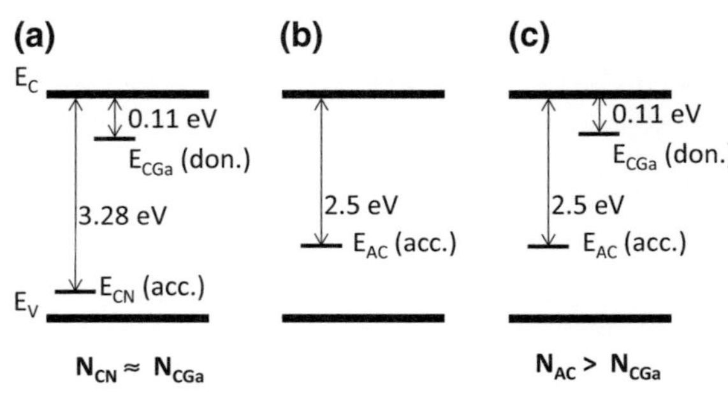

Fig. 8.2 Sketch of trap levels assumed to model carbon doping according to the **a** autocompensation model, **b** the dominant acceptor model, and **c** the partial autocompensation model

with optimized buffer [9]. The extracted E_{AUID} is indeed not consistent with any of the C-related E_{CN}, E_{CGa} or E_{AC} states. It might be related to one of the majority carrier traps associated with edge dislocations in n-type GaN [16]. For these reasons, it has been included in the GaN layers regardless of whether model (A), (B), or (C) is adopted for C doping.

8.2.1 Pulsed Mode Behavior

Buffer conductivity compensation relies on the controlled generation of intrinsic defects during growth or the introduction of acceptor impurities like Fe and C. In all cases, the device is made potentially more prone to trap-related effects, generally leading to dispersion between DC and pulse-mode characteristics. Moreover, in devices having an insulated-gate structure, threshold-voltage instabilities are commonly observed in response to the gate–source voltage switching, as a result of electron trapping/detrapping effects at the interface between gate dielectric and barrier and/or in the gate dielectric bulk. Electrons can, in principle, be provided by the channel and/or the gate depending on the applied bias and the gate dielectric conductivity properties [17–19].

Devices considered here have both C-doped buffer and insulated gate. It is of great importance for optimization of this category of GaN HEMT technologies to be able to distinguish buffer-related and dielectric-related trap effects from each other. This calls for an accurate understanding/modelling of the role that C-related trap states can play under pulse-mode operation.

8.3 Non-optimized Buffer Technology

In this section, we show that pulse-mode threshold-voltage (V_{TH}) instabilities can actually arise in insulated-gate, C-doped AlGaN/GaN HEMTs also as a result of buffer trapping/detrapping phenomena. We investigate the underlying physical mechanisms and provide the instruments to distinguish the observed buffer-related V_{TH} instabilities from those linked to dielectric and interface traps.

Figure 8.3 shows pulse-mode drain current (I_D) versus drain–source voltage (V_{DS}) characteristics at a gate–source voltage (V_{GS}) of −5.2 V, measured by sequentially using different baselines ($V_{GS,BL}$, $V_{DS,BL}$). Drain and gate are pulsed simultaneously with source and substrate grounded. Pulse width and period are 5 and 500 μs, respectively. The first pulsed curve is obtained from a (0, 0 V) baseline. Repeated pulse-mode measurements all taken from this "harmful" quiescent condition result in no appreciable dispersion between different curves (not shown). This curve can therefore be adopted as a reference for subsequent measurements using baselines that are instead able to induce dynamic dispersion effects. The second pulsed I_D–V_{DS} curve has a baseline of (−0, 400 V) and results in a significant I_D increase as a result of a

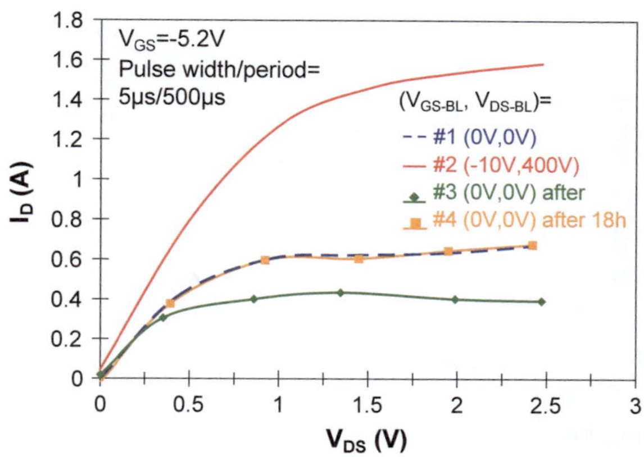

Fig. 8.3 Experimental pulse-mode output characteristics at a gate–source voltage (V_{GS}) of −5.2 V obtained by applying the baseline sequence #1 to #4. Device is a large-periphery, packaged HEMT featuring gate length of 3 μm, gate–source spacing, gate–drain spacing, and source-connected field-plate overhang of 1.5, 12, and 6 μm, respectively. Reprinted with permission [8]

negative V_{TH} shift of about 0.75 V. A (0, 0 V) baseline measurement taken immediately after leads to a positive V_{TH} shift of almost 1 V, thus moving the I_D–V_{DS} curve under the initial (0, 0 V) curve. Another (0, 0 V) baseline measurement taken 18 h after the previous measurement eventually results in a full recovery of the I_D–V_{DS} curve to its initial state.

Similar behavior, characterized by a double V_{TH} shift in negative and positive directions followed by recovery to initial state, is also observed after a short-term (order of seconds) stress under large-V_{DS}, off-state bias conditions. This is illustrated by Fig. 8.4, showing several DC transfer characteristics at $V_{DS} = 0.5$ V, taken at different times prior to and after application of a quiescent bias of (−10, 100 V) for 30 s (Fig. 8.4a) and corresponding I_D values at a V_{GS} of −5.5 V as a function of acquisition time (Fig. 8.4b). Point A ($I_D \approx 0.9$ A) represents the pre-stress state. As can be noted, at short times after removal of the off-state bias (point B), I_D largely overshoots the pre-stress value, this behavior being related to a V_{TH} shift in negative direction; see Fig. 8.4a. The drain current then decreases and reaches a minimum (indicated as C) that is below the pre-stress value after a transient of 100–200 s. This corresponds to a positive V_{TH} shift; see Fig. 8.4a. Afterward, I_D starts to increase back and recovers to a value that is slightly larger than the pre-stress value after 30–60 min (point D).

Before describing the way our device-level modelling approach is able to reproduce and explain the observed V_{TH} instabilities, it is useful to discuss the spectrum of possible underlying mechanisms on the basis of previous literature and commonly accepted understanding of trap effects in GaN HEMTs.

Considering first the negative V_{TH} shift, negative charge must have been removed from somewhere under the gate during the off-state phase, so that, when the device is switched to above-threshold conditions, V_{TH} is dynamically more negative and I_D overshoots the DC value. This has to be followed by negative charge rebuild leading to I_D recovery to DC level through a slow decreasing transient. Mechanisms that are in principle able to cause this process are as follows. (N1) Electron emission from channel traps during the off-phase due to 2DEG depletion, followed by 2DEG

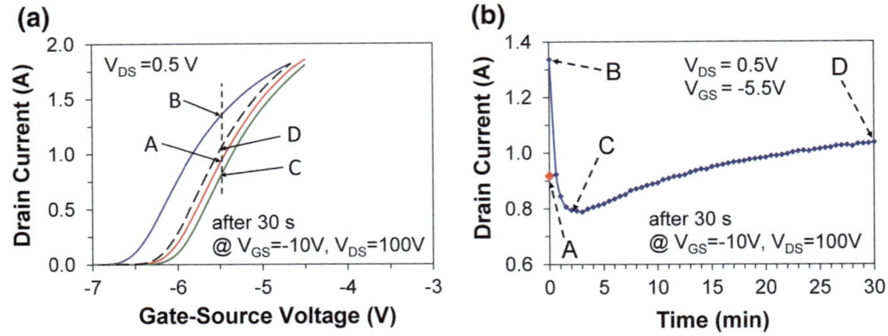

Fig. 8.4 **a** DC transfer characteristics at a drain–source voltage (V_{DS}) of 0.5 V acquired at different times prior to and after 30-s stress at $V_{GS} = -10$ V and $V_{DS} = 100$ V and **b** corresponding drain current values at a gate–source voltage of −5.5 V as a function of acquisition time. For sake of clarity, not all DC curves used to construct the I_D versus time plot shown in (**b**) are plotted in (**a**). Same device as in Fig. 8.3. Reprinted with permission [8]

reformation and electron trapping into channel traps. (N2) Electron emission from barrier and/or barrier/dielectric interface traps during the off-state phase, followed by electron capture into same traps.

Mechanism (N1) is a quite basic mechanism always present in GaN HEMTs unless trap density in the GaN channel is negligible at least in the region close to the barrier interface. Its predominant role in analyzed devices is supported by observation that, as will be shown in the next section, the negative V_{TH} shift vanishes in devices having optimized buffer and same processing. Mechanism (N2) is instead less likely to have a role during measurements like those considered, where V_{GS} is pulsed from sub to slightly above V_{TH}. It would rather require a positive V_{GS} during the on-phase to enable significant channel electron injection over the AlGaN/GaN barrier.

Turning our attention to the positive V_{TH} shift, negative charge must, in this case, accumulate somewhere under the gate during the off-state phase, so that, when the device is switched to above-threshold conditions, V_{TH} is dynamically less negative and I_D is initially smaller than the DC value. This must be followed by negative charge removal, resulting in I_D recovery to DC level through a slow increasing transient. Mechanisms that are in principle able to cause this process are as follows: (P1) Buffer trap charge-up during the off-state phase owing to source–drain leakage current, followed by buffer trap discharge. (P2) Electron trapping into barrier and/or surface traps due to gate leakage current during the off-state phase, followed by electron re-emission from same traps. (P3) Electron trapping into oxide traps due to gate leakage current during the off-state phase, followed by electron re-emission from same traps.

(P1) is the mechanism typically invoked to explain buffer-related, current-collapse effects in GaN HEMTs [20]. It can be held responsible also for the I_D drop observed at high V_{GS}, related to drain access resistance increase and clearly shown in Fig. 8.4 (curve B). In C-doped devices, trap charge/discharge can in principle occur not only by electron capture/emission into/from electron traps in the channel and/or buffer layers, as commonly assumed, but also through hole emission/capture from/into hole traps in the C-doped buffer. In the latter case, acceptor states C_N and/or A_C (see Fig. 8.2) can be involved. The dominant role played by mechanism (P1) in our devices is supported [like for (N1)] by observation that V_{TH} instabilities are not present in devices with optimized buffer and same processing (see next section). Specifically, as will be shown later, simulations suggest that C-related hole traps A_C in the buffer are responsible for this mechanism in the devices under consideration. Mechanisms (P2) and (P3) are instead unlikely to play a dominant role due to very small gate leakage in these devices [$<10^{-10}$ A/mm up to the highest (lowest) V_{DS} (V_{GS}) adopted]. In addition, as far as (P3) mechanism is concerned, a typical signature for it is the presence of "counterclockwise" hysteresis in DC I_D–V_{GS} curves [17], which is, however, not observed in devices under consideration.

Results of device simulations are compared with corresponding measurements in Fig. 8.5. Measured data are the same already shown in Fig. 8.4b. Simulation data are obtained by adopting model (C) for C doping. As can be noted, simulations are able to reproduce experiments with good accuracy. Interpretation of experimental data provided by simulations can be summarized as follows.

(a) During application of the large-V_{DS}, off-state quiescent bias and acceptor traps A_{UID} in the UID channel under the gate are neutral (no trapped electrons) as a result of 2DEG channel depletion. C-related acceptor traps A_C in the C-doped buffer (both under the gate and in the drain access region) are instead negatively charged (no trapped holes), owing to the large and positive V_{DS} that leads to enhanced buffer leakage current. The negative V_{TH} shift induced by the empty channel traps dominates and I_D immediately after off-state bias removal (point B) is higher than prior to its application (point A).
(b) A_{UID} traps in the channel capture electrons during the turn-on transient as a result of 2DEG channel reformation, and this explains the initial drain current decrease (transient from point B to point C in Fig. 8.5) and the associated positive V_{TH} shift. Owing to a longer time constant, buffer traps are still negatively charged, so that I_D reaches a smaller value (point C) than the pre-stress value (point A) and V_{TH} shifts to a less negative value than pre-stress case.
(c) A_C traps, instead, capture holes in the weakly p-type buffer during the turn-on transient, as the excess electron buffer leakage current has been suppressed. This happens with a very long time constant in the order of 10^2–10^3 s owing to the small hole density in the C-doped buffer. This leads to the final, slow I_D recovery to the pre-stress value (transient from point C to point D in Fig. 8.5) and the correlated, final negative V_{TH} shift.

In summary, simulations suggest that V_{TH} instabilities in these devices result from the combined effect of electron traps in the UID channel (A_{UID}) and hole traps in the C-doped buffer (A_C), respectively, causing mechanism (N1) and (P1) to take place simultaneously but with distinct time constants.

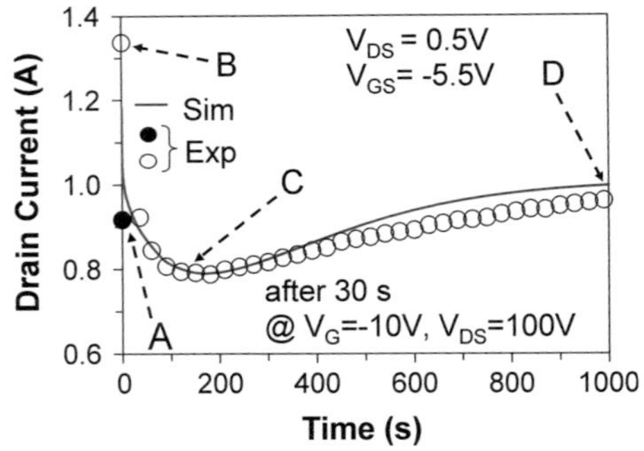

Fig. 8.5 Measured (*symbols*) and simulated (*solid line*) drain current (I_D) values as a function of time prior and after 30-s stress at $V_{GS} = -10$ V and $V_{DS} = 100$ V. Same device as in Figs. 8.3 and 8.4. Reprinted with permission [8]

As far as alternative models for C doping are concerned, same above considerations and capability to account for V_{TH} instabilities apply to model (B) as well. This is because C-doping impact on pulse-mode behavior is mainly related to acceptor traps A_C which are common to both models (B) and (C). Discrimination between which of models (B) and (C) and explanation why model (C) is more suited to describe the actual behavior of devices with non-optimized buffer considered in this section has actually been achieved by means of AC capacitance simulation. This is shown in Sect. 8.4.1. Model (A) is instead unable to explain observed V_{TH} instabilities. As will be shown in the next section, model (A) results in limited trap effects and, particularly, to negligible V_{TH} instabilities, which is consistent with experimental data from optimized buffer devices.

8.4 Optimized Buffer Technology

Double-pulse output characteristics of a HEMT representative of the optimized buffer technology are shown in Fig. 8.6. Pulse width and period are 1 and 100 μs, respectively. In all tested devices (10 from different wafer cells), dispersion of pulsed characteristics with respect to the (0, 0 V) baseline reference curve is negligible for $V_{DS,BL}$ up to 20 V and remains <10 % for $V_{DS,BL} = 50$ V. Dynamic $R_{DS,ON}$ increase is <5 %. V_{TH} instabilities like those shown in Figs. 8.3 and 8.4 are not present in these devices.

Figure 8.7a shows experimental and simulated I_D transients following the application of a (−10, 50 V) quiescent bias for 100 s followed by turn-on pulsing to (0, 5 V). Simulation results are reported for models (A) and (B) for C doping. All curves are normalized to their final values ($t = 100$ s). As can be noted, the slow component of the experimental I_D waveform has a very small amplitude (≈2 % of the total I_D change) and a relatively long time constant in the order of tens of seconds. Such slow response is governed by the A_{UID} trap having an energy $E_{AUID} = 0.37$ eV and an electron capture cross section of 1.1×10^{-21} cm^2, as extracted from temperature-dependent drain current transients (not shown). We assumed this trap to be located

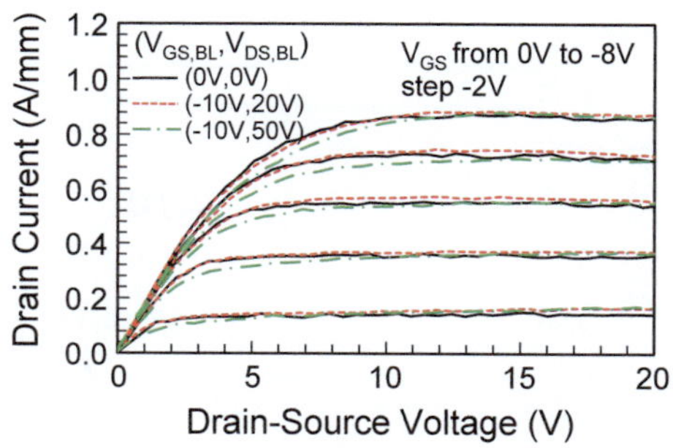

Fig. 8.6 Experimental pulse-mode output characteristics for different baselines ($V_{GS,BL}$, $V_{DS,BL}$). Device is a small-periphery HEMT with gate length (L_G) of 1 μm, gate–source spacing, gate–drain spacing, and source-connected field-plate overhang of 1.5, 12, and 6 μm, respectively. Reprinted with permission [9]

in the UID channel, since, due to the insulated-gate structure and the negligible gate leakage current, modulation of surface and barrier traps is unlikely to play a role. Moreover, current collapse emerges in these devices only for large $V_{DS,BL}$, indicating that it is mainly related to drain-lag effects and thus pointing to channel or buffer location as for responsible traps [20].

The same procedure adopted to extract E_{AUID} experimentally has been applied to simulated temperature-dependent I_D versus time waveforms (as they were experimental data). Experimental and simulated Arrhenius plots are shown in Fig. 8.7b.

Significant qualitative/quantitative differences between predictions of model (A) and (B) are evident when comparing simulations with experimental data in Fig. 8.7.

When model (A) is adopted, (1) both amplitude and time constant of simulated I_D transients are comparable with the experimental ones; see Fig. 8.7a; (2) the activation energy extracted from simulated I_D transient is equal to the experimental value E_{AUID}; see Fig. 8.7b, as the (small) current-collapse effects in the simulated device are due to A_{UID} traps in the UID channel; (3) C-related trap states have no appreciable impact on simulated I_D waveform.

On the contrary, by adopting model (B), (1) predicted current-collapse effects are much larger than those observed experimentally (≈40 % of the total I_D change); see Fig. 8.7a; (2) the activation energy extracted from simulated I_D transient is different from the experimental value E_{AUID}; see Fig. 8.7b. It instead corresponds to energy of A_C traps when referred to the valence band edge [$E_{AC} - E_V = E_G - E_{AC} = (3.4 - 2.5)$ eV $= 0.9$ eV], thus meaning that current-collapse effects would in this case be governed by E_{AC} states in the C-doped buffer acting as hole traps, consistently with simulations reported in [21] where model (B) is actually adopted to describe C doping.

In summary, these devices with optimized buffer are characterized by no dynamic V_{TH} instability and by small current-collapse effects, the latter resulting from the effect of UID channel traps A_{UID}. This behavior is reproduced by simulations when model (A) is adopted, and explained as a consequence of the perfect autocompensation of C_N–C_{Ga} states.

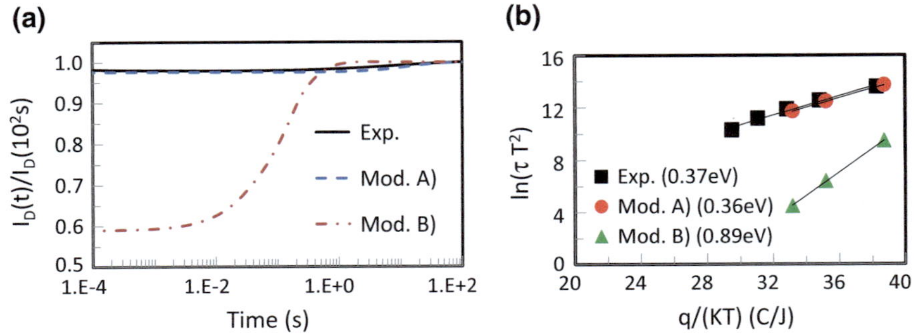

Fig. 8.7 **a** Experimental room-temperature drain current (I_D) transients following the application of a (−10, 50 V) quiescent bias for 100 s followed by turn-on pulsing to (0, 5 V), compared to simulated waveforms obtained by assuming models (A) and (B) for carbon doping. All curves are normalized to their final values (at $t = 100$ s). **b** Arrhenius plots obtained from experimental and simulated temperature dependent I_D versus time transients. Same device as in Fig. 8.6

8.4.1 AC Capacitances

Figure 8.8 shows experimental and simulated off-state drain–source capacitances (C_{DS}) as a function of V_{DS}, for representative devices of the non-optimized and optimized buffer technologies. C_{DS} decreases with increasing V_{DS} as a result of the gradual 2DEG depletion at the AlGaN/GaN interface, proceeding from the drain-end edge of the gate toward the drain contact. C_{DS} versus V_{DS} curves exhibit distinct roll-off points, labeled as R1 in Fig. 8.8, in correspondence of the V_{DS} values for which 2DEG depletion reaches the source-connected field-plate (SCFP) edge. As can be noted by comparing experimental C_{DS} curves in Fig. 8.8a, b, this roll-off voltage is a sensitive function of C doping, with device processing and geometry being the same for the two technologies.

By considering first the behavior of the non-optimized technology device in Fig. 8.8a, as can be noted, only model (C) allows the $C_{DS}(V_{DS})$ curve to be accurately reproduced by simulations. On the contrary, models (A) and (B) result in marked overestimation and underestimation of the roll-off voltage (point R1), respectively. Moreover, model (B) predicts the presence of a second roll-off point (labeled as R2) at about 70 V, corresponding to the V_{DS} value at which 2DEG depletion has reached the drain contact. This condition is instead reached in measured data only for $V_{DS} > 200$ V (outside the measurement range in Fig. 8.8a).

The underlying physics pointed out by simulations is as follows. The net charge associated with C-related trap states in the buffer has a strong impact on 2DEG depletion and specifically on the V_{DS} values required to reach R1 and R2 points. Negative charge (contributed by C-related acceptors) makes 2DEG depletion "easier," leading to smaller V_{DS} values required to reach R1 and R2 points. Positive charge (contributed by C-related donors) has the opposite effect. In the case of model (B), carbon doping is modelled by means of A_C acceptors only. This makes the C_{DS}

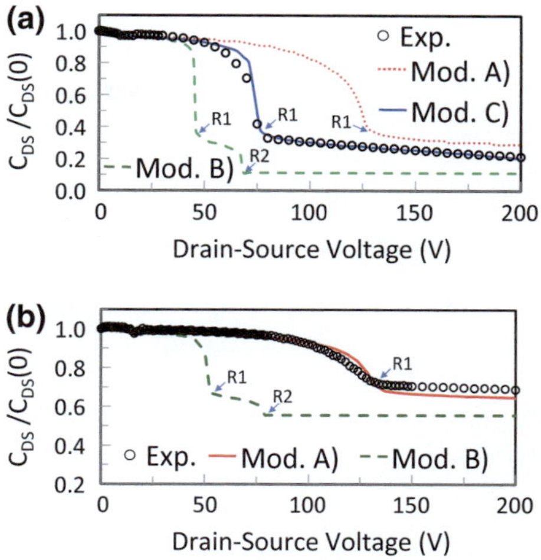

Fig. 8.8 Experimental drain–source AC capacitance (C_{DS}) as a function of drain–source voltage (V_{DS}) under off-state conditions, as compared with simulated values. Curves are normalized to their zero-V_{DS} value. Device is a HEMT of the non-optimized (**a**) and optimized (**b**) buffer technology with gate–drain spacing and source-connected field-plate overhang 12 and 6 μm, respectively

roll-off to take place at smaller V_{DS} and more abruptly than in experiments. In model (A), on the other hand, the negative charge associated with C_N acceptors is perfectly balanced by the positive one related to C_{Ga} donors, and this nullifies the contribution of the C-doped buffer to 2DEG depletion. In this case, C_{DS} roll-off voltages are underestimated by simulations. By adjusting the donor-to-acceptor density ratio N_{CGa}/N_{AC}, simulations implementing model (C) are, instead, able to accurately reproduce the experimental $C_{DS}(V_{DS})$ curve. Best fitting is achieved for $N_{CGa}/N_{AC} = 0.95$. This value was actually used in all simulations adopting model (C) shown in Sect. 8.2.1.

Turning now our attention to the $C_{DS}(V_{DS})$ behavior in the optimized buffer technology in Fig. 8.8b, it is model (A), in this case, that allows the $C_{DS}(V_{DS})$ curve to be reproduced, whereas model (B) results in significant underestimation of roll-off V_{DS} values corresponding to R1 and R2 points. These results confirm those shown in Sect. 8.2.1 on pulse-mode behavior of the optimized buffer technology.

In summary, the drain–source capacitance is found to be a sensitive function of C doping, suggesting that its monitoring can be adopted as a fast technique to assess C doping and related buffer compensation properties during growth/device optimization. By considering both AC capacitance results and pulse-mode behavior, model (C) is the most suited C-doping model, among the considered ones, for describing devices with non-optimized C doping. On the other hand, it is model (A) that is able to provide pulse-mode and AC predictions fully consistent with experimental data for devices of the optimized buffer technology.

Figure 8.9 shows an example of the model accuracy on a normally-on GaN HEMT. The simulation model is considered to be satisfactory only when it can predict accurately the DC characteristics, the AC characteristics, and the transient behavior, all at the same time.

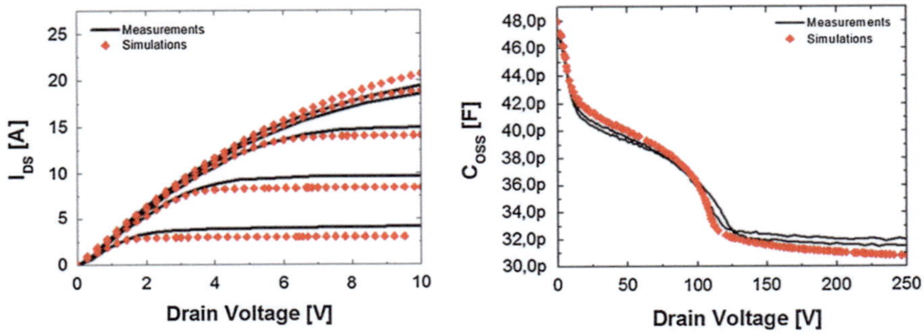

Fig. 8.9 Experimental (*solid*) and simulated (*symbols*) *IV* curves and *CV* curves for a normally-on GaN HEMT

8.4.2 Off-state Breakdown

Understanding the physical mechanisms limiting the off-state breakdown voltage in AlGaN/GaN HEMTs is a difficult task owing to structure and technology complexities, like the presence of field plates and back-barrier layers and the possible role played by intrinsic defects and/or extrinsic impurities used for compensating the unintentional GaN conductivity. Correlating the HEMT breakdown characteristics with the high-voltage behavior of more simple, lateral, and vertical test structures, as well as adopting device simulation as an interpretative tool, can be useful in view of a more rapid and effective technology optimization.

Devices adopted for this study are from the non-optimized buffer technology. In the simulations, model (C) has therefore adopted to account for C doping, in agreement with previous results from pulse-mode and AC capacitance behavior of this technology. In lateral ohmic-to-ohmic structures, isolation is achieved by means of Ar implantation. The latter has been modelled in the simulations by following [22].

Room-temperature experimental and simulated breakdown characteristics of lateral isolation test structures are illustrated by Fig. 8.10, where one of the ohmic contact and the substrate contact are grounded, while a positive voltage V_{OHM} is applied to the other ohmic contact. Lateral isolation structures having different isolation implant lengths (L_{ISOL}) and therefore different ohmic-to-ohmic distances are considered. As can be noted, simulations are able to reproduce quantitatively the breakdown-voltage scaling with L_{ISOL}. Same satisfactory agreement is achieved at different temperatures (not shown).

Simulations suggest that the physical mechanisms underlying breakdown in these lateral isolation structures are as follows: (i) lateral, ohmic-to-ohmic punch-through for $L_{ISOL} \leq 4$ μm and (ii) vertical, ohmic-to-substrate breakdown for $L_{ISOL} \geq 8$ μm. It is worth emphasizing that the detailed description of trap states in the GaN channel and buffer layers resulting from our previous analysis of trap-related effects in this technology was instrumental to achieving a quantitative breakdown prediction like that shown in Fig. 8.10. As a matter of fact, channel and buffer traps (along with the

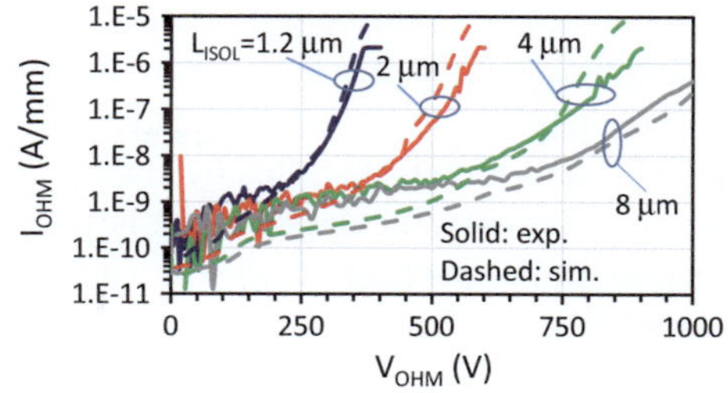

Fig. 8.10 Experimental (*solid*) and simulated (*dashed*) *IV* curves in lateral, ohmic-to-ohmic isolation structures for different isolation implant lengths (L_{ISOL}). One of the ohmic contact and the substrate contact are grounded while a positive voltage V_{OHM} is applied to the other ohmic contact

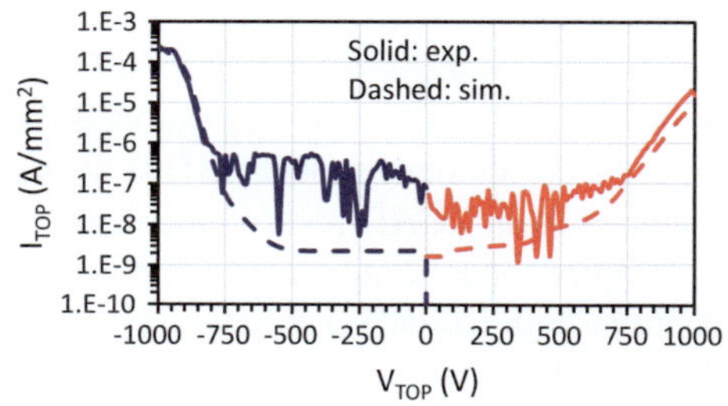

Fig. 8.11 Experimental (*solid*) and simulated (*dashed*) *IV* curves in a vertical, ohmic-to-substrate structure where the voltage V_{TOP} is applied to the top ohmic contact and the substrate is grounded

isolation implant) impact the pre-breakdown leakage current level as well as have an influence on the electric field under the ohmic contacts and, as a consequence of this, on their carrier injection properties.

The analysis of vertical, ohmic-to-substrate structures is illustrated by Fig. 8.11 for both cases when positive and negative voltages are applied to the top ohmic contact (V_{TOP}) with respect to the grounded substrate. A good agreement between measurements and simulations is achieved also for these vertical structures. Apparent discrepancies in the pre-breakdown regime are mainly related to noise, masking the actual leakage current level in the measured curves. As can be noted, vertical breakdown experimentally takes place at ≈950 and ≈850 V for positive and negative V_{TOP}, respectively (assuming a 10^{-5} A/mm^2 breakdown definition). Underlying breakdown mechanisms suggested by simulations are as follows.

When a positive voltage is applied to the top ohmic contact, high-field carrier generation in the GaN buffer layers where the electric field tends to accumulate at high voltages has a crucial role, electron injection from the substrate being inhibited by the low-resistivity p-type doping. In our simulations, we used band-to-band tunneling as high-field carrier generation mechanism in order to reproduce the experimental breakdown curves shown in Fig. 8.11. Thermally activated injection and/or transport mechanisms (like Schottky emission or Poole–Frenkel transport) have been ruled out as limiting mechanisms in these devices, owing to the observed negligible temperature dependence of vertical breakdown curves (not shown). When a negative voltage is applied to the top ohmic contact, breakdown is governed by a combination of electron injection from the top ohmic contact and high-field carrier generation in the buffer layer where most of the applied voltage drops.

The physical mechanisms leading to three-terminal, off-state breakdown in HEMT devices are suggested by simulations to be the same as in lateral, ohmic-to-ohmic isolation structures, i.e., (i) lateral, source–drain punch-through for “short” gate-to-drain spacings (L_{GD}) and (ii) vertical, drain-to-substrate breakdown for “long” L_{GD} devices.

8.5 Spice Model Development and Calibration

Second step in the development of a GaN virtual prototyping consists in the development and calibration of Spice models that can be used in application circuit simulations. A proprietary model [23] has been developed and calibrated with device data obtained from TCAD and/or electrical characterization of real fabricated devices. This Spice model can be used for both normally-on and normally-off concepts and contains both L1 and L3 model levels. As far as the L3 model is concerned, accurate thermal characterization of packaged transistors has been carried out in order to extract their thermal characteristics (R_{TH}, Z_{TH}). Developed model has been extensively compared with simulated and characterization data and has proven to be reliable and robust in all the relevant application simulations.

Figure 8.12 shows an example of a typical Spice model skeleton in the case of a pGaN normally-off approach [24]. The model contains the typical device capacitances (C_{GS}, C_{GD}, C_{DS}) that are described by nonlinear capacitances dependent on the voltage of the internal drain, gate and source terminals. Nonlinear capacitances are implemented as voltage-dependent charge sources, and they account for the different contributions coming from the intrinsic device, interconnects and packaging.

Moreover, independent resistors placed at the different nodes of the device equivalent circuit allow to account for both the device and package (interconnect, bonding, …) contribution to each of the total terminal resistances.

Special care, in the case of a normally-off pGaN concept, must be taken in order to correctly model the non-negligible gate leakage that can be experimentally observed when the transistor is switched in on-state and its gate voltage overcomes a certain critical value which is imposed by the particular technology fabrication steps used for the definition of the pGaN gate module [24, 25]. The gate-to-source and gate-to-drain components of the total leakage current coming from the pseudo-pn junction at the gate terminal is modelled by two independent voltage-controlled current generators in series with two gate resistors which, again, can account for both the device and package contributions.

The calibrated compact models have been extensively used for the following System-level assessment of the performance of both normally-on and normally-off GaN device concepts.

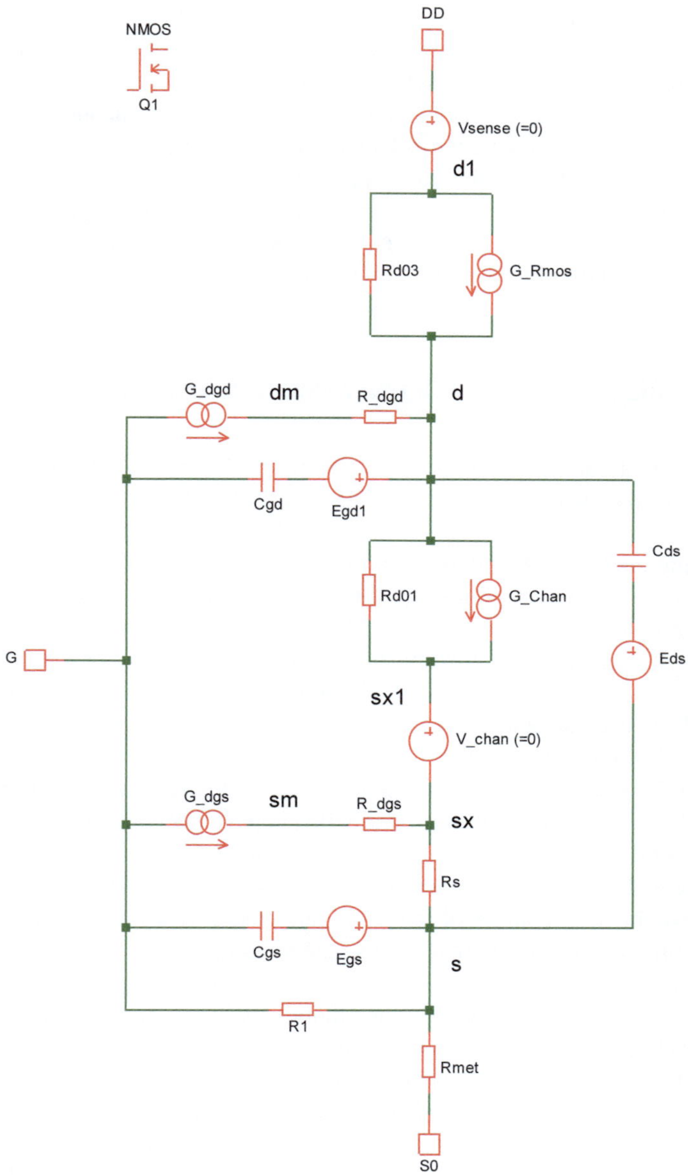

Fig. 8.12 Example of a GaN Spice model for a normally-off pGaN concept

It is worth mentioning that the compact model here described and used afterward for the application simulations do not include intentionally any dynamic trap-related effect, like a description of the current-collapse and/or of the dynamic R_{DSON} effect. Main reason for this is twofold: First, the analyzed devices are optimized devices that suffer from very small dynamic effects, so that their impact on the overall system-level performance can be considered negligible; second, in order to have a reliable and fast approach, we did not want to add unnecessary complexity. The effect of trap-related effects like, threshold voltage shift and R_{DSON} increase, can still be studied and quantified in our approach by artificially tuning the Spice model parameters and assessing their individual impact on the overall system performance. However, it is worth mentioning that several compact model approaches which include also trapping effects have been already proposed, and interested readers can refer to published data by several different groups [26, 27].

8.6 Application Board Characterization and Simulations

Third important component of the GaN VP approach is represented by the reliable modelling of the switching characteristics of GaN HEMTs in relevant application tests.

The calibrated TCAD input decks previously described have been used to simulate both the DC characteristics and the AC characteristics of normally-off GaN HEMT with pGaN gate module [24, 25] and also normally-on GaN devices with insulated gate. In particular, the transfer characteristics (I_{DS}–V_{GS}), the output characteristics (I_{DS}–V_{DS}) and the body diode behavior versus gate voltage have been simulated and used to calibrate the DC currents. Device capacitances (C_{GS}, C_{DS} and C_{GD}) have been simulated for a device in off-state conditions, i.e., zero voltage applied to the gate electrode and a drain voltage ranging from zero up to 600 V. Moreover, the device capacitances C_{GS} and C_{GD} have been also simulated, for the normally-off devices, as a function of the gate voltage from 0 V up to the full turn-on voltage and for different values of the drain-to-source voltage in order to fully characterize the AC behavior of the input gate diode, which is a typical feature of normally-off GaN transistors with pGaN gate module.

Simulated electrical characteristics were used to calibrate Spice models that have been used afterward in real application simulations. It is worth noticing that a pure device-level calibration has been shown not to be sufficient to have a reliable GaN VP. On the other side, a full-system approach must be followed that allows to correctly link the device characteristics with the system-level performance of the GaN device inserted in real switch-mode power supply applications.

For this reason real application boards have been designed, fabricated and characterized. Figure 8.13 shows a normally-off application board comprising a low-side pGaN device, a high-side SiC diode and an integrated driving scheme. Also visible are the voltage stabilization capacitances and the current shunt resistor used to measure the current flowing in the device. Not shown is instead the large inductor that allows to have an inductive switching current between the low-side and the high-side device. For this particular case, the GaN HEMT is packaged in a ThinPaK package, while the SiC diode is mounted in a T0220 package. Similar boards where

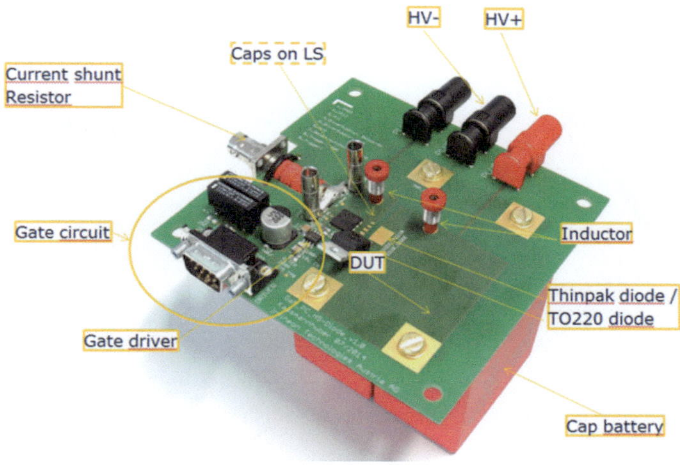

Fig. 8.13 Inductive load switching application board that has been designed and characterized for normally-off pGaN devices. An integrated asymmetric driving scheme has been used in order to minimize parasitics and to maximize performances

both GaN HEMT and SiC diode are packaged in a ThinPaK package have been also realized and studied.

In order to minimize the board parasitics and to consequently reduce the risks of possible voltage and current spikes on the gate of our normally-off GaN components, we have decided to have a special driving scheme that is integrated on the board. This integrated approach allows to have a total gate loop inductance which is lower than 10–20 pF and also allows an asymmetric driving scheme which should be favorable for the overall system-level performance and should help minimizing risks of unwanted spurious turn-on effects. This approach could be particularly useful for normally-off GaN HEMTs due to their intrinsic electrical characteristics, such as a low threshold voltage and a Q_{GD}/Q_{GS} ratio that can hardly be minimized at the levels today possible with Silicon technology. Here, values below unity can be achieved and this allows a drastic reduction of risks of unwanted capacitive spurious turn-on. Those effects depend, indeed, on the ratio between the total charge Q_{GD} and the total charge Q_{GS} and also they depend on the value of the total gate resistance (transistor + driver) used to drive the semiconductor device. In GaN technologies, unfortunately, a high ratio between the two charges is generally achieved especially because of the long metal field plates used to minimize current-collapse effects and, in general, all related trapping effects. Having an asymmetric driving scheme can, therefore, alleviate those parasitic effects by introducing some voltage headroom that acts as a safety margin against spurious turn on effects. Clearly, the extra complexity and costs in this more complex driving scheme must be accounted for.

The inductive load switching application board, shown in Fig. 8.13, has been afterward fully simulated with the Q3D simulation tool from Ansys [28]. The Q3D software is a 2-D and 3-D parasitic extraction tool used mainly by engineers

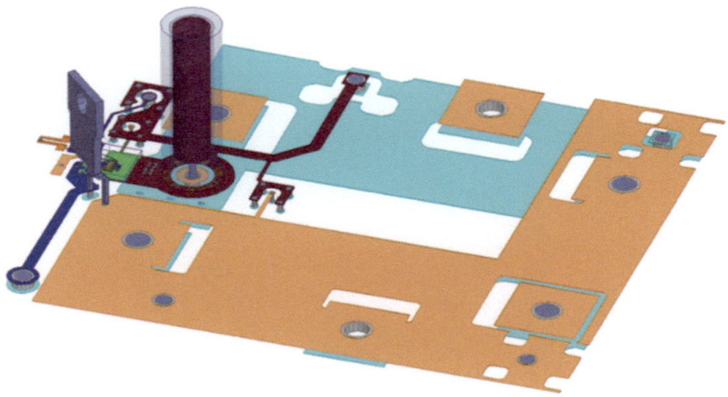

Fig. 8.14 Simulated normally-off application board with the Q3D tool

designing electronic packaging and power electronic components. The tool uses method of moments and performs 2-D and 3-D electromagnetic field simulations required for the extraction of resistance, capacitance, and inductance RLC parameters from a design. In our example, Q3D allows to extract a full matrix of parasitic inductances, resistance, and capacitances which are introduced by the PCB, by interconnects, by the packages and, in general, by the particular board design. This RLC parasitic network is afterward included in our application simulations, and it represents, in our opinion, the most efficient and accurate approach to quantify all the parasitic elements introduced during the application board design, and it also represents one of the most relevant requirements in order to have a reliable and predictive virtual prototyping platform.

A picture of the simulated Q3D board is depicted in Fig. 8.14 and fully considers all the interconnect levels and all the details concerning the device packages and other external components.

The RLC matrix extracted from the Q3D simulation can be afterward conveniently imported in a Spice circuit simulator and used for assessing the switching performance of GaN devices in real applications. For this particular example, we have used the SIMetrix tool [29] due to its simplicity and robustness demonstrated for our switch-mode power supply simulations. The parasitic RLC matrix obtained from Q3D can be imported in SIMetrix as a new model and contains a certain number of input and output ports which depend on the particular board complexity and configuration.

The application simulation environment comprising the calibrated Spice models on TCAD and real measurements, the Q3D parasitic board component, and the full integrated driving scheme can be therefore compared with the measured switching performance of the considered GaN devices.

A schematic representation of the GaN VP is sketched in Fig. 8.15, and it outlines the different components of the chain. One of the most important features of the developed approach is that it allows to have a fast and reliable prediction of the

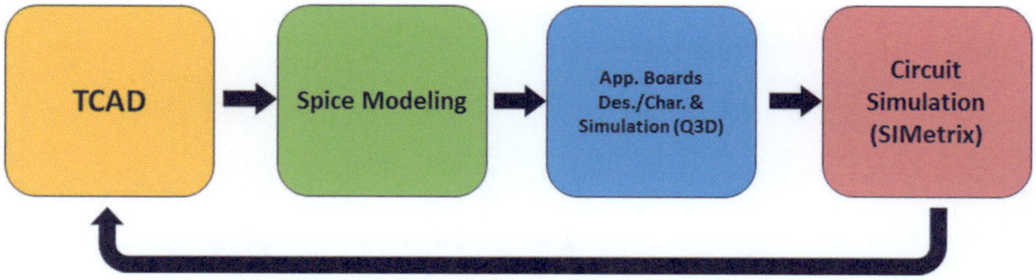

Fig. 8.15 Schematic description of the GaN Virtual Prototype approach comprising of device simulations, compact model calibration and application board design, characterization and simulation

overall system-level performance of GaN HEMTs and can assist both technologists and designers. The feedback loop is the key point that link technology and application and allows to optimize the device concepts and the different technology steps depending on the particular application that the customer is targeting. This approach has proven to be very effective especially with disruptive technologies where the lack of design and technology experience and especially the immature status of the technology largely increase the learning cycle when a pure experimental approach is being followed. On the other side, having the support of a predictive simulation environment allows not only to gain a detailed understanding of the GaN system properties but also to identify and, possibly solve earlier, the key challenges that are still open before the GaN introduction into the market, with clear advantages in term of technology development time and resources needed.

8.6.1 Normally-off pGaN Transistors

The GaN VP approach previously described has been used in the rest of this chapter to predict the switching performance of GaN devices in real end customer switch-mode power supplies. In particular, a classic power-factor correction (PFC) circuit is considered where a low-side GaN normally-off transistor is switched against a SiC diode. An active PFC switch is basically an AC/DC converter, which controls the current supplied to the consumer via a pulse width modulation (PWM). The PWM controls the turn-on and turn-off switching transitions of the power switch, which separates the intermediate DC voltage in constant pulse sequences. This pulse sequence is then filtered and smoothened by an intermediate capacitor, which generates the DC output voltage.

We have simulated the active PFC shown in Fig. 8.13, and simulated waveforms obtained from our GaN VP have been compared with measured data. Figure 8.16 shows a comparison between the simulated and measured output voltage of the PFC, considering a target operation at 400 V and a load current of 1 A. Voltage and current

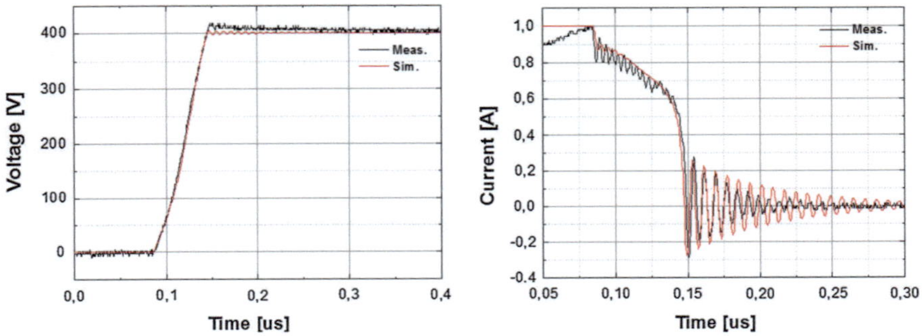

Fig. 8.16 Simulated and measured turn-off transition of a pGaN device at 400 V off-state voltage and for a load current of 1 A

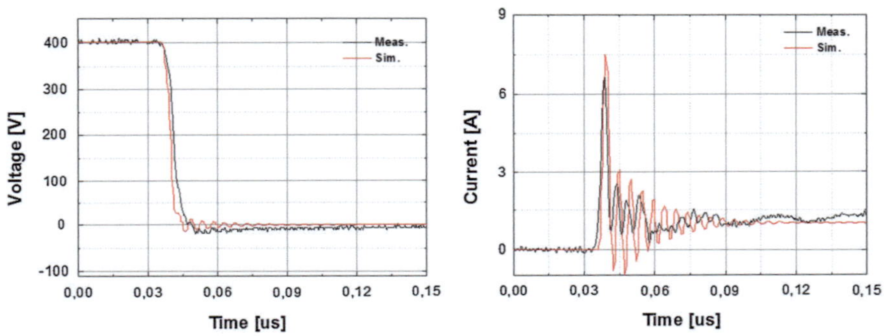

Fig. 8.17 Simulated and measured turn-on transition of a pGaN device at 400 V off-state voltage and for a load current of 1 A

waveforms are separately considered. A turn-off transition is considered in the first case. As clearly visible in the plots, the normally-off GaN VP is capable of accurately predicting both waveforms, including also the oscillation caused by the parasitics. Indeed, both the oscillation amplitude and frequency can be precisely modelled.

Figure 8.17 shows instead a comparison between the simulated and measured switching performance, at the same bias conditions as before, this time for a turn-on transition. Again, the normally-off GaN VP is capable of accurately predicting both waveforms, including also the oscillation caused by the parasitics. The current peaks and oscillations during the turn-on transition are accurately predicted.

Thanks to the full-system approach, we are able to obtain a very high accuracy in predicting the current and voltage waveforms. The very detailed GaN modelling at device level and also the full-board simulation by Q3D and the extraction of the RLC matrix of parasitics introduced during packaging and PCB layout, allow not only to accurately predict the steady-state values of voltages and currents after turn-on and turn-off transitions but also to predict the voltage overshoots, current

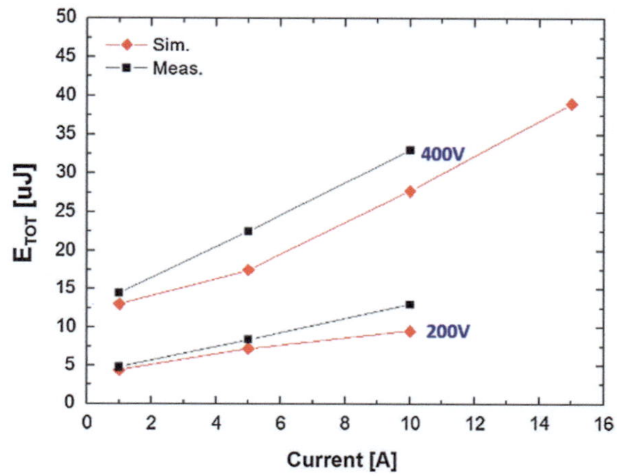

Fig. 8.18 Total switching energy of the normally-off pGaN transistor. Comparison between measurements and simulations is shown

spikes, and oscillations caused by the "unwanted" RLC parasitic elements present in the packages and in the application board.

As a consequence, we should be able to reasonably predict the overall switching performance of GaN devices in real industrial applications, since voltage and current waveforms can be very precisely estimated. Indeed, Fig. 8.18 summarizes the overall system performance of our pGaN devices inserted in a classic PFC circuit. The total switching losses, obtained as the sum of the switch-on and switch-off energy losses, are plotted as obtained from the direct characterization of our board and from the GaN VP approach. The comparison is made for two different target operation voltages, i.e., 200 and 400 V, and for a load current ranging from 0.1 A up to a maximum of 16 A.

A maximum mismatch between measurements and simulations of about ~20 % is found, which is also in the same order of the characterization uncertainty. This gives high confidence on the system modelling strategy previously described and especially on its capability of predicting the system-level performance of even disruptive and immature technologies, such as GaN technology for power applications.

It is worth noticing that a possible root cause of the small mismatch between measurements and simulations has been identified in the asymmetric driving scheme used to counterbalance possible effects of unwanted capacitive spurious turn-on. Indeed, as it will be shown in the next section, the same analysis performed on a normally-on GaN transistor with conventional driving scheme shows a better accuracy and almost no difference between simulated and measured data.

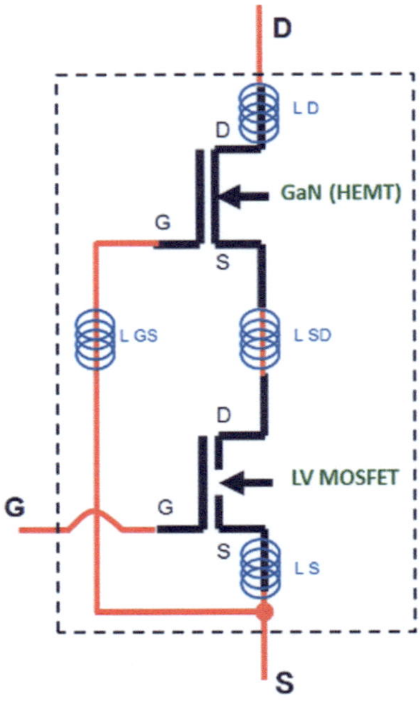

Fig. 8.19 Schematic of the GaN cascode approach, comprising a high-voltage GaN HEMT and a low-voltage Si MOSFET

8.6.2 Normally-on HEMT: Cascode Design

Very similarly to the investigation conducted for the normally-off pGaN transistors, we have designed and characterized also application boards for normally-on GaN transistors in cascode configuration. The cascode approach allows to have a three-terminal enhancement transistor having at disposal a normally-on GaN technology and consists in the co-packaging of the GaN device in series with a normally-off Si transistor. Moreover, the realization of a high-voltage cascode approach requires the use of only a low-voltage Si transistor, e.g., a 20- to 30-V rated device. Figure 8.19 shows a schematic representation of the GaN cascade, where the gate of the high-voltage GaN HEMT is connected to the source of the LV MOSFET, while the source of the GaN HEMT and the drain of the Si MOSFET are tied together. The gate of the Si MOSFET is used as overall input terminal, and the switching on and off of the GaN device is obtained only indirectly via the switching of the Si transistor. Also highlighted in the schematic are the parasitic inductances introduced during assembling.

Clear advantages of this approach are the robustness of the input device, i.e., a consolidated and well-known Si device, the possibility of using standard driving scheme with clear advantage in terms of time to market and easy of usage. However, several drawbacks can be also identified in the cascode approach. First, the presence of the low-voltage (LV) MOSFET implies some performance loss due to the extra capacitances that must be charged and discharged during switching and due to the

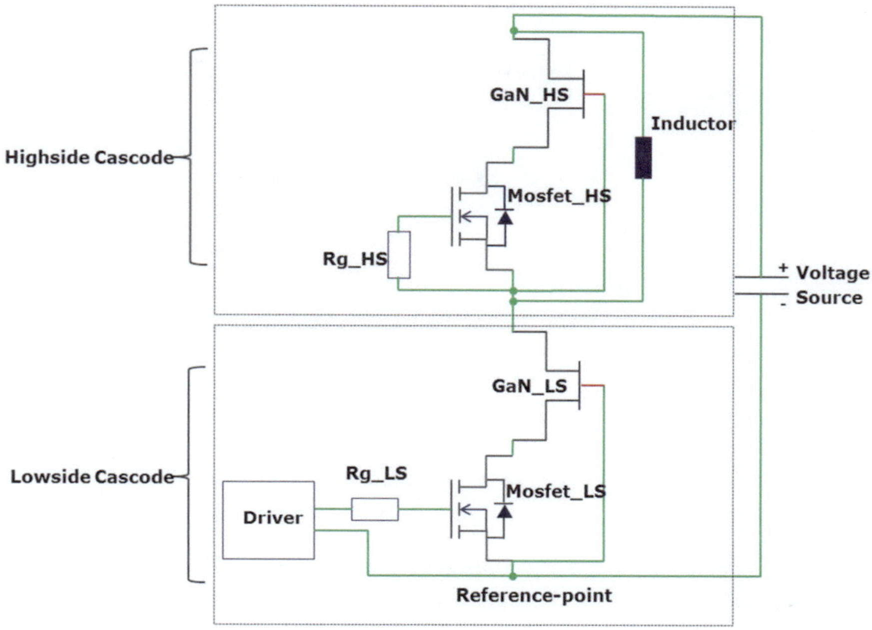

Fig. 8.20 Schematic of the half-bridge GaN cascode board

series resistances which add to the total R_{DSON}. Second, this concept is particularly suited for high-voltage operation (600 V) but loses progressively all its advantages when the operating voltage is scaled down. Finally, one of the most serious issues with the cascode approach is represented by the risks of unwanted oscillation and loss of controllability which can cause the avalanche breakdown in the Si device. In particular, during turn-off, the voltage in the middle terminal of the cascode can rise to a value which is higher than the maximum rated voltage of the LV MOSFET. At this point, the LV MOSFET enters avalanche conditions with consequent performance degradation of the cascode and possible reliability issues for both components. Avalanche issues are strongly linked to the parasitic inductances introduced during assembling, as represented in Fig. 8.19 [30–32].

To study the behavior of the GaN cascode approach, we have performed extensive TCAD simulations of insulated-gate normally-on transistors. Details on the device concepts and technology have been already described in Sect. 8.2. Moreover, we have designed, fabricated, and characterized cascode application boards. In particular, Fig. 8.20 shows the schematic of the cascode board which has been designed in order to get more insight into the behavior of GaN devices in totem-pole PFC applications [33, 34].

Indeed, the board is designed in such a way that the low-side GaN transistor is switching against another high-side GaN transistor. And both are mounted in cascode configuration. A large inductor defines the load current and a rated voltage operation

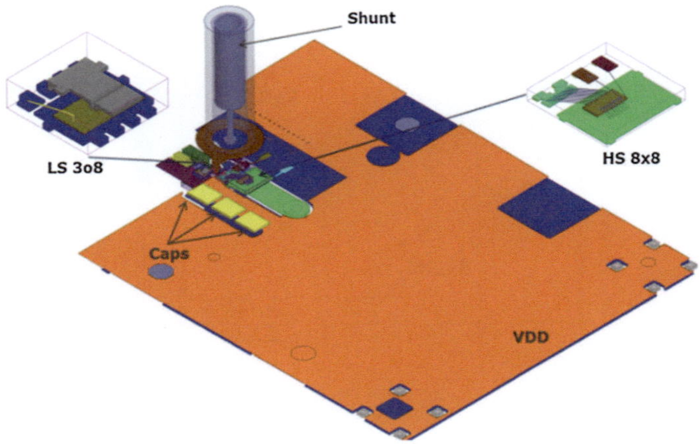

Fig. 8.21 Application board for the GaN cascode approach

at 400 V has been considered. Two different drivers are also considered, one for the low-side cascode and one for the high-side cascade. However, for the present investigation, the high-side device is kept in off-state condition, and the main emphasis of the study has been to assess the potential risk of avalanche of the low-side cascode transistor. The lesson learnt in this case can afterward be applied also to the high-side case.

As previously explained, the unwanted avalanche of the low-voltage silicon transistor during application causes the contemporary loss of the stability of the cascode with consequent large increase in energy losses and also a rapid deterioration of the reliability of both the LV Si MOSFET and the HV GaN HEMT. For these reasons, it is mandatory to optimize the GaN HEMT, the LV MOSFET, and the package in such a way to prevent spurious turn-on and avalanche effects. Clearly, performance must also be optimized at the same time.

An important factor in determining the stability of the cascode is represented by the parasitic inductances and resistances introduced during packaging. In particular, it has been identified that a very sensitive element that affects the overall performance and stability of the GaN cascode approach is represented by the parasitic inductance between the source of the GaN HEMT and the drain of the LV MOSFET. For this reason, we have designed a special application board of the cascode approach which makes use of two discrete packages for the two individual component of the cascade. In particular, the GaN HEMT is packaged in the Infineon standard ThinPaK 8 × 8, whereas the LV MOSFET is packaged in the standard Infineon S308 package.

Figures 8.21 and 8.22 show different views of the cascode board. Clearly visible are also the stabilization capacitances and the shunt used to measure the load current.

More importantly, the two components of the cascade, for both high side and low side, are mounted on the opposite side of the PCB. In this way, the parasitic inductance (L_{SD}) between the source of the GaN and the drain of the LV MOSFET is

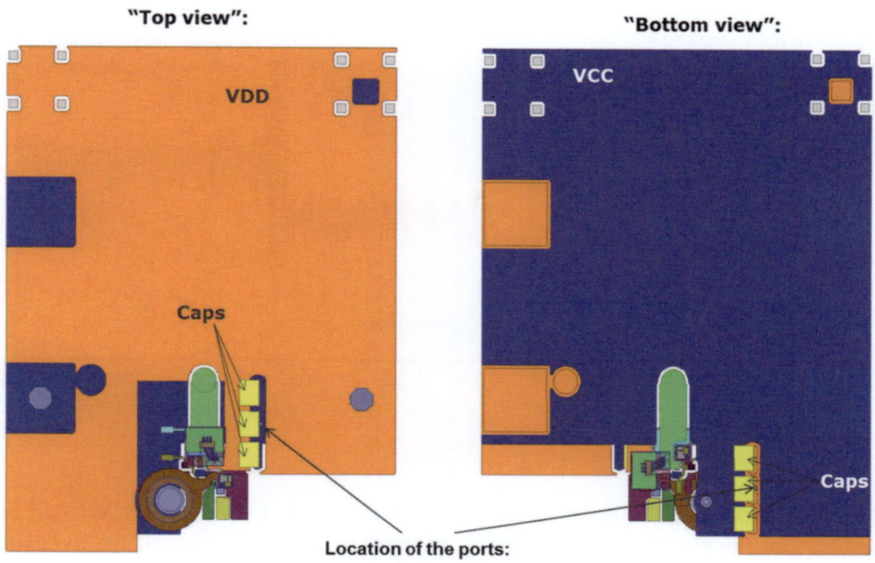

Fig. 8.22 Top view and bottom view of the application board for the GaN cascode approach

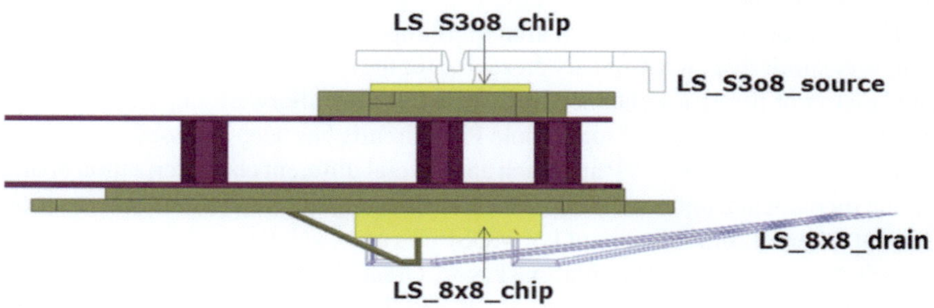

Fig. 8.23 Front view of the application board for the GaN cascode approach with emphasis on the low-side cascade. The GaN HEMT and the LV MOSFET are mounted on the opposite sides of the PCB

determined mainly by three components: the parasitic inductance of the two packages, ThinPaK and S308, and the thickness of the PCB.

We have also fabricated boards with different thicknesses in order to arbitrarily vary the parasitic L_{SD} and study its impact on the overall stability of the cascode approach. Details on the PCB thickness and on the mounting of the GaN and Si chips are shown in Fig. 8.23.

The disadvantages of the proposed approach, with respect to a fully integrated one, is clearly the larger parasitics introduced. However, the main advantage is represented by the flexibility of the approach (individual components can be easily

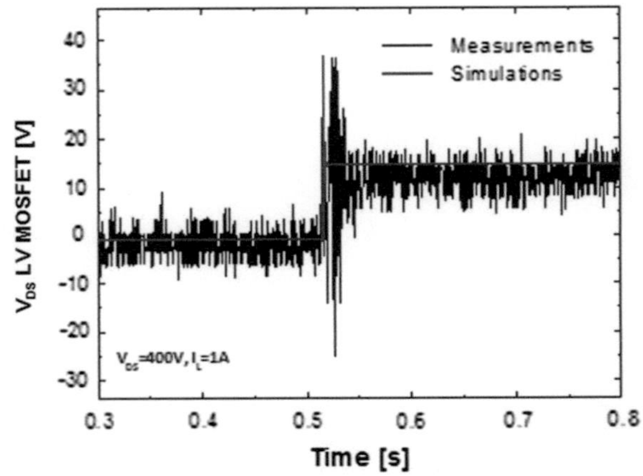

Fig. 8.24 Drain-to-source voltage of the LV MOSFET during a turn-off transition

changed and different combinations of LV MOSFET and HV GaN HEMT can be tested) and the possibility to arbitrarily change the absolute value of the parasitic inductance L_{SD}. It is worth mentioning that the main purpose of our approach was not to fully exploit the GaN performance in totem-pole PFC, rather the design of a fast and reliable tool that could be used for the fundamental understanding of the main issues of the cascode design and to draw guidelines towards its optimization.

The cascode boards have been afterward imported and simulated with the Q3D tool, very similarly to what already presented for the normally-off case. The extracted RLC matrix of parasitics obtained from Q3D has been afterward imported in SIMetrix, and the performance of totem-pole PFC circuits has been assessed.

We have used the GaN VP approach and several different characterization boards for the fundamental understanding of the cascode properties and for its optimization. Several different combinations of LV MOSFETs and GaN devices have been studied as well as the impact of different packages. Great attention has been given to the possible risks of avalanche of the LV MOSFET and also to possible voltage and current spikes on the different terminals of the GaN HEMT during transitions. Figure 8.24 shows, for example, the drain-to-source voltage of the LV MOSFET during a turn-off transition. Simulated and measured data are reported for comparison. A 30-V rated device was considered in this case. As clearly visible, if the cascode is not properly optimized, the V_{DS} of the LV MOSFET can exceed the maximum rated voltage and the Si device enters in the avalanche region with consequent performance degradation and instabilities of the cascode system. The detailed optimization of the GaN cascode is outside the scope of this chapter. Interested readers can refer to several published data by different groups [30–32].

To conclude, we have also designed and tested normally-on boards to assess the properties of the normally-on GaN technology, in a very similar way to what has been already presented for normally-off and for cascaded devices.

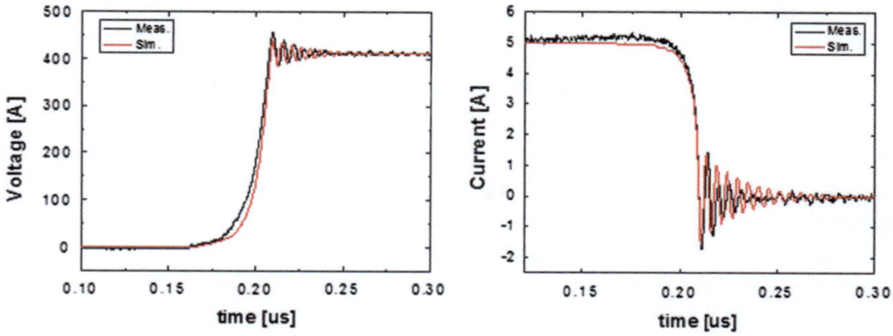

Fig. 8.25 Simulated and measured turn-on transition of a normally-on GaN device at 400 V off-state voltage and for a load current of 5 A

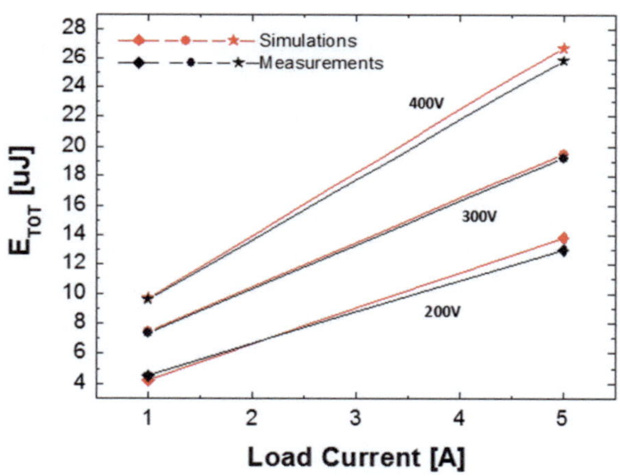

Fig. 8.26 Total switching energy of the normally-off pGaN transistor. Comparison between measurements and simulations is shown

Figures 8.25 and 8.26 summarize the main outcomes of the comparison between the VP approach and the direct measurements of the performance of normally-on GaN transistors inserted in power factor correction circuits. The GaN VP is capable to accurately predict both current and voltage waveforms during turn-on and turn-off transitions, and this allows a very reliable estimation of the total switching losses in PFC applications. Indeed, Fig. 8.26 shows the measured and simulated total switching losses ($E_{ON} + E_{OFF}$) for a load current ranging from 1 A up to 5 A and for three different output voltages, 200, 300, and 400 V.

8.7 Conclusions

We have presented a system-level modelling approach for GaN HEMT technologies that are capable to efficiently link and mimic one to one the different steps of manufacturing of a typical semiconductor product: from technology development, to assembling and to final end customer application. The GaN virtual prototype approach presented is composed by four main steps: (i) TCAD device simulations; (ii) compact model development and calibration; (iii) application boards design and simulation with the Q3D Ansys tool; and (iv) Spice application circuit simulations.

We believe that due to the very complex nature of GaN-on-silicon technology, a full-system approach is needed where technology, packaging, and application are all optimized at the same time. Our proposed approach has proven to be fast, robust, and reliable on several different technologies and allows to accurately predict the system-level performance of GaN-based transistors in switch-mode power supply applications.

We have presented extensive comparison between measurements and simulations and the GaN VP has been applied to three different case studies: normally-off pGaN technology, GaN normally-on plus LV silicon MOSFET cascode technology, and insulated-gate normally-on GaN technology. In all three cases, excellent agreement between predictions and real measurements has been shown, and the GaN VP has demonstrated its capabilities to be predictive and accurate starting from the device-level performance up to the system-level switching performance.

In general, the GaN VP can assist both technologists and designers and should be used, in our opinion, as a general guideline tool for GaN system optimization with clear advantages in terms of decision making, budget resources, and fastening GaN products introduction into the market.

References

1. Treu M, Vecino E, Pippan M et al (2012) The role of silicon, silicon carbide and gallium nitride in power electronics. In IEDM Technical Digest, pp 147–150
2. Curatola G, Kassmanhuber A, Yuferev S, Franke J, Pozzovivo G, Lavanga S, Prechtl G, Detzel T, Haeberlen O (2014) GaN virtual prototyping: from traps modeling to system-level cascode optimization. In Proceedings of the 44th European solid-state device research conference (ESSDERC 2014), Venezia (Italy), 22–26 Sept 2014, pp 337–340
3. Haffouz S, Tang H, Bardwell JA et al (2005) AlGaN/GaN field effect transistors with C-doped GaN buffer layer as an electrical isolation template grown by molecular beam epitaxy. Solid-State Electr. 49:802–807
4. Kato S, Satoh Y, Sasaki H et al (2007) C-doped GaN buffer layers with high breakdown voltages for high power operation AlGaN/GaN HFETs on 4-in Si substrates by MOVPE. J Cryst Growth 298:831–834
5. Fang Z-Q, Claflin B, Look DC et al (2010) Deep traps in AlGaN/GaN heterostructures studied by deep level transient spectroscopy: Effect of carbon concentration in GaN buffer layers. J Appl Phys 108:063706

6. Bahat-Treidel E, Brunner F, Hilt O et al (2010) AlGaN/GaN/GaN: C back-barrier HFETs with breakdown voltage of over 1 kV and low RON × A. IEEE Trans Electr Dev 57:3050–3058
7. Ramdani MR, Chmielowska M, Cordier Y et al (2012) Effect of carbon doping on crystal quality, electrical isolation and electron trapping in GaN based structures grown silicon substrates. Solid-State Electr. 75:86–92
8. Meneghesso G, Silvestri R, Meneghini M, Cester A, Zanoni E, Verzellesi G, Pozzovivo G, Lavanga S, Detzel T, Haberlen O, Curatola G (2014) Threshold voltage instabilities in D-mode GaN HEMTs for power switching applications. In Proceedings of the 52nd IEEE international reliability physics symposium (IRPS 2014), Waikoloa (HI, USA), 1–5 June 2014, pp 6C.2.1–6C.2.5
9. Verzellesi G, Morassi L, Meneghesso G, Meneghini M, Zanoni E, Pozzovivo G, Lavanga S, Detzel T, Häberlen O, Curatola G (2014) Influence of buffer carbon doping on pulse and AC behavior of insulated-gate field-plated power AlGaN/GaN HEMTs. IEEE Electr Dev Lett 35(4):443–445
10. Curatola G, Huber M, Daumiller I, Haeberlen O, Verzellesi G (2015) Off-state breakdown characteristics of AlGaN/GaN MIS-HEMTs for switching power applications. In Proceedings of the 2015 IEEE conference on electron devices and solid-state circuits, Singapore, June 2015
11. Synopsys Inc. (2012) Sentaurus device user guide, Version G-2012.06
12. Ambacher O, Smart J, Shealy JR et al (1999) Two-dimensional electron gases induced by spontaneous and piezoelectric polarization charges in N- and Ga-face AlGaN/GaN heterostructures. J Appl Phys 85:3222–3233
13. Wright AF (2002) Substitutional and interstitial carbon in wurtzite GaN. J Appl Phys 92:575–2585
14. Amstrong A, Poblenz C, Green DS et al (2006) Impact of substrate temperature on the incorporation of carbon-related defects and mechanism for semi-insulating behavior in GaN grown by molecular beam epitaxy. Appl Phys Lett 88:082114
15. Lyons JL, Janotti A, Van de Walle CG (2010) Carbon impurities and the yellow luminescence in GaN. Appl Phys Lett 97:152108
16. Johnstone D (2007) Summary of deep level defect characterization in GaN and AlGaN. Proc SPIE 6473:64730
17. Tapajna M, Jurkovic M, Valik L, Hascik S, Gregusov D, Brunner F, Cho E-M, Kuzmik J (2013) Bulk and interface trapping in the gate dielectric of GaN based metal-oxide-semiconductor high-electron-mobility transistors. Appl Phys Lett 102:243509
18. Lagger P, Schiffman A, Pobegen G, Pogany D, Ostermaier C (2013) Very fast dynamics of threshold voltage drifts in GaN-based MIS-HEMTs. IEEE Electr Dev Lett 34:1112–1114
19. Van Hove M, Kang X, Stoffels S, Wellekens D, Ronchi N, Venegas R, Gennes K, Decoutere S (2013) Fabrication and performance of Au-free AlGaN/GaN-on-silicon power devices with Al_2O_3 and Si_3N_4/Al_2O_3 gate dielectrics. IEEE Trans Electr Dev 60(10):3071–3078
20. Binari SC, Klein PB, Kazior TE (2002) Trapping effects in GaN and SiC microwave FETs. Proc IEEE 90:1048–1058
21. Uren MJ, Möreke J, Kuball M (2012) Buffer design to minimize current collapse in GaN/AlGaN HFETs. IEEE Trans Electr Dev 59:3327–3333
22. Umeda H, Takizawa T, Anda Y, Ueda T, Tanaka T (2013) High-voltage isolation technique using Fe ion implantation for monolithic integration of AlGaN/GaN transistors. IEEE Trans Electr Dev 60(2):771–775
23. www.infineon.com
24. Uemoto Y et al (2007) Gate injection transistor (GIT)—a normally-off AlGaN/GaN power transistor using conductivity modulation. IEEE Elect Dev 54:3393–3399
25. www.epc-co.com
26. Radhakrishna U, Piedra D, Zhang Y, Palacios T, Antoniadis D (2013) High voltage GaN HEMT compact model: experimental verification, field plate optimization and charge trapping, IEEE Electronic Device Meeting (IEDM)
27. Koudymov A, Shur MS, Simin G (2007) Compact model of current collapse in heterostructure field-effect transistors. IEEE Elec Dev Lett 28:332–335

28. www.ansys.com
29. www.simetrix.co.uk
30. Liu Z, Huang X, Lee FC, Li Q (2014) Package parasitic inductance extraction and simulation model development for the high-voltage cascode GaN HEMT. IEEE Trans Power Electron 29
31. Huang X, Li Q, Liu Z, Lee FC (2013) Analytical loss model of high voltage GaN HEMT in cascode configuration, ECCE Conference, pp 3587–3594
32. Huang X, Dum W, Liu Z, Lee FC, Li Q (2014) Avoiding Si MOSFET avalanche and achieving true zero-voltage-switching for cascode devices, ECCE conference, pp 106–112
33. Lingxiao X, Zhiyu S, Boroyevich D, Mattavelli P (2015) GaN-based high frequency totem-pole bridgeless PFC design with digital implementation. APEC 2015:759–766
34. Zhou L, Wu Y, Honea J, Wang Z (2015) High-efficiency true bridgeless totem pole PFC based on GaN HEMT: design challenges and cost-effective solution. PCIM Europe 2015:1–8

Chapter 9
Performance-Limiting Traps in GaN-Based HEMTs: From Native Defects to Common Impurities

Isabella Rossetto, Davide Bisi, Carlo de Santi, Antonio Stocco, Gaudenzio Meneghesso, Enrico Zanoni and Matteo Meneghini

A significant effort has been recently spent with the aim of investigating the properties of the defects that limit the dynamic performance of GaN-based power transistors and of correlating the results obtained through the different deep-level characterization techniques with the microscopic origin of the defect. A correct identification is important, since it provides useful information about the improvements needed in the growth process that could enhance the performance of the devices.

Several experimental and theoretical issues make the comparison between different papers difficult. The results of deep-level investigation can be significantly dependent on the technique used for the analysis, mainly due to the specific physical quantities and the procedures used for the excitation of the deep level and for the detection of the trap-sensitive parameter. Moreover, the electric field at the trap location plays an important role, since it may lead to an underestimation of the activation energy (E_a) through the emission barrier lowering caused by the Poole–Frenkel effect. The presence of a capture barrier implies a further problem, related to the different level of trap filling during temperature- or field-dependent measurements. For this reason, in the following, we will group together reports that may reasonably refer to the same deep level.

If we take into account all the possible growth techniques and used materials, the list of possible defect types may be ascribed to the following categories:

I. Rossetto · D. Bisi · C. de Santi · A. Stocco (✉) · G. Meneghesso
E. Zanoni · M. Meneghini
Department of Information Engineering, University of Padova, Padua, Italy
e-mail: rossett6@dei.unipd.it

M. Meneghini et al. (eds.), *Power GaN Devices*,
Power Electronics and Power Systems, DOI 10.1007/978-3-319-43199-4_9

- Native defects: These are the defects that may be present in bulk GaN due to its crystalline structure. They can be summarized in: vacancies, i.e., missing nitrogen (V_N) or gallium (V_{Ga}) atoms in their usual position in the lattice; antisites, when a nitrogen atom occupies the expected location of a gallium atom (N_{Ga}) or vice versa (Ga_N); and interstitials, caused by a nitrogen (N_i) or gallium (Ga_i) atom in a position where no element should be present.
- Impurities: These are foreign atoms inside the gallium nitride. They can be in interstitial or substitutional position if they are located at a typically unoccupied location of the GaN crystal or at the place of a Ga or N atom in the lattice, respectively. They may be expected dopant atoms (magnesium, silicon, iron, carbon), elements used in the growth chamber in precursors or flow gases (carbon, hydrogen), or impurities incorporated from the atmosphere (hydrogen, oxygen). Although every element may have functional benefits (such as the dopants), it creates also unwanted allowed energy levels inside the forbidden bandgap. When studying impurities, it is important to point out the difference between deep levels specifically caused by the actual impurity species and intrinsic defects whose concentration is enhanced by the presence of the external atom.
- Extended defects: These are larger defects, created by more than a single atom, and their actual structure may be complex. They may be two- or three-dimensional extensions of native defects, or originate from particular configurations of the crystalline lattice, as in the case of stacking faults and edge or screw dislocations. Extended defects may also be composed by clusters involving native defects and/or impurities and by complexes or other arrangements of multiple defects (even other extended defects).
- Surface defects: They may be point or extended defects located at the outermost layer of the device. Here, at the interface between the lattice and a non-crystalline material, the periodicity structure is totally and abruptly stopped, leading to the formation of dangling bonds, bonds with impurities, and loop bonds. The analysis of these defects is difficult since the band diagram formulation does not apply when periodicity is absent. However, the concentration of surface defects may be typically lowered through proper passivation and/or surface treatment.

In the following, we will describe the main deep levels associated with these defects, taking into account data and interpretations from more than 80 papers on different devices, growth methods, and characterization techniques.

Nitrogen vacancies are one of the most common defects and are reported to behave as mid-shallow donors, covering almost the entire range from $E_C - 0.089$ eV to $E_C - 0.26$ eV [1–15]. Even though they are present typically as point defects, there are also reports of an extended defect behavior at $E_C - 0.19$ eV [6], $E_C - 0.23$ eV [6, 9], and $E_C - 0.25$ eV [4, 16] likely due to a connection with dislocations. Two additional levels at $E_C - 0.35$ eV [5] and $E_C - 0.613$ eV [7] are probably related to V_N complexes. The level at $E_C - 0.24$ eV was tentatively associated with either nitrogen or gallium vacancies by Honda et al. [12]; the comparison with further works reported in the literature on the topic suggests the first hypothesis.

The reports on nitrogen antisites are coherent and place them as electron traps in the narrow range from $E_C - 0.5$ eV to $E_C - 0.664$ eV [3, 6, 9, 12, 15–20]. No extended defect behavior has been described.

Few papers identify nitrogen interstitials. Three levels are referenced at $E_C - 0.76$ eV [21], $E_C - 0.89$ eV [20], and $E_C - 1.2$ eV [5].

Gallium vacancies are another commonly reported defect in GaN, probably due to the lower formation energy compared to interstitials and substitutional atoms. They introduce several allowed energy levels. The one at $E_C - 0.24$ eV [12] was tentatively associated with either nitrogen or gallium vacancies, but this level is clearly linked to nitrogen in other papers as previously described; therefore, we may rule it out as gallium-related. Another set of data is grouped around $E_C - 0.62$ eV [13, 18] and may relate to a gallium vacancy–oxygen complex, as an additional level at $E_C - 1.118$ eV [18], or to dislocations, which seems to be the most likely hypothesis through comparison with other papers. Other complexes with hydrogen create a deep level at $E_C - 2.49$ eV [22, 23] and another level at $E_C - 2.62$ eV [11, 14, 24, 25], which is possibly related to hydrogen. A last level associated with gallium vacancies is at $E_C - 2.85$ eV [26].

Dislocations, due to their extended nature and to the formation of complexes, are known to induce a broad range of deep levels in the upper half of the energy gap. Three series of activation energies at $E_C - 0.18$ eV [2, 6], $E_C - 0.24$ eV [2, 6, 9], and $E_C - 0.27$ eV [2] probably refer to the same broad level, possibly involving nitrogen vacancies [6, 9]. Another group of levels is in the range from $E_C - 0.59$ eV to $E_C - 0.642$ eV [2, 14, 18]. A single level is reported at $E_C - 0.8$ eV [16] which may be related to intrinsic defects arranged along dislocations, but its origin and its possible characteristic as AlGaN-specific defect are still unclear. The set of levels at $E_C - 1$ eV [5], $E_C - 1.02$ eV [27], and $E_C - 1.118$ eV [18] may also refer to the same broad deep level caused by dislocations.

Figure 9.1 reports a graphical description of the allowed energy levels introduced in the energy gap by intrinsic defects and dislocations. Zero energy corresponds to the upper border of the valence band, the thick black line to the lower border of the conduction band, blue lines to deep levels identified with acceptable certainty, while red levels are tentatively ascribed to their specific cause.

Hydrogen is typically present in complexes involving other defects. A level at $E_C - 0.62$ eV [27] has been related to the formation of Mg–H complexes. Two reported levels grouped at $E_C - 2.48$ eV [23] and $E_C - 2.62$ eV [11] may both involve hydrogen in complexes with gallium vacancies; therefore, they may be part of a same broad deep level centered around $E_C - 2.55$ eV. Other deep levels at $E_C - 0.14$ eV [28], $E_C - 0.49$ eV [28], and $E_C - 0.578$ eV [29] were tentatively ascribed to either hydrogen or carbon impurities.

Few reports are available on oxygen-related levels, since oxygen impurities have very low concentration when proper growth conditions and insulation from the atmosphere are ensured. Oxygen is known to behave as donor at $E_C - 0.44$ eV [30] when in nitrogen substitutional position (O_N). Some levels associated with V_{Ga}–oxygen complexes or dislocations are reported at $E_C - 0.599$ eV [18], $E_C - 0.642$ eV [18], and $E_C - 1.118$ eV [18].

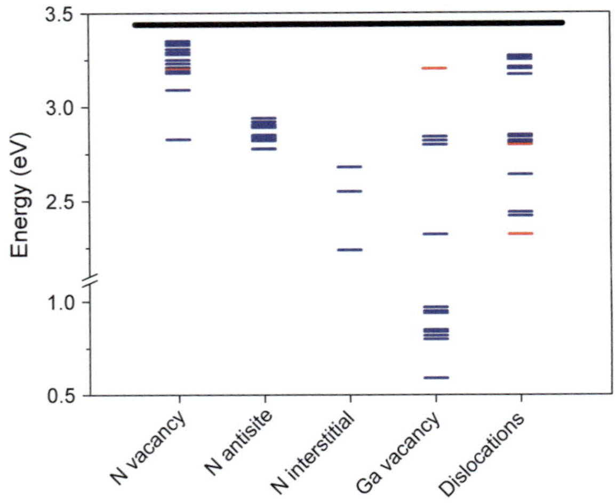

Fig. 9.1 Allowed energy levels inside the forbidden GaN gap caused by intrinsic defects, dislocations, and their complexes. Zero energy corresponds to the upper border of the valence band, the *thick black line* to the lower border of the conduction band, *blue lines* to deep levels identified with acceptable certainty, while red levels are tentatively ascribed to their specific cause

Magnesium is an important element used in order to achieve p-type doping and is often associated with hydrogen due to the strong Mg–H bonds. A defect related to the formation of Mg–H complexes was reported at E_C – 0.62 eV [27], which is probably the same state as the level at E_C – 0.597 eV reported in [31]. Another electron trap caused by Mg is located at E_C – 0.355 eV [1]. The actual acceptor level is at nearly E_C – 3.2 eV [11, 32]. Therefore, the defect reported at E_C – 3.22 eV [25] as either C- or Mg-related is probably induced by magnesium.

Silicon is commonly used as shallow donor dopant (ionization energy 14 ÷ 30 meV in GaN). It creates other deep levels at E_C – 0.37 eV [2], E_C – 0.4 eV [2], and E_C – 0.59 eV [33].

Carbon is often used in gallium nitride to compensate the intrinsic n-type conductivity. The actual role of carbon is quite complicated due to its double acceptor and donor (amphoteric) nature, and the possible consequent self-compensation mechanism. The acceptor behavior is obtained through carbon in nitrogen substitutional position (C_N), causing a band in the region from E_C – 3.24 eV to E_C – 3.31 eV [1, 11, 14, 23, 26, 34, 35]. The additional level reported at E_C – 3.22 eV [25] as either C- or Mg-related is likely related to magnesium, since it falls slightly outside the carbon band and closely matches other magnesium levels [11, 32]. Another level caused by carbon-related defects and ascribed to carbon substitutionals is located at E_C – 2.58 eV, as supported by hybrid functional calculations [12, 36]. Carbon in interstitial positions (C_i) causes a deep level in the range from E_C – 1.28 eV to E_C – 1.35 eV [11, 14, 25, 37]. C-related donor is located at E_C – 0.4 eV [12], tentatively related to carbon in gallium substitutional position (C_{Ga}). Other donor levels related to carbon or hydrogen are located at E_C – 0.14 eV [28], E_C – 0.49 eV [28], and E_C – 0.578 eV [29].

Iron is another species used to obtain (semi-)insulating GaN. One level at E_C – 0.34 eV [38] was related to $Fe^{3+/2+}$ deep acceptor, and no specific transition is given for levels E_C – 0.397 eV [17] and E_C – 0.4 eV [39], which are likely the

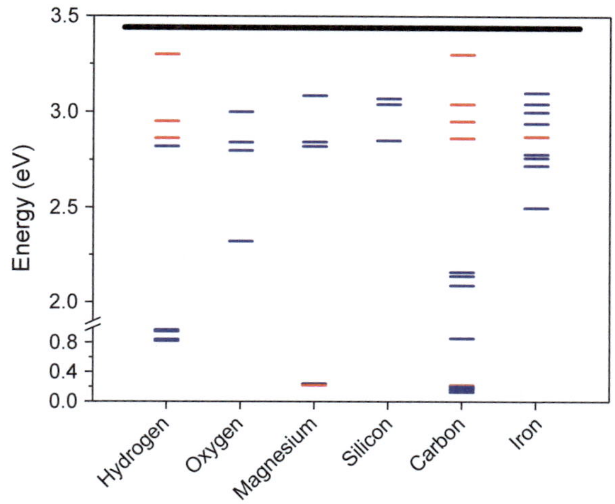

Fig. 9.2 Allowed energy levels inside the forbidden GaN gap caused by intrinsic defects, dislocations, and their complexes. Zero energy corresponds to the upper border of the valence band, the thick *black line* to the lower border of the conduction band, *blue lines* to deep levels identified with acceptable certainty, while *red* levels are tentatively ascribed to their specific cause

same level. A Fe-induced level was located at $E_C - 0.5$ eV [24] and another tentatively at $E_C - 0.57$ eV [40]. A defect related to iron is found between $E_C - 0.66$ eV and $E_C - 0.72$ eV [41], whose actual activation energy depends on the dopant concentration. A last level was detected at $E_C - 0.94$ eV [39].

Figure 9.2 describes all the deep levels related to impurities in a single graphical description, as is shown in Fig. 9.1 for intrinsic defects and dislocations.

By comparing Figs. 9.1 and 9.2, a deep level at $E_V + 2.85$ eV can be noticed for almost every impurity. A similar level was also attributed to vacancies, N antisites, and dislocations. We can assume it may be a native defect whose concentration is increased by the presence of impurities, probably related to extended defects and/or complexes.

Surface states are more difficult to describe due to the previously described theoretical limitations. A deep level ($E_C - 0.12$ eV [15]) was detected in n-GaN possibly induced by high-energy electrons. Other reports describe results obtained on AlGaN/GaN HEMTs; therefore, they refer to traps at the AlGaN surface. Described traps exhibit capture and emission characteristics compatible with deep levels at $E_C - 0.57$ eV [37], $E_C - 2.3$ eV [34], $E_V + 0.57$ eV [42], $E_V + 2.3$ eV [43], and $E_V + 0.578$ eV [43].

The complete signature of a defective state is composed not only by the activation energy, which describes the temperature dependency of the capture/emission process, but also by the typical response time window, usually related to the apparent capture cross section. The Arrhenius plot in Fig. 9.3 summarizes and compares these two quantities for the previously described papers. By comparing the experimental data with this database, it is possible to extract useful information on the likely physical cause of the detected performance variation.

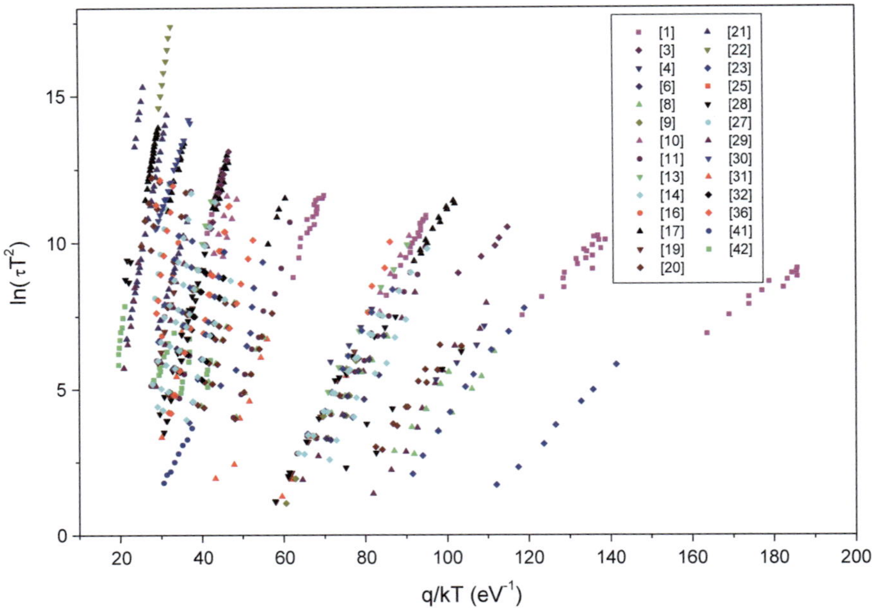

Fig. 9.3 Summary of the signatures of the collected deep levels

9.1 Surface-Related Trapping

One of the key issues affecting the dynamic properties of the GaN high-electron-mobility transistors is the charge trapping related to surface states. During high-field off-state conditions, parasitic electrons could be injected from the gate metal to the epitaxial surface, where they are readily available to be trapped at surface states. Surface states include donor-like states required for the formation of the 2DEG at the AlGaN/GaN interface [44]. If negative charge is trapped at the surface, the surface potential decreases and the ungated region depletes, suppressing the 2DEG carrier concentration and promoting current collapse. Discovered in the early 2000s [45–47], this phenomenon was historically referred to as "virtual gate" [46]. An effective solution to drastically mitigate the surface-related current collapse is the introduction of a passivation layer (usually silicon nitride) on top of the ungated access regions [48–51].

In the following, we report on the signature of surface-related trapping by means of pulsed I−V and drain current transient spectroscopy.

Devices under test belong to a split experiment wafer implementing both non-passivated and passivated devices. Devices under test were grown by metal-organic chemical vapor deposition (MOCVD) on a silicon carbide substrate. The epitaxial structure consists of a 20-nm $Al_{0.36}Ga_{0.64}N$ barrier layer, a 0.7-nm AlN layer, a 10-nm GaN channel layer, a 1-μm-thick $Al_{0.04}Ga_{0.96}N$ back-barrier layer, and a

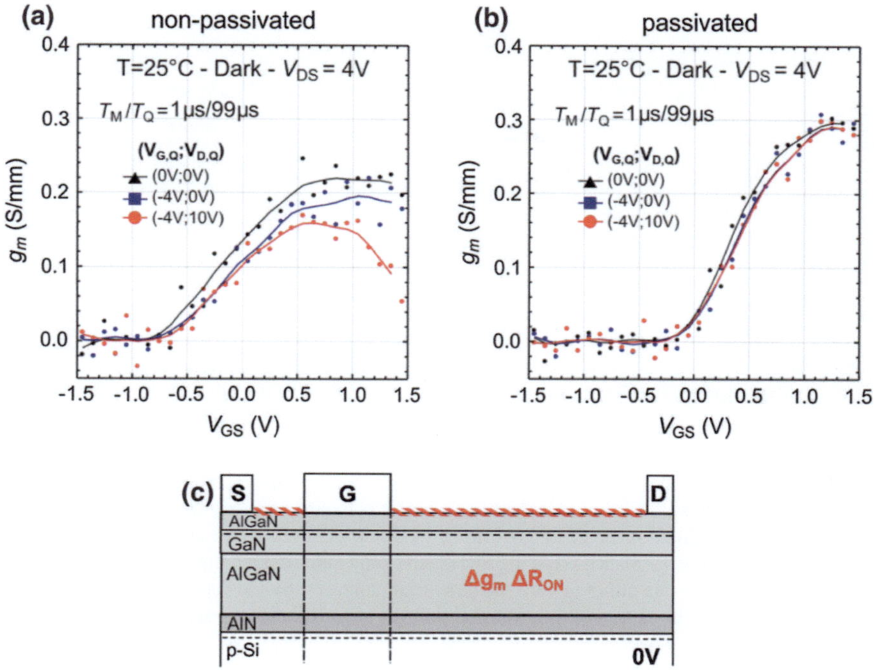

Fig. 9.4 Pulsed g_m-V_G measured on **a** non-passivated and **b** passivated HEMTs. When exposed to off-state quiescent bias point, non-passivated devices feature noticeable transconductance peak variation and no V_{TH} shift

300-nm $Al_{0.09}Ga_{0.91}N$ layer. Gate length is 1 µm, and gate–drain and gate–source distances are equal to 2 µm and 1 µm, respectively.

When exposed to off-state bias (see pulsed measurements in Fig. 9.15), non-passivated devices experience a measurable reduction of the transconductance, with negligible shift of the threshold voltage. This behavior suggests that the dominant charge-trapping mechanism takes place in the access regions. Since passivated devices feature negligible dispersion effects (Fig. 9.4), we can confirm that the current dispersion in non-passivated devices can be ascribed to surface-related charge trapping.

Drain current transient spectroscopy in Fig. 9.5 discloses useful information about surface trap signature. When the non-passivated HEMTs are switched on after 100 s at the filling bias (V_{GQ}; V_{DQ}) = (−4 V; 10 V), the drain current features a significant recovery. This recovery is related to the emission of the carriers trapped at the surface of the devices, which determines a gradual reduction of the virtual gate effect responsible for the drain current decrease.

The analysis of the detrapping kinetics indicates that carrier emission is a slow process with a dominant time constant around 100 ms and that the recovery of the current can be fitted by the following stretched exponential function

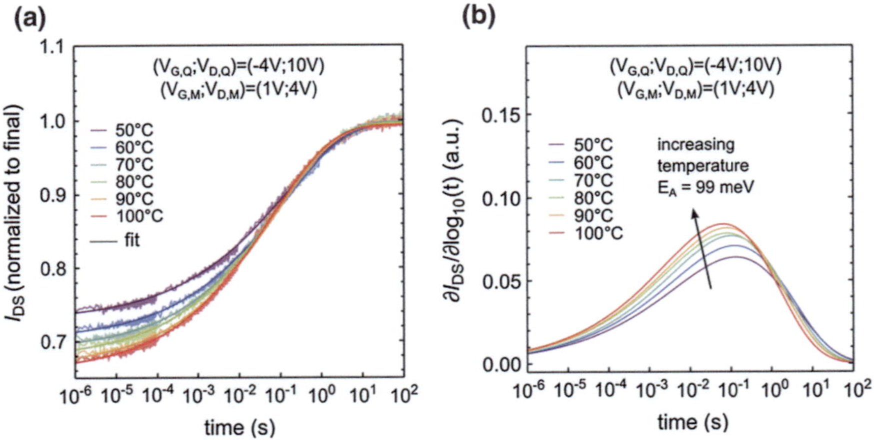

Fig. 9.5 **a** Drain current transient and **b** related spectroscopy acquired at multiple temperatures on a non-passivated HEMT. Highly stretched exponential behavior and very low thermal activation suggest that the dominant mechanism governing the detrapping of surface states and the recovery of pristine electrical performance is not the SRH thermionic emission but the transport mechanisms from the surface trap state either toward the gate/drain contacts (by means of surface conduction) or toward the 2DEG (by means of trap-assisted transport mechanisms through the AlGaN barrier)

$$I_{\mathrm{DS}}(t) = I_{\mathrm{DS,final}} - Ae^{-\left(\frac{t}{\tau}\right)^{\beta}}$$

where the fitting parameters A, τ, and β are, respectively, the amplitude, the typical time constant, and the non-exponential stretching factor of the detected charge emission processes. In this case study, a relatively stretched exponential kinetics with a small stretching factor $\beta \approx 0.3$ was detected. Both results are consistent with a current collapse mechanism related to the presence of surface traps [46]. In addition, temperature-dependent measurements indicated that the detrapping of electrons from surface states is only weakly thermally activated, with activation energy of 99 meV (see Fig. 9.5b). The slow detrapping transients and the low activation energy indicate that thermal emission is not the main mechanism responsible for the release of the electrons trapped at surface states. The emission of the electrons trapped at the surface can be slowed down by other factors: Once the device is switched on (after the filling pulse), the trapped electrons leave the surface, possibly via hopping toward the contacts, through the surface, or through trap states [52–54]. This mechanism may act as a bottleneck for the release of electrons trapped in the gate–drain access region during the detrapping phase, i.e., when the gate–drain electric field is relatively low: as a consequence, the detrapping kinetics can be long, and the time constants may have a relatively low temperature dependence.

9.2 Impact of Iron Doping

The poor carrier confinement in the 2DEG is one of the most representative issues which affect gallium nitride (GaN) high-electron-mobility transistors (HEMTs); this is due to the sensitivity of field effect transistors (FETs) to short-channel effects and to the gallium nitride intrinsic n-type doping. The enhancement of the carrier confinement in the 2DEG, as well as the reduction of leakage components (e.g., vertical leakage current and punch-through currents), grants significant improvements in the device high-power and high-frequency performance and in the pinch-off properties. With this aim, several growth techniques have been reported: heterostructure growth on semi-insulating substrates; use of a hetero-junction structure below the 2DEG (double-heterostructure); and use of a single heterostructure with a highly resistive or semi-insulating GaN layer. The latter can be achieved with the introduction of deep acceptor states through unintentional doping (u.i.d., e.g., with the use of threading dislocations and/or with unintentional impurities) or with the deliberate introduction of acceptor dopants, namely iron (Fe) and carbon (C) [55–57].

A controlled iron doping of the GaN buffer demonstrated a significant impact on the carrier confinement, leading to an important suppression of the leakage current and to a better control of the pinch-off [58–61]. Bougrioua et al. [61] indicated a technique for the growth of highly resistive GaN layers on sapphire or SiC substrates using a Fe modulation doping, demonstrating good performance in terms of reduced dislocation density, 2 DEG mobility, and pinch-off properties. Uren et al. [60] further demonstrated that a proper design of the iron doping (e.g., Fe concentration, Fe doping depth, shape of Fe doping profile) determines negligible change in the RF performance with respect to the undoped samples.

Nevertheless, the use of an iron-doped GaN buffer has a non-negligible impact on charge-trapping effects, thus limiting the device dynamic performance [59]. In an early study, Desmaris et al. [59] claimed that intentionally iron-doped structures suffered from severe drain lag ascribable to the introduction of trapping centers in the GaN buffer layer. The worsening of the dynamic performance induced by the iron doping mainly consists in the increase of the current collapse (i.e., dynamic decrease of the drain current) and in the dynamic shift of the threshold voltage when the devices are submitted to a negative gate voltage and/or to a high drain voltage [62].

9.2.1 Properties of Deep Level E2 and Impact of Iron Doping

Iron doping is commonly associated with a deep level (hereafter E2) located in the buffer layer. E2 trap causes an allowed energy state at $E_C - 0.53–0.7$ eV, with an apparent capture cross section between 1×10^{-13} and 1×10^{-15} as reported in [40, 41, 62, 63]. The activation energy and the cross section do not change with a different iron doping amount [41, 62]. Consistent results were reported with different evaluation techniques, namely dynamic transconductance frequency sweep [41], capacitance [17], and current deep-level spectroscopy [62].

Meneghini et al. [62] used pulsed measurements (i.e., simultaneous pulses to the gate and drain terminal from a trapping condition or a quiescent bias point) to show that trapping effects corresponding to the iron doping indicate deep levels located mostly under the gate than in the gate–drain access regions. The current collapse is indeed sensitive to an applied negative gate bias and/or a positive drain bias, and it is mainly ascribed to a dynamic threshold voltage shift, more prominent in the saturation than in the knee voltage region, rather than to on-resistance dynamic increase. Silvestri et al. [41] demonstrated similar results by means of simulations and dynamic transconductance frequency sweep evaluation; measurements are taken in the subthreshold region at low drain voltage levels, thus avoiding the ohmic regime and suggesting that the detected traps are located in the area under the gate. Silvestri et al. [41] further indicated that the E2 trap level is located in the buffer, due to its correlation with the iron doping concentration and due to the absence of a correspondence between the frequency of the E2 trap peak and the applied gate bias.

A slightly discordant hypothesis is reported by Cardwell et al. [40] after analyses performed with macro-scale resistance transients (RTs), nanometer-scale surface potential transients (SPTs), and atomic force microscope (AFM) analysis. Cardwell et al. [40] confirm, by means of the surface potential feedback signal detected with the AFM probe, that the E2 trap is located in the buffer layer; this is further confirmed by the dependence of the trap signal on the iron doping buffer compensation. Moreover, the detection of the E2 trap in devices with and without SiN passivation and/or with a barrier layer grown with a different material confirms that E2 is not located in the AlGaN barrier or in the surface [40]. Cardwell et al. [40] suggest that the E2 level is located near the gate edge in the drain access region. This hypothesis is suggested by the trap concentration measured through the resistance transient amplitude when different bias points are applied, and it is further supported by the simulation of the vertical electric fields in the GaN buffer and in the AlGaN barrier layer calculated in consistent bias points.

The E2 trap level is found to be significantly influenced by the iron doping concentration [17, 40, 41, 62]. Initial studies were performed, by using electrical and optical measurements, on semi-insulating GaN films where the lower part was compensated by iron doping [39, 64]. Polyakov et al. [64] detected a trap level with energy $E_C - 0.5$ eV in semi-insulating GaN films partially compensated with iron doping (only 0.5 μm of the layer in close proximity to the sapphire substrate). The activation energy was ascribed to the depth of the dominant trap level pinning the Fermi level, while the concentration of the corresponding traps (measured by C–V analysis) was found to be significantly higher in the Fe-doped portion of the film. Carrier capture in partially doped GaN films was further studied by Aggerstam et al. [39] by time-resolved photoluminescence (PL). The comparison of the PL transients in samples with a different iron compensation indicated a strong dependence between luminescence decay and iron concentration, thus suggesting the importance, in case considered, of Fe centers as predominant non-radiative recombination channels.

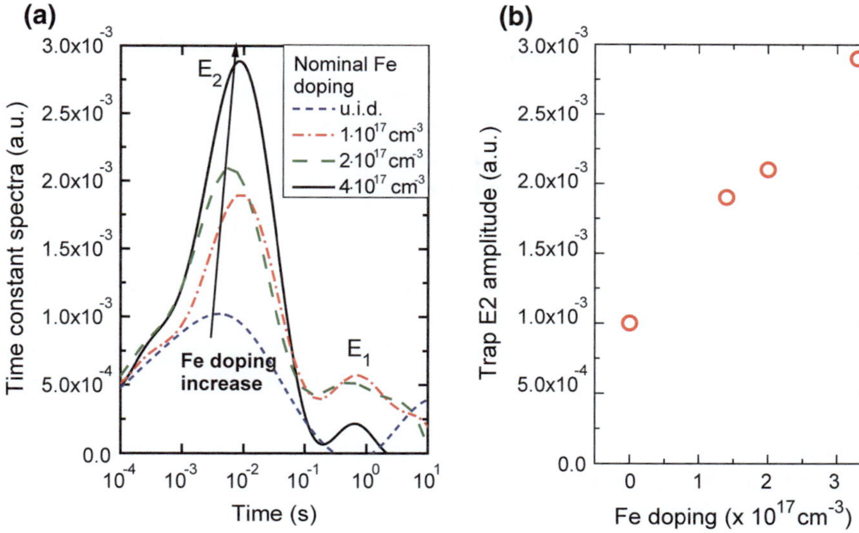

Fig. 9.6 **a** Time constant spectra measured through drain current transients in devices with iron-doped buffer layer; **b** E2 trap amplitude plotted as a function of iron concentration. Time constant spectra are measured extrapolated from the drain current transient by using the stretched exponential technique

Figure 9.6 shows the time constant spectra extrapolated from the drain current transient by using the stretched exponential technique [65] and the corresponding trap amplitude plotted as a function of the iron doping concentration measured by secondary ion mass spectroscopy (SIMS) on the tested devices. Figure 9.6 shows, as demonstrated by Meneghini et al. [62], that the E2 trap amplitude is well correlated with the iron concentration in the GaN buffer. Silvestri et al. [41] reported consistent results, demonstrating a significant variation of the trap conductance peak with the iron doping compensation. Consistent results were furthermore revealed by Cardwell et al. [40] through the comparison of trap amplitudes measured by macro-scale resistance transients in devices with a different Fe doping compensation.

The E2 trap level is furthermore found to be significantly sensitive to the iron doping profile. Chini et al. [66] demonstrated that the iron doping decay in the GaN buffer and the corresponding concentration in the proximity of the heterointerface plays a significant role in the generation of trapping effects. Therefore, measurement (e.g., SIMS) and simulation of the iron doping profile can, in principle, predict trapping effects related to the iron doping.

Conversely, Uren et al. [67] report by means of numerical simulations only a small variation in the current dispersion for varying the iron doping profile. This result was ascribed to the fact that the iron doping compensation leads to a weakly n-type buffer as a consequence of the Fermi level pinning in the upper half of the bandgap.

9.2.2 Origin of the Trap E2

The origin of the trap E2 has been subject of intense studies. Silvestri et al. [41] preliminary suggested that E2 is located in the buffer and that it is caused by iron impurities, due to the dependence of the E2 amplitude on the iron doping concentration and to the absence of a dependence between the E2 peak frequency and the gate bias applied. The correspondence between the different iron doping concentration and the magnitude of the trap-sensitive signal was further studied by a simulation of the GaN conduction band with different iron doping and the corresponding depth at which the Fe level crosses the Fermi level.

A more recent study by Meneghini et al. [62] provided further insight on the origin of the trap E2 and demonstrated that this deep level can be present even in samples without iron doping. The level E2 was ascribed to an intrinsic defect of GaN or unintentional impurities, whose concentration increases with the amount of iron in the buffer. This hypothesis is based on the detection of the E2 trap level on the devices grown with different techniques and/or with no iron doping buffer compensation [1, 3, 6, 11, 20, 29, 32, 68–70]: several studies [20, 69, 70] indeed ascribe the E2 trap level to nitrogen antisite defects, while further studies [70] relate E2 to an intrinsic GaN defect. Figure 9.7 compares E2 trap level detected on different studies. The correspondence between the E2 trap amplitude and the iron doping buffer compensation further confirms the E2 level location in the GaN buffer.

The filling time influences the trap amplitude of E2, with a logarithmic trend; according to Meneghini et al. [62], the linearly arranged defects originate from point defects clustering along dislocations. Similar inferences were provided by Rudzinski et al. [71]. By means of defect selective etching (to visualize the defects) and scanning electron microscope (to evaluate the defect density), it was reported that iron influences the density of the threading dislocations and the number of defects created during the growth. Results were furthermore provided by Mei et al. [72] with a detailed analysis, by means of optical measurements, of the impact of the iron doping on the device morphology. The presence of a high number of threading dislocations in the iron-doped GaN layer was confirmed by high-resolution transmission electron microscopy (HRTEM) performed at different growth stages; consistently, atomic force microscope (AFM) revealed, in the iron-doped GaN layer, a rough feature with spiral hillocks.

The E2 trap capture seems to be influenced by the bias conditions that involve the GaN buffer. Meneghini et al. [62] demonstrate that two main phenomena contribute to the E2 capture mechanism (Fig. 9.7). During the off state, the trapping is influenced by the gate leakage current; during the semi-on state, a strong contribution is given by the injection of hot electrons from the 2DEG into the buffer, leading to a significant increase of the current collapse (i.e., the dynamic decrease of the drain current) and of the trap amplitude associated with the deep level E2 (Fig. 9.8). The second mechanism, in agreement with the distribution of the electric field, is more prominent near the gate edge on the drain side. The contribution of the hot electrons (Fig. 9.8) determines a sublinear relation

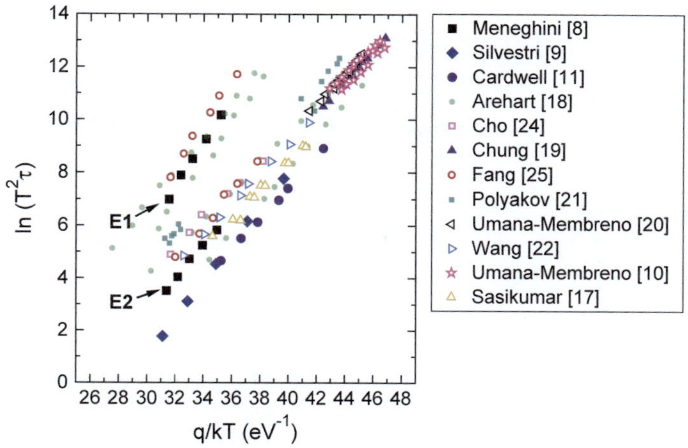

Fig. 9.7 Arrhenius plot of the deep levels E1 and E2 detected by Meneghini et al. [62]. Activation energy of 0.82 and 0.63 eV is extrapolated, respectively. Comparison of the deep levels detected in Meneghini et al. [62] with previous works reported in the literature concerning devices with and without intentionally iron doping

between the trapping effects (demonstrated by the current collapse increase) and the corresponding dc drain current.

Experimental evidence of charge-trapping mechanisms promoted in the semi-on state, possibly related to hot electrons, was reported also by Wang et al. [73] and Meneghini et al. [74]. Wang et al. [73] indicate results consistent with [62], although measured on devices with carbon buffer doping. In [73] the current collapse, consistently with the on-resistance variation, increases with the quiescent bias point. The bell-shaped behavior of the current collapse variation suggests the influence of the hot electrons. For gate voltage values near the threshold voltage, the electrons start to form the channel. The strong correlation between the increase of the drain current and of the current collapse suggests that an additional trapping process can be generated when the electrons, which may have high kinetic energy levels, start to flow in the channel and are injected into the trap states. Meneghini et al. [62] report a sublinear dependence between the drain current and the current collapse variation measured in semi-on state; the results are explained by considering that the effect of the hot electrons depends not only on their number but also on their average energy, which may decrease as a consequence of the decrease of the accelerating field and of the increase of the electron scattering.

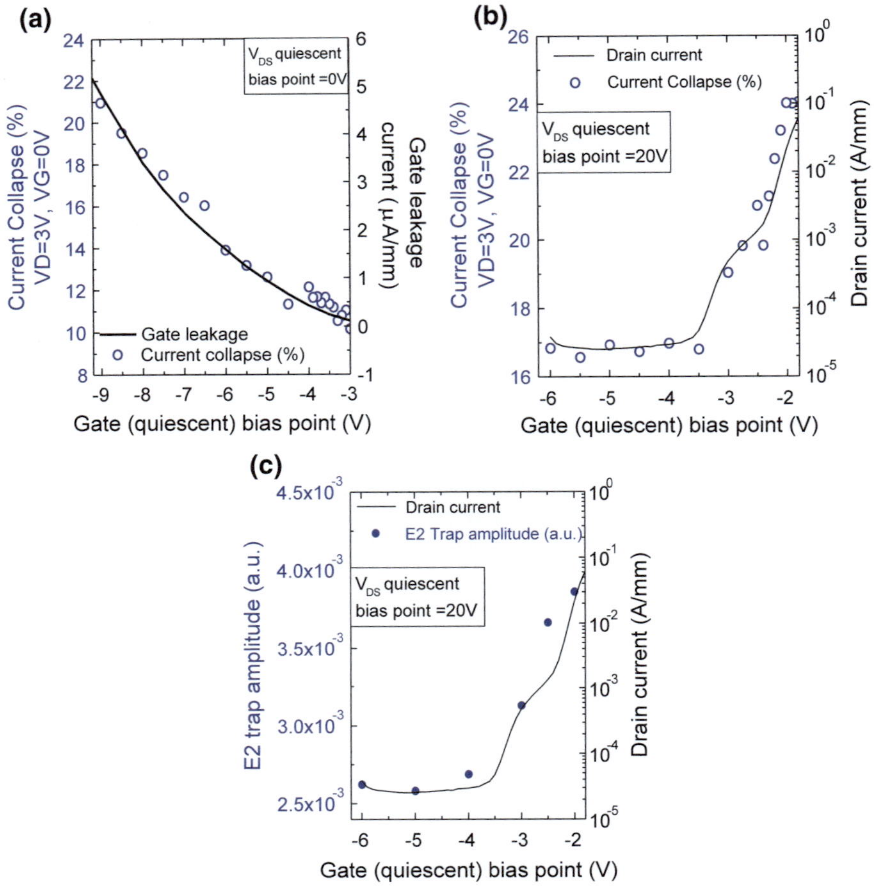

Fig. 9.8 Dependence between the dynamic performance of devices with iron doping buffer compensation and the leakage current measured in off state (**a**) or the drain current evaluated in off and semi-on state (**b**). Enhancement of the trap E2 signal measured in off and in semi-on state (**c**). The current collapse is defined as the ratio of the current variation in two quiescent bias points to the current measured in the first one

9.2.3 Impact of Electrical Stress on Trapping Mechanisms

The impact of a dc stress on the variation of trapping mechanisms in devices with a different iron doping buffer compensation was studied in detail [75, 76].

Meneghini et al. [75] clarified the influence of a negative bias dc stress on the leakage current and of the increase of the trapping mechanisms in devices with a different iron doping compensation. The reverse bias stress induces a significant worsening of both the static and dynamic performance of device, represented by a significant increase in the dc leakage current and in the current collapse, respectively.

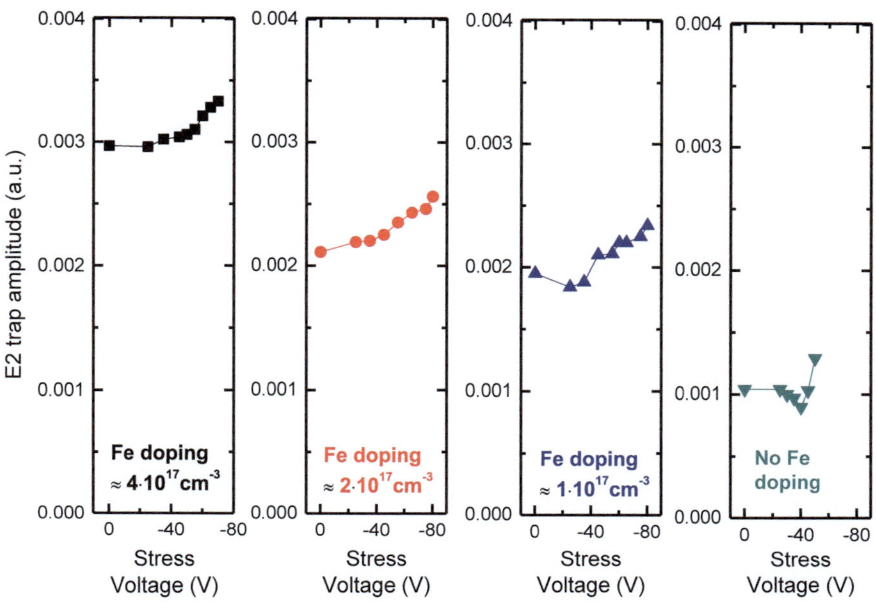

Fig. 9.9 Time constant spectra monitored during a negative gate bias stress in devices with a different iron doping buffer compensation

Figure 9.9 shows the time constant spectra calculated by the stretched exponential technique [65] on the drain current transients monitored after defined steps during the stress. The properties of the traps E1 and E2, namely activation energy and apparent capture cross section, demonstrate a negligible change after the negative bias stress (Fig. 9.10). Nevertheless, the stress induces an increase in the amplitude of both traps, thus indicating an enhancement of the signal associated with the traps that were already detected on fresh devices (Fig. 9.9).

Meneghini et al. [75] suggest that the signal E1, probably located in the AlGaN barrier layer, rises due to the increase in preexistent defects, ascribed to converse piezoelectric effect or to the high applied electric field. Conversely, the deep level E2 is located in the buffer; charge mechanisms can be due to electrostatic effect or to the gate leakage current. The increase or the generation of leakage paths after a negative bias stress induces an increase of the current collapse. The corresponding increase of the trap signal of E2 therefore suggests that the negative bias stress leads to a higher concentration of electrons trapped in the E2 trap states located in the buffer.

The influence of an RF test on the trapping effects in devices with different buffer doping, namely iron and carbon doping, was investigated by Bisi et al. [76]. After 24-h RF tests, the devices experienced a variation of the current collapse. A good correlation between the variation of the current collapse and of the RF output power was demonstrated Bisi et al. [76]. In Bisi et al. [76], the authors shows that the worsening of the RF output power (P_{OUT}) is more evident in the devices with worse

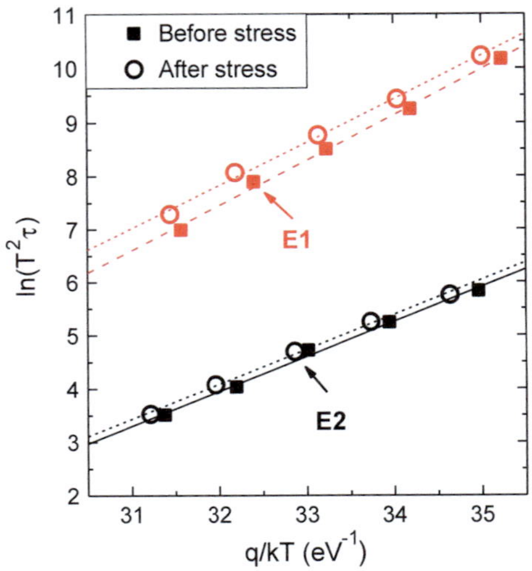

Fig. 9.10 Activation energy measured before and after a negative bias stress in the devices with iron doping buffer compensation

pinch-off properties and enhanced short-channel effects. The devices which experience the higher P_{OUT} degradation are indeed characterized by a worse performance in terms of subthreshold slope, drain-induced barrier lowering effects, and source-to-drain leakage current. The good correlation demonstrated between the RF output power performance and the initial subthreshold current suggests that in the devices with non-optimal buffer compensation, the degradation of the P_{OUT} is caused by the increase of the trapping effects and of the preexisting defect density.

9.3 Impact of Carbon Doping

Carbon is an effective agent to compensate the unintentional doping of III-nitride materials, achieving highly insulating buffer layers for GaN-based electronic devices.

In metal-organic chemical vapor deposition (MOCVD), the carbon impurity commonly exists in the materials due to the use of metal-organic precursors. The incorporation of carbon in the MOCVD III-N layers can be effectively controlled by tuning the growth conditions, including pressure, temperature, III/V ratio, growth rate, and Al content [77–82].

Contrary to iron and magnesium, carbon shows no segregation or memory effect in the growth system and in the epitaxial stack, and it allows a precise control of the doping profile and a sharp transition to the unintentionally doped GaN channel layer.

From a static point of view, carbon doping allows the control of the resistivity of GaN films [79], the reduction of lateral and vertical buffer leakage current [82–85], the increase of the breakdown voltage [80, 84, 86], and the suppression of the short-channel effects [76] of GaN-based high-electron-mobility transistors (HEMTs).

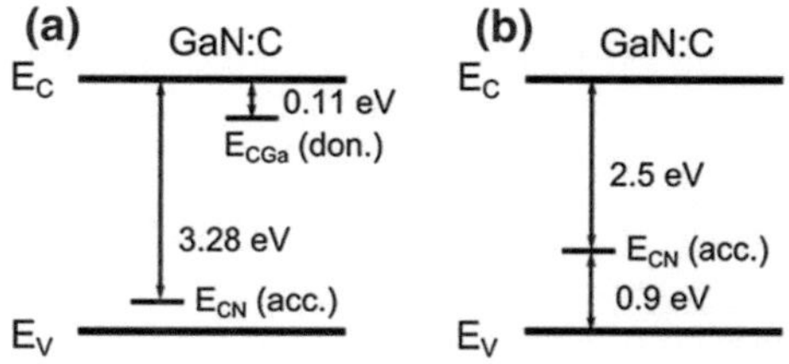

Fig. 9.11 Schematic representation of the energy levels assumed in the C-doped GaN layer according to the **a** self-compensation model (see Ref. [92]) and **b** the dominant deep acceptor model (see Ref. [36])

As a drawback, carbon introduces deep levels and promotes parasitic effects related to charge trapping, current collapse, and the degradation of dynamic performance. The degradation of the transconductance (g_m) and of the on-resistance (R_{ON}) of C-doped AlGaN/GaN HEMTs was reported by several research works [67, 87–91].

In this section, we analyze the theoretical and experimental evidence of trap states related to carbon impurities, we report on the effects of these deep levels on the dynamic performance of GaN-based HEMTs, and we present the solutions proposed by multiple research groups to mitigate the current collapse and dynamic R_{ON} increase in GaN-based transistors.

As predicted by first-principle calculations [36, 92], carbon in GaN has an amphoteric nature. Depending on the Fermi level, carbon can be incorporated either in nitrogen substitutional position (C_N), behaving as acceptor state, in gallium substitutional (C_{Ga}), or in interstitial position (C_i), behaving as donor state.

Two scenarios have been proposed to explain the semi-insulating behavior of carbon-doped GaN. In the early 2000s, Wright et al. proposed the self-compensation of acceptor species (C_N) and donor species (e.g., C_{Ga} or C_i) which incorporate in GaN with similar formation energies [92] (Fig. 9.11a). A decade later, Lyons et al. proposed that carbon in nitrogen substitutional position ($\mathrm{C_N}$) could behave as deep acceptor with the (0/-) transition level at $E_V + 0.9$ eV [36, 93] (Fig. 9.11b); if incorporated with sufficient concentration, it would compensate the unintentionally n-type doping, pin the Fermi level at $E_V + 0.9$ eV, and achieve semi-insulating behavior, at room temperature, with small hole concentration and negligible p-type conductivity.

Both donor-like and acceptor-like deep levels associated with carbon impurities have been detected in experimental research works. With deep-level optical spectroscopy (DLOS) and deep-level transient spectroscopy (DLTS), Armstrong et al. reported the deep levels at $E_C - 0.11$ eV, $E_C - 1.35$ eV, $E_C - 2.05$ eV, $E_C - 3.0$ eV, and $E_C - 3.28$ eV and $E_V + 0.9$ eV whose concentration monotonically tracks the concentration of carbon doping in MBE GaN [94]. In light of the theoretical prediction of Wright et al. [92], the levels $E_C - 0.11$ eV and $E_C - 3.28$ eV were tentatively ascribed to C_{Ga} and C_N defects, respectively. The level $E_C - 1.35$ eV was found to be sensitive to the threading dislocation density and tentatively ascribed to carbon interstitial incorporated preferentially in along threading dislocations. $E_C - 0.11$ eV, $E_C - 1.35$ eV, and $E_C - 3.28$ eV were detected also in a previous works on MOCVD GaN [70, 95].

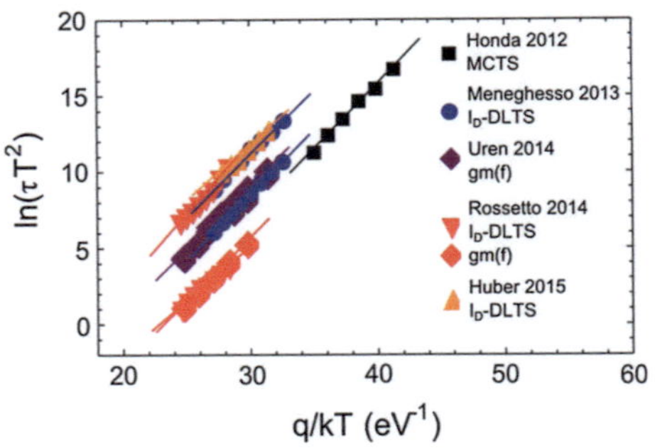

Fig. 9.12 Deep-level signature with apparent activation energies ranging from 0.71 to 0.94 eV identified to be the primary cause of current collapse and dynamic R_{ON} increase in high-electron-mobility transistors equipped with carbon-doped buffer

By means of transmission electron microscopy (TEM) and X-ray diffraction (XRD), Wickenden et al. reported that MOCVD GaN films grown with lower growth pressure exhibit not only increasing carbon concentrations, but also decreasing grain size and increasing density of threading edge dislocations [79]. Based on DLTS, Fang et al. detected both electron traps and hole traps in AlGaN/GaN Schottky barrier diodes (SBD) with intentionally carbon-doped buffers [96]. The overall density of electron traps was higher in the devices with higher carbon concentration, and the DLTS signal exhibited a strong dependence on filling pulse width; authors suggested their association with extended defects.

More recently, Honda et al. reported a hole trap at $E_V + 0.86$ eV, whose concentration was directly influenced by carbon concentration in MOCVD GaN [12]. That work provided further experimental evidence of the possible existence of the deep acceptor state C_N at $E_V + 0.9$ eV theorized in [93].

Deep-level signatures with apparent activation energies ranging from 0.71 to 0.94 eV (Arrhenius plot in Fig. 9.12) were identified to be the primary cause of current collapse and dynamic R_{ON} increase in high-electron-mobility transistors with carbon-doped buffer [87, 89–91, 97]. Rossetto et al. [89] and Huber et al. [90] explicitly pointed out increasing trap density with increasing carbon doping concentration.

Reference	Method	E_A (eV)	σ_c (cm^2)
Honda et al. [12]	MCTS	0.86	1.6×10^{-13}
Uren et al. [97]	g_m(f)	0.83	1.0×10^{-14}
Meneghesso et al. [87]	I_D–DLTS	0.85	4.0×10^{-14}
		0.83	1.2×10^{-15}
Rossetto et al. [89]	I_D–DLTS	0.71	1.7×10^{-14}
		0.94	1.8×10^{-14}
Rossetto et al. [89]	g_m(f)	0.83	3.5×10^{-13}
Bisi et al. [91]	I_D–DLTS	0.84	4.0×10^{-14}
Huber et al. [90]	I_D–DLTS	0.73/0.83	5.5×10^{-16}

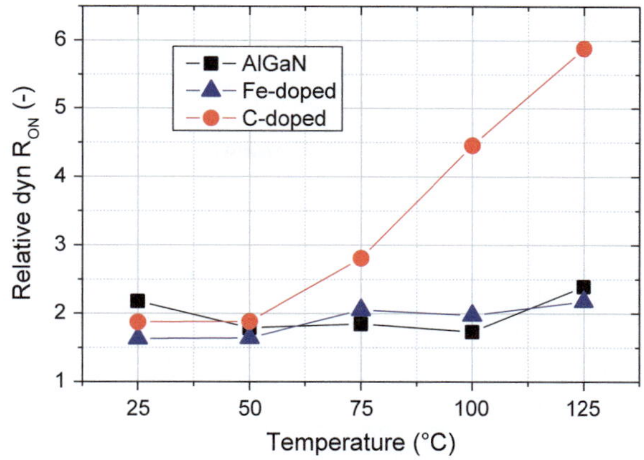

Fig. 9.13 Relative dynamic R_{ON} from switching transients at 200 V as function of temperature. Contrary to devices with iron-doped GaN or u.i.d. AlGaN back-barriers, the dynamic R_{ON} increase of carbon-doped devices rises with increasing temperature. Data from Rossetto et al. [88], courtesy of E. Bahat Treidel et al. FBH

The net negative charge stored during high-voltage off-state conditions by the slowly responding deep levels in the carbon-doped buffer could act as virtual back gate, reducing the 2DEG concentration and promoting drain lag and bias history-dependent dynamic R_{ON} increase.

By investigating the effects of the temperature on the dynamic performance of GaN HEMTs, Bahat-Ttreidel et al. experimentally pointed out that conversely to iron-doped GaN or u.i.d. AlGaN back-barriers, the dynamic R_{ON} increase of carbon-doped devices is positively dependent on temperature [88], Fig. 9.13. Similar evidence is reported by Bisi et al. [98], where the relative dynamic R_{ON} increase of carbon-doped AlGaN/GaN/AlGaN MIS-HEMTs is worsened from 2 to 169 %, if base plate temperature is raised from 40 (Fig. 9.14a) to 160 °C (Fig. 9.14b).

Positive temperature-dependent R_{ON} increase could represent a serious issue for GaN-based electronics, which is supposed to be operated in high-temperature conditions. Thanks to stress/recovery transient measurements, we reported the kinetics of the temperature-dependent charge trapping and detrapping in AlGaN/GaN/AlGaN MIS-HEMTs with a carbon concentration of ~10^{18} cm^{-3} in the AlGaN buffer [98]. Since the observed effect is completely recoverable, it can be ascribed to charge-trapping effects, whereas no permanent degradation is induced.

Devices under tests were grown on 150-mm p-type (111) silicon substrate; the epitaxial structure consists of 150-nm-thick AlN interlayer, 2-μm-thick AlGaN back-barrier composed of three layers with 70, 40, and 18 % Al concentration, respectively, a 150-nm undoped GaN channel layer, a 10-nm $Al_{0.25}Ga_{0.75}N$ barrier layer capped by 10 nm of in situ Si_3N_4, and 15 nm of Al_2O_3 atomic layer deposition (ALD) [99].

Time-resolved stress/recovery R_{ON} transients were acquired by means of sampled transient measurements: During the 1-ks-long stress, the device was repeatedly biased with short pulses in the linear region (V_G; V_{DS}) = (0 V; 0.5 V) during which the instantaneous I_{DS} is acquired, and the related instantaneous R_{ON} is extrapolated; during the 1-ks-long recovery, the device was continuously biased in the linear region (V_G; V_{DS}) = (0 V; 0.5 V).

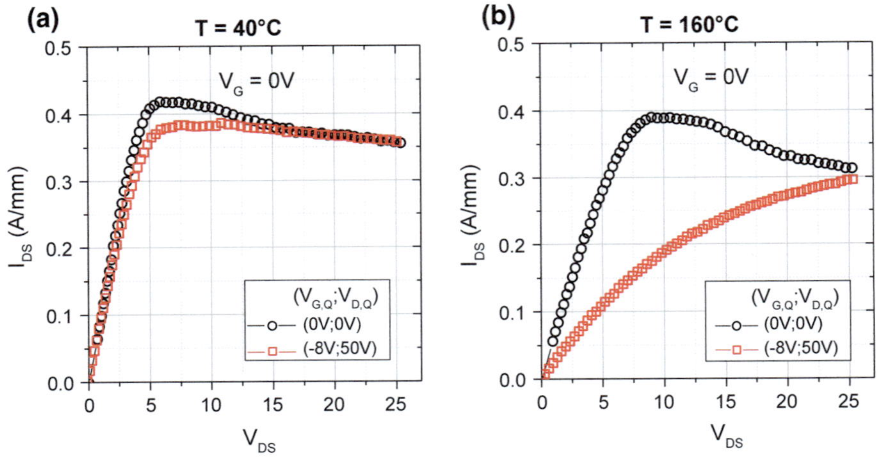

Fig. 9.14 Pulsed I_D–V_D characteristics of an AlGaN/GaN/AlGaN MIS-HEMT acquired at **a** 40 °C and **b** 160 °C. When exposed to the off-state quiescent bias point, the devices under test experience an increase of the dynamic R_{ON}. This effect is positive temperature-dependent, thus of great concern for high-temperature operations. Data from Bisi et al. [98]

Figure 9.15a and b depicts the stress/recovery transients acquired applying an off-state stress bias at (V_G; V_{DS}; V_B) = (−8 V; 25 V; 0 V). It can be noticed that temperature influences both the rate of the R_{ON} recovery and the rate of the R_{ON} increase. This explains why the device experiences remarkably higher current collapse and dynamic R_{ON} increase when exposed to higher temperature. The activation energy of R_{ON} increase and R_{ON} recovery are 0.90 and 0.95 eV, respectively.

To confirm that the observed current collapse was related to charge trapping in the buffer and not at the surface, auxiliary back-gating measurements were taken. During the stress phase, a negative potential was applied to the substrate terminal ($V_B = -25$ V) and no potential difference was applied between source, gate, and drain terminals ($V_G = V_D = V_S = 0$ V). Under back-gating, surface trapping is supposed to be negligible because no bias is applied between gate and drain terminals, and the formed 2DEG would screen the superficial layers from the field effect induced by back-gating.

Results reveal that back-gating stress leads to similar stress/recovery transients, with similar kinetics, similar timings, and similar thermal activation than those acquired during off-state stress (Fig. 9.15c, d). This indicates that the observed charge-trapping process is localized in the buffer region (Fig. 9.16) and is the dominant cause promoting current dispersion during off-state regimes.

As predicted by Uren et al. [67], the charge trapping leading to the dynamic R_{ON} increase related to buffer trapping is not located at the gate edge, but it is distributed over the entire gate–drain access region. Among the consequences, the dynamic R_{ON} increase is proportional to the gate–drain length (L_{GD}), as experimentally proved in Meneghini et al. [35, 100].

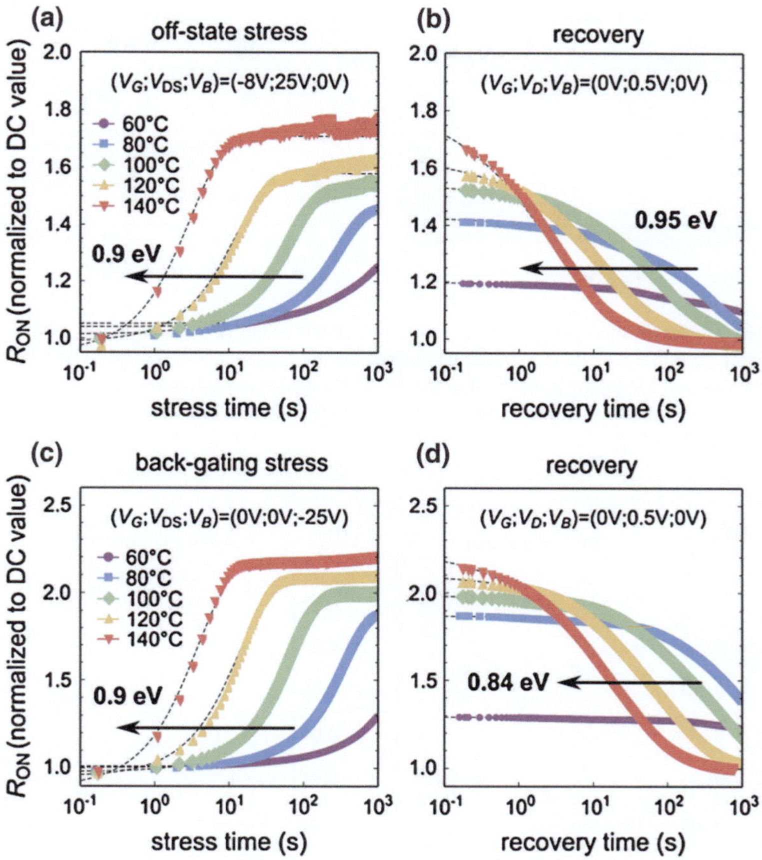

Fig. 9.15 Stress and recovery transients performed in off-state and back-gating stress. Both charge trapping and charge detrapping are thermally activated. Off-state stress and back-gating stress lead to similar stress/recovery transients, suggesting unique charge-trapping mechanism localized in the buffer. Data from Bisi et al. [98]

As is shown in Fig. 9.15, the magnitude of the R_{ON} increase during back-gating is higher (up to 120 % of DC value) than during off state (up to 80 %). During off-state stress, the high vertical electric field between the formed 2DEG and the substrate is confined in the gate–drain access region, whereas during back-gating stress, it affects the entire source-to-drain extension. By performing I_D–V_G measurement prior to and after 1-ks-long back-gating stress (Fig. 9.17), it can be noticed that the device experienced not only the dynamic R_{ON} increase, but also a positive V_{TH} shift (+1.43 V), proving that the buffer charge trapping promoted by back-gating bias happens also below the gate contact, within the intrinsic device region.

The observed R_{ON} increase could be explained by two different mechanisms. The first one considers the involvement of electron traps. When the devices are submitted to high drain bias, electrons can be injected from the silicon substrate to the buffer

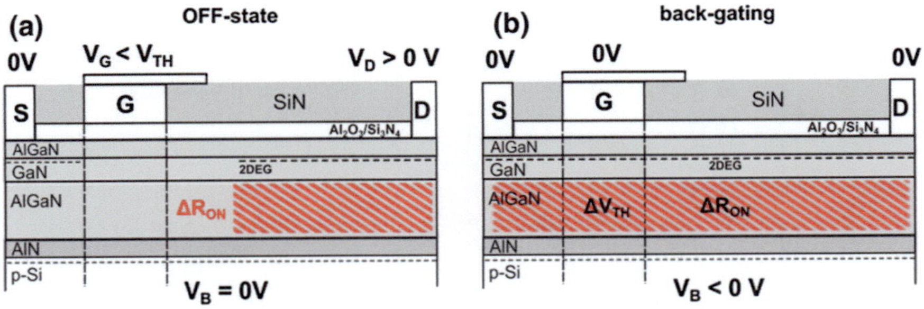

Fig. 9.16 **a** In the off state, the dominant charge-trapping mechanism is promoted by vertical drain-to-substrate potential, is localized in the gate–drain access region within the epitaxial buffer/back-barrier layers, and causes the dynamic R_{ON} increase. **b** In back-gating stress, charge trapping affects the entire source-to-drain length, promoting not only the dynamic R_{ON} increase, but also the shift of the threshold voltage

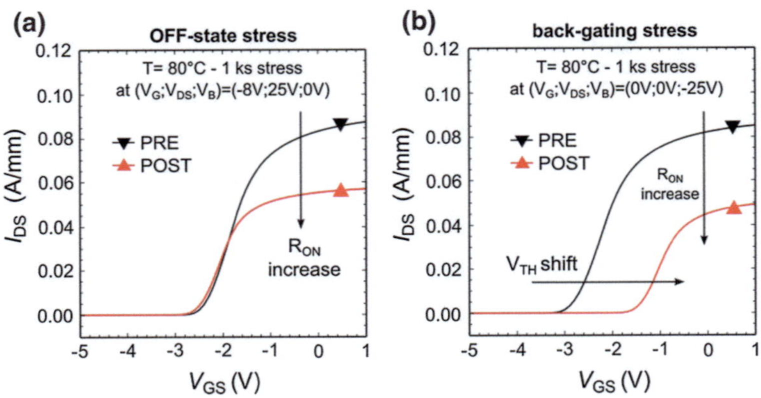

Fig. 9.17 I_D–V_G characteristics (V_{DS} = 1 V) acquired prior to and after **a** off-state stress and **b** back-gating stress. When exposed to back-gating stress, the device experiences not only the dynamic R_{ON} increase, but also a positive V_{TH} shift of +1.43 V, proving that the buffer charge trapping happens also within the intrinsic device region, below the gate contact. Data from Bisi et al. [98]

through the highly defective nucleation layer [101, 102] and could be possibly trapped at deep levels in the buffer. Under these assumptions, the strong temperature dependence of the R_{ON} increase would be related to the availability of free electrons, hence to the thermally activated substrate leakage current and/or to the thermally activated capture cross section of the involved traps [103]. Conversely, the activation energy of the R_{ON} recovery would account for the thermal emission from the trap state to the conduction band. Possible relation between the parasitic vertical leakage current and the buffer charge trapping is reported in [104–106].

The second hypothesis considers the involvement of deep acceptors. Assuming the model predicted by Lyons et al. [36], the interface between a C-doped buffer and a unintentionally doped GaN channel could be modeled as a p–n junction with the Fermi level of the weakly p-type C-doped GaN pinned in the proximity of the deep acceptor state ($E_V + 0.9$ eV) [67, 107]. During high-voltage off-state conditions, the gate–drain access region is exposed to high drain potential, and the depletion width would extend within the GaN buffer. The slow response of the deep acceptors to change their charge state and the low conductivity of the buffer would lead to strong delays to reach the electrostatic steady state during on–off–on transitions. The strong temperature dependence of the R_{ON} increase would account for the ionization of the deep acceptor state (0.9 eV) by the thermal emission of trapped holes, whereas the strong temperature dependence of the R_{ON} recovery would account for the availability of free holes required to neutralize the deep acceptors and restore the original charge balance.

The solutions proposed to prevent and mitigate the carbon-related dynamic R_{ON} increase involve the optimization of both the epitaxial structure and the device architecture. Carbon-related dynamic R_{ON} increase could be suppressed by optimizing the carbon concentration, by using a u.i.d. GaN channel layer, by increasing the vertical spacing between the carbon-doped buffer and the 2 DEG [80, 105], and by implementing a superlattice buffer [108].

A different solution, suggested by Uren et al. [97] indicates that the suppression of the dynamic R_{ON} dispersion could be favored in devices with a high density of active defects, where the reverse leakage current through the depletion region would allow the floating weakly p-type buffer to remain in equilibrium with the 2 DEG.

Maintaining the introduction of a source of holes to avoid the negative charging of deep trap states is a concept proposed also in Kaneko et al. [109]. The authors proposed a GaN gate injection transistor (GIT) with an additional *p*-GaN region beside the drain electrode. Though featuring slightly higher off-state leakage current and similar breakdown voltage, the device proposed in [109] shows remarkably less dynamic R_{ON} increase than conventional GITs; the authors speculated that the holes injected in the off state from the auxiliary *p*-GaN drain effectively release the trapped electrons, preventing the current collapse and the dynamic R_{ON} degradation.

9.4 Trapping Mechanisms in Metal Insulator Semiconductor High-Electron-Mobility Transistors (MIS-HEMTs)

GaN high-electron-mobility transistors designed for power applications require high performance in terms of low leakage current, high breakdown voltage, low dynamic on-resistance, and stable threshold voltage. An effective solution to suppress the parasitic gate leakage current is the insertion of a dielectric layer under the gate contact, i.e., the use of a metal insulator semiconductor high-electron-mobility transistor (MIS-HEMTs) [99, 110–114]. This approach is very interesting for the power applications since it allows the implementation of normally off devices [115–119].

Nevertheless, the performance of these devices is limited by trapping effects due to the presence of deep levels originating from defects, threading dislocations, impurities (intentional and unintentional), or surface states [98]. The presence of a dielectric layer under the gate causes peculiar trapping mechanisms in GaN MIS-HEMT structures, which are mainly located under the gate in the dielectric or at the dielectric/III-nitride interface [98, 120–124]. The presence of charge trapping under the gate promotes detrimental threshold voltage (V_{TH}) instabilities, especially when submitted to positive gate bias [98, 120, 124, 125]. Although several dielectric materials have been studied (e.g., SiO_2, Al_2O_3, Si_3N_4, HfO_2) [126], the intrinsic defectiveness of the dielectric layer and that of the corresponding interface with the III-nitride epitaxial structure still represent a significant issue concerning the optimization of these devices.

9.4.1 Origin of the Trapping Induced by Positive Gate Bias

The mechanisms that cause the threshold voltage instabilities in GaN MIS-HEMTs are still under discussion.

In the early studies, Mizue et al. [122] provided a detailed description of the interface states and the charge-trapping mechanisms at the dielectric/AlGaN interface by means of capacitance–voltage (C–V) measurements and simulations. As a peculiar feature in MIS-HEMT structures, a C–V curve with two steps is observed; the two steps in the C–V profile correspond to the accumulation of electrons at the AlGaN/GaN and Al_2O_3/AlGaN interfaces. The fixed charges and the different densities of the interface states were studied by simulations in the Al_2O_3 gate dielectric. The C–V slope in forward bias is found to decrease with the increase of the density of the interface states, thus achieving a trend more similar to the experimental results. The different slope of the C–V curves (obtained with an increase of the interface state density) is ascribed to the electron trapping at the dielectric/AlGaN interface and to the consequent suppression of the potential modulation in the AlGaN layer due to the negatively charged interface states. The electron capture mechanisms at the interface (or in the dielectric) remain unchanged when the device is submitted to a high reverse gate bias. The simulation of the band diagram (and the corresponding position of the Fermi and valence band of the AlGaN at the dielectric/AlGaN interface) determines the negligible influence of the gate voltage on the electron occupation when a negative bias is applied [120, 122]. Similar analyses were discussed by Johnson et al. [127] on HfO_2/AlGaN structures; they suggest that the traps are localized in or near the interface, due to the absence of a shift in the gate leakage current onset when the charge is trapped.

A detailed study of the mechanisms which determine the V_{TH} instabilities was further reported by Lagger et al. [129]. They confirmed that when submitted to positive gate bias, MIS-HEMTs show a V_{TH} shift with no degradation of the

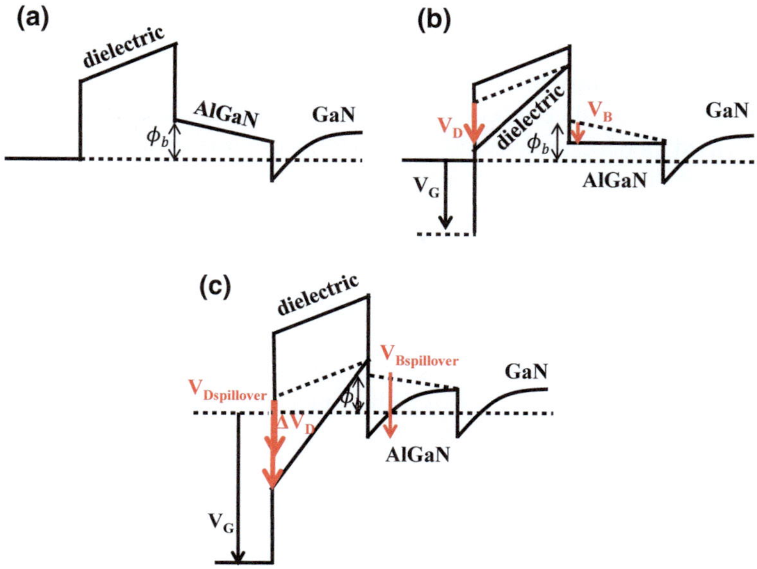

Fig. 9.18 Band diagram variation with different positive gate bias levels. **a** At the thermal equilibrium, there is no net electron flow between the channel and the interface. **b** At positive gate bias levels, according to the change of the initial potential barrier, an electron flow may be observed via the barrier toward the III-N interface. **c** At high positive gate bias levels (called spillover condition in Lagger et al. [128]), a second channel is created at the dielectric interface. The electrons provided by the second channel may get trapped in defect states

transconductance, due to the trapping at the dielectric/III-N interface away from the 2DEG. The analysis of the V_{TH} shift during capture/recovery tests indicates dependence with the gate bias and a broad distribution of the time constants with a negligible dependence on the temperature.

The dependence on the gate bias was explained by distinguishing between three different regimes [128]. A similar distinction is described in Fig. 9.18. In Fig. 9.18c, hereafter called spillover condition (or regime), a second electron channel is formed at the dielectric/AlGaN interface. If a higher gate voltage level is applied in the condition shown in Fig. 9.18c, there is no change in the voltage drop ($V_D - V_B$); the electrons trapped at the dielectric/AlGaN interface (or dielectric) are provided by the second channel and are considered responsible for the increase of the ΔN_{it} (namely the number of trapped electrons at the interface).

Lagger et al. [124] further explained, according to the results in SiO_2/AlGaN MIS structures, the evidence of a broad distribution for the capture/emission time constants. Several hypotheses were discussed. (i) According to Shockley–Read–Hall, a broad distribution of time constants can be explained by a broad and homogeneous distribution of individual defects at the dielectric/AlGaN interface. (ii) Spatial distribution of border traps in the dielectric (silicon dioxide in Lagger et al. [124]): if we consider the contributions to the time constant in a tunneling process from the dielectric/AlGaN interface to the bulk defects (in the case considered) in the oxide, a broad distribution can be due to a uniform distribution of tunneling distances to the individual dielectric defects. (iii) There is a distribution of activation energies due to the lattice relaxation effects (as a consequence of the disordered materials used for the gate dielectric). (iv) There is the presence of multistate defects and statistical distribution of the energy barriers for their state transition. (v) The time constant is influenced by the contribution of the AlGaN barrier layer, due to the fact that the barrier acts as a rate limiter. This contribution can be distributed due to the nature of localized leakage paths in the AlGaN. (vi) The contribution of the barrier layer in the time constant can be influenced by the accumulation of electrons in the dielectric/AlGaN interface during the stress.

Rossetto et al. [130] and Meneghesso et al. [131] clarified that independently from the gate insulator used, the V_{TH} shift is well correlated with the gate forward leakage and bias. This result suggests that the trapping is promoted by the injection of electrons in the gate insulator with forward gate bias.

Several techniques have been used to study the trap levels in the dielectric or at the dielectric/AlGaN interface. With the aim of studying the fast/slow trap time constants and the corresponding activation energies, threshold voltage transients have been used [129–135]. Lagger et al. [129] further suggest, due to the absence of the transconductance variation, a correspondence between the drain current and the V_{TH} transients. As a consequence of the broad distribution of the time constants, Lagger et al. [129] suggested the use of the CET (capture emission time) maps, widely used for the positive bias temperature instability in CMOS devices [136].

A fruitful discussion has been provided, too, in order to determine the density distribution of interface states. The most widely adopted techniques are dynamic conductance dispersion technique [121, 137–139], frequency and/or temperature capacitance voltage measurements [140–142], photocapacitance light intensity measurement [143], and photo-assisted C–V analysis [120].

Capriotti et al. [144] presented an analysis of the limitations of the conductance method, aimed at the calculation of the interface state density; the discussion was supported by experimental measurements, by simulations with different interface state density (D_{it}), and by the description of a method to improve a model (lumped element model) already described by Yang et al. [141]. The model was applied to the gate stack in the so-called spillover regime (i.e., when a second channel at the dielectric/AlGaN interface is formed). The purpose of the work is to discuss the relation between the trapping mechanisms and the intrinsic response of the gate stack, thus demonstrating the limitations of the measurement techniques currently used (e.g., the conductance method). The interdependence revealed between the trapping effects and the intrinsic response of the gate stack explains that these two mechanisms (phenomena) can lead to similar effects. The relation described above was studied by means of a simulation of the C–V curves with an increasing interface state density (D_{it}) at different frequencies. The results are furthermore confirmed by simulations of the conductance–frequency curves for D_{it} ranging from 1×10^9 to 1×10^{14}. Limitations of this method are demonstrated by the not linear relation of the conductance peak (value and number of peaks measured with the conductance–frequency curves) with D_{it} and by the possible underestimation of the density of the defects.

9.4.2 *Analysis of Fast and Slow Trapping Mechanisms*

A significant effort was spent in order to provide an analysis of fast and slow trapping mechanisms.

Wu et al. [133] provided a detailed analysis of slow traps and the corresponding threshold voltage (V_{TH}) instabilities after gate forward bias in MIS-HEMTs with an in situ bilayer (SiN/Al_2O_3). Activation energy of 0.7 eV was calculated by means of capture and recovery V_{TH} transients at several ambient temperatures (Shockley–Read–Hall theory was considered due to the nature of bulk-like defects). Similar trap states were already reported in the literature [145] and ascribed to MIS structure. On the basis of simulated band diagrams, Wu et al. [133] identified the corresponding level as a donor-like state located at the interface between the AlGaN and the gate dielectric.

Lansbergen et al. [135] presented an analysis of the distinct fast components in the V_{TH} drift and a discussion concerning the possible optimization by means of the III-V surface treatment. Fast traps that are detected by AC measurements show no dependence on the employed gate dielectric and are characterized by an emission/capture time constant ranging from 10 μs to 10 ms; the trapping effects cannot be detected by the DC measurements as a consequence of the longer integration time used. A comparison between SiN and Al_2O_3 gate insulator demonstrates that the dielectric has no impact on the detected fast trapping mechanisms. Evidence of fast trapping was also reported by Ma et al. [137] in MIS-HEMTs with an Al_2O_3 gate insulator by means of a dynamic capacitance dispersion technique.

9.4.3 *Materials and Deposition Techniques for the Improvement of Trapping Effects*

Several materials and deposition techniques have been proposed for the optimization of a MIS-HEMT structure and the reduction of the trapping effects. According to the band offset energy, required for the suppression of the leakage current, several materials were taken into account in the early studies, namely SiO_2, SiN_x, Al_2O_3, HfO_2, and further high-k dielectrics. Several works studied the performance of the SiO_2 gate insulator [146, 147] because of its large bandgap (despite the low permittivity); by conductance analysis, a low state density was demonstrated by Gaffey et al. [146] (5×10^{10} eV^{-1} at $E_C - 0.8$ eV with a capture cross section of 2.4×10^{-17}). Improvements in MIS-HEMT structures employing SiO_2 bilayers (SiO_2/Al_2O_3) have been demonstrated [148].

Another solution is provided by SiN/GaN structures [32, 149]. SiN reports a low permittivity and a lower bandgap (or band lineup) [150] with respect to further dielectrics (Al_2O_3, SiO_2); however, the possibility of in situ deposition [151] makes it a very interesting solution. Bilayer gate insulators (SiN/Al_2O_3) demonstrated improvements in terms of trapping effects and threshold voltage instabilities [99, 152].

Al_2O_3 demonstrates good performance, as a consequence of the large bandgap, of relatively high permittivity, and of the high breakdown field [113, 137, 153, 154]. Low state density (1×10^{11} at $E_C - 0.8$ eV) has been reported [155]. Several studies discussed the evidence of a high density of fixed charges in the Al_2O_3 films, whose origin still remain unclear [156–161]. Choi et al. [159] suggested that transition levels close to the conduction band are determined by oxygen vacancies; they further ascribed the fixed charge to the presence of Al vacancies and interstitials.

HfO_2 [127, 139, 157, 162] and high-k dielectrics were furthermore analyzed in detail, demonstrating good performance.

The impact of the design of the gate dielectric stack on the trapping effects was discussed by Lagger et al. [128]. The threshold voltage instabilities were found to be influenced by the thickness of the barrier layer and of the dielectric and by the used dielectric material. More specifically, Lagger et al. [128] explain that the V_{TH} instabilities are improved by decreasing the barrier thickness or by increasing the dielectric thickness. The dependence on the barrier thickness is justified by its influence on the gate voltage required for the creation of a second channel between the barrier and the dielectric interface (spillover voltage). The V_{TH} drift finally differs for the employed dielectric.

Several works reported in the literature finally demonstrated that the trapping effects can be improved by the surface treatment [135, 163] or by a different deposition technique. Plasma-enhanced atomic layer deposition has been tested in detail for the SiN gate insulator [116, 130, 131, 140, 164, 165].

The impact of a different deposition technique on SiN MIS-HEMTs has been reported by Rossetto et al. [130] and Meneghesso et al. [131].

Figure 9.19 reports a frequency-dependent C–V measurement performed in large capacitors. The impact of a different gate insulator (namely Al_2O_3 and SiN) and of the deposition technique (plasma-enhanced atomic layer deposition and rapid thermal chemical vapor deposition) is studied.

C–V measurement from depletion to accumulation results in threshold voltage shift and demonstrates that devices experience frequency-dependent capacitance dispersion. The latter can be ascribed to (i) the influence of the AlGaN layer on the frequency response to the gate bias [144], and/or (ii) the charge (de)trapping at border traps by means of tunneling mechanisms [166], and/or (iii) parasitic conduction mechanisms and the leakage current [167]. The comparison among different structures demonstrated that both the deposition technique and the dielectric insulator can influence the trapping mechanisms.

The dielectric material and the deposition technique show a severe impact on the onset of a broad conductive loss which follows an exponential trend with the gate

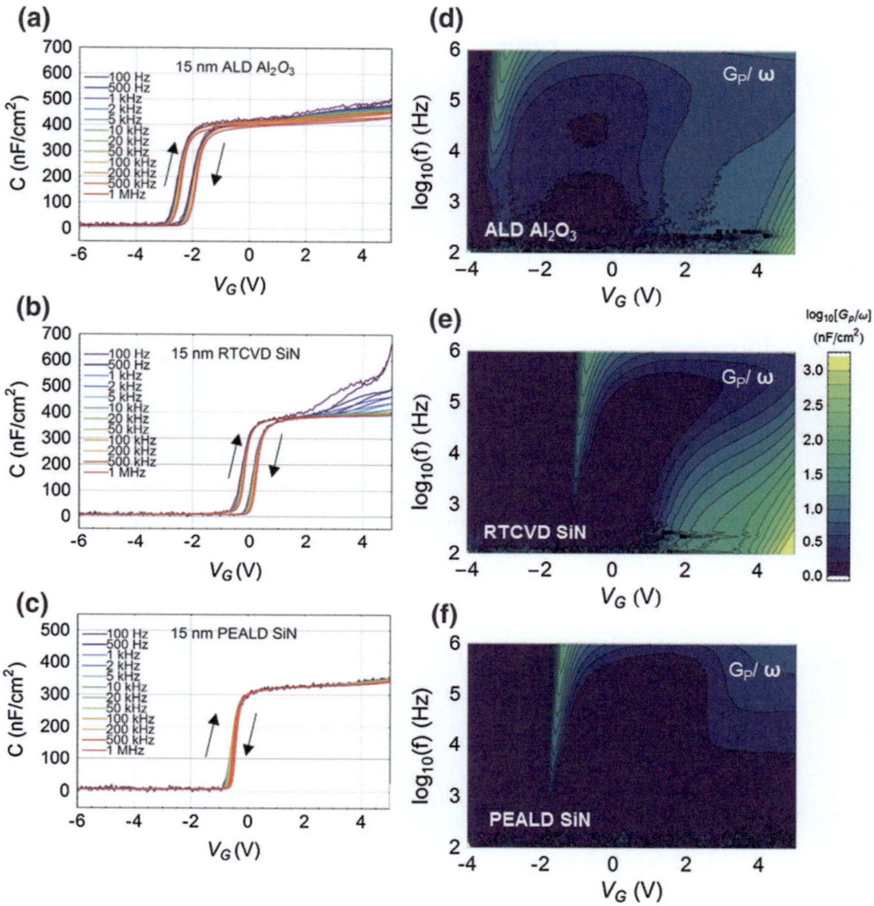

Fig. 9.19 **a–c** C–V–f measurements and **d–f** normalized parallel conductance as a function of frequency and gate voltage for the three tested wafer. Reprinted with the permission from Ref. [131] © 2015 Elsevier Ltd

voltage. This effect, correlated with the presence of high dc leakage components, is significantly lowered in devices with a PEALD SiN gate insulator.

Consistent results were revealed by pulsed measurements. The deposition technique and/or the gate dielectric used have a strong impact on the leakage gate current and on the severity of the threshold voltage instabilities induced by the forward gate bias; the trapping effects mainly consist in the threshold voltage shift with a negligible transconductance peak decrease. In all the cases (namely PEALD SiN, RTCVD SiN, ALD Al_2O_3), a good correlation is demonstrated by the V_{TH} shift and the gate leakage current (Fig. 9.20). The devices with PEALD SiN insulator demonstrate a lower sensitivity toward the gate positive bias-induced trapping effects, as shown in Fig. 9.20; moreover, in PEALD SiN devices, a negligible impact of the reverse bias

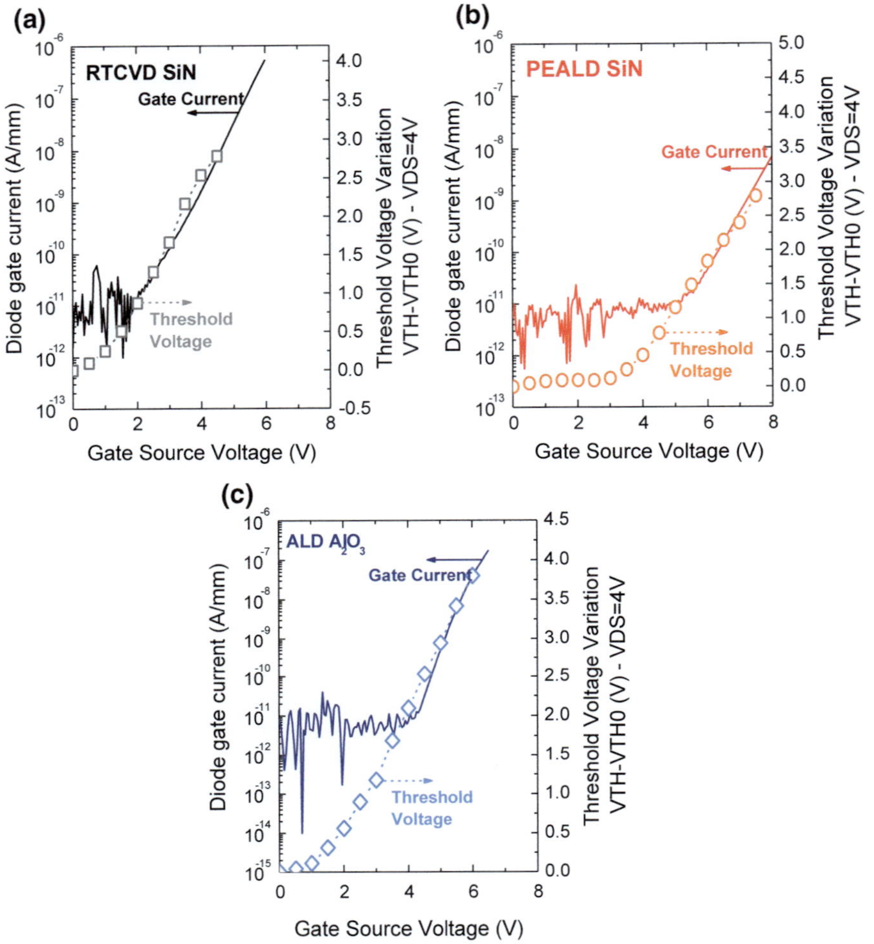

Fig. 9.20 Correlation between the threshold voltage dynamic shift (*void circles*) and the gate leakage forward current (*solid line*) for different dielectric materials (SiN, Al_2O_3) and deposition techniques (plasma-enhanced atomic layer deposition and rapid thermal chemical vapor deposition)

is demonstrated. This improvement is partially ascribed to the reduction of the trapping effects and partially to the lowering of the leakage current and its correspondence with the V_{TH} instabilities.

Results are confirmed by the capture/emission drain current (V_{TH}) transients. PEALD SiN devices, indeed, demonstrate significant improvements and lower sensitiveness toward the forward gate bias with respect to RTCVD SiN and ALD Al_2O_3 devices. The capture/recovery transients further show the presence of slow traps not thermally activated.

References

1. Umana-Membreno GA, Dell JM, Hessler TP, Nener BD, Parish G, Faraone L, Mishra UK (2002) 60Co gamma-irradiation-induced defects in n-GaN. Appl Phys Lett 80(23):4354–4356
2. Soh CB, Chua SJ, Lim HF, Chi DZ, Liu W, Tripathy S (2004) Identification of deep levels in GaN associated with dislocations. J Phys Condens Matter 16(16):6305–6315
3. Chung HM, Chuang WC, Pan YC, Tsai CC, Lee MC, Chen WH, Chen WK, Chiang CI, Lin CH, Chang H (2000) Electrical characterization of isoelectronic In-doping effects in GaN films grown by metalorganic vapor phase epitaxy. Appl Phys Lett 76(7):897–899
4. Park YS, Lee M, Jeon K, Yoon IT, Shon Y, Im H, Park CJ, Cho HY, Han M-S (2010) Deep level transient spectroscopy in plasma-assisted molecular beam epitaxy grown Al 0.2 Ga 0.8 N/GaN interface and the rapid thermal annealing effect. Appl Phys Lett 97:112110
5. Look DC, Fang Z-Q, Claflin B (2005) Identification of donors, acceptors, and traps in bulk-like HVPE GaN. J Cryst Growth 281(1):143–150
6. Cho HK, Kim KS, Hong CH, Lee HJ (2001) Electron traps and growth rate of buffer layers in unintentionally doped GaN. J Cryst Growth 223(1–2):38–42
7. Johnstone D, Biyikli S, Dogan S, Moon YT, Yun F, Morkoc H (2005) Comparison of deep levels in GaN grown by MBE, MOCVD, and HVPE. Proc SPIE—Light Diodes Res Manuf Appl IX 5739:7–15
8. Fang Z, Look DC, Kim W, Fan Z (1998) Deep centers in n-GaN grown by reactive molecular beam epitaxy. Appl Phys Lett 72(18):2277–2279
9. Cho HK, Kim CS, Hong CH (2003) Electron capture behaviors of deep level traps in unintentionally doped and intentionally doped n-type GaN. J Appl Phys 94(3):1485–1489
10. Choi KJ, Jang HW, Lee JL (2003) Observation of inductively coupled-plasma-induced damage on n-type GaN using deep-level transient spectroscopy. Appl Phys Lett 82:1233–1235
11. Arehart AR, Corrion A, Poblenz C, Speck JS, Mishra UK, DenBaars SP, Ringel SA (2008) Comparison of deep level incorporation in ammonia and rf-plasma assisted molecular beam epitaxy n-GaN films. Phys Status Solidi Curr Top Solid State Phys 5(6):1750–1752
12. Honda U, Yamada Y, Tokuda Y, Shiojima K (2012) Deep levels in n-GaN doped with carbon studied by deep level and minority carrier transient spectroscopies. Jpn J Appl Phys 51:04DF04-1–04DF04-4
13. Cho HK, Khan FA, Adesida I, Fang Z-Q, Look DC (2008) Deep level characteristics in n-GaN with inductively coupled plasma damage. J Phys D Appl Phys 41:155314
14. Arehart AR, Homan T, Wong MH, Poblenz C, Speck JS, Ringel SA (2010) Impact of N- and Ga-face polarity on the incorporation of deep levels in n-type GaN grown by molecular beam epitaxy. Appl Phys Lett 96:242112
15. Chen S, Honda U, Shibata T, Matsumura T, Tokuda Y, Ishikawa K, Hori M, Ueda H, Uesugi T, Kachi T (2012) As-grown deep-level defects in n-GaN grown by metal-organic chemical vapor deposition on freestanding GaN. J Appl Phys 112:053513
16. Kindl D, Hubík P, Krištofik J, Mareš JJ, Vyborny Z, Leys MR, Boeykens S (2009) Deep defects in GaN/AlGaN/SiC heterostructures. J Appl Phys 105:093706
17. Umana-Membreno GA, Parish G, Fichtenbaum N, Keller S, Mishra UK, Nener BD (2007) Electrically active defects in GaN layers grown with and without Fe-doped buffers by metal-organic chemical vapor deposition. J Electron Mater 37(5):569–572
18. Stuchlíková L, Šebok J, Rybár J, Petrus M, Nemec M, Harmatha L, Benkovská J, Kováč J, Škriniarová J, Lalinský T, Paskiewicz R, Tlaczala M (2010) Investigation of deep energy levels in heterostructures based on GaN by DLTS. In: Proceedings of the 8th international conference on advanced semiconductor devices and microsystems (ASDAM), pp 135–138
19. Hacke P, Detchprohm T, Hiramatsu K, Sawaki N, Tadatomo K, Miyake K (1994) Analysis of deep levels in n-type GaN by transient capacitance methods. J Appl Phys 76(1):304–309

20. Fang ZQ, Polenta L, Hemsky JW, Look DC (2000) Deep centers in as-grown and electron-irradiated n-GaN. In: IEEE semiconducting and semi-insulating materials conference (SIMC), pp 35–42
21. Asghar M, Muret P, Beaumont B, Gibart P (2004) Field dependent transformation of electron traps in GaN p–n diodes grown by metal–organic chemical vapour deposition. Mater Sci Eng, B 113(3):248–252
22. Calleja E, Sánchez FJ, Basak D, Sánchez-García MA, Muñoz E, Izpura I, Calle F, Tijero JMG, Sánchez-Rojas JL, Beaumont B, Lorenzini P, Gibart P (1997) Yellow luminescence and related deep states in undoped GaN. Phys Rev B 55(7):4689–4694
23. Zhang Z, Hurni CA, Arehart AR, Yang J, Myers RC, Speck JS, Ringel SA (2012) Deep traps in nonpolar m-plane GaN grown by ammonia-based molecular beam epitaxy. Appl Phys Lett 100:052114
24. Polyakov AY, Smirnov NB, Govorkov AV, Shlensky AA, Pearton SJ (2004) Influence of high-temperature annealing on the properties of Fe doped semi-insulating GaN structures. J Appl Phys 95(10):5591–5596
25. Hierro A, Ringel SA, Hansen M, Speck JS, Mishra UK, DenBaars SP (2000) Hydrogen passivation of deep levels in n–GaN. Appl Phys Lett 77(10):1499–1501
26. Henry TA, Armstrong A, Kelchner KM, Nakamura S, Denbaars SP, Speck JS (2012) Assessment of deep level defects in m-plane GaN grown by metalorganic chemical vapor deposition. Appl Phys Lett 100:082103
27. Fang ZQ, Look DC, Kim DH, Adesida I (2005) Traps in AlGaNGaNSiC heterostructures studied by deep level transient spectroscopy. Appl Phys Lett 87:182115
28. Lee WI, Huang TC, Guo JD, Feng MS (1995) Effects of column III alkyl sources on deep levels in GaN grown by organometallic vapor phase epitaxy. Appl Phys Lett 67:1721–1723
29. Wang CD, Yu LS, Lau SS, Yu ET, Kim W (1998) Deep level defects in n-type GaN grown by molecular beam epitaxy. Appl Phys Lett 72(10):1211–1213
30. Caesar M, Dammann M, Polyakov V, Waltereit P, Bronner W, Baeumler M, Quay R, Mikulla M, Ambacher O (2012) Generation of traps in AlGaN/GaN HEMTs during RF-and DC-stress test. IEEE Int Reliab Phys Symp Proc CD 6.1–CD 6.5
31. Hacke P, Nakayama H, Detchprohm T, Hiramatsu K, Sawaki N (1996) Deep levels in the upper bandgap region of lightly Mg-doped GaN deep levels in the upper band-gap region of lightly Mg-doped GaN. Appl Phys Lett 68(10):1362–1364
32. Sasikumar A, Arehart A, Kolluri S, Wong MH, Keller S, Denbaars SP, Speck JS, Mishra UK, Ringel SA (2012) Access-region defect spectroscopy of DC-stressed N-polar GaN MIS-HEMTs. IEEE Electron Device Lett 33(5):658–660
33. Chen XD, Huang Y, Fung S, Beling CD, Ling CC, Sheu JK, Lee ML, Chi GC, Chang SJ (2003) Deep level defect in Si-implanted GaN n(+)-p junction. Appl Phys Lett 82(21):3671–3673
34. Sasikumar A, Arehart A, Ringel SA, Kaun S, Wong MH, Mishra UK, Speck JS (2012) Direct correlation between specific trap formation and electric stress-induced degradation in MBE-grown AlGaN/GaN HEMTs. IEEE Int Reliab Phys Symp Proc 2C 3.1–2C 3.6
35. Meneghini M, Bisi D, Marcon D, Stoffels S, Van Hove M, Wu T-L, Decoutere S, Meneghesso G, Zanoni E (2014) Trapping and reliability assessment in D-mode GaN-based MIS-HEMTs for power applications. IEEE Trans Power Electron 29(5):2199–2207
36. Lyons JL, Janotti A, Van de Walle CG (2014) Effects of carbon on the electrical and optical properties of InN, GaN, and AlN. Phys Rev B 89:035204
37. Arehart AR, Sasikumar A, Via GD, Winningham B, Poling B, Heller E, Ringel SA (2010) Spatially-discriminating trap characterization methods for HEMTs and their application to RF-stressed AlGaN/GaN HEMTs. Tech Dig Int Electron Devices Meet (IEDM) 464–467
38. Heitz R, Maxim P, Eckey L, Thurian P, Hoffmann A, Broser I, Pressel K, Meyer BK (1997) Excited states of Fe^{3+} in GaN. Phys Rev B 55(7):4382–4387
39. Aggerstam T, Pinos A, Marcinkevičius S, Linnarsson M, Lourdudoss S (2007) Electron and hole capture cross-sections of Fe acceptors in GaN: Fe epitaxially grown on sapphire. J Electron Mater 36(12):1621–1624

40. Cardwell DW, Sasikumar A, Arehart AR, Kaun SW, Lu J, Keller S, Speck JS, Mishra UK, Ringel SA, Pelz JP (2013) Spatially-resolved spectroscopic measurements of $E_c - 0.57$ eV traps in AlGaN/GaN high electron mobility transistors. Appl Phys Lett 102:193509
41. Silvestri M, Uren MJ, Kuball M (2013) Iron-induced deep-level acceptor center in GaN/AlGaN high electron mobility transistors: Energy level and cross section. Appl Phys Lett 102:073501
42. Verzellesi G, Mazzanti A, Canali C, Meneghesso G, Chini A, Zanoni E (2003) Study on the origin of dc-to-RF dispersion effects in GaAs- and GaN-based heterostructure FETs. Reliab Work Compd Semicond 155–156
43. Auret FD, Meyer WE, Wu L, Hayes M, Legodi MJ, Beaumont B, Gibart P (2004) Electrical characterisation of hole traps in n-type GaN. Phys Status Solidi C Conf 201(10):2271–2276
44. Ibbetson JP, Fini PT, Ness KD, DenBaars SP, Speck JS, Mishra UK (2000) Polarization effects, surface states, and the source of electrons in AlGaN/GaN heterostructure field effect transistors. Appl Phys Lett 77(2):250–252
45. Binari SC, Ikossi K, Roussos JA, Kruppa W, Park D, Dietrich HB, Koleske DD, Wickenden AE, Henry RL (2001) Trapping effects and microwave power performance in AlGaN/GaN HEMTs. IEEE Trans Electron Devices 48(3):465–471
46. Vetury R, Zhang NQQ, Keller S, Mishra UK (2001) The impact of surface states on the DC and RF characteristics of A1GaN/GaN HFETs. IEEE Trans Electron Devices 48(3):560–566
47. Meneghesso G, Verzellesi G, Pierobon R, Rampazzo F, Chini A, Mishra UK, Canali C, Zanoni E (2004) Surface-related drain current dispersion effects in AlGaN-GaN HEMTs. IEEE Trans Electron Devices 51(10):1554–1561
48. Green BM, Chu KK, Chumbes EM, Smart JA, Shealy JR, Eastman LF (2000) The effect of surface passivation on the microwave characteristics of undoped AlGaN/GaN HEMT' s. IEEE Electron Device Lett 21(6):268–270
49. Vertiatchikh AV, Eastman LF, Schaff WJ, Prunty T (2002) Effect of surface passivation of AIGaN/GaN heterostructure field-effect transistorl. Electron Lett 38(8):388–389
50. Koley G, Tilak V, Eastman LF, Spencer MG (2003) Slow transients observed in AlGaN/GaN HFETs: Effects of SiN_x passivation and UV illumination. IEEE Trans Electron Devices 50(4): 886–893
51. Tilak V, Green B, Kim H, Dimitrov R, Smart J, Schaff WJ, Shealy JR, Eastman LF (2000) Effect of passivation on AlGaN/GaN HEMT device performance. In: 2000 IEEE international symposium on compound semiconductors. Proceedings of the IEEE twenty-seventh international symposium on compound semiconductors (Cat. No.00TH8498), pp 357–363
52. Hasegawa H, Akazawa M (2009) Current collapse transient behavior and its mechanism in submicron-gate AlGaN/GaN heterostructure transistors. J Vac Sci Technol B Microelectron Nanom Struct 27:2048–2054
53. Kotani J, Tajima M, Kasai S, Hashizume T (2007) Mechanism of surface conduction in the vicinity of Schottky gates on AlGaN/GaN heterostructures. Appl Phys Lett 91:093501
54. Sabuktagin S, Moon YT, Dogan S, Baski AA, Morkoç H (2006) Observation of surface charging at the edge of a Schottky contact. IEEE Electron Device Lett 27(4):211–213
55. Heikman S, Keller S, Denbaars SP, Mishra UK (2002) Growth of Fe doped semi-insulating GaN by metalorganic chemical vapor deposition. Appl Phys Lett 81(3):439–441
56. Uren MJ, Nash KJ, Balmer RS, Martin T, Morvan E, Caillas N, Delage SL, Ducatteau D, Grimbert B, De Jaeger JC (2006) Punch-through in short-channel AlGaN/GaN HFETs. IEEE Trans Electron Devices 53(2):395–398
57. Uren MJ, Kuball M (2014) GaN transistor reliability and instabilities. In: 2014 10th International Conference on advanced semiconductor devices microsystems (ASDAM), Smolenice
58. Lee W, Ryou JH, Yoo D, Limb J, Dupuis RD, Hanser D, Preble E, Williams NM, Evans K (2007) Optimization of Fe doping at the regrowth interface of GaN for applications to III-nitride-based heterostructure field-effect transistors. Appl Phys Lett 90:093509

59. Desmaris V, Rudziñski M, Rorsman N, Hageman PR, Larsen PK, Zirath H, Rödle TC, Jos HFF (2006) Comparison of the DC and microwave performance of AlGaN/GaN HEMTs grown on SiC by MOCVD with Fe-doped or unintentionally doped GaN buffer layers. IEEE Trans Electron Devices 53(9):2413–2417
60. Uren MJ, Hayes DG, Balmer RS, Wallis DJ, Hilton KP, Maclean JO, Martin T, Roff C, Mcgovern P, Benedikt J, Tasker PJ (2006) Control of short-channel effects in GaN/AlGaN HFETs. In: Proceedings of the 1st European microwave integrated circuits conference, pp 65–68
61. Bougrioua Z, Azize M, Lorenzini P, Laügt M, Haas H (2005) Some benefits of Fe doped less dislocated GaN templates for AlGaN/GaN HEMTs grown by MOVPE. Phys Status Solidi Appl Mater Sci 202(4):536–544
62. Meneghini M, Rossetto I, Bisi D, Stocco A, Chini A, Pantellini A, Lanzieri C, Nanni A, Meneghesso G, Zanoni E (2014) Buffer traps in Fe-doped AlGaN/GaN HEMTs: investigation of the physical properties based on pulsed and transient measurements. IEEE Trans Electron Devices 61(12):4070–4077
63. Umana-Membreno GA, Parish G, Fichtenbaum N, Keller S, Mishra UK, Nener BD (2008) Electrically active defects in GaN layers grown with and without Fe-doped buffers by metal-organic chemical vapor deposition. J Electron Mater 37(5):569–572
64. Polyakov AY, Smirnov NB, Govorkov AV, Pearton SJ (2003) Electrical and optical properties of Fe-doped semi-insulating GaN templates. Appl Phys Lett 83(16):3314–3316
65. Bisi D, Meneghini M, De Santi C, Chini A, Dammann M, Bruckner P, Mikulla M, Meneghesso G, Zanoni E (2013) Deep-level characterization in GaN HEMTs-Part I: advantages and limitations of drain current transient measurements. IEEE Trans Electron Devices 60(10): 3166–3175
66. Chini A, Di Lecce V, Soci F, Bisi D, Stocco A, Meneghini M, Meneghesso G, Zanoni E, Gasparotto A (2012) Experimental and numerical correlation between current-collapse and Fe-doping profiles in GaN HEMTs. IEEE Int Reliab Phys Symp Proc CD 2.1–CD 2.4
67. Uren MJ, Moreke J, Kuball M (2012) Buffer design to minimize current collapse in GaN/AlGaN HFETs. IEEE Trans Electron Devices 59(12):3327–3333
68. Polyakov AY, Smirnov NB, Govorkov AV, Markov AV, Sun Q, Zhang Y, Yerino CD, Ko TS, Lee IH, Han J (2010) Electrical properties and deep traps spectra of a-plane GaN films grown on r-plane sapphire. Mater Sci Eng B Solid-State Mater Adv Technol 166:220–224
69. Haase D, Schmid M, Kürner W, Dörnen A, Härle V, Scholz F, Burkard M, Schweizer H (1996) Deep-level defects and n-type-carrier concentration in nitrogen implanted GaN. Appl Phys Lett 69(17):2525–2527
70. Armstrong A, Arehart AR, Moran B, DenBaars SP, Mishra UK, Speck JS, Ringel SA (2004) Impact of carbon on trap states in n-type GaN grown by metalorganic chemical vapor deposition. Appl Phys Lett 84(3):374–376
71. Rudziński M, Desmaris V, Van Hal PA, Weyher JL, Hageman PR, Dynefors K, Rödle TC, Jos HFF, Zirath H, Larsen PK (2006) Growth of Fe doped semi-insulating GaN on sapphire and 4H-SiC by MOCVD. Phys Status Solidi Curr Top Solid State Phys 3(6):2231–2236
72. Mei F, Fu QM, Peng T, Liu C, Peng MZ, Zhou JM (2008) Growth and characterization of AlGaN/GaN heterostructures on semi-insulating GaN epilayers by molecular beam epitaxy. J Appl Phys 103:094502
73. Wang M, Yan D, Zhang C, Xie B, Wen CP, Wang J, Hao Y, Wu W, Shen B (2014) Investigation of surface- and buffer-induced current collapse in GaN high-electron mobility transistors using a soft switched pulsed I-V measurement. IEEE Electron Device Lett 35(11): 1094–1096
74. Meneghini M, Bisi D, Marcon D, Stoffels S, Van Hove M, Wu TL, Decoutere S, Meneghesso G, Zanoni E (2014) Trapping in GaN-based metal-insulator-semiconductor transistors: Role of high drain bias and hot electrons. Appl Phys Lett 104:143505

75. Meneghini M, Rossetto I, Bisi D, Stocco A, Cester A, Meneghesso G, Zanoni E, Chini A, Pantellini A, Lanzieri C (2014) Role of buffer doping and pre-existing trap states in the current collapse and degradation of AlGaN/GaN HEMTs. IEEE Int Reliab Phys Symp Proc 6C 6.1–6C 6.7
76. Bisi D, Chini A, Soci F, Stocco A, Meneghini M, Pantellini A, Nanni A, Lanzieri C, Gamarra P, Lacam C, Tordjman M, Meneghesso G, Zanoni E (2015) Hot-electron degradation of AlGaN/GaN high-electron mobility transistors during RF operation: correlation with GaN buffer design. IEEE Electron Device Lett 36(10):1011–1014
77. Parish G, Keller S, Denbaars SP, Mishra UK (2000) SIMS investigations into the effect of growth conditions on residual impurity and silicon incorporation in GaN and $Al_xGa_{1-x}N$. J Electron Mater 29(1):15–20
78. Koleske DD, Wickenden AE, Henry RL, Twigg ME (2002) Influence of MOVPE growth conditions on carbon and silicon concentrations in GaN. J Cryst Growth 242:55–69
79. Wickenden AE, Koleske DD, Henry RL, Twigg ME, Fatemi M (2004) Resistivity control in unintentionally doped GaN films grown by MOCVD. J Cryst Growth 260:54–62
80. Brunner F, Bahat-Treidel E, Cho M, Netzel C, Hilt O, Würfl J, Weyers M (2011) Comparative study of buffer designs for high breakdown voltage AlGaNGaN HFETs. Phys Status Solidi Curr Top Solid State Phys 8(7–8):2427–2429
81. Chen JT, Forsberg U, Janzén E (2013) Impact of residual carbon on two-dimensional electron gas properties in AlxGa1-xN/GaN heterostructure. Appl Phys Lett 102:193506
82. Gamarra P, Lacam C, Tordjman M, Splettstösser J, Schauwecker B, Di Forte-Poisson MA (2015) Optimisation of a carbon doped buffer layer for AlGaN/GaN HEMT structures. J Cryst Growth 414:232–236
83. Poblenz C, Waltereit P, Rajan S, Heikman S, Mishra UK, Speck JS (2004) Effect of carbon doping on buffer leakage in AlGaN/GaN high electron mobility transistors. J Vac Sci Technol B Microelectron Nanom Struct 22(3):1145–1149
84. Wang WZ, Selvaraj SL, Win KT, Dolmanan SB, Bhat T, Yakovlev N, Tripathy S, Lo GQ (2015) Effect of carbon doping and crystalline quality on the vertical breakdown characteristics of GaN layers grown on 200-mm silicon substrates. J Electron Mater 44(10): 3272–3276
85. Kim D-S, Won C-H, Kang H-S, Kim Y-J, Kim YT, Kang IM, Lee J-H (2015) Growth and characterization of semi-insulating carbon-doped/undoped GaN multiple-layer buffer. Semicond Sci Technol 30:035010
86. Bahat-treidel E, Brunner F, Hilt O, Cho E, Würfl J, Tränkle G (2010) AlGaN/GaN/GaN: C back-barrier HFETsWith breakdown voltage of over 1 kV and low RON×A. IEEE Trans Electron Devices 57(11):3050–3058
87. Meneghesso G, Meneghini M, Zanoni E (2013) GaN-based power HEMTs, parasitic, reliability and high field issues. 224th ECS Meet
88. Bahat-Treidel E, Hilt O, Brunner F, Pyka S, Wurfl J (2014) Systematic study o f GaN based power transistors' dynamic on-state resistance at elevated temperatures. Int Symp Compd Semicond
89. Rossetto I, Rampazzo F, Meneghini M, Silvestri M, Dua C, Gamarra P, Aubry R, Di Forte-Poisson MA, Patard O, Delage SL, Meneghesso G, Zanoni E (2014) Influence of different carbon doping on the performance and reliability of InAlN/GaN HEMTs. Microelectron Reliab 54:2248–2252
90. Huber M, Silvestri M, Knuuttila L, Pozzovivo G, Andreev A, Kadashchuk A, Bonanni A, Lundskog A (2015) Impact of residual carbon impurities and gallium vacancies on trapping effects in AlGaN/GaN metal insulator semiconductor high electron mobility transistors. Appl Phys Lett 107:032106
91. Bisi D, Stocco A, Rossetto I, Meneghini M, Rampazzo F, Chini A, Soci F, Pantellini A, Lanzieri C, Gamarra P, Lacam C, Tordjman M, Di Forte-Poisson MA, De Salvador D, Bazzan M, Meneghesso G, Zanoni E (2015) Effects of buffer compensation strategies on the electrical performance and RF reliability of AlGaN/GaN HEMTs. Microelectron Reliab 55:1662–1666

92. Wright AF (2002) Substitutional and interstitial carbon in wurtzite GaN. J Appl Phys 92(5): 2575–2585
93. Lyons JL, Janotti A, Van De Walle CG (2010) Carbon impurities and the yellow luminescence in GaN. Appl Phys Lett 97:152108
94. Armstrong A, Arehart AR, Green D, Mishra UK, Speck JS, Ringel SA (2005) Impact of deep levels on the electrical conductivity and luminescence of gallium nitride codoped with carbon and silicon. J Appl Phys 98:053704
95. Armstrong A, Arehart AR, Moran B, DenBaars SP, Mishra UK, Speck JS, Ringel SA (2003) In 30th international symposium on compound semiconductor. San Diego, USA, p 42
96. Fang ZQ, Claflin B, Look DC, Green DS, Vetury R (2010) Deep traps in AlGaN/GaN heterostructures studied by deep level transient spectroscopy: effect of carbon concentration in GaN buffer layers. J Appl Phys 108:063706
97. Uren MJ, Silvestri M, Casar M, Hurkx GAM, Croon JA, Sonsky J, Kuball M (2014) Intentionally carbon-doped AlGaN/GaN HEMTs: necessity for vertical leakage paths. IEEE Electron Device Lett 35(3):327–329
98. Bisi D, Meneghini M, Van Hove M, Marcon D, Stoffels S, Wu TL, Decoutere S, Meneghesso G, Zanoni E (2015) Trapping mechanisms in GaN-based MIS-HEMTs grown on silicon substrate. Phys Status Solidi Appl Mater Sci 212(5):1122–1129
99. Van Hove M, Boulay S, Bahl SR, Stoffels S, Kang X, Wellekens D, Geens K, Delabie A, Decoutere S (2012) CMOS process-compatible high-power low-leakage AlGaN/GaN MISHEMT on silicon. Electron Device Lett IEEE 33(5):667–669
100. Bisi D, Meneghini M, Marino FA, Marcon D, Stoffels S, Van Hove M, Decoutere S, Meneghesso G, Zanoni E (2014) Kinetics of buffer-related R_{ON}-increase in GaN-on-Silicon MISHEMTs. IEEE Electron Device Lett 35(10):1004–1006
101. Pérez-Tomàs A, Fontseré A, Llobet J, Placidi M, Rennesson S, Baron N, Chenot S, Moreno JC, Cordier Y (2013) Analysis of the AlGaN/GaN vertical bulk current on Si, sapphire, and free-standing GaN substrates. J Appl Phys 113:174501
102. Zhou C, Jiang Q, Huang S, Chen KJ (2012) Vertical leakage/breakdown mechanisms in AlGaN/GaN-on-Si devices. In: Proceedings of the 24th international symposium on power semiconductor devices ICs 3–7 June 2012. Bruges, Belgium, pp 1132–1134
103. Tanaka K, Ishida M, Ueda T, Tanaka T (2013) Effects of deep trapping states at high temperatures on transient performance of AlGaN/GaN heterostructure field-effect transistors. Jpn J Appl Phys 52:04CF07–1–04CF07–5
104. Meneghini M, Vanmeerbeek P, Silvestri R, Dalcanale S, Banerjee A, Bisi D, Zanoni E, Meneghesso G, Moens P (2015) Temperature-dependent dynamic R_{ON} in GaN-based MIS-HEMTs: role of surface traps and buffer leakage. IEEE Trans Electron Devices 62(3):782–787
105. Moens P, Vanmeerbeek P, Banerjee A, Guo J, Liu C, Coppens P, Salih A, Tack M, Caesar M, Uren MJ, Kuball M, Meneghini M, Meneghesso G, Zanoni E (2015) On the impact of carbon-doping on the dynamic Ron and off-state leakage current of 650 V GaN power devices. Proc Int Symp Power Semicond Devices ICs 37–40
106. Kwan MH, Wong K, Lin YS, Yao FW, Tsai MW, Chang Y, Chen PC, Su RY, Wu C, Yu JL, Yang FJ, Lansbergen GP, Wu H, Lin M, Wu CB, Lai Y, Hsiung C, Liu P, Chiu H, Chen C, Yu CY, Lin HS, Chang M, Wang S, Chen LC, Tsai JL, Tuan HC, Kalnitsky A (2014) CMOS-compatible GaN-on-Si field-effect transistors for high voltage power applications. Electron Devices Meet (IEDM) IEEE Int 17.6.1–17.6.4
107. Uren MJ, Caesar M, Karboyan S, Moens P, Vanmeerbeek P, Kuball M (2015) Electric field reduction in C—doped AlGaN/ GaN on Si high electron mobility transistors. Electron Device Lett IEEE 36(8):826–828
108. Stoffels S, Zhao M, Venegas R, Kandaswamy P, You S, Novak T, Saripalli Y, Van Hove M, Decoutere S (2015) The physical mechanism of dispersion caused by AlGaN/GaN buffers on Si and optimization for low dispersion. 2015 IEEE Int Electron Devices Meet 35.4.1–35.4.4

109. Kaneko S, Kuroda M, Yanagihara M, Ikoshi A, Okita H, Morita T, Tanaka K, Hikita M, Uemoto Y, Takahashi S, Ueda T (2015) Current-collapse-free operations up to 850 V by GaN-GIT utilizing hole injection from drain. Proc Int Symp Power Semicond Devices ICs 41–44
110. Liu ZH, Ng GI, Arulkumaran S, Maung YKT, Teo KL, Foo SC, Sahmuganathan V (2010) Improved linearity for low-noise applications in 0.25-μm GaN MISHEMTs using ALD Al_2O_3 as gate dielectric. Electron Device Lett IEEE 31(8):803–805
111. Wu T, Marcon D, Zahid MB, Van Hove M, Decoutere S, Groeseneken G (2013) Comprehensive investigation of on-state stress on DMode AlGaN/GaN MIS-HEMTs. Reliab Phys Symp (IRPS), 2013 IEEE Int 3C.5.1–3C.5.7
112. Khan MA, Simin G, Yang J, Zhang J, Koudymov A, Shur MS, Gaska R, Hu X, Tarakji A (2003) Insulating Gate III-N heterostructure field-effect transistors for high-power microwave and switching applications. IEEE Microw Theory Tech 51(2):624–633
113. Imada T, Motoyoshi K, Kanamura M, Kikkawa T (2011) Reliability analysis of enhancement-mode GaN MIS-HEMT with gate-recess structure for power supplies. IEEE Int Integr Reliab Work Final Rep 38–41
114. Kordoš P, Heidelberger G, Bernát J, Fox A, Marso M, Lüth H (2005) High-power SiO2/AlGaN/GaN metal-oxide-semiconductor heterostructure field-effect transistors. Appl Phys Lett 87:143501
115. Hahn H, Benkhelifa F, Ambacher O, Brunner F, Noculak A, Kalisch H, Vescan A (2015) Threshold voltage engineering in GaN-based HFETs: A systematic study with the threshold voltage reaching more than 2 V. IEEE Trans Electron Devices 62(2):538–545
116. Choi W, Ryu H, Jeon N, Lee M, Cha H-Y, Seo K-S (2014) Improvement of Vth instability in normally-off GaN MIS-HEMTs employing PEALD-SiN_x as an interfacial layer. Electron Device Lett IEEE 35(1):30–32
117. Saito W, Takada Y, Kuraguchi M, Tsuda K, Omura I (2006) Recessed-gate structure approach toward normally electronics applications. IEEE Trans Electron Devices 53(2):356–362
118. Oka T, Nozawa T (2008) AlGaN/GaN recessed MIS-Gate HFET with high-threshold-voltage normally-off operation for power electronics applications. IEEE Electron Device Lett 29(7): 668–670
119. Xu Z, Wang J, Cai Y, Liu J, Yang Z, Li X, Wang M, Yu M, Xie B, Wu W, Ma X, Zhang J, Hao Y (2014) High temperature characteristics of GaN-based inverter integrated with enhancement-mode (E-Mode) MOSFET and depletion-mode (D-Mode) HEMT. IEEE Electron Device Lett 35(1):33–35
120. Yatabe Z, Hori Y, Ma W-C, Asubar JT, Akazawa M, Sato T, Hashizume T (2014) Characterization of electronic states at insulator/AlGaN interfaces for improved insulated gate and surface passivation structures of GaN-based transistors. Jpn J Appl Phys 53(10):2014
121. Zhu JJ, Ma XH, Hou B, Chen WW, Hao Y (2014) Investigation of trap states in high Al content AlGaN/GaN high electron mobility transistors by frequency dependent capacitance and conductance analysis. AIP Adv 4:037108
122. Mizue C, Hori Y, Miczek M, Hashizume T (2011) Capacitance–voltage characteristics of Al2O3/AlGaN/GaN structures and state density distribution at Al2O3/AlGaN interface. Jpn J Appl Phys 50:021001
123. Ganguly S, Verma J, Li G, Zimmermann T, Xing H, Jena D (2011) Presence and origin of interface charges at atomic-layer deposited Al 2O3/III-nitride heterojunctions. Appl Phys Lett 99:193504
124. Lagger P, Reiner M, Pogany D, Ostermaier C (2014) Comprehensive study of the complex dynamics of forward bias-induced threshold voltage drifts in GaN based MIS-HEMTs by stress/recovery experiments. IEEE Trans Electron Devices 61(4):1022–1030
125. Guo A, Del Alamo JA (2015) Positive-bias temperature instability (PBTI) of GaN MOSFETs. IEEE Int Reliab Phys Symp Proc 6C51–6C57
126. Eller BS, Yang J, Nemanich RJ (2013) Electronic surface and dielectric interface states on GaN and AlGaN. J Vac Sci Technol A Vac Surf Film 31(5):050807

127. Johnson DW, Lee RTP, Hill RJW, Wong MH, Bersuker G, Piner EL, Kirsch PD, Harris HR (2013) Threshold voltage shift due to charge trapping in dielectric-gated AlGaN/GaN high electron mobility transistors examined in Au-free technology. IEEE Trans Electron Devices 60(10):3197–3203
128. Lagger P, Steinschifter P, Reiner M, Stadtmuller M, Denifl G, Naumann A, Muller J, Wilde L, Sundqvist J, Pogany D, Ostermaier C (2014) Role of the dielectric for the charging dynamics of the dielectric/barrier interface in AlGaN/GaN based metal-insulator-semiconductor structures under forward gate bias stress. Appl Phys Lett 105:033512
129. Lagger P, Ostermaier C, Pobegen G, Pogany D (2012) Towards understanding the origin of threshold voltage instability of AlGaN/GaN MIS-HEMTs. Tech Dig Int Electron Devices Meet (IEDM) 299–302
130. Rossetto I, Meneghini M, Bisi D, Barbato A, Van Hove M, Marcon D, Wu TL, Decoutere S, Meneghesso G, Zanoni E (2015) Impact of gate insulator on the dc and dynamic performance of AlGaN/GaN MIS-HEMTs. Microelectron Reliab 55:1692–1696
131. Meneghesso G, Meneghini M, Bisi D, Rossetto I, Wu TL, Van Hove M, Marcon D, Stoffels S, Decoutere S, Zanoni E (2015) Trapping and reliability issues in GaN-based MIS HEMTs with partially recessed gate. Microelectron Reliab 58:151–157
132. Lagger P, Schiffmann A, Pobegen G, Pogany D, Ostermaier C (2013) Very fast dynamics of threshold voltage drifts in GaN-based MIS-HEMTs. IEEE Electron Device Lett 34(9):1112–1114
133. Wu TL, Marcon D, Ronchi N, Bakeroot B, You S, Stoffels S, Van Hove M, Bisi D, Meneghini M, Groeseneken G, Decoutere S (2015) Analysis of slow de-trapping phenomena after a positive gate bias on AlGaN/GaN MIS-HEMTs with in-situ Si_3N_4/Al_2O_3 bilayer gate dielectrics. Solid State Electron 103:127–130
134. Capriotti M, Alexewicz A, Fleury C, Gavagnin M, Bethge O, Visalli D, Derluyn J, Wanzenbock HD, Bertagnolli E, Pogany D, Strasser G (2014) Fixed interface charges between AlGaN barrier and gate stack composed of in situ grown Si_3N4 and Al2O3 in AlGaN/GaN high electron mobility transistors with normally off capability. Appl Phys Lett 104:113502
135. Lansbergen GP, Wong KY, Lin YS, Yu JL, Yang FJ, Tsai CL, Oates AS (2014) Threshold voltage drift (PBTI) in GaN D-MODE MISHEMTs: Characterization of fast trapping components. IEEE Int Reliab Phys Symp Proc 6C 4.1–6C 4.6
136. Grasser T (2012) Stochastic charge trapping in oxides: From random telegraph noise to bias temperature instabilities. Microelectron Reliab 52:39–70
137. Ma XH, Zhu JJ, Liao XY, Yue T, Chen WW, Hao Y (2013) Quantitative characterization of interface traps in Al2O3/AlGaN/GaN metal-oxide-semiconductor high-electron-mobility transistors by dynamic capacitance dispersion technique. Appl Phys Lett 103:033510
138. Lu X, Ma J, Jiang H, May K (2014) Lau, "Characterization of in situ SiNx thin film grown on AlN/GaN heterostructure by metal-organic chemical vapor deposition". Appl Phys Lett 104:032903
139. Sun X, Saadat OI, Chang-Liao KS, Palacios T, Cui S, Ma TP (2013) Study of gate oxide traps in HfO2/AlGaN/GaN metal-oxide-semiconductor high-electron-mobility transistors by use of ac transconductance method. Appl Phys Lett 102:103504
140. Huang S, Jiang Q, Yang S, Tang Z, Chen KJ (2013) Mechanism of PEALD-Grown AlN passivation for AlGaN/GaN HEMTs: compensation of interface traps by polarization charges. IEEE Electron Device Lett 34(2):193–195
141. Yang S, Tang Z, Wong KY, Lin YS, Lu Y, Huang S, Chen KJ (2013) Mapping of interface traps in high-performance Al 2 O 3/AlGaN/GaN MIS-heterostructures using frequency- and temperature-dependent C–V techniques. Tech Dig Int Electron Devices Meet (IEDM) 6.3.1–6.3.4
142. Shih HA, Kudo M, Suzuki TK (2012) Analysis of AlN/AlGaN/GaN metal-insulator-semiconductor structure by using capacitance-frequency-temperature mapping. Appl Phys Lett 101:043501

143. Matys M, Adamowicz B, Hashizume T (2012) Determination of the deep donor-like interface state density distribution in metal/Al2O3/n-GaN structures from the photocapacitance-light intensity measurement. Appl Phys Lett 101:231608
144. Capriotti M, Lagger P, Fleury C, Oposich M, Bethge O, Ostermaier C, Strasser G, Pogany D (2015) Modeling small-signal response of GaN-based metal-insulator-semiconductor high electron mobility transistor gate stack in spill-over regime: Effect of barrier resistance and interface states. J Appl Phys 117:024506
145. Okino T, Ochiai M, Ohno Y, Kishimoto S, Maezawa K, Mizutani T (2004) Drain current DLTS of AlGaN-GaN MIS-HEMTs. IEEE Electron Device Lett 25(8):523–525
146. Gaffey B, Guido LJ, Wang XW, Ma TP (2001) High-quality oxide/nitride/oxide gate insulator for GaN MIS structures. IEEE Trans Electron Devices 48(3):458–464
147. Kanechika M, Sugimoto M, Soejima N, Ueda H, Ishiguro O, Kodama M, Hayashi E, Itoh K, Uesugi T, Kachi T (2007) A vertical insulated gate AlGaN/GaN heterojunction field-effect transistor. Jpn J Appl Phys 46(21):L503–L505
148. Kambayashi H, Nomura T, Ueda H, Harada K, Morozumi Y, Hasebe K, Teramoto A, Sugawa S, Ohmi T (2013) High quality SiO2/Al2O3 gate stack for GaN MOSFET. Jpn J Appl Phys 52:04CF09
149. Tang Z, Jiang Q, Lu Y, Huang S, Yang S, Tang X, Chen KJ (2013) 600-V normally off SiN_x/AlGaN/GaN MIS-HEMT with large gate swing and low current collapse. IEEE Electron Device Lett 34(11):1373–1375
150. Robertson J, Falabretti B (2006) Band offsets of high K gate oxides on III-V semiconductors. J Appl Phys 100:014111
151. Derluyn J, Boeykens S, Cheng K, Vandersmissen R, Das J, Ruythooren W, Degroote S, Leys MR, Germain M, Borghs G (2005) Improvement of AlGaN/GaN high electron mobility transistor structures by in situ deposition of a Si3N4 surface layer. J Appl Phys 98:054501
152. Van Hove M, Kang X, Stoffels S, Wellekens D, Ronchi N, Venegas R, Geens K, Decoutere S (2013) Fabrication and Performance of Au-Free AlGaN/GaN-on-silicon Power Devices with Al_2O_3 and Si_3N_4/Al_2O_3 Gate Dielectrics. IEEE Trans Electron Devices 60(10):3071–3078
153. Liu S, Yang S, Tang Z, Jiang Q, Liu C, Wang M, Chen KJ (2014) Al2O3/AlN/GaN MOS-channel-HEMTs with an AlN interfacial layer. Electron Device Lett IEEE 35:723–725
154. Kanamura M, Ohki T, Ozaki S, Nishimori M, Tomabechi S, Kotani J, Miyajima T, Nakamura N, Okamoto N, Kikkawa T, Watanabe K (2013) Suppression of threshold voltage shift for normally-Off GaN MIS-HEMT without post deposition annealing. Proc Int Symp Power Semicond Devices ICs 411–414
155. Hori Y, Mizue C, Hashizume T (2010) Process conditions for improvement of electrical properties of Al 2 O 3/n-GaN structures prepared by atomic layer deposition. Jpn J Appl Phys 49(8):080201
156. Esposto M, Krishnamoorthy S, Nath DN, Bajaj S, Hung TH, Rajan S (2011) Electrical properties of atomic layer deposited aluminum oxide on gallium nitride. Appl Phys Lett 99:133503
157. Son J, Chobpattana V, McSkimming BM, Stemmer S (2012) Fixed charge in high-k/GaN metal-oxide-semiconductor capacitor structures. Appl Phys Lett 101:102905
158. Choi M, Janotti A, Van De Walle CG (2013) Native point defects and dangling bonds in α-Al2O3. J Appl Phys 113:044501
159. Choi M, Lyons JL, Janotti A, Van de Walle CG (2013) Impact of native defects in high-k dielectric oxides on GaN/oxide metal-oxide-semiconductor devices. Phys Status Solidi Basic Res 250(4):787–791
160. Liu X, Kim J, Yeluri R, Lal S, Li H, Lu J, Keller S, Mazumder B, Speck JS, Mishra UK (2013) Fixed charge and trap states of in situ Al2O3 on Ga-face GaN metal-oxide-semiconductor capacitors grown by metalorganic chemical vapor deposition. J Appl Phys 114:164507
161. Weber JR, Janotti A, Van de Walle CG (2011) Native defects in Al2O3 and their impact on III-V/Al2O3 metal-oxide-semiconductor-based devices. J Appl Phys 109:033715

162. Fontserè A, Pérez-Tomás A, Godignon P, Millán J, De Vleeschouwer H, Parsey JM, Moens P (2012) Wafer scale and reliability investigation of thin HfO2·AlGaN/GaN MIS-HEMTs. Microelectron Reliab 52:2220–2223
163. Lin Y, Wong K, Lansbergen GP, Yu JL, Yu CJ, Hsiung CW, Chiu HC, Liu SD, Chen PC, Yao FW, Su RY, Chou CY, Tsai CY, Yang FJ, Tsai CL, Tsai CS, Chen X, Tuan HC, Kalnitsky A (2014) Improved trap-related characteristics on SiNx/AlGaN/GaN MISHEMTs with surface treatment. In: Proceedings of the 26th international symposium power semiconductor devices IC's, pp 293–296
164. Ronchi N, De Jaeger B, Van Hove M, Roelofs R, Wu T, Hu J, Kang X, Decoutere S (2015) Combined plasma-enhanced-atomic-layer-deposition gate dielectric and in situ SiN cap layer for reduced threshold voltage shift and dynamic ON-resistance dispersion of AlGaN/GaN high electron mobility transistors on 200 mm Si substrates Combined plasma-e. Jpn J Appl Phys 54:04DF02
165. Choi W, Seok O, Ryu H, Cha HY, Seo KS (2014) High-voltage and low-leakage-current gate recessed normally-Off GaN MIS-HEMTs with dual gate insulator employing PEALD-SiN_x/RF-sputtered-HfO_2. IEEE Electron Device Lett 35(2):175–177
166. Yuan Y, Wang L, Yu B, Shin B, Ahn J, McIntyre PC, Asbeck PM, Rodwell MJW, Taur Y (2011) A distributed model for border traps in Al_2O_3—inGaAs MOS devices. IEEE Electron Device Lett 32(4):485–487
167. Tang K, Negara A, Kent T, Droopad R, Kummel AC, Mcintyre PC (2015) Border trap analysis and reduction in AL D-high-k InGaAs gate stacks. Compound Semiconductors, St. Barbar (CA)

Chapter 10
Cascode Gallium Nitride HEMTs on Silicon: Structure, Performance, Manufacturing, and Reliability

Primit Parikh

10.1 Motivation and Configuration of the Cascode GaN HEMT

Due to the polarization charges for the AlGaN–GaN heterostructure system [1], GaN HEMTs by nature are depletion mode (d-mode) or normally on devices. While d-mode devices have been successful for RF applications [1], in power switching applications "normally off" devices are mostly preferred for safety and ease of control in the absence of input or control (gate) voltage, when the GaN power device and the associated circuit is off. Unlike the well-established SiO_2–Si oxide insulator-semiconductor system, the lack of a suitable native oxide for the AlGaN–GaN system has hindered the development of enhancement-mode (e-mode) or normally off devices. Although intrinsic e-mode GaN devices have been demonstrated [2, 3], a reliable high-voltage device that performs in a very stable manner in high-voltage applications has been elusive.

A cascode GaN HEMT combines a d-mode high-voltage GaN HEMT with a low-voltage Si MOSFET device to achieve an effective normally off high-voltage device (intrinsically even a Si superjunction transistor is combination of inbuilt low voltage normally off "gating" region and a high voltage normally on "drift/blocking" region). By combining the well-established normally off gating functionality of the Si device with the high-voltage capability of the GaN HEMT, the cascode has enabled a highly reliable, very-high-performance GaN product to the market [4, 5] (Fig. 10.1).

P. Parikh (✉)
Transphorm Inc., Goleta, CA 93117, USA
e-mail: pparikh@transphormusa.com

M. Meneghini et al. (eds.), *Power GaN Devices*,
Power Electronics and Power Systems, DOI 10.1007/978-3-319-43199-4_10

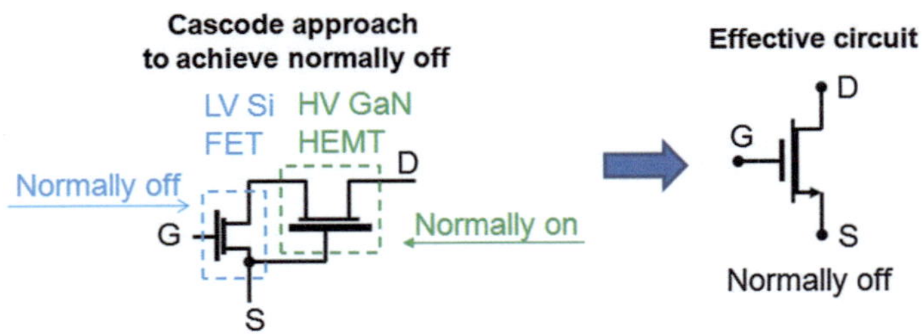

Fig. 10.1 Schematic of the cascode high-voltage GaN HEMT

10.2 Functionality and Benefits of Cascode GaN HEMT

The operation of the cascode GaN HEMT effectively utilizes the key properties of the Si "driver" FET's positive threshold voltage and the GaN HEMT's high off-state blocking. Referring to Fig. 10.2, note that the drain-source voltage of the Si FET is the gate-source voltage of the GaN HEMT. When zero gate voltage is applied to the Si FET, the Si FET turns off first, and its drain to source voltage (VDS1) increases. Since this is the gate-source voltage (VGS2) of the GaN HEMT, it will turn off as soon as this voltage is more negative than its threshold voltage. The HEMT will then block a high off-state voltage (VDS2). When a positive gate voltage is applied to the Si FET (>VTH1), it turns on first, which reduces VDS1 and thus VGS2 to near zero. Hence, the GaN HEMT turns on as its gate-source voltage is now higher than its intrinsic negative threshold voltage. An appropriate Si FET should be chosen whose off-stage voltage is more negative than the GaN HEMT threshold voltage to effectively turn off the GaN HEMT under all conditions.

One of the biggest benefits of the cascode GaN HEMT is the ability to use standard commercial silicon MOSFET drivers. This presents both significant ease of use as well as a cost benefit (due to standard driver products) versus using customized (i.e., higher cost) drivers for today's single-chip e-mode GaN devices that need delicate turn-on/maximum gate voltage protection.

Since the low-voltage Si FET is fast (very low Q_g and Q_{rr}), the total gate charge and reverse recovery charge of the normally off cascode device is very small when compared to a 600-V Si superjunction (SJ) MOSFET, as shown in Fig. 10.3. All-in-all the cascode GaN offers ultra-low resistance, a minimal Miller plateau (a feature of the cascode configuration itself), and fast switching operation in a diode-free bridge (explained in Sect. 10.3).

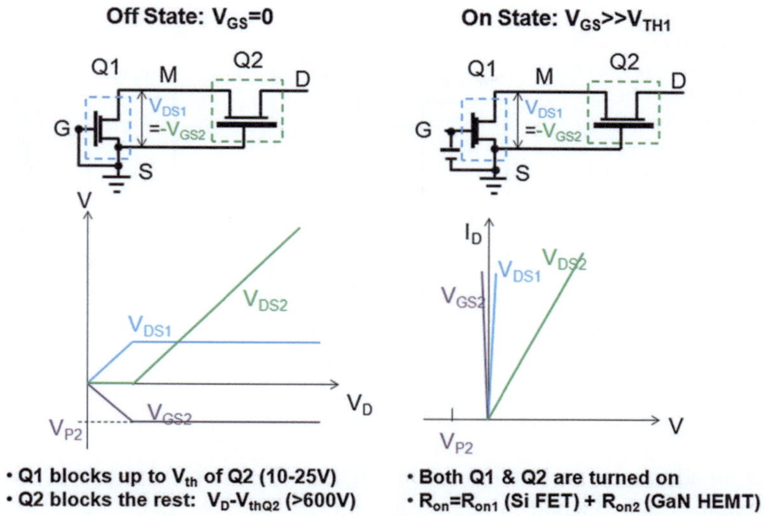

Fig. 10.2 Cascode GaN HEMT off-/on-state operation: *Q1* Si FET, *Q2* GaN HEMT

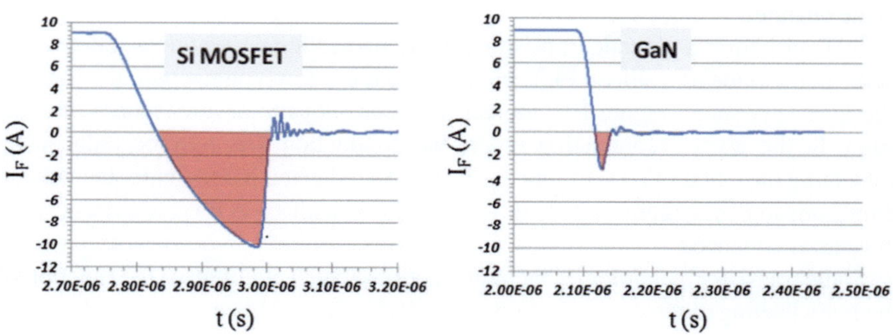

Fig. 10.3 Reverse recovery charge test result for a Si MOSFET and a GaN HEMT with similar on-resistance, showing a 20× reduction in Qrr

10.3 Key Applications and Performance Advantage of Cascode GaN HEMTs

10.3.1 Diode-Free Half-Bridge Architecture

Bridge circuits are among the most ubiquitous building blocks in power electronic converter and inverter circuits. The cascode GaN HEMT, capable of 3-quadrant operation with 20 times less reverse recovery charge versus the best Si SJ MOSFETs, is a key enabler for diode-free hard-switched bridges. A particular advantage of GaN HEMTs in bridge circuits is that they can carry the freewheeling current without the

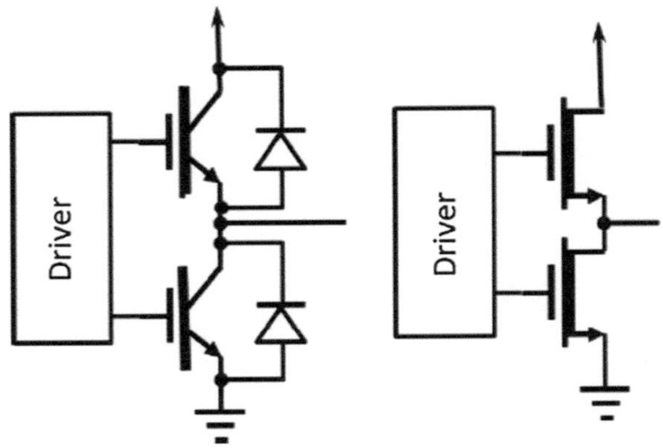

Fig. 10.4 Comparison of a traditional half-bridge with the antiparallel diodes (*left*) with a GaN diode-free half-bridge (*right*) with only two GaN transistors (each can be a cascode or e-mode normally off)

need of an additional antiparallel diode. Figure 10.4 compares a traditional high-voltage half-bridge to a half-bridge made with GaN devices. In the traditional half-bridge, each switch (shown here as an IGBT) is paired with a freewheeling diode. Because the HEMT channel exists in pure, undoped GaN, there is no parasitic p-n junction to contribute loss, and bidirectional flow of majority carriers can be realized in the channel.

In Transphorm's cascode transistors, the freewheeling current does indeed flow in the body diode of a silicon FET, but because it is a low-voltage part, the injected charge is very small. Figure 10.2 indicates the current path for three modes of operation. In the reverse conducting mode, the conduction loss may be reduced by enhancing the silicon FET (driving $V_{gs} > V_{th}$). As indicated in the figure, the voltage drop from source to drain decreases by about 0.8 V with a 5 A reverse current when the gate is enhanced.

Some transistor technologies include junctions which could serve the function of the freewheeling diode, the body diode of a MOSFET being one example. However, since the reverse recovery charge for these devices is much larger than for a GaN transistor with similar ratings, switching losses will be very high and the diode-free operation will not be practical (Fig. 10.5).

Detailed analysis of the operation, switching, and loss minimization for the GaN HEMT cascode device is also described in a recent article by the Center of Power Electronics (CPES) from Virginia Tech University [6].

10.3.2 Gate-Drive Considerations

Because high slew rates on the order of 100 V/ns are normal for GaN transistors, the high-side gate driver in bridge applications must have good common-mode transient immunity. Apart from that consideration, there are no special requirements for the high or low gate driver used with cascode GaN switches. As with any

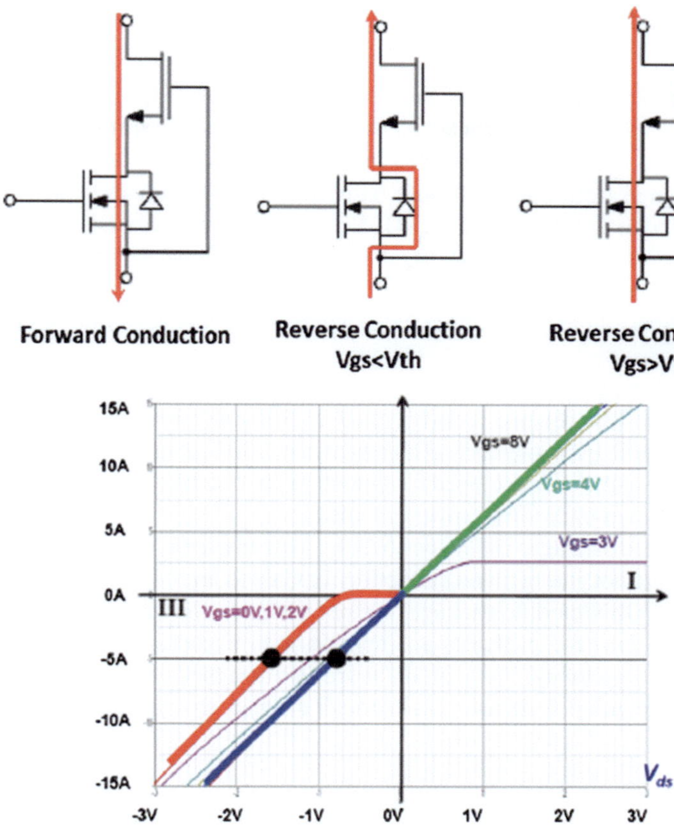

Fig. 10.5 Current paths in the cascode GaN switch for three operating modes, and the corresponding I-V characteristic

insulated-gate power transistor, the gate-drive current should be consistent with the desired turn-on time and total gate charge. The transistor is fully on with about +8 V, gate to source. Since many gate-drive chips have an undervoltage protection threshold of 8 or 9 V, Transphorm evaluation boards often use 12-V auxiliary supplies to avoid nuisance trips. The use of a negative gate voltage in the off-state is not mandatory, but is a viable option for increasing noise margin. A further consideration for switching the cascode GaN HEMT in applications is that a series gate resistor is not generally as helpful as a small ferrite bead. Selecting a gate driver with a lower drive current can be appropriate for reducing dI/dt. Drivers with 0.5 A output current have been used with good result, for example.

10.4 Products in the Market

Compared to similarly rated state-of-the-art Si superjunction (SJ) MOSFETs, Transphorm's GaN devices offer significant reduction in gate charge, on-resistance, output capacitance, and reverse recovery charge. The on-resistances of products from Transphorm range from 250 to 30 mΩ based on the device/module size and are presently offered in TO-220, TO-247, PQFN packages as well as customer modules (Fig. 10.6). The achieved FOM of $R_{on}*Q_g$ ~1nVs and $R_{on}*Q_{rr}$ ~8.5 nVs is a significant enhancement over mature Si SJ MOSFETs. In addition, the cascode device has +2.1 V gate threshold and maximum gate swings of ±18 V, easily driven by low-cost MOSFET drivers.

As noted before, the hybrid GaN HEMT is also capable of 3-quadrant operation and has 20 times less reverse recovery charge versus the best Si SJ MOSFETs, making a diode-free hard-switched bridge possible. 600 V 2-in-1 modules with on-resistances of 30 mΩ were developed for operation up to several hundred kHz, about 10 times higher than traditional power modules.

While advanced packages should be considered as any semiconductor technology develops, they are not gating market adoption for GaN today. The myth that conventional TO2XX/PQFN/module packages are not adequate for GaN is untrue. Transphorm products in these packages are reliable with exceptional performance in end-user applications today. These products are also designed into end-user systems, having met the required system-level electrical, thermal, and mechanical reliability requirements.

Specification	Discretes				Module
Product	TPH3006PS/PD	TPH3002PS/PD	TPH3006LS/LD	TPH3002LS/LD	TPH3215M

	Source Tab (PS) Drain Tab (PD)		Source Dap (LS) Drain Dap (LD)		Custom Module
Package	TO220	TO220	PQFN88	PQFN88	Custom Module
RDS(ON)Typ. (OHM)	0.15	0.29	0.15	0.29	0.031 (31mohm)
ID25°C (A)	17	9	17	9	70
Co(er) (pF)	56	36	56	36	400
Co(tr) (pF)	110	63	110	63	640
Og (ns)	6.2	6.2	6.2	6.2	28
Trr (ns)	30	30	30	30	80
Qrr (nC)	54	29	54	29	260
Vgs(V) (Gate Voltage)	+/- 18	+/- 18	+/- 18	+/- 18	+/- 18

Fig. 10.6 Early generation GaN HEMT 600-V products available in the market (from Transphorm Inc.) that are both JEDEC-qualified and demonstrate multi-million hours intrinsic lifetime (current products: www.transphormusa.com)

10.5 Applications and Key Performance Benefits

Powered by higher performance, reliability-proven product, GaN HEMTs are being adopted in variety of applications ranging from power supplies to PV inverters for their higher efficiency, low loss, and small form factor (compact) power conversion system enablement. Some key examples are described below.

10.5.1 Totem-Pole Power Factor Correction Circuit (PFC)

The diode-free operation coupled with the ultra-low recovery charge lends itself well to a true-bridgeless PFC application—the totem-pole PFC. Pushed by the demands of very-high-efficiency titanium class power supplies [7] for telecom and data center power supply needs, the losses of the conventional diode bridge rectifier input stage-based PFCs are not acceptable. Among the various ways of achieving bridge-less PFC applications [8], the totem-pole PFC represents a simple and true bridge-less implementation. However, due to the inadequacies of existing power switching MOSFETs, practical demonstrations have been limited, an obstacle that has been removed by the advent of high-performance GaN HEMTs. GaN HEMT-based totem-pole PFCs have been achieved with stunning efficiencies of 99 %, significantly outperforming a corresponding silicon-based implementation as shown in Fig. 10.7 [9].

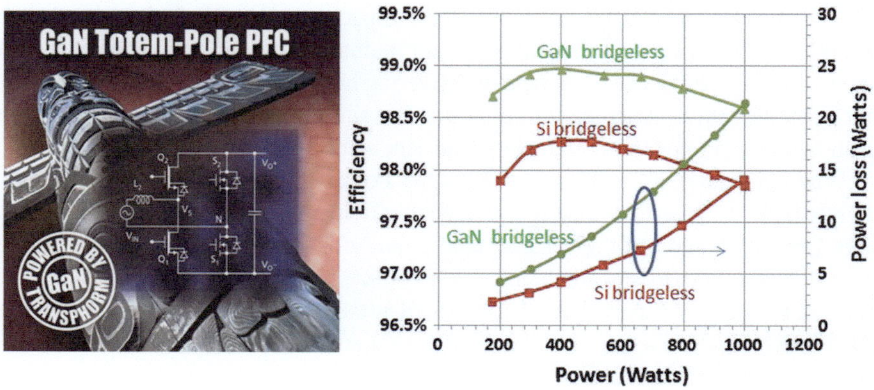

Fig. 10.7 Totem-pole PFC enabled by diode-free GaN and the resulting performance for a 1-kW application circuit implementation. Higher power 3-kW applications have also been commercialized successfully [10]

10.5.2 PV Inverters

While most of the attention in PV industry in the last decade has been focused on solar cells and PV modules, one of the overlooked part of the PV installation bill of materials has been the DC–AC inverter. Traditional IGBT- or MOSFET-based inverters typically operate in the sub-20-kHz pulse width modulation range and are typically both lower efficiency and bulky in size. GaN with its attributes of an ideal switch enables simple circuits which otherwise would need additional "shrubbery" in the form of more complex circuit functions/components in conjunction with a traditional silicon MOSFET- or IGBT-based switch. A traditional textbook boost-inverter stage with low-loss, high-frequency switching of GaN HEMTs enables a significant volume reduction, loss reduction, and a value in system cost due to reduction of the system components such as casing, heat sinks, and passive components such as inductors/filters.

The high voltage GaN HEMTs have been successfully used in an end application for 4.5-kW PV inverter, commercialized by Yaskawa Electric in Japan. Peak efficiencies of more than 98 % were achieved at the system level with simultaneous loss and system size reduction of 40 %. Fan-free, quite operation was also achieved. For this higher power application, the GaN HEMT switches were packaged in half-bridge modules for optimum performance (Fig. 10.8).

10.5.3 All-in-One Power Supplies with GaN AC–DC PFC and Full-Bridge Resonant Switching LLC DC–DC Converter

Figures 10.9 and 10.10 show the schematic and the reference board for a 200-W power supply that is significantly smaller and more efficient than a similar rating state-of-the-art Si-based power supply. The compact-size board is designed to switch at 200 kHz in order to shrink the size and maintain the high efficiency that best showcases GaN devices' advantage in delivering both small size and high efficiency not possible with existing silicon solutions. For simplicity, the cascode GaN device is shown as a single circuit element in the schematic.

With universal AC input, the All-in-One Power Supply Evaluation Board can deliver up to 20 A at 12 V output with a peak efficiency of 95.4 % from a 230-V AC line. One key benefit of the cascode configuration is the ability to drive the device with standard gate drivers and being able to utilize industry leading controllers. The All-in-One Power Supply Evaluation Board uses ON semiconductor control ICs (NCP4810, NCP1654, NCP1397, and NCP432) and three Transphorm 600-V GaN high-electron-mobility transistors (HEMTs) in both the PFC and bridge configuration.

Fig. 10.8 GaN HEMT power modules have enabled 4.5-kW commercial PV inverters for residential use (picture and product courtesy of Yaskawa Electric)

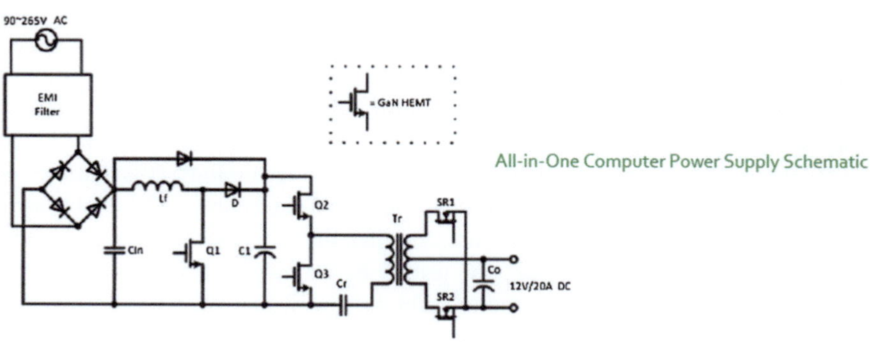

Fig. 10.9 GaN HEMT application in an AC–DC power supply with resonant DC–DC switching stage

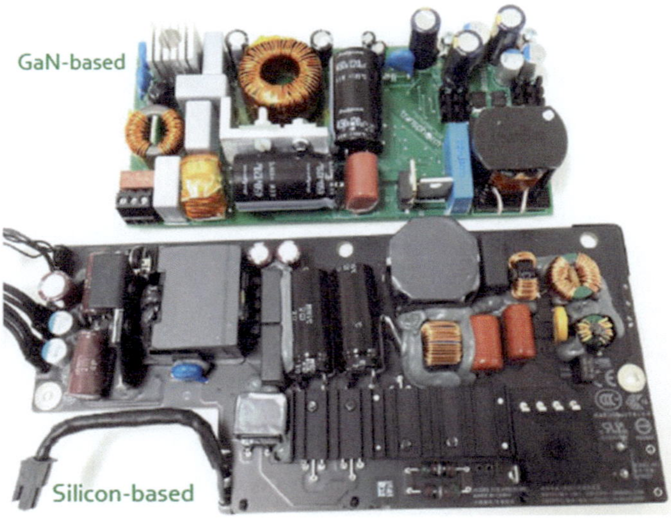

Fig. 10.10 GaN-based 250-W power supply showing significant size reduction versus a Si MOSFET-based version. The cascode GaN allows use of industry-standard drivers and controllers

10.6 Qualification and Reliability of Cascode GaN HEMTs

A key mission for a new technology like GaN-based power conversion solutions to properly serve the market is to develop a comprehensive understanding of product reliability, so that the risks associated are no greater than those associated with silicon-based devices. From its inception, and along with industry leading performance, Transphorm has supported comprehensive reliability testing of its products and has achieved industry firsts in qualifying 600/650-V GaN products for the marketplace. After utilizing standard JEDEC qualification testing prior to commercializing its GaN power devices, Transphorm has continued to expand its testing to determine quality and FIT levels and long-term reliability of its products, to ensure that the quality of GaN devices will meet customer expectations for reliability. As the JEDEC tests were originally developed on silicon technology, it is appropriate to examine the assumptions that underlie the tests and determine the level of protection that these tests offer for GaN products. Further, the testing should also go beyond the minimum requirements of JEDEC testing by running tests on a much larger number of devices than the minimum required (Fig. 10.11). We will also examine the impact that these extended tests have on the projected FIT rates for our GaN devices.

Beyond initial quality, we have used accelerated testing to predict how long the devices will last. High-temperature testing commonly in the field has been used to predict device lifetime due to temperature-related degradation as the devices are passing current at low voltages. The high-voltage rating of the parts is related to the blocking portion of operation. High field testing has been used to evaluate this portion

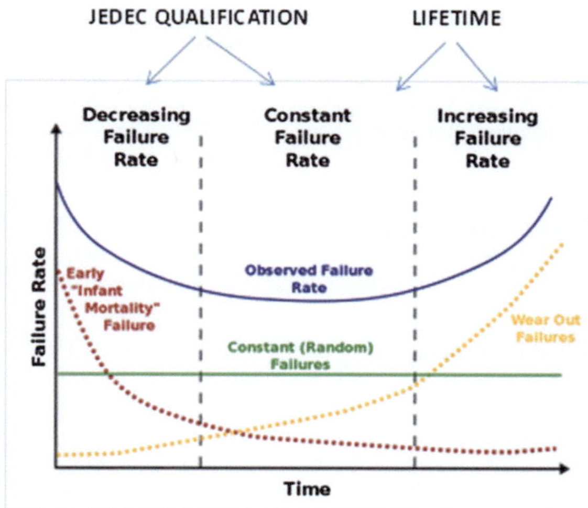

Fig. 10.11 Both JEDEC and long-term testings have been appropriately combined to ensure reliable operation of GaN devices

where there is no current and the voltages are high. Additionally, the transition between the two operating conditions has been tested by operating the devices for extended periods of time at maximum operating conditions.

10.6.1 JEDEC Qualification

JEDEC testing typically utilizes relatively large numbers of devices and applies accelerated stress to those devices per accepted standards. The tests are designed to give a lot of information about infant mortality and statistical failure and have also been correlated to longer term lifetimes in established technologies such as silicon. Such correlation needs to be verified/extended for new devices like GaN. However, it is a minimum necessity to ensure passing under JEDEC qualification conditions to ensure a basic level of reliability for GaN. The conditions and the results of testing are shown in Table 10.1.

In addition to these electrical tests above, there are the standard mechanical tests, die attach, and bond wire strength typical for any semiconductor products. By passing this suite of tests, we have good assurance that our products are free from any of the "typical" defects that can have a negative impact on reliability.

Both the JEDEC qualifications focused on the first two phases of the bathtub curve, extended qualification and the wear-out/intrinsic testing that address the later phases of the bath tub curve has been completed for the GaN HEMT devices.

Table 10.1 JEDEC testing summary of GaN HEMTs based on JESD 47 standard (Recently testing extended to 650V)

Test	Symbol	Conditions	Sample	Result/pass
High-temperature reverse bias	HTRB	TJ = 150 °C V_{DS} = 480 V 1000 h	3 lots 77 parts per lot 231 total parts	0 fails
Highly accelerated temp and humidity test	HAST	130 °C, 85 % RH 33.3 PSI, Bias = 100 V, 96 h	3 lots 77 parts per lot 231 total parts	0 fails
Temperature cycle	TC	−55 °C/150 °C 2 cycles/h 1000 cycles	3 lots 77 parts per lot 231 total parts	0 fails
Power cycle	PC	25 °C/150 °C ΔT = 125 °C 5000 cycles	3 lots 77 parts per lot 231 total parts	0 fails
High-temperature storage life	HTSL	150 °C 1000 h	3 lots 77 parts per lot 231 total parts	0 fails

10.6.2 Extended Qualification/Reliability Testing

The more parts we sample in any qualification test, the more likely we are to sample a part that fails. The industry-standard test which has been influenced very heavily by automotive quality requirements is to test three lots, with 77 parts in each lot, and passing the test with zero failed parts (3 × 0/77), with a total sample size of 231. This testing scheme satisfies the <3 % lot tolerance percent defective (LTPD) quality level as per JESD47, and all Transphorm products must meet this standard before being released to production. So far this standard has served the semiconductor industry well, gating the entry of new products into the field, but it is also common practice to run larger samples over extended period of time to gain a better understanding of defect levels.

Transphorm has completed HTRB testing on over 2000 parts for 1000 h without failure. By testing many lots over an extended period of time, Transphorm tests the intrinsic capability of its device and also tests the stability of the production operation over time. The 2000 parts were samples in batches of ~50 parts in each lot during the course of 12 months of production time. To have a 95 % probability of passing that test, the defect levels need to be below 0.003 %. This represents more than an order of magnitude improvement in quality over the standard JEDEC qualification scheme. These extended samples are large enough to detect very low levels of defects, and continuous testing and continuous improvement will help ensure that it remains that way. Further, a detailed FIT rate estimation of GaN devices has also been done [5].

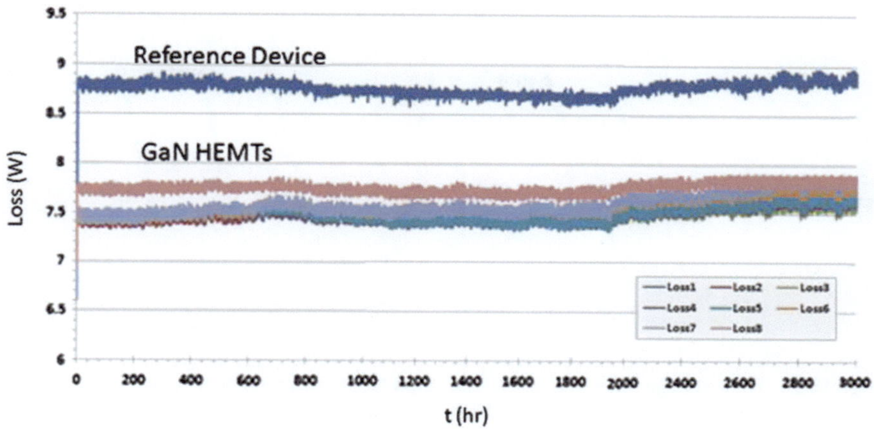

Fig. 10.12 Loss plot for GaN HEMTs along with a reference device all operating under 300 kHz, 175 C, 400 W, 1:2 boost switching conditions for the HTOL test. No change observed in performance and key parameters (before/after) following 3000 h of operation

10.6.3 Operating and Intrinsic Lifetime Testing

During normal operation, the devices may be exposed to many of the JEDEC test conditions simultaneously. High-temperature operating life test (HTOL) mimics the hard switching conditions in applications and provides a window into possible interactions affecting reliability. We ran the test on standard parts operating as a main switch in a boost converter. The devices were run at 175 °C, which is higher than the 150 °C reported in the data sheet. The higher temperature provides a minor acceleration of the test, but higher temperatures also resulted in the degradation of external components in the test circuit, thereby limiting the maximum junction temperature to 175 °C. The conversion plot of Fig. 10.4 shows conversion loss of the devices over the life of the HTOL. Degradation of the contacts and external components are the reason for the increase in conversion loss after 2000 h; the devices showed no significant change in performance when measured after the HTOL. While this test does not predict lifetime, the GaN devices are robust for extended times at the maximum rated temperature in actual operating conditions (Fig. 10.12).

Lifetime projections which depend on understanding and modeling the wear-out process are based upon accelerated testing to failure. The operating conditions of the GaN power switch allow for separation of the major stress factors. In the off-state, the device is blocking a large voltage across the switch with no current flowing in both normal operation and in accelerated voltage testing, allowing for high field testing and lifetime projections without the complications of current and/or temperature extremes. In the on-state, the device is conducting current with a small voltage across it that is similar at both nominal and elevated temperature operations needed

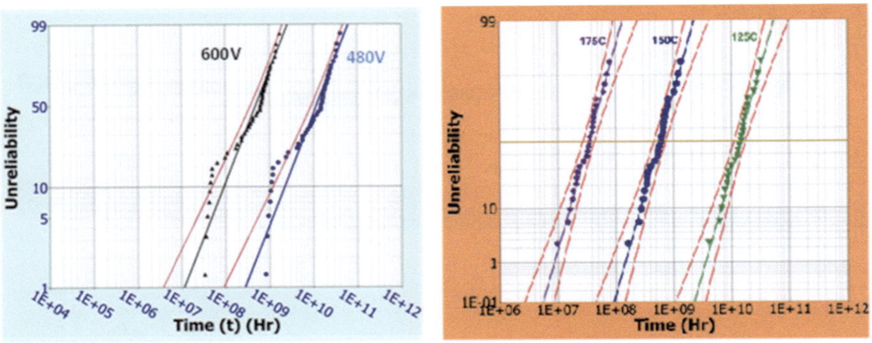

Fig. 10.13 High-voltage off-state (*left*)—and high-temperature on-state (*right*)—accelerated testing at multiple voltages (HVOS) and multiple temperatures (HTOS) to determine intrinsic lifetimes under these wear-out mechanisms

for lifetime predictions. The combined understanding from on-state and off-state testing provides confidence that the wear-out predictions are accurate.

We have shown the first report of GaN high field-related lifetime of >1X10^8 h at 600 V. The high voltages might seem to represent a significant reliability concern, but device design has limited the electric field strength to levels similar to the RF GaN devices that have been studied even more extensively.

Standard 600-V production parts (cascode-configured GaN HEMTs with rated current of 15 Amps) were used for the high field lifetime testing. Three sets of devices were biased in the off-state at high drain voltages of 1050 V, 1100 V, and 1150 V (the inherent high robustness of Transphorm GaN allows for testing at higher than a kilovolt). Device temperature was set at 82 °C to mimic expected use. In each plot, all devices (including multiple sample sets) tested at accelerated conditions are projected back to a use condition using the physical parameters determined from the plots above. The Weibull plot combines all of the devices into a single set and allows a more detailed understanding of the variability of the process as well as predicting device lifetime. The results in Fig. 10.13 (left) are from high field testing of 3 separate sample sets per above. High field lifetime at 600 V (or 480 V as shown) can be directly taken from the plot, not only at 50 % failures (>1 × 10^8 h), but also at low percentages such as 10 % (~1 × 10^8 h) or 1 % (~1 × 10^7 h). The relatively steep slope of the multiple sample sets shows the small variability of the process wafer to wafer as well as with a wafer. The 95 % confidence limits provide confidence that the projected lifetime for 1 % failures is >1 × 10^6 h. Additionally, the first failures do not form a significant tail. The lack of a tail indicates that the FiT rate will remain low for the lifetime of the devices.

Similar to the field-related use plot, the temperature-related use plot of Fig. 10.13 (right) shows median lifetime of >2 X10^7 h at the peak-rated junction temperature of 175 °C. Additional temperatures show the advantages of limiting device temperature on device reliability. The steep slope of the fit line and the narrow 95 % confidence limits show the small variability of the process. Device and test time availability and an activation energy (1.8 eV) that matches reported values contribute to the limited sample sets, but also to our confidence in the validity of the results.

GaN reliability life cycles have been experimentally investigated to address the perceived reliability risk that has been a major impediment to the widespread adoption of GaN. Through JEDEC-style testing and HTOL, the initial quality and robustness of the design and process is sufficient for insertion in user applications. The projected mean lifetimes for both the on-state and off-state are greater than $1X10^7$ hrs at nominal operating conditions, which exceeds known requirements. FiT analysis from HTRB testing further predicts a low rate of failures over the lifetime of the parts that is supported by the small variability seen in the highly accelerated testing. The quality and reliability of Transphorm's GaN power devices have been shown to be excellent. At the very least, the reliability of GaN is indistinguishable from the reliability of silicon.

10.7 Manufacturing Excellence

In order to satisfy mass market, it is required to have a manufacturing process and supply chain in place capable of high-quality mass production. The GaN products described here have gone through such a stringent process already and are now commercially available in high volumes.

GaN epi-layers are manufactured on 6-in silicon substrates in a multi-wafer MOCVD production platform and are developed to have a low-defect density, high 2-DEG mobility, and high charge density such that both high breakdown and low-channel resistance is achieved. Depletion-mode GaN HEMTs were then manufactured in wafer fabrication facilities available in standard silicon process lines, using gold-free and other high-volume silicon manufacturing steps. Transphorm's GaN HEMTs have been commercially released from our high-volume foundry partner, Fujitsu Semiconductor Limited's Aizu-Wakamatsu wafer fabrication facility. The same wafer fabrication also supplies silicon-based devices to top-tier automotive manufacturers.

The devices have been designed and fabricated with significant margin for reliability. For example, the breakdown voltage of our packaged 600-V-rated devices is typically in excess of 1000 V, and the high-temperature leakage is lower than competing Si CoolMOS devices. Contrary to the popular myth, dislocations do not impact the leakage current levels or reliability in otherwise well-designed and optimally manufactured lateral GaN devices. This fact has been well proven now with the ultra-low leakages achieved in operation, and the stringent long-term reliability at high voltages that has been experimentally demonstrated. Further, all the GaN devices described here are also rated for a spike-voltage tolerance of 750 V (Fig. 10.14).

It is important to achieve these characteristics not just on a few wafers but across the entire wafer and repeatable over all wafers that are manufactured. This has been

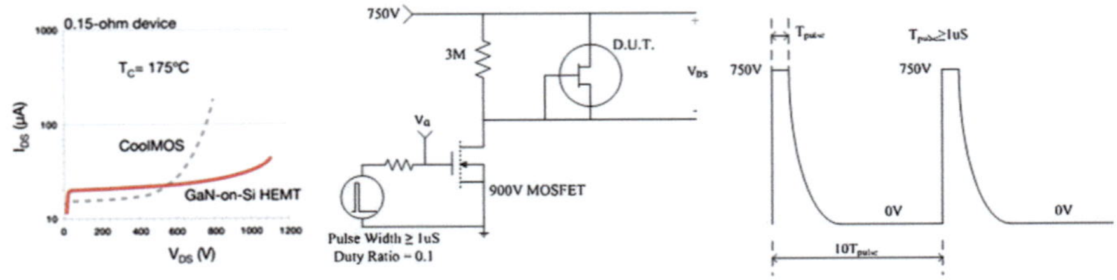

Fig. 10.14 600 V GaN HEMT with low leakage and V_{bd} > 1000 V along with a standard spike rating of 750 V

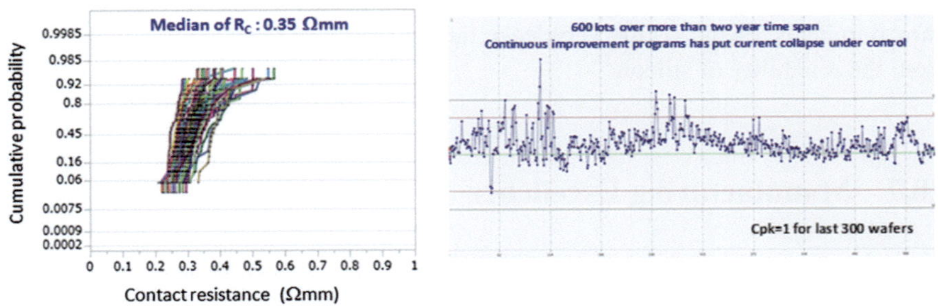

Fig. 10.15 Example of stringent statistical process control and reproducibility shown by on-wafer resistance distribution and a control chart of dynamic on-resistance across 300 wafers

systematically achieved by optimized processing and epitaxial growth. The GaN HEMTs are subject to stringent statistical manufacturing monitoring and control. Several hundred parameters ranging from MOCVD epi-wafer conditions to wafer processing parameters to unit cell and in-process device monitors to final wafer and product parameters are under SPC charting in automated quality systems. Just one example is shown below in Fig. 10.15 illustrating comprehensive on-wafer contact resistance measurement (left) and packaged final product dynamic on-resistance over 300 wafers with process capability >1.

10.8 On Single-Chip e-Mode GaN

While cascode-based GaN HEMTs offer a highly reliable, high-performance product in the market with effective normally off operation, single-chip e-mode GaN is also attractive for reasons such as a relatively simpler package due to elimination of one die (the silicon driver FET). Low-voltage sub-300-V e-mode single-chip devices have been offered commercially by EPC Corporation [2]. Other companies such as Panasonic [11] have also announced sampling of higher voltage/600-V class e-mode GaN devices but comprehensive JEDEC qualification and extended reliability data with lifetimes and corresponding activation energies is not yet available. Higher voltage single-chip e-mode GaN devices have also been reported from Transphorm, developed under the US DOE ARPA-E-funded Solar ADEPT program [12]. To ensure successful field insertion of the high-voltage GaN e-mode devices, not only superior performance but ease of use (std. drivers), stable circuit operation, and superior long-term reliability is a key requirement that companies offering this solution must demonstrate.

10.9 Future Outlook

10.9.1 Next-Generation Products

While the early-generation GaN products have successfully demonstrated key benefits in power conversion applications while outperforming silicon-based MOSFETs and IGBTs (that have neared or are nearing the limits of their physical performance), new-generation GaN devices are poised to further grow the gap between GaN- and traditional silicon-based solutions. Significant advances in figure of merit (resistance, capacitance, and chip size) are expected along with packaging innovations that will continue to offer superior value/cost performance to continue taking an increasing share of power electronics applications from the entrenched silicon solutions while being the forerunner for new emerging applications.

10.9.2 Intellectual Property Considerations

As the GaN power market expands, and very similar to the history witnessed in the GaN LED market explosion in the late 1990s/early 2000s, companies with the strongest IP along with an ability to supply products to the market will dominate. IP across the value chain, ranging from GaN-on-Si epi-growth; GaN device design, and fabrication; high-speed/high-frequency packages for GaN as well as patents for circuits; and utilization of GaN in applications will become critical in the rapid commercialization phase for GaN products that is now imminent. Companies aspiring to produce or market these GaN products must be able to successfully navigate the existing and continuously growing IP portfolios of the leading players in GaN domain.

10.9.3 In Summary

High-voltage GaN is redefining power conversion, providing cost-competitive and easy-to-embed solutions that can reduce costly energy loss by as much as 50 %;

shrink size; and simplify the design and manufacturing of power supplies and adapters, PV inverters, motor drives, and electric vehicles. The high voltage GaN (successfully commercialized by Transphorm) has established the next power conversion platform—demonstrating breakthrough performance and introducing the world's first 600/650-V GaN products with its EZ-GaN™ platform.

We look forward to an exciting phase in power electronics, as GaN, the new power conversion platform, is widely adopted by the industry. Get ready to reimagine power conversion for an energy-efficient world.

Acknowledgments The author would like to acknowledge Transphorm Inc. and its entire world-wide team, partners, investors and customers from inception till date for the exciting progress that has resulted in the first commercialization of the high-voltage GaN HEMT. Special acknowledgment to team members whose input and material has been incorporated in this chapter—Dr. Yifeng Wu (Circuits and Applications), Dr. Jim Honea (Diode-free Cascode Operation), Dr. Kurt Smith and Ron Barr (Reliability and Lifetime), Dr. Toshi Kikkawa (Manufacturing Results) and to Dr. Peter Smith for his help on detailed proofreading of this chapter.

References

1. Mishra UK, Parikh P, Wu Y-F (2002) AlGaN/GaN HEMTs—an overview of device operation and applications. In: Proceedings of IEEE
2. Lidow A, Strydom J, de Rooji M, Ma Y (2012) GaN transistors for efficient power conversion perfect paperback, 20 Jan 2012
3. Marcon D, Van Hove M, De Jaeger B, Posthuma N, Wellekens D, You S, Kang X, Li T, Wu MW, Stoffels S, Decoutere S (2015) Direct comparison of GaN-based e-mode architectures (recessed MISHEMT and p-GaN HEMTs) processed on 200 mm GaN-on-Si with Au-free technology. In: Proceedings of SPIE 9363, gallium nitride materials and devices, Mar 2015
4. Parikh P (2014) Commercialization of high 600 V GaN-on-silicon power devices. In: International solid state device and materials, Tsukuba, Japan
5. Smith K, Barr R (2016) Reliability lifecycle of GaN power devices. http://www.transphormusa.com/wpcontent/uploads/2016/02/reliability-lifcycle-at-transphorm.pdf
6. Cheng X, Liu Z, Li Q, Lee FC (2013) Evaluation and application of 600 V GaN HEMT in cascode structure. IEEE Trans Power Electron 29(5)
7. Definition of the 80 Plus Power Supply Standard, https://en.wikipedia.org/wiki/80_Plus
8. Huber L, Yungtaek J, Jovanovic MM (2008) Performance evaluation of bridgeless PFC boost rectifiers. IEEE Trans Power Electron 23(3):1381–1390
9. Mishra U, Zhou L, Wu Y-F (2013) True-bridgeless totem-pole PFC based on GaN HEMTs. In: PCIM conference and trade show, Nuremburg
10. Transphorm Inc. Application Resources, http://www.transphormusa.com/wp-content/uploads/2016/02/TDPS2800E2C1_AppNote_rev1.1.pdf
11. Panasonic Corporation news release, A 600 V gallium nitride (GaN) power transistor with its stable switching operations, Osaka, Japan, 19 Mar 2013
12. Heidel T (2014) Achieving high efficiency, high power density, high reliability and low cost. In: ARPA-E, APEC 2014 conference and trade show. http://www.apec-conf.org/wp-content/uploads/2014/03/06_ARPA-E_Initiatives_in_Power_Electronics_APEC2014_Plenary.pdf

Chapter 11
Gate Injection Transistors: E-mode Operation and Conductivity Modulation

Tetsuzo Ueda

This chapter describes an enhancement-mode (E-mode) GaN transistor named as Gate Injection Transistor (GIT) and various technologies to improve the performances. The operation principle of the GIT as well as its state-of-the-art DC and switching performances are described after summarizing reported E-mode GaN power transistors. Status of the reliability including current collapse is summarized, followed by results of the application of the GIT to practical power switching circuits targeting at high efficiencies. Future technologies to improve the performances and to extract the potential are also described.

11.1 Operation Principle of GIT

Lateral GaN power switching transistors are very promising owing to the inherent high sheet carrier density at AlGaN/GaN hetero-interface caused by the material's unique polarization-induced electric field [1, 2]. The low parasitic capacitances give strong expectation for the extremely high-speed switching that cannot be achieved by existing Si power devices. The high carrier concentration together with high electron mobility can contribute to serve very low series resistances; however, this has been an obstacle to enable an E-mode operation which is strongly desired for safe operations of power switching systems. Novel gate structure had been strongly

T. Ueda (✉)
Panasonic Corporation, Automotive & Industrial Systems Company,
Industrial Business Development Center, 1006 Oaza-Kadoma,
Kadoma-shi, Osaka 571-8506, Japan
e-mail: ueda.tetsuzo@jp.panasonic.com

M. Meneghini et al. (eds.), *Power GaN Devices*,
Power Electronics and Power Systems, DOI 10.1007/978-3-319-43199-4_11

desired to deplete the sheet charges underneath the gate for the E-mode operation. Although various approaches have been proposed so far, most of them have their technical issues that are difficult to be overcome for practical use by forecastable efforts. Figure 11.1 is a summary of reported technical approaches including a GIT to enable E-mode operations of GaN power transistors [3–6]. While these have unique gate structures with its positive threshold voltages, there remains unstable operation and/or not reproducible fabrication. In contrast, the proposed GIT which has p-type gate to deplete the gate region is free from the instability of the gate, thus very promising for the commercialization of GaN power devices in the near future. The GIT itself has a structure very similar to conventional junction field effect transistor (JFET), except for the experimentally found conductivity modulation at higher gate voltages. The conductivity modulation that is a unique feature caused by injection of holes at the AlGaN/GaN hetero-junction increases the drain current enabling lower on-state resistances together with the E-mode operation. Figure 11.2 shows a band diagram of GIT at the gate region in which two-dimensional electron gas (2DEG) underneath the gate is depleted by high-potential barrier of p-AlGaN. Figure 11.3 summarizes the basic operation principle of GIT by illustrations [3]. At the zero gate bias, the channel under the gate is fully depleted enabling E-mode operation. By increasing the gate voltage up to the built-in voltage of the gate pn-junction, the drain current is increased just by the control of the potential at 2DEG. The operation is based on what is observed in conventional unipolar field effect transistors. Once the applied gate voltage exceeds the built-in voltage typically at around 3.5 V for the GIT, holes start being injected from the p-type gate onto the 2DEG. The injected holes produce equal numbers of electrons at the 2DEG to maintain the neutrality of charges. Note that the injection of the electrons from the channel to the gate is suppressed by the hetero-barrier at AlGaN/GaN. The electrons produced by the above conductivity modulation are moved to the drain side by the applied drain bias taking advantage of the high mobility, while the injected holes stay around the gate because of the two order of magnitude lower mobility than that of electrons. The device operation enables significant increase of the drain current maintaining low gate current. Figure 11.4 shows the I_{ds}–V_{gs} characteristics of a fabricated GIT in which the second peak of the transconductance g_m at higher gate voltage is a sign of the proposed conductivity modulation. The electroluminescence (EL) images at various gate voltages as shown in Fig. 11.5 are also the proofs of the proposed operation principle. At the gate voltage over 5 V, the place of the light emission is moved from the gate edge at the drain side to that at the source side [7]. This indicates that the recombination of holes and electrons occurs by hole injection at higher gate voltages.

11.2 DC and Switching Performances of GIT

Additional feature of GIT is that the lateral device has been fabricated on cost-effective Si substrate. So far, it has been fabricated on 6-inch Si(111) substrate by well-established epitaxial growth technologies to overcome the lattice and thermal

	p-gate HFET (GIT) (3)	MIS-HFET (4)	F-doped Gate (5)	Cascoded (6)
Structure	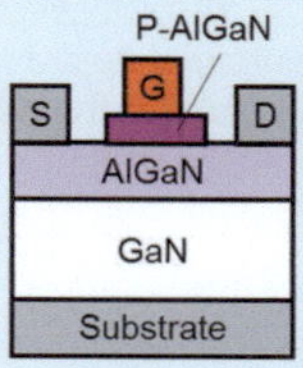	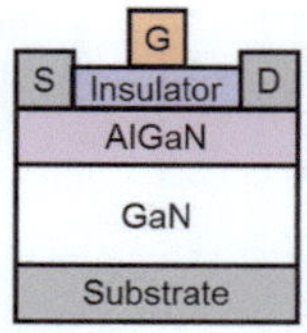	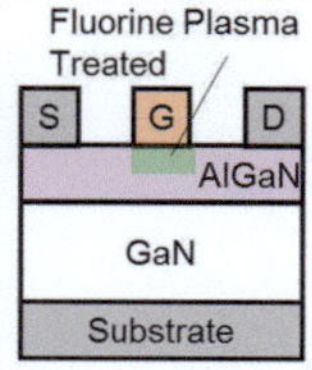	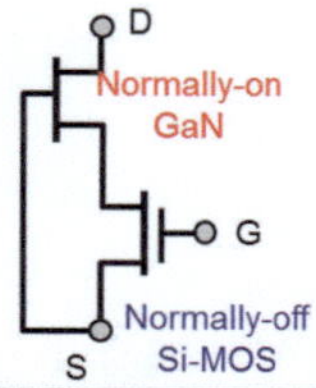
Advantages	✓ Stable Vth ✓ Highly reliable	✓ Low gate current ✓ High drain current	✓ Low gate current ✓ Controllability of Vth	✓ High Vth ✓ Gate drive compatible with that for Si devices
Challenges		✓ Controllability of Vth ✓ Stability of MIS interface	✓ Stability of F-doped region	✓ High total fabrication cost ✓ Integration is impossible ✓ Limited high speed driving

Fig. 11.1 Summary of technical approaches for E-mode GaN-based transistors

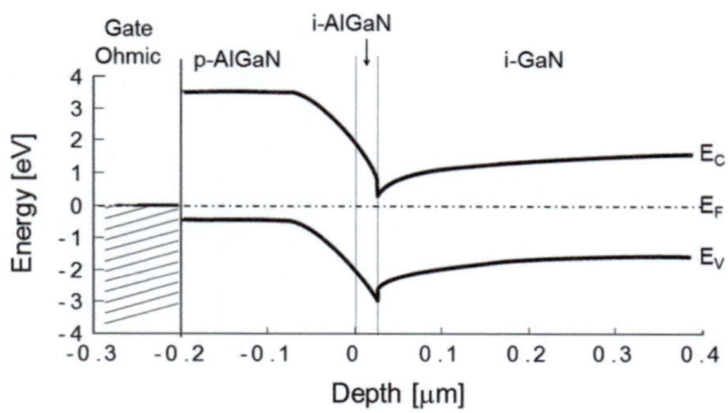

Fig. 11.2 Band diagram at the gate region of a GIT in which p-AlGaN gate is placed over AlGaN/GaN hetero-junction

mismatches between GaN and Si [8, 9]. Competitiveness on costs over existing Si power devices has been anticipated from the uniform and reproducible crystal growth on the large diameter substrates. Figure 11.6 shows a cross-sectional scanning electronic microscope (SEM) image of the fabricated GIT on Si. After the epitaxial growth of p-AlGaN/AlGaN/GaN hetero-structure on Si, the p-gate is selectively formed by dry etching. The employed gate and source/drain metals are Pd and Ti/Al, respectively. The gate length L_g employed so far for 600 V GITs is typically 2.0 μm.

I_{ds}–V_{ds} characteristics of the fabricated GIT at on-state and off-state are summarized in Fig. 11.7. The GIT with a 600 V rating blocking voltage has a breakdown voltage over 1000 V considering its current Collapse and reliability as it is mentioned in the following subchapter. The measured specific on-state resistance R_{on}A of GIT is as low as 2.3 mΩ cm^2 with the breakdown voltage of 1260 V. Breakdown voltages of GITs on Si are increased by the extension of L_{gd} as shown

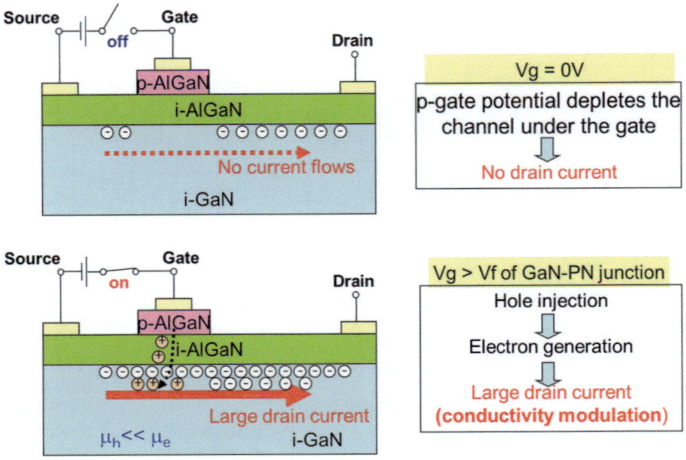

Fig. 11.3 Cross-sectional illustrations describing the operation principle of GIT

Fig. 11.4 I_{ds}–V_{gs} and g_m–V_{gs} characteristics of a fabricated GIT

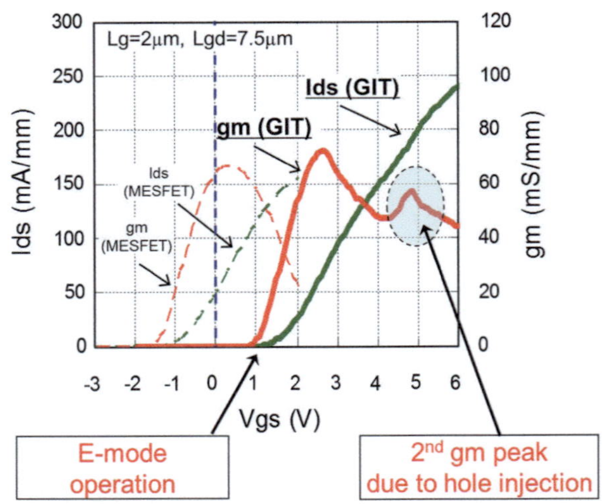

in Fig. 11.8. At longer L_{gd}, the breakdown voltages are saturated to a certain value that is determined by the thickness of GaN on Si. This is because vertical electric fields are dominant at high drain voltages on conductive Si substrates. The device with wider gate width shows an on-state resistance Ron of 65 mΩ and the rating drain current of 15 A, while the peak current exceeds 30 A. The gate charge Q_g of GIT with the rating blocking voltage of 600 V is 11 nC and the $R_{on}Q_g$ as a figure of merit for high-speed switching is 700 mΩ nC that is one-thirteenth of that by state-of-the-art super junction Si MOSFETs. These results indicate that the GIT has a great potential at high-frequency operation enabling very compact

Fig. 11.5 EL images taken over a fabricated GIT at various gate voltages

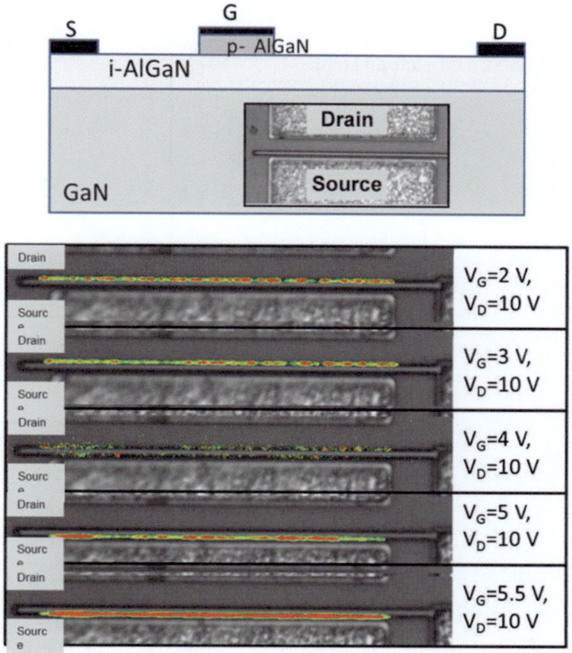

Fig. 11.6 Cross–sectional SEM image of the fabricated GIT on Si

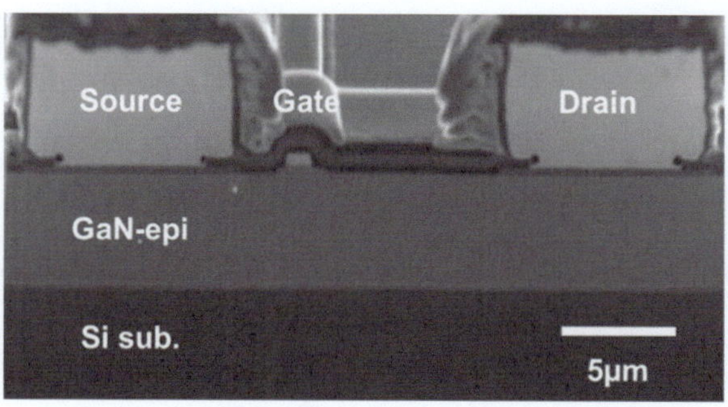

switching systems. Table 11.1 is the summary of DC performances of the currently available 600 V GITs aiming at the mass production.

Smaller switching loss is a very attractive feature of GaN devices as mentioned earlier. Switching waveforms are measured for fabricated GITs, where the effect of packaging is examined. Parasitic inductances are effectively reduced by flip–chip bonding as compared to those by conventional TO-220 package with long lead frame. Figure 11.9 shows simulated parasitic inductances by flip–chip and TO-220. Based on the simulations, the parasitic inductance is reduced down to 2 nH by flip–chip bonding from 24 nH by TO-220. The switching waveforms by fabricated GITs are also shown in Fig. 11.9, indicates that the reduction of the parasitic inductances

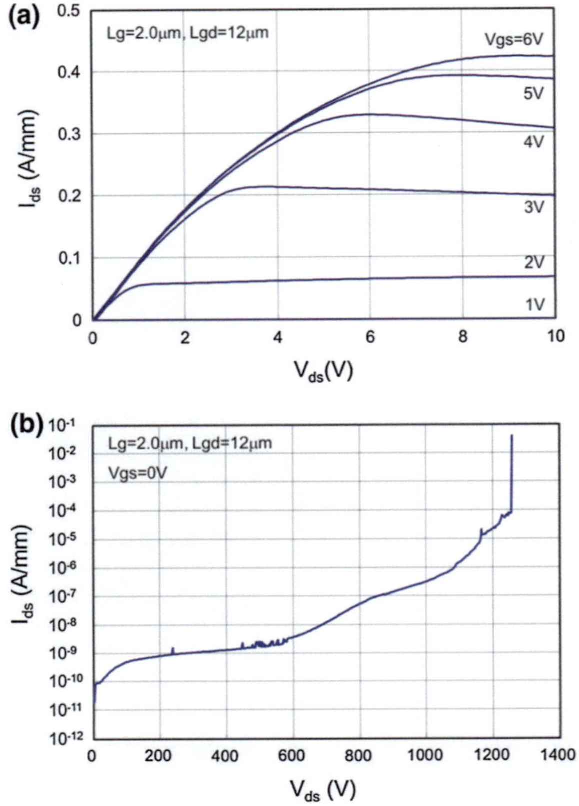

Fig. 11.7 I_{ds}–V_{ds} characteristics of fabricated GIT on Si, **a** on-state and **b** off-state

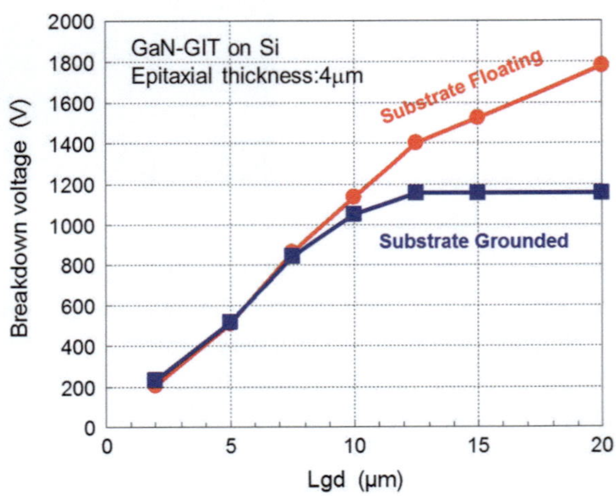

Fig. 11.8 Breakdown voltages of fabricated GITs on Si for various L_{gd} in which the effect of grounding Si substrates are examined

Table 11.1 Summary of DC performances of currently available GITs

Threshold voltage V_{th}	1.2 V (E-mode)
Blocking voltage BV_{ds}	600 V
Rating current (continuous) I_d	15 A
On-state resistance R_{on}	65 mΩ
Gate charge Q_g	11 nC

enables extremely high-speed switching [10]. High dV/dt of 170 V/ns is achieved by flip–chip-bonded GITs, so that 600 V switching with the time period in the order of ns is possible by the GIT. These results show a great potential of GITs for high–frequency switching system with very compact size.

11.3 State-of-the-Art Reliability of GIT

In order to achieve stable operation of GaN power switching transistors, current collapse has been the most serious technical issue to be overcome. Carrier trapping in the GaN transistor during the switching operations causes a serious reduction of the drain current resulting in increase of on-state resistances [11]. Figure 11.10 schematically explains the current Collapse in I_{ds}–V_{ds} characteristics and in the switching waveform. The increase of the on-state resistance raises the temperature, so that it might lead to the failure of the device. It is also noted that the increase of the on-state resistance has been more serious at higher drain voltage for the off-state. Another point of discussion on current collapse besides how to eliminate it is that how the phenomena is characterized since it is strongly affect by the measurement condition such as time duration of the off-state and timing of measurement after turning on. Figure 11.11 shows the circuit diagram for GITs to be tested. Multi-pulsed switching with inductive load is employed for the test as shown in the figure. As compared to the conventional resistive load, the load lines for the inductive give more serious bias conditions since high current is flown at high drain current. The collapse has been worsened by the inductive switching implying that the condition is satisfactorily good to see whether it is fully eliminated or not. So far, improvements of the quality of interfaces and crystal as well as the modification of the device structure have successfully eliminated the above-mentioned current collapse [12]. Currently available GITs exhibit the switching operation free from current collapse up to the drain voltage of 850 V with the measurement using inductive load and multi-pulses as shown in Fig. 11.12. Increase of the dynamic on-state resistances is presumably due to the increase of the temperature caused by the applied multi-pulses of the current. Thus, the GIT on Si at present serves only a viable alternative for practical applications owing to the E-mode and current collapse-free operations.

More detailed reliability tests have been carried out for the GITs as summarized in Table 11.2. These results satisfy JEDEC qualification standards and no failure is

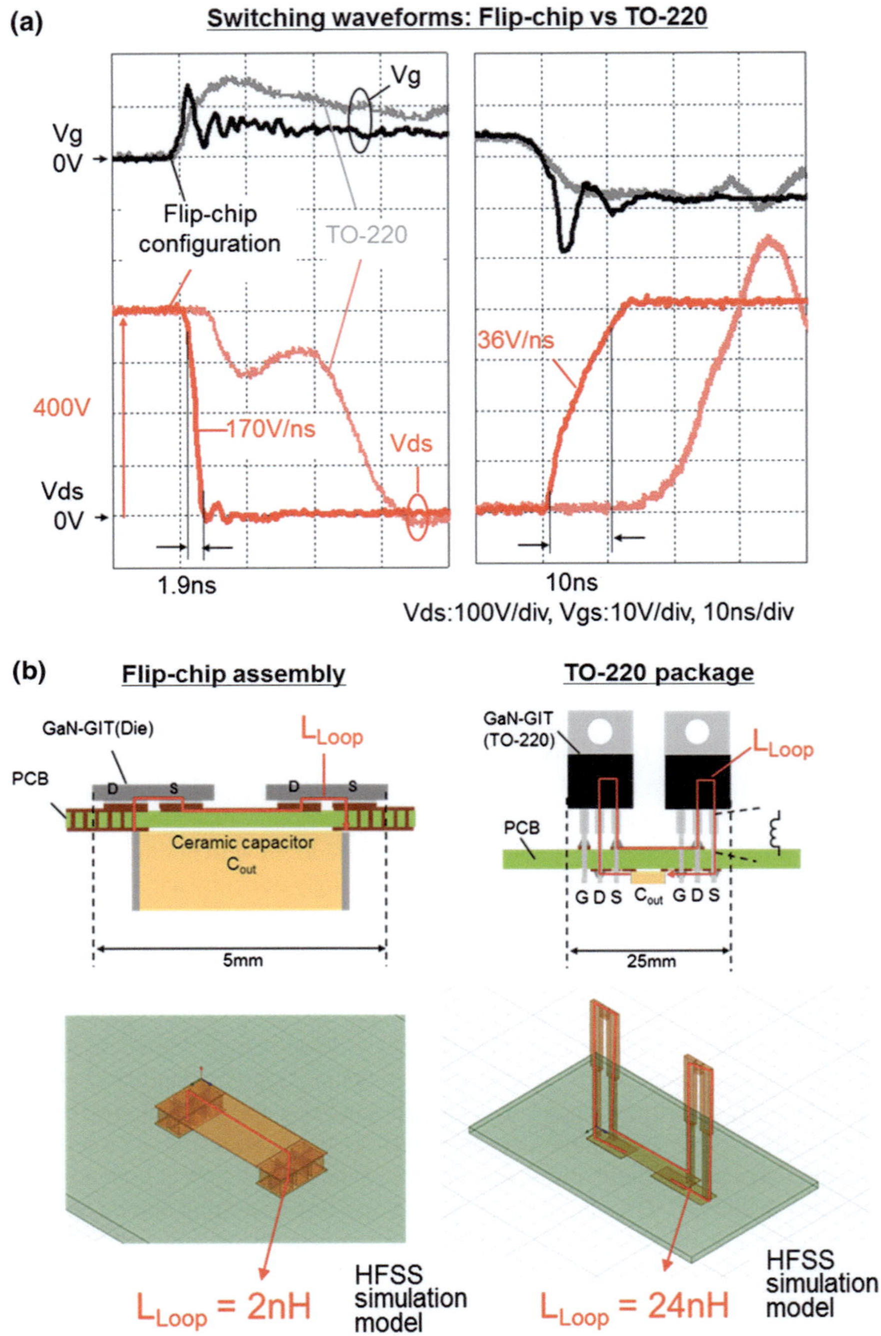

Fig. 11.9 Switching waveforms of GITs with **a** flip–chip assembly and **b** TO-220 packaging, where the simulation results on parasitic inductances are also shown

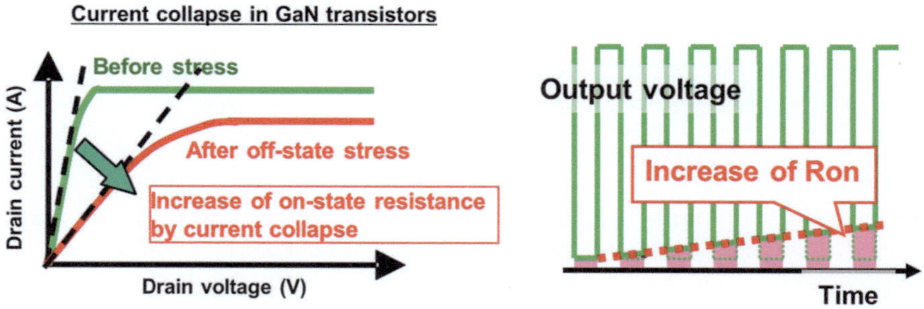

Fig. 11.10 Schematic illustrations to explain Current Collapse of GaN power transistors

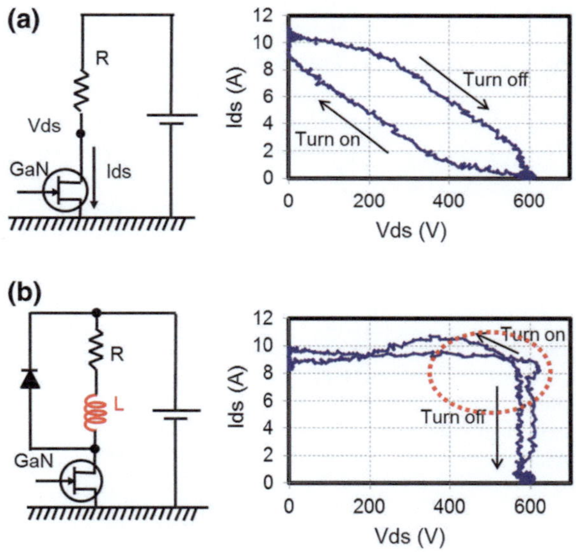

Fig. 11.11 Circuit diagram for measurements of current collapse using **a** conventional resistive load and **b** inductive load for GITs

observed in over 10,000 devices by high temperature reverse bias (HTRB) tests equivalent to the lifetime of 20 years. The fabricated GITs on Si are reliable enough for practical switching applications.

11.4 Applications of GIT to Practical Switching Systems

GaN power devices have a potential to achieve far better performances of power switching systems than those by conventional Si power devices owing to the superior material properties. Followings describe highly efficient operations of power switching systems using E-mode GITs. Inverter systems and various power supply circuits are described as the promising applications of GaN.

Table 11.2 Summary of reliability test results on state-of-the-art GITs PKG :T0220D

	Test items	Test condition	Test results (failure/tested)
1.	Temperature humidity reverse bias test	Ta = 85 °C, RH = 85 %, $V_{ds} = 480$ V, $t = 1000$ h	0/231 (77 × 3 lot)
2.	High temperature reverse bias test	Ta = 150 °C, $V_{ds} = 480$ V, $t = 1000$ h	0/10 k (500 × 20 lot)
3.	High temperature gate bias test (forward)	Tj = 150 °C, $V_{gs} = 4.0$ V, $t = 1000$ h	0/231 (77 × 3 lot)
4.	High temperature gate bias test (reverse)	Tj = 150 °C, $V_{gs} = 12$ V, $t = 1000$ h	0/231 (77 × 3 lot)
5.	High temperature storage test	Ta = 150 °C, $t = 1000$ h	0/231 (77 × 3 lot)
6.	Temperature cycling test	−65 to 150 °C, 100 cyc, 30 min each, gaseous phase	0/231 (77 × 3 lot)
7.	Solder heat resistance test	Ta = 260 °C, $t = 10$ s	0/231 (77 × 3 lot)
8.	ESD test (HBM.CDM)	HBM:C = 100 pF, $R = 1.5$ kΩ, ±2000 V, CDM = ±2000 V	0/90 (30 × 3 1ot)

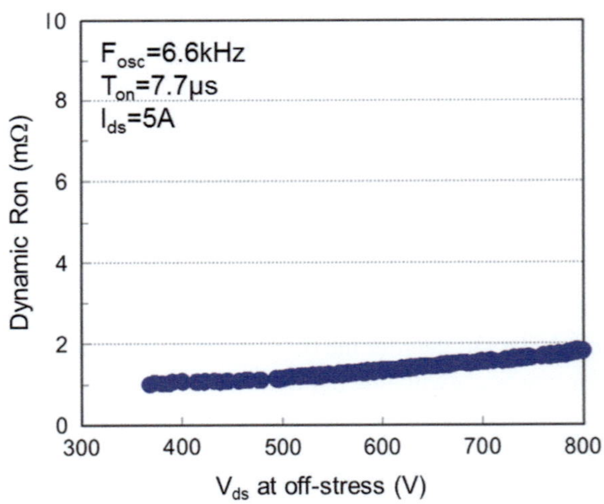

Fig. 11.12 Dynamic on-state resistances of fabricated GITs plotted as a function of the applied off-state drain voltages

GaN can achieve lower operating loss in an inverter system than that by conventional insulating gate bipolar transistor (IGBT) with voltage offset in the forward current–voltage characteristics, since GaN is free from the offset. Figure 11.13 summarizes the advantage of GaN in the inverter system as compared to IGBT. By further reducing the on-state resistances of GaN, far better efficiency is expected

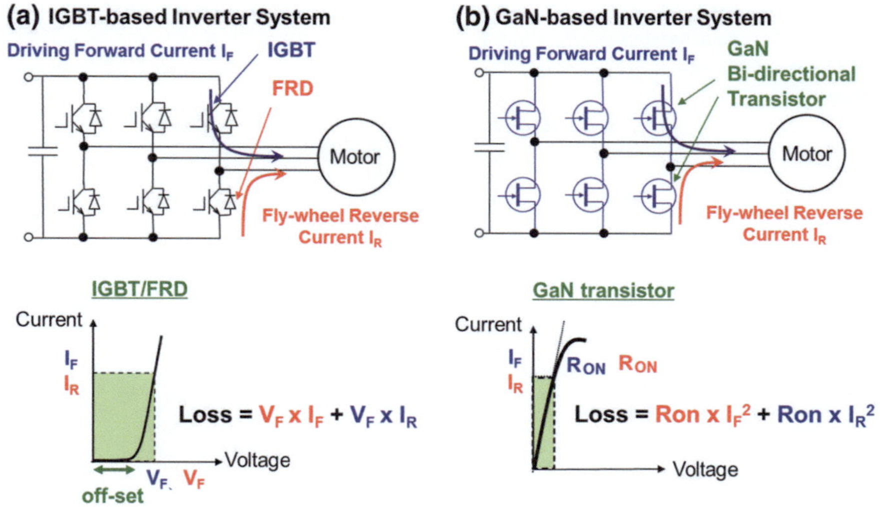

Fig. 11.13 Operations of **a** a conventional IGBT inverter system and **b** GaN-based one with the description of forward and reverse conduction losses

especially at low output power. Bidirectional operation of the lateral GIT is an additional advantage so that the fast recovery diodes (FRDs) connected in parallel with IGBTs can be eliminated. Reverse conduction of the GIT at the off-state exhibits the current voltage characteristics as if the reverse diode with the built-in voltage as small as the threshold voltage of 1.5 V. These current–voltage characteristics of a GIT for both forward and reverse directions are summarized in Fig. 11.14, indicating that the turning on for the gate at the reverse current flow further reduce the conduction loss. So-called recovery characteristics of the reverse GIT is measured as shown in Fig. 11.15, which confirms significantly small recovery loss by GIT as compared to the large loss by conventional Si FRD. Figure 11.16 shows the conversion efficiency of an inverter using GITs on Si substrates for various output power [13]. The efficiencies for conventional IGBT-based inverter are also plotted for comparison. The efficiencies by GITs are higher than those by IGBTs, where the highest value of 99.3 % at 1.5 kW is confirmed. The low on-state resistance and reduced switching loss by GITs remarkably reduce the operating loss in a wide range of the output power. At smaller output where the efficiency is affected by the voltage offset, the improvement of the efficiency by GaN is more significant. The operating loss at 500 W is reduced in a GIT inverter by 60 % from that in IGBT-based one as shown in Fig. 11.17.

In addition to inverters, power supplies are promising applications of GaN power transistors, since high-frequency operation would reduce the size of the system by possible use of smaller passive components. Detailed circuit diagram of typical power supply is summarized in Fig. 11.18, in which, GaN transistors with various blocking voltages can be applied. A PFC (Power Factor Correction) circuit and an

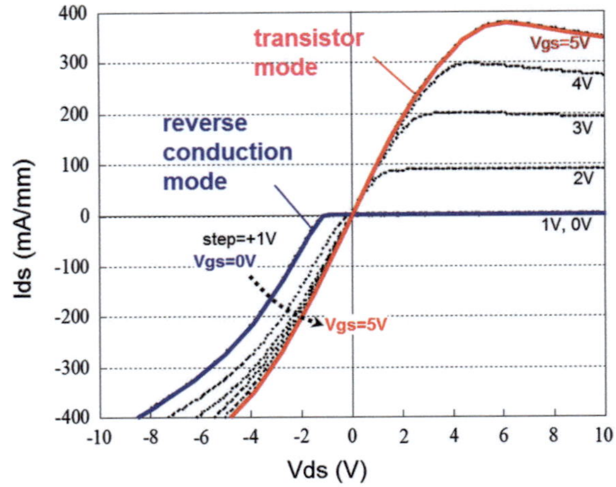

Fig. 11.14 Forward and reverse I_{ds}–V_{ds} characteristics of GIT for various gate voltages

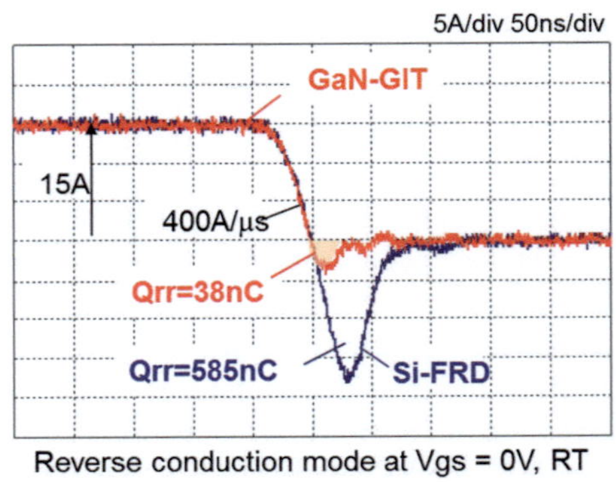

Fig. 11.15 Recovery characteristics of reverse conduction diode by a fbricated GIT with those by a conventional Si FRD

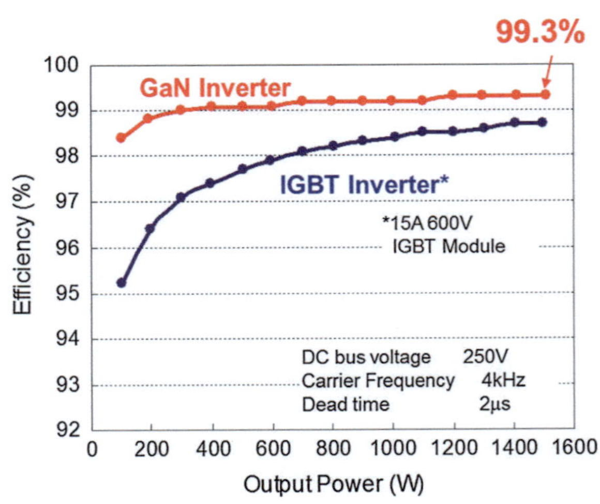

Fig. 11.16 Power conversion efficiency of an inverter system using GITs for various output power as compared to those by conventional IGBTs

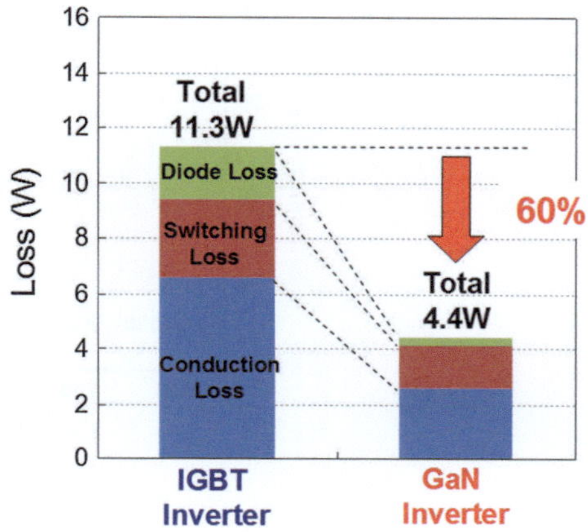

Fig. 11.17 Details of operating losses or inverter systems using GITs and IGBTs at the output power of 500 W

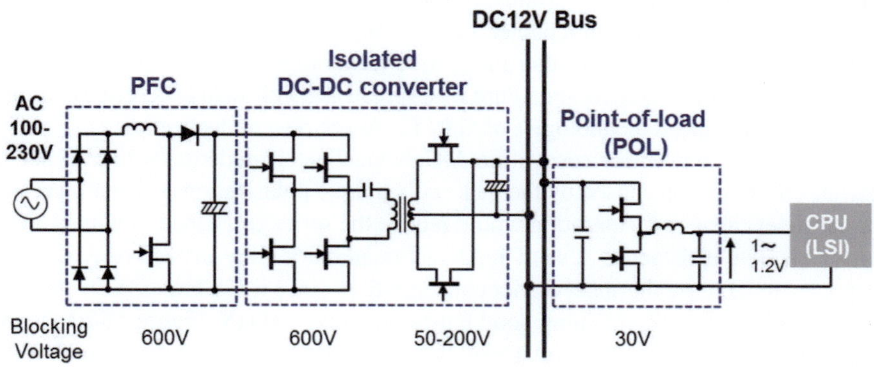

Fig. 11.18 Circuit diagram of a typical power supply system

isolated DC–DC converter with a transformer are the applications for 600 V GaN devices. A resonant LLC converter is a typical circuit topology of the isolated DC–DC converter for which GITs are applied to increase the operating frequency and thus to make the system smaller. Figure 11.19 shows the measured efficiency of the fabricated LLC converter as a function of the output power. The results exhibit successful operation at 1 MHz by GITs with high efficiency of 96.4 % at 1 kW output. Since existing Si power devices cannot be operated at such high frequencies, the performances demonstrate a great potential of GITs for power supply circuits.

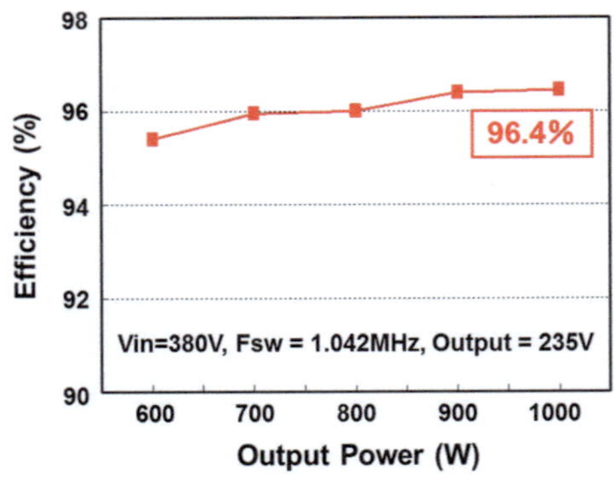

Fig. 11.19 Operating efficiencies of a fabricated resonant LLC converter using GITs at 1 MHz for various output powers

11.5 Advanced Technologies of GIT for Future Power Electronics

GaN power devices are very promising for practical switching applications as it is described in the previous sub-chapter. Considering competing situation of GaN with existing Si power devices, further innovative technologies on GaN power devices should also be considered for the future widespread use. Followings describe demonstrations of advanced technologies of GaN for future power electronics.

Lateral GaN power devices feature AlGaN/GaN hetero-junction with high sheet carrier concentration caused by the material's unique polarization. Further increase of the sheet carrier concentration would reduce the series resistances leading to the lower on-state resistances. Use of InAlGaN quaternary alloy instead of AlGaN is expected to increase the sheet carrier concentration, since it can increase the magnitude of polarization maintaining good lattice matching to GaN. Figure 11.20 shows band gap energy of InAlGaN material systems plotted as a function of the lattice constants. Use of slightly strained InAlGaN over GaN more than double the sheet carrier concentration from conventional AlGaN/GaN. This InAlGaN/GaN heterostructure with the sheet carrier concentration of 3×10^{13} cm^{-2} is applied for GIT with p-type gate, of which a schematic cross section is shown in Fig. 11.21. The fabricated GIT employing InAlGaN/GaN exhibits smaller on-state resistance together with increased maximum drain current as shown in Fig. 11.22 [14]. Note that the E-mode operation is maintained with the threshold voltage of 1.1 V. The reduction of series resistance would be a great help for the reduction of the conduction loss in practical switching operations.

A unique feature of the lateral GaN transistor is that multi-GaN power devices can be integrated into one chip while conventional vertical Si power devices cannot. The integration helps to reduce the parasitic inductances, and thus enable faster switching. Peripheral circuits such as gate drivers also can be integrated, which enables a high performance and small power switching system by a single chip.

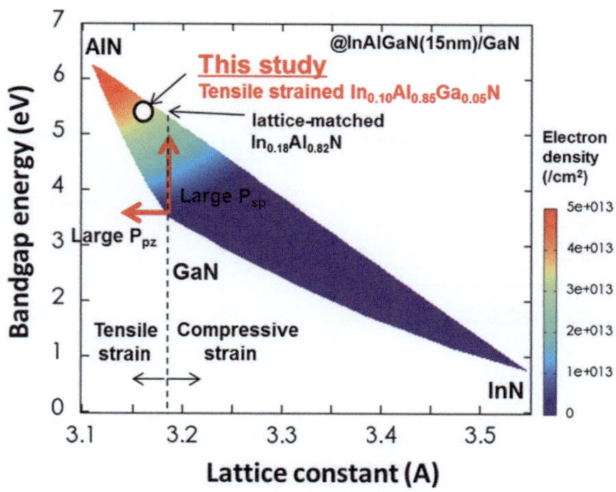

Fig. 11.20 Bandgap energies of InAlGaN quaternary alloy system plotted as a function of the lattice constants

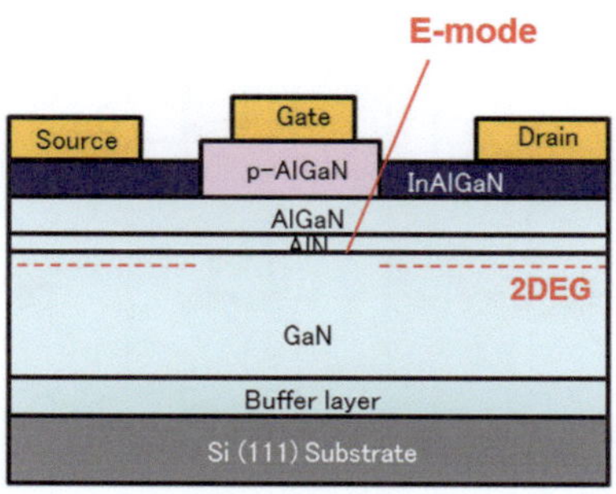

Fig. 11.21 Schematic cross section of the fabricated p-gate GIT employing InAlGaN/GaN heterostructure with high sheet carrier concentration

One example of the integration is a single chip inverter IC using lateral GITs. GaN-based inverter can be fabricated just by six transistors free from the fast recovery diodes that are required for conventional IGBT-based one [15].

Another example is the integration of gate drivers with GITs for a DC–DC converter [16]. Here, two GITs consisting of a point of load (POL) are integrated with gate drivers as a single-chip POL. Note that the POL converts the DC voltage from 12 to 1.8 V and the integrated GITs have the blocking voltage of 30 V and the gate lengths shortened down to 0.5 Ωm [17]. The reduction of the gate length successfully reduces the $R_{on}Q_g$ down to 19 mΩ nC that is half of that by conventional Si MOSFETs. The gate drivers consist of a DCFL (Direct Coupled FET Logic) with a buffer amplifier enabling low-power consumption in the gate driving. The circuit diagram and the schematic cross section of the integrated devices are shown in

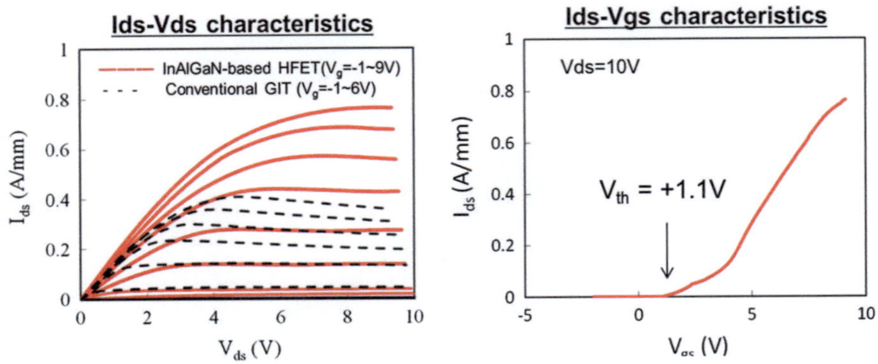

Fig. 11.22 I_{ds}–V_{ds} and I_{ds}–V_{gs} characteristics of the GIT fabricated over InAlGaN/GaN, where I_{ds}–V_{ds} of the conventional GIT on AlGaN/GaN is also shown

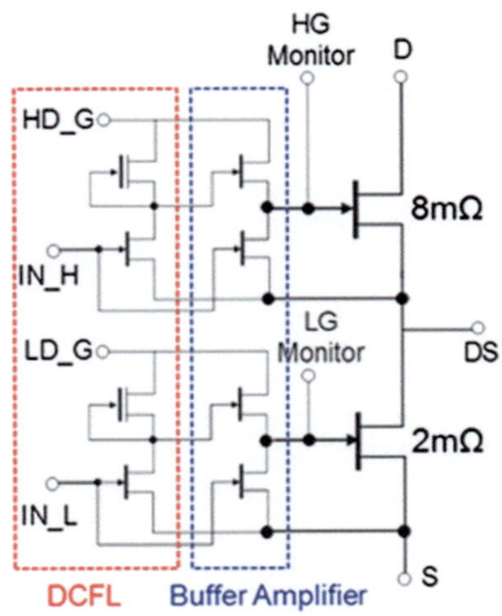

Fig. 11.23 Circuit diagram of the DC–DC converter (POL) IC with gate drivers

Figs. 11.23 and 11.24, respectively. Depletion mode (D-mode) GaN hetero-junction field effect transistors (HFETs) are also fabricated by a part of the total processing for the E-mode GIT. Figure 11.25 shows a photograph of the fabricated POL IC with high-speed gate drivers. Successful conversions at 1–3 MHz are confirmed, where the peak efficiency at 2 MHz is as high as 88.2 %. The reduction of the parasitic inductances is an advantage of the integration so that highly efficient operations are achieved together with the reduced system size.

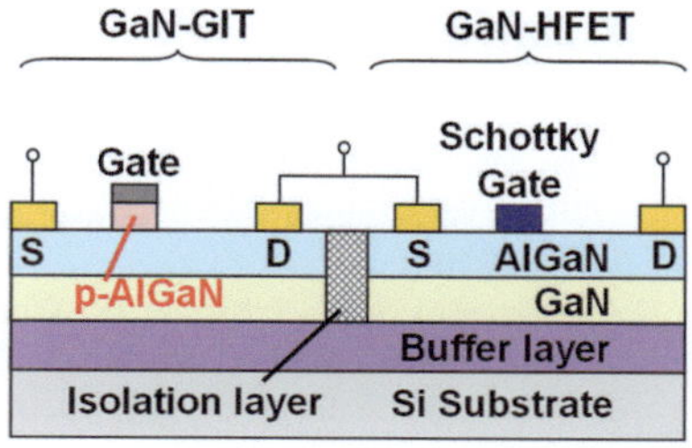

Fig. 11.24 Schematic cross section of the GaN devices integrated as a part of the gate driver in a POL IC

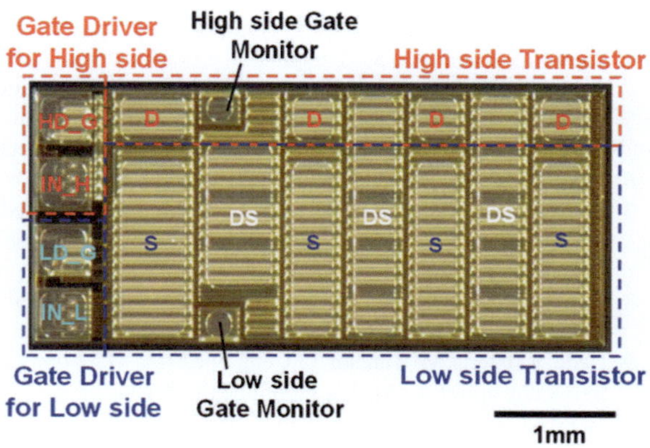

Fig. 11.25 Photograph of the fabricated POL IC in which two GITs and two gate drivers are integrated into a single chip

11.6 Summary

State-of-the-art technologies of E-mode GaN-based GITs on Si are summarized. GIT serves an E-mode operation by p-type gate over AlGaN/GaN hetero-structure with an experimentally found conductivity modulation by injection of holes from the gate that increases the drain current. Current Collapse, which had been the most serious technical issue, has been successfully eliminated in GITs up to 850 V. Various power circuits such as inverters and isolated DC–DC converters are examined by using the GITs demonstrating far better performances than those by Si power devices. The demonstrated technologies on the GITs including the advanced technologies are extracting the full potential of GaN and to be applicable to future highly efficient power switching systems. The GITs with these technologies are very promising for future energy-efficient power electronics.

Acknowledgments The author would like to gratefully acknowledge Dr. Tsuyoshi Tanaka and other members of Panasonic Corporation for their technical advices and help throughout the work. He also would like to acknowledge Dr. Daisuke Ueda, Kyoto Institute of Technology, for his technical advices. The work is partly supported by the New Energy and Industrial Technology Development Organization (NEDO), Japan, under the Strategic Development of Energy Conservation Technology Project and the Strategic Development of Energy Saving Innovative Technology Development Project.

References

1. Khan MA, Van Hove JM, Kuznia JN, Olson DT (1991) High electron mobility GaN/AlxGa1-xN heterostructures grown by low-pressure metalorganic chemical vapor deposition. Appl Phys Lett 58:2408
2. Ambacher O, Smart J, Shealy JR, Weimann NG, Chu K, Murphy M, Schaff WJ, Eastman LF, Dimitrov R, Wittmer L, Stutzmann M, Rieger W, Hilsenbeck J (1999) Two-Dimensional electron gases induced by spontaneous and piezoelectric polarization charges in N- and Ga-face AlGaN/GaN heterostructures. J Appl Phys 85:3222
3. Uemoto Y, Hikita M, Ueno H, Matsuo H, Ishida H, Yanagihara M, Ueda T, Tanaka T, Ueda D (2007) Gate Injection Transistor (GIT)—A normally-off AlGaN/GaN power transistor using conductivity modulation. IEEE Trans Electron Device 54:3393
4. Sugiura S, Kishimoto S, Mizutani T, Kuroda M, Ueda T, Tanaka T (2008) Normally-off AlGaN/GaN MOSHFETs with HfO2 gate oxide. Phys Status Solidi C 5:1923
5. Cai Y, Zhou Y, Chen KJ, Lau KM (2005) High-performance enhancement-mode AlGaN/GaN HEMTs using fluoride-based plasma treatment. IEEE Electron Dev Lett 26:435
6. http://www.irf.com/product-info/ganpowir/GaNAPEC.pdf
7. Meneghini M, Scamperle M, Pavesi M, Manfredi M, Ueda T, Ishida H, Tanaka T, Ueda D, Meneghesso G, Zanoni E (2010) Electron and hole-related luminescence processes in gate injection transistors. Appl Phys Lett 97:033506
8. Ishida M, Ueda T, Tanaka T, Ueda D (2013) GaN on Si technologies for power switching Devices. IEEE Trans Electron Device 60:3053
9. Ueda T, Ishida M, Tanaka T, Ueda D (2014) GaN transistors on Si for switching and high-frequency applications. Jpn J Appl Phys 53:100214
10. Morita T, Handa H, Ujita S, Ishida M, Ueda T (2014) 99.3 % Efficiency of boost-up converter for totem-pole bridgeless PFC using GaN gate injection transistors. PCIM Europe, Nuremberg, Germany, May 2014
11. Tanaka K, Ishida M, Ueda T, TanakaT (2013) Effects of deep trapping states at high temperatures on transient performance of AlGaN/GaN heterostructure field-effect transistors. Jpn J Appl Phys 52:04CF07
12. Kaneko S, Kuroda M, Yanagihara M, Ikoshi A, Okita H, Morita T, Tanaka K, Hikita M, Uemoto Y, Takahashi S, Ueda T (2015) Current-collapse-free operations up to 850 V by GaN-GIT utilizing hole injection from drain. In: International symposium on power semiconductor devices and ICs (ISPSD) 2015, Hong Kong, May 2015
13. Morita T, Tamura S, Anda Y, Ishida M, Uemoto Y, Ueda T, Tanaka T, Ueda D (2011) 99.3 % Efficiency of three-phase inverter using GaN-based gate injection transistors. In: Proceedings of 26th IEEE applied power electronics conference and exposition (APEC 2011), Fort Worth, USA, p 481, March 2011
14. Kajitani R, Tanaka K, Ogawa M, Ishida H, Ishida M, Ueda T (2014) A novel high-current density GaN-based normally-off transistor with tensile strained quaternary InAlGaN barrier. Extended abstract of international conference on solid state devices and materials, Tsukuba, Japan, E-3-2, September 2014
15. Uemoto Y, Morita T, Ikoshi A, Umeda H, Matsuo H, Shimizu J, Hikita M, Yanagihara M, Ueda T, Tanaka T, Ueda D (2009) GaN monolithic inverter IC using normally-off gate injection transistors with planar isolation on Si substrate. IEEE IEDM technical digest, Baltimore, USA, p 165, Dec 2009
16. Ujita S, Kinoshita Y, Umeda H, Morita T, Tamura S, Ishida M, Ueda T (2014) A compact GaN-based DC-DC converter IC with high-speed gate drivers enabling high efficiencies. International symposium on power semiconductor devices and ICs (ISPSD), Wikoloa, USA, B1L-A-1, June 2014
17. Umeda H, Kinoshita Y, Ujita S, Morita T, Tamura S, Ishida M, Ueda T (2014) Highly efficient low-voltage DC-DC converter at 2-5 MHz with high operating current using GaN gate injection transistors. PCIM Europe, Nuremberg, Germany, p 45, May 2014

Chapter 12
Fluorine-Implanted Enhancement-Mode Transistors

Kevin J. Chen

12.1 Introduction: Fluorine in III-Nitride Heterostructures: Robust V_{th} Control

The fluorine plasma ion implantation approach was discovered when large positive shift in V_{th} was observed in AlGaN/GaN HEMTs as F ions were incorporated into the barrier layer by carbon tetrafluoride (CF_4) plasma treatment, whose major role is identified as providing F ions with enough energy to penetrate into the subsurface layer of III-nitride epitaxial layers grown along the [0001] direction [1, 2]. The effectiveness of fluorine implantation technique in GaN and related compounds stems from the strong electronegativity of F element and the intrinsic III-nitride wurtzite crystal structure. In AlGaN/GaN heterostructures, because of the very tight lattice structure (with an in-plane lattice constant of a ~3.2 Å), the implanted F ions tend to be stabilized at the interstitial sites by the repulsive forces from the neighboring atoms (Al, Ga, or N). Since group-VII F has the strongest electronegativity among all the chemical elements, a single F ion at the interstitial site tends to capture a free electron and becomes a negative fixed charge. These negative fixed charges subsequently modulate the local potential and deplete the 2DEG in the channel, as shown in Fig. 12.1b. The amount of the V_{th} shift can be robustly controlled by the implantation time, as shown in Fig. 12.2. The fluorine plasma implantation technique also leads to the first demonstration of E-mode GaN-on-Si HEMTs [3].

The RF plasma power is another parameter that needs to be adjusted during process optimization. While this power needs to exceed a lower bound (in the range of ~100–250 W depending on the specific plasma systems used) to introduce

K.J. Chen (✉)
Department of Electronic and Computer Engineering,
The Hong Kong University of Science and Technology,
Clear Water Bay, Kowloon, Hong Kong
e-mail: eekjchen@ust.hk

M. Meneghini et al. (eds.), *Power GaN Devices*,
Power Electronics and Power Systems, DOI 10.1007/978-3-319-43199-4_12

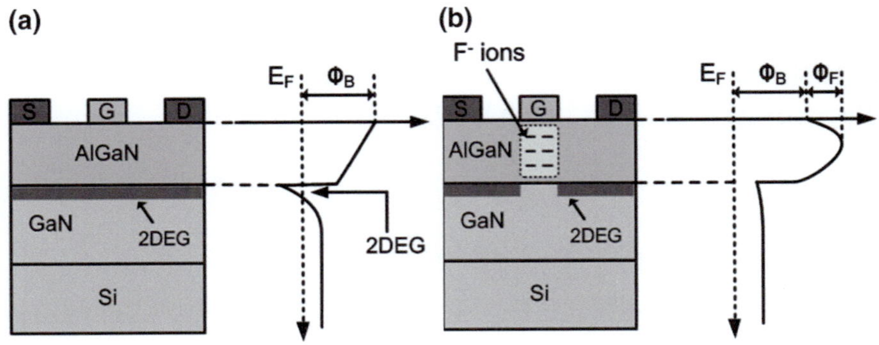

Fig. 12.1 Cross section and the conduction band profiles of standard **a** normally on and **b** normally off AlGaN/GaN HEMTs. The normally off HEMT features F^- ions incorporated in the AlGaN barrier by plasma ion implantation

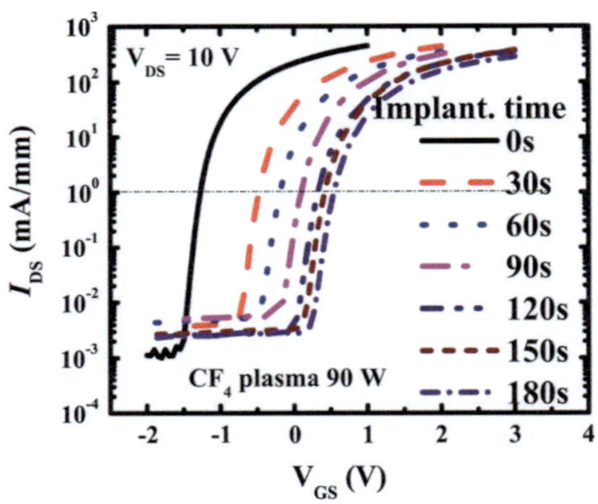

Fig. 12.2 Transfer characteristics of AlGaN/GaN HEMTs with different fluorine plasma ion implantation times

significant V_{th} shift, it is preferred to keep this power as low as possible in order to minimize the plasma-induced lattice damages and the amount of F ions that can penetrate to the channel region. Annealing at 400 °C has been proven to be effective in removing majority of the plasma-induced damages. The nature of the ion implantation process, however, indicates that there is a small amount of F ions in the 2DEG channel, presenting themselves as impurities that could lead to slight mobility degradation. With process optimization, the 2DEG mobility degradation can be reduced to 10–20 %.

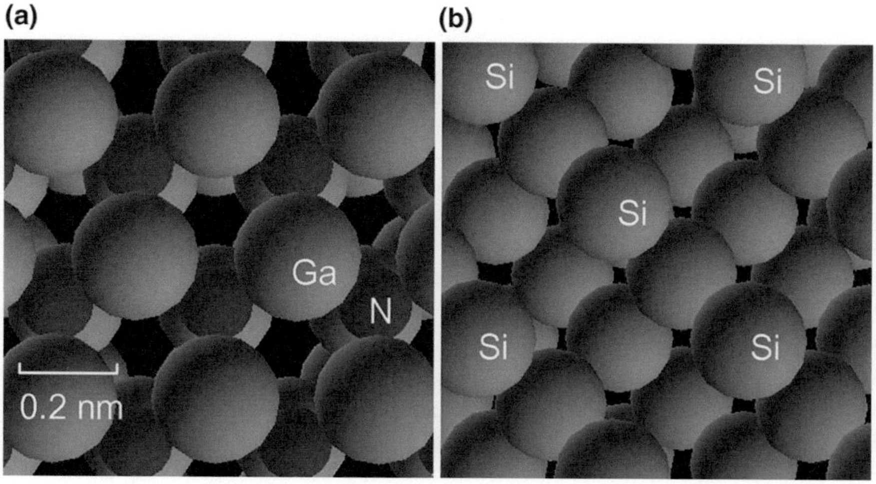

Fig. 12.3 Top view of **a** Ga-faced (0001) GaN wurtzite crystal, and **b** (100) Si crystal. The same physical scale is used in (**a**) and (**b**). Even Si features larger lattice constant (5.43 Å), there is little opening along the incident direction

12.2 Physics Mechanism of Fluorine Implantation

12.2.1 Atomistic Modeling of F Plasma Ion Implantation

To provide a theoretical framework for understanding the physical mechanisms and developing a modeling tool for process design and optimization, it is necessary to develop an approach capable of predicting the ion or dopant distribution profiles and the effects of the ion–solid interactions.

Despite the high atomic densities of GaN and related compound, low-energy (0.1–1.0 keV) F ions are able to penetrate to a shallow depth in AlGaN/GaN heterostructures, partially as a result of the narrow opening channels along the [0001] GaN, as shown in Fig. 12.3. A molecular dynamics (MD) simulation framework [4] was established to study the F plasma implantation into III-nitride material system, because the MD simulation takes into account the ion–lattice interactions on an atomistic scale and includes lattice structure information that is absent in other more traditional implantation simulation tools such as Stopping and Range of Ions in Matter (SRIM) which are based on Monte Carlo method. The MD simulation also provides time and space evolution of atomic coordinates, resembling a real-time observation of the F ion implantation process.

The substrate structure under study is a standard C-plane AlGaN/GaN HEMT structure with a 20-nm AlGaN barrier layer. F^- ions are implanted by CF_4 plasma into the region under the gate. The regular wurtzite lattice structure is applied to model both AlGaN and GaN layers in the MD simulation. Because of its small thickness (20 nm), the AlGaN barrier is assumed to be strained without any lattice relaxation. Thus, the piezoelectric charge polarization and the slight variation in the vertical lattice constant, as a result of the strained AlGaN layer, are both considered in MD simulations. It was found that the charge polarization has negligible effect on the fluorine dopant profiles.

According to published results [5], F^- and CF_3^+ ions are the main ions existing in CF_4 plasma. However, after CF_4 plasma implantation, no obvious change has been observed in the carbon concentration according to SIMS (second ion mass spectroscopy) measurement and the detected carbons are already in presence in the as-grown sample. There are two possible causes for this phenomenon, (1) F^- is much lighter than CF_3^+, with a mass ratio of 9:33. With the same acceleration time determined by the RF signal applied across the two electrodes driving the plasma, majority of F^- ions travel enough distance to reach the sample, which is located at the bottom electrode. On the contrary, only small fraction of CF_3^+ can reach the sample surface before the electric field reverses its direction and pulls the ions away from the sample; (2) In the plasma system, the bottom electrode is grounded, and the top electrode is capacitively coupled to the RF source. Thus, a negative self-bias in the order of several hundreds of volts is induced at the top electrode, which attracts CF_3^+ and pulls them away from the sample. It should be noticed that the absence of additional C atoms in the sample is preferred because C impurities present adverse effects to AlGaN/GaN HEMT [6].

With MD simulation, the 1-D profile of F ion in AlGaN/GaN heterostructure is calculated and plotted in Fig. 12.4, along with the experimental results measured by SIMS. The simulation results with a (100) Si substrate are also calculated as reference. Excellent agreement between simulation and measurement is obtained in the case of F ion in AlGaN/GaN heterostructure. It can be seen that F ion implantation in AlGaN/GaN could get a smaller full width at half maximum (FWHM) compared to Si. This can be explained by the higher density of $Al_{0.3}Ga_{0.7}N$ (5.19 g/cm^3) compared to Si (2.33 g/cm^3).

In the region near the substrate surface, i.e., in the AlGaN layer, the F concentration exhibits a Gaussian-like profile. F ions can penetrate deeper into AlGaN/GaN heterojunction than (100) Si, despite higher atomic density in AlGaN/GaN, indicating a strong channeling effect for the F ions in III-nitride wurtzite lattice structure.

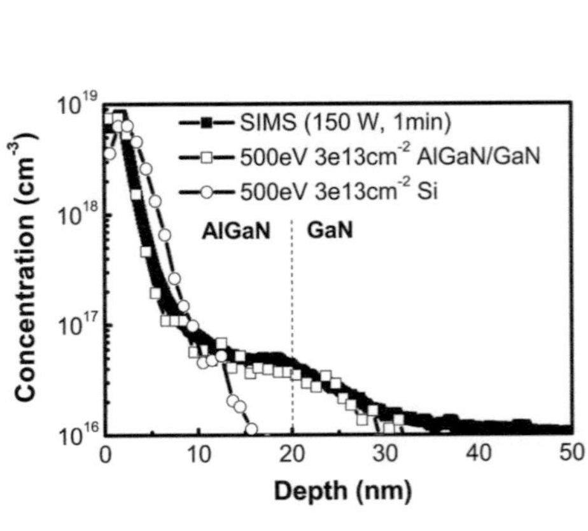

Fig. 12.4 Simulated F profiles in AlGaN/GaN and Si by MD simulation. The implantation energy of F ion is 500 eV and the dose of F is 3×10^{13} cm^{-2}. The F profile in an implanted AlGaN/GaN sample was measured by SIMS and plotted for comparison

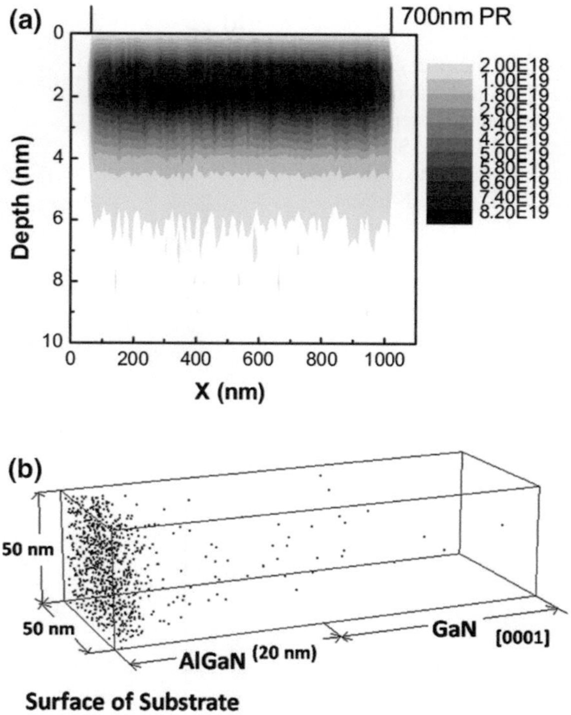

Fig. 12.5 **a** 2-D and **b** 3-D F ion distribution profiles by molecular dynamics simulations. In (**a**), the gate length is 1 μm and the height of the protection mask (photoresist) is 700 nm. In (**b**), a 50 × 50 × 40 nm volume was considered

From the top views of both III-nitride wurtzite lattice and (100) Si crystal structure, it can be seen that there are many channels available in GaN for incident fluorine ions to go through, while the cubic lattice of silicon does not provide these open channels in the direction of implantation.

2-D and 3-D F distribution profiles have been simulated, as shown in Fig. 12.5. The simulation tool considers F ions with all possible incident angles. It should be noted that the F ions penetrating into the AlGaN/GaN structure are capable of inducing vacancies along the implantation path. The long-term high-temperature (400 °C) annealing indicates no shift in V_{th} [7], suggesting weak correlation between these vacancies and the threshold voltage.

12.2.2 Stability of F Ions in AlGaN/GaN Heterostructures

Unlike in Si or other compound semiconductors in which the F ions have shown rather poor thermal stability due to F's small atomic size and large space in the interstitial regions of these materials [8, 9], much better thermal and electrical stabilities have been reported for F ions in AlGaN/GaN heterostructures [7, 10]. The F ions' stability is revealed by the potential energy profiles of F ions in AlGaN/GaN,

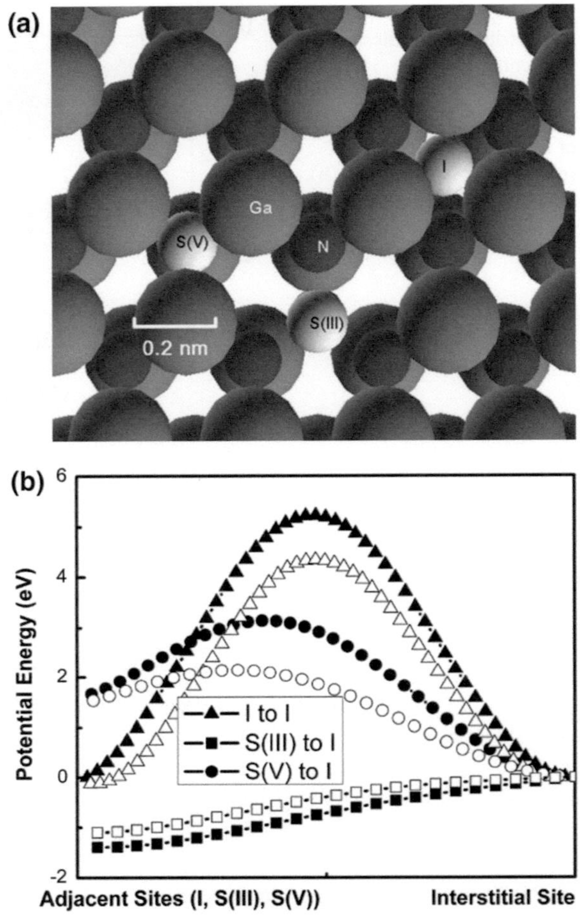

Fig. 12.6 **a** Top view of (0001) GaN with three possible sites that can accommodate fluorine ions. **b** The potential energy profiles from I, S(III) and S(V) sites to their nearest I site of fluorine ion in $Al_{0.25}Ga_{0.75}N$ (*open symbols*) and GaN (*solid symbols*). The $Al_{0.25}Ga_{0.75}N$ is strained and lattice matched to GaN

also calculated by MD simulation [11]. The three locally stable sites that can physically accommodate fluorine ions are (Fig. 12.6a): interstitial site I, substitutional group-III cation site S(III), and substitutional group-V anion site S(V). Since majority (>90 %) of the implanted F ions are initially located at the I sites after implantation, the potential energy profiles between an I site and these three possible nearby sites along the minimum energy path are calculated and shown in Fig. 12.6b.

Several conclusions can be drawn from Fig. 12.6 as follows:

- Interstitial F ions will be stabilized at the I site unless there is an S(III) site nearby.
- An F ion exhibits higher potential energy at the S(V) site compared to an I site since the Ga atoms around an N vacancy provide strong repulsive force to F ions and also help prevent interstitial F ions from moving into N vacancies.
- Around a Ga vacancy, the N atoms are smaller and the repulsive force to an F ion is significantly weaker. Thus, it is easier for an interstitial fluorine ion to move into a nearby Ga vacancy when available.

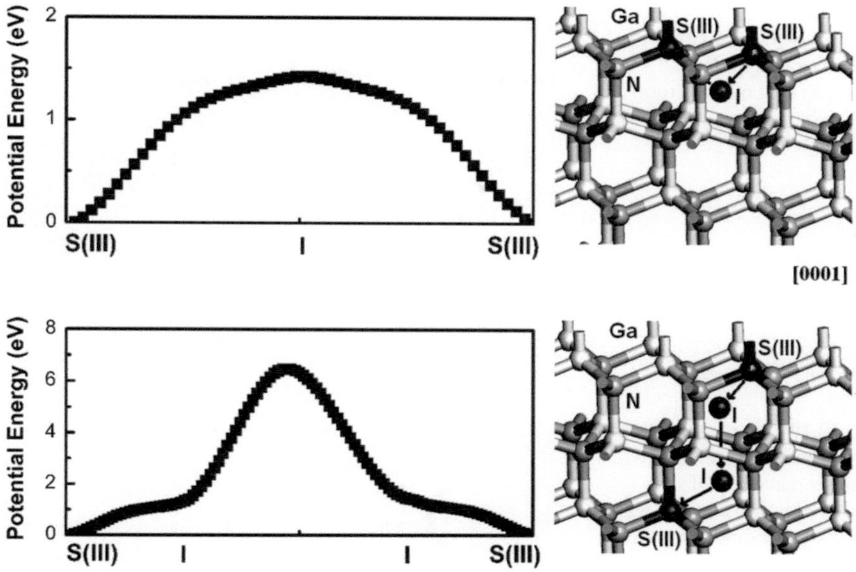

Fig. 12.7 Potential energy profiles through S(III)-I-S(III) and S(III)-I-I-S(III) paths. The movement paths are shown on the *right*. The *black*, *dark gray*, and *light gray balls* are F sites, N atoms, and Ga atoms, respectively

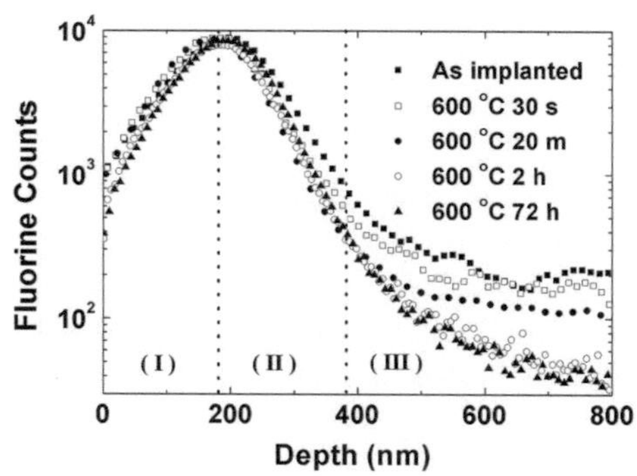

Fig. 12.8 Depth profiles of fluorine after post-implantation annealing at 600 °C. This experiment was conducted on a sample implanted by 180 keV F ions with a dose of $1 \times 10^{15}/\text{cm}^2$, so that the peak F concentration is deep inside the bulk

When there are no continuous S(III) vacancy chains, the F ions will be trapped at the S(III) sits, as depicted in Fig. 12.7. The thermal stability of F ions in GaN is highlighted in Fig. 12.8 and the F–Ga vacancy interactions are revealed by positron annihilation spectroscopy experiment, as shown in Fig. 12.9 [12].

Excellent thermal stability has been reported in F-implanted normally off GaN HEMTs as long as the ambient temperature does not exceed the limit of the gate

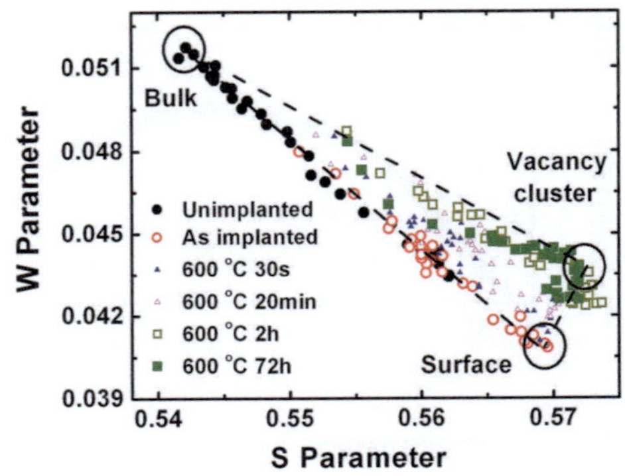

Fig. 12.9 S-E curve of the fluorine-implanted samples after annealing in N_2 atmosphere at 600 °C from 30 s to 72 h

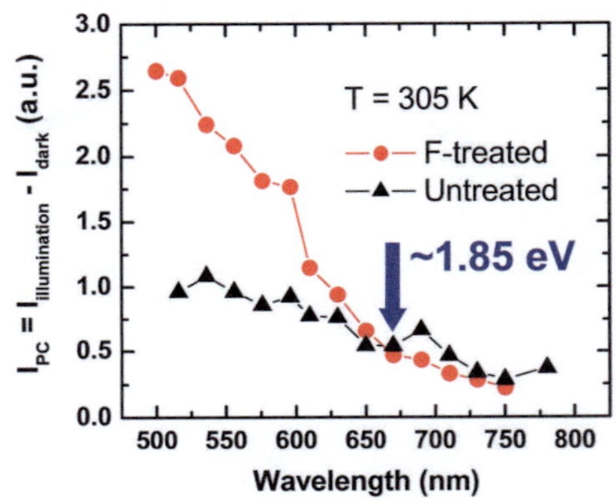

Fig. 12.10 Photoconductive current between two ohmic contacts made to 2DEG, measured in two samples with and without F-implant

metal (e.g., 500 °C for Ni) [7] or gate dielectric (>800 °C for SiN_x). The stability of the gate metal or gate dielectric is critical since it provides the cover that prevents F ions from escaping through the surface under high-temperature thermal excitation. It is observed that F ions can be stable under high electric field stress as long as large number of defects and dislocations are not introduced, e.g., by the inverse piezo-electric effect.

12.2.3 Electron Binding Energy Around F Ions

To study the binding energy of electrons associated with F ions, photoconductivity measurements were conducted using a Xe lamp together with a monochromator. The photocurrents as a function of the excitation wavelength are plotted in Fig. 12.10. The photocurrent increases slowly as the wavelength decreases (photon energy increases) from 750 nm in both the F-implanted and unimplanted samples. As the wavelength falls below 670 nm (corresponding to a 1.85 eV photon energy), the F-implanted sample starts to exhibit a much higher increasing rate compared to the unimplanted sample, indicating activation of electrons from certain deep levels. Thus, the electron binding energy associated with the F ions in AlGaN/GaN is ~1.85 eV [13].

12.3 Fluorine-Implanted Enhancement-Mode GaN MIS-HEMTs

12.3.1 GaN MIS-HEMTs

Many of the early normally off GaN FETs are HEMT devices with Schottky gate, which has relatively small threshold voltage (<1 V) and small gate swing that is limited by the Schottky-gate forward turn-on voltage (e.g., <3 V). In power-switching applications, large positive threshold voltage (e.g., >3 V) and large gate swing (e.g., >10 V) are required to prevent the power switches from faulty turn-on by electromagnetic interference and to become compatible with the gate drive designs currently used for Si-based power transistors. MIS-HEMT (metal–insulator–semiconductor HEMT) with a gate dielectric inserted between the gate metal and group-III-nitride surface provides the capability of scaling up the threshold voltage and gate swing. Various dielectric materials (e.g., Al_2O_3, SiO_2, and SiN_x) have been employed to convert Schottky gate to MIS gate [14, 15]. A few reports have shown GaN E-mode devices with large gate swings [16]. Meanwhile, low current collapse, or low dynamic on-resistance (R_{on}), is another indispensable requirement for achieving high-efficiency power conversion under high-voltage switching conditions, whereas effective device passivation still demands more effort [17].

Device Structure and Fabrication

The studied sample features a 21-nm barrier layer (with 2-nm GaN cap, 18 nm $Al_{0.25}Ga_{0.75}N$, and 1 nm AlN) and a 4-μm GaN buffer/transition layer, grown on a 4-inch p-type (111) Si substrate. The schematic cross section of a fabricated E-mode MIS-HEMT is shown in Fig. 12.11. An AlN/SiN_x passivation stack structure was deposited with plasma-enhanced atomic layer deposition (PE-ALD) of 4 nm AlN and plasma-enhanced chemical vapor deposition (PECVD) of 50 nm SiN_x in sequence [18]. Planar device electrical isolation was realized by standard fluorine ion implantation. The gate window was opened with low-power dry etching of the passivation stack. The gate region was then subjected to CF_4 plasma at an RF power of 200 W for 250 s with photoresist as the plasma implantation mask. After removing photoresist and carrying out an *ex situ* surface cleaning process with remote NH_3/Ar/N_2 plasma at 300 °C in the PE-ALD system to effectively remove the detrimental Ga–O bonds at the surface [19], the second SiN_x thin film of 17 nm thickness was immediately grown by PECVD and deployed as the gate insulator to reduce gate leakage. Gate electrode with 1-μm footprint and 0.5-μm overhang was formed with Ni/Au deposited upon the second SiN_x.

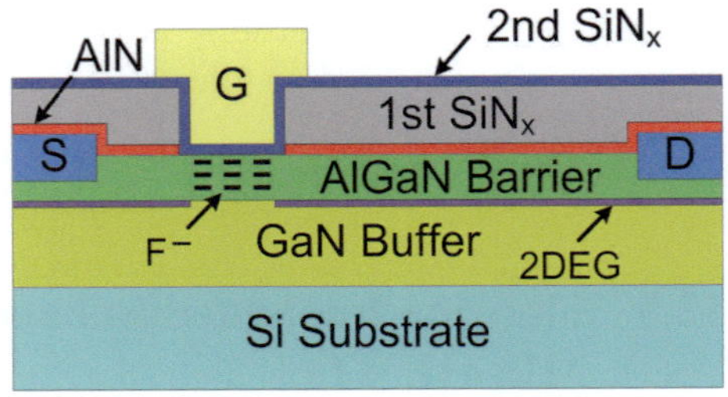

Fig. 12.11 Schematic cross section of a fabricated E-mode MIS-HEMT

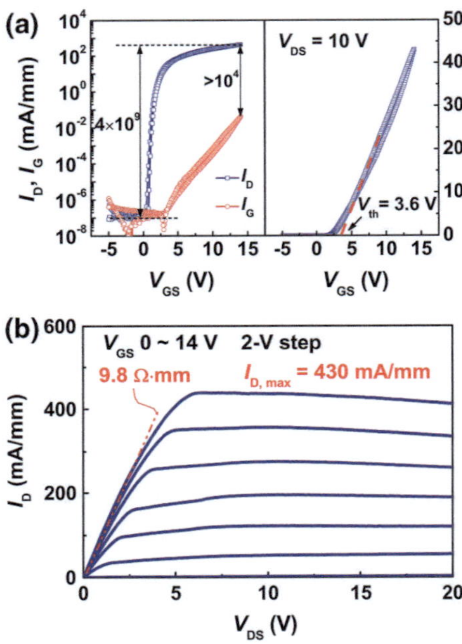

Fig. 12.12 **a** DC I_D–V_{GS} and I_G–V_{GS} characteristics measured with gate bias positive/negative sweeping at a drain bias of 10 V. **b** DC I_D–V_{DS}

Measurement Results and Analysis

Device characterization was carried out on devices with $L_{GS} = 1$ μm, $(W/L)_G = 10$ μm/1 μm, and $L_{GD} = 15$ μm (unless otherwise specified). Figure 12.12 shows typical DC transfer and output characteristics of an MIS-HEMT. The threshold voltage Vth, determined by the linear extrapolation method (i.e., the gate bias intercept of the linear extrapolation of drain current at the point of peak transconductance), is extracted to be +3.6 and +1.2 V for the MIS-HEMTs and the Schottky-gate control HEMTs fabricated on the same wafer, respectively. The positive shift of V_{th} in the MIS-HEMT is primarily a result of the reduced gate-to-channel capacitance by the insertion of the SiN_x gate dielectric.

The trap density at SiN_x/barrier interface is estimated to be ~3×10^{12} cm^{-2} using a pulsed I_D–V_{GS} measurement [20]. The device exhibits an ON/OFF current ratio of 4×10^9 and a drive current of 430 mA/mm. The gate leakage is effectively suppressed

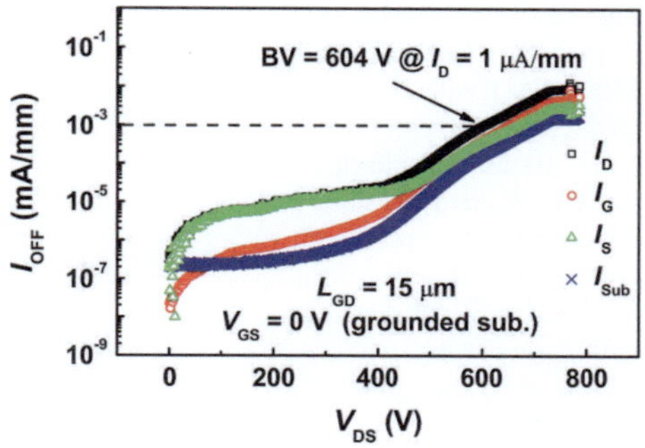

Fig. 12.13 Off-state breakdown/leakage characteristics with $V_{GS} = 0$ V and the substrate connected to the ground. A breakdown voltage of 604 V is achieved at a drain leakage current of 1 μA/mm

by the SiN_x at positive gate bias, enabling a large gate swing of 14 V and a low R_{on} of 9.8 Ω mm. The DC-specific R_{on} is calculated to be 2.1 mΩ cm², taking into account a 1.5 μm transfer length for each ohmic contact (i.e., source and drain) in active area calculation.

The off-state breakdown/leakage characteristics of an MIS-HEMT are shown in Fig. 12.13. Gate and source electrodes are biased at 0 V and the substrate is grounded. It can be observed that the leakage current is mainly originated from the source injection current at a drain bias V_{DS} up to 450 V, when the off-state gate leakage has been substantially suppressed by the gate insulator SiN_x. When V_{DS} continues to increase, the gate leakage and vertical substrate leakage will both increase to be comparable to the source injection current as V_{DS} approaches 600 V. The breakdown voltage (BV), defined as the drain bias at a drain leakage current of 1 μA/mm, is measured to be 604 V.

Current collapse evaluation is performed using AMCAD-pulsed *I–V* measurement system. Figure 12.14a shows the pulsed I_D–V_{DS} characteristics of a normally off device with L_{GS} = 1 μm, $(W/L)_G$ = 2 × 50 μm/1 μm, and L_{GD} = 10 μm. The quiescent bias point is set at (V_{GSQ}, V_{DSQ}) = (0 V, 60 V). The pulse width and pulse period are 500 ns and 1 ms, respectively. The DC output curves are plotted for reference. Effective suppression of current collapse by AlN/SiN_x passivation is demonstrated with little difference between the DC and pulsed drain current in the linear region. The elimination of self-heating effect is believed to be the reason for the higher drain current obtained in the pulsed *I–V* measurement. For high-voltage switching measurement, a slow-switching on-wafer testing was conducted using Agilent B1505A power device analyzer/curve tracer. After a high drain bias stress in the off-state with V_{GS} = 0 V, the device is turned on when the dynamic R_{on} is measured at V_{GS} = 12 V and V_{DS} = 1.5 V with a transient on-state current of ~ 100 mA/mm, as shown in Fig. 12.14b. The off-to-on switching time is 0.1 s when the drain stress voltage V_{DS} is lower than 200 V and 2.7 s when V_{DS} is higher than 200 V. The device has slight dynamic R_{on} degradation of only 76 % for an off-state drain bias stress of 650 V.

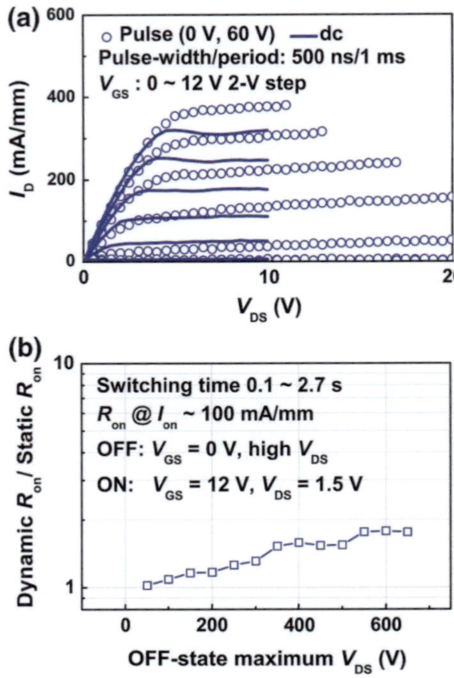

Fig. 12.14 **a** Pulsed I_D–V_{DS} characteristics of an E-mode MIS-HEMT with $L_{GS} = 1\ \mu m$, $(W/L)_G = 2 \times 50\ \mu m/1\ \mu m$, and $L_{GD} = 10\ \mu m$. **b** Ratio of dynamic R_{on} and DC static R_{on} obtained by low-speed high-voltage switching measurement. The device is with $L_{GD} = 15\ \mu m$, $(W/L)_G = 10\ \mu m/1\ \mu m$, and $L_{GD} = 15\ \mu m$

12.3.2 GaN MIS-HEMTs with Partially Recessed Fluorine-Implanted Barrier

For the GaN transistors with metal–insulator–semiconductor (MIS) structure, their high-temperature stability can be hindered by the challenges of V_{th} instability originating from the thermal electron emission of trap states at the dielectric/III-N interface [21, 22]. To address this issue, a thinned barrier layer is proposed to bring the deep interface traps below the Fermi level at pinch-off so that they become inactive [22]. In this work, the normally off MIS-HEMTs featuring a partially recessed (Al)GaN barrier were realized by a fluorine plasma implantation/etch technique. The partially recessed barrier leads to improved thermal stability, while the fluorine implantation can convert the device from D-mode to E-mode without completely removing the barrier and sacrificing the high mobility heterojunction channel [1].

Device Fabrication

The schematic cross section of the normally off MIS-HEMT is shown in Fig. 12.15. Both the fluorine ion implantation and gate recess were carried out using CF_4 plasma [23]. By properly adjusting the power level of the RF source driving the fluorine plasma, we are able to obtain two desirable results: (1) a well-controlled slow dry etching for gate recess; and (2) effective shallow implantation of fluorine ions into the AlGaN barrier. Fluorine plasma implantation at a higher RF power level of 200 W

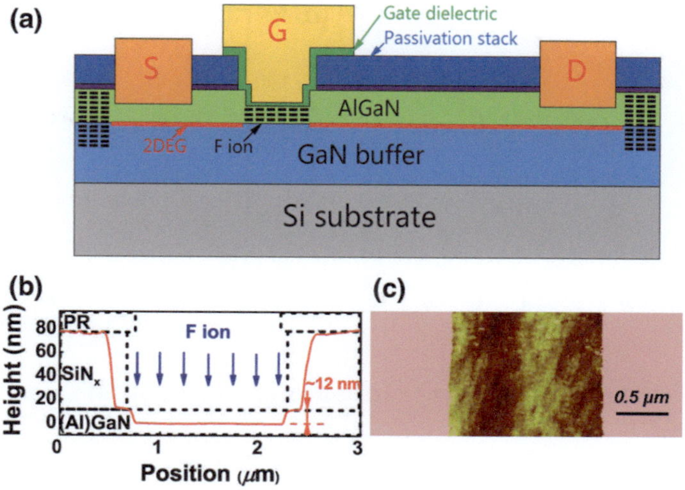

Fig. 12.15 **a** Cross section of a fabricated E-mode MIS-HEMT. **b** Side wall profile of the gate region after F-implantation (*solid line in red*) and the outline of gate region before F-implantation. **c** Surface morphology of the recessed gate region

resulted in a well-controlled slow-etching process with an etching rate of 2 nm/min. Meanwhile, a lower RF power of 150 W only induced insignificant etching of the barrier layer [1]. After 6 min of F-implantation/etch, a recess depth of ~12 nm and a smooth etched surface were obtained. After removing another 2 nm AlGaN by a digital etching [24], 20 nm Al_2O_3 was deposited by ALD with an *in situ* nitridation process [19].

Results and Discussion

The partially recessed F-implanted E-mode MIS-HEMT exhibits a threshold voltage (V_{th}) of +0.6 V at a drain current of 10 μA/mm, a maximum drive current of 730 mA/mm, an on-resistance of 7.07 Ω mm (Fig. 12.16), and a hysteresis of ~ 0.3 V between the up- and down-V_{GS}-sweep with a relatively fast sweeping rate (0.7 V/s).

The temperature (T)-dependent transfer and output characteristics of an MIS-HEMT were characterized (Fig. 12.17). When temperature increases from 25 to 200 °C, an increase in 3 orders of magnitude is observed in the OFF-state drain leakage due to increased buffer leakage, while the drain current exhibits an decrease (e.g., from 240 mA/mm to 200 mA/mm at $V_{GS} = 4$ V). By using a current criteria of I_{DS} of 10 μA/mm, V_{th} shifted by 0.5 V toward negative direction when T increased from 25 to 200 °C.

According to a recent report [22], the thermally stable V_{th} of MIS-HEMTs with recessed gate is mainly attributed to the thin barrier thickness (Fig. 12.18a). As shown in figure inset, the V_{th} shift with temperature increasing from 25 to 200 °C depends on the barrier layer thickness (t_{BR}). As t_{BR} is reduced, the thermally induced V_{th} shift can be significantly suppressed. The recessed barrier buries the deep interface traps

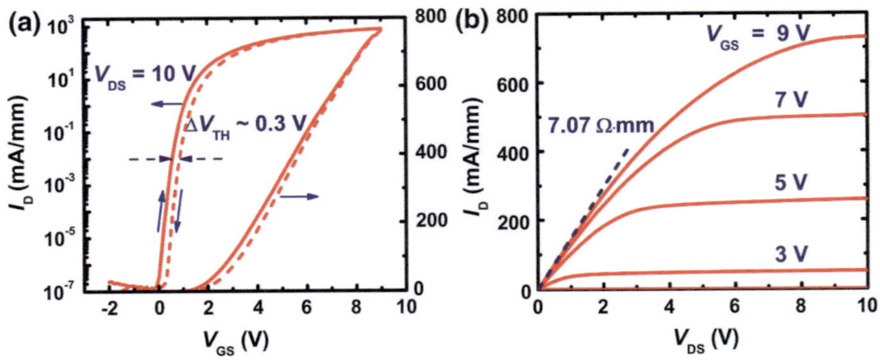

Fig. 12.16 (a) Transfer and (b) output characteristics of a MIS-HEMT with L_G = 1.5 μm and L_{GD} = 10 μm

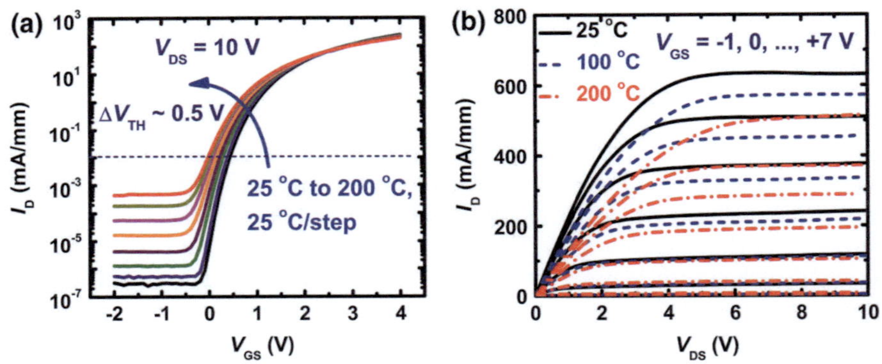

Fig. 12.17 Temperature (T)-dependent **a** transfer and **b** output characteristics of a MIS-HEMT with L_G = 1 μm and L_{GD} = 2 μm at T increasing from 25 to 200 °C

below the Fermi level at pinch-off (i.e., $\Delta E_3 < \Delta E_2 < \Delta E_1$) so that they become inactive and do not participate in the thermally sensitive emission (Fig. 12.18b). Stable charge states of the interface traps would lead to more stable V_{th}.

12.3.3 GaN Smart Power ICs

Although the current focus of GaN power device technologies is on discrete power devices, higher level integration of GaN power electronics could lead to many application-relevant benefits including reduced cost and improved reliability. In a more complete power converter module, besides the core power components (i.e., switches and rectifiers), intelligent control units are also needed to achieve precise

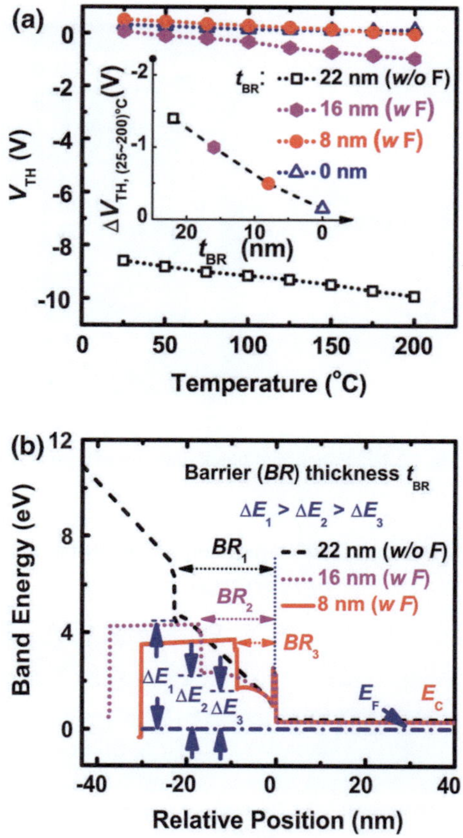

Fig. 12.18 **a** *T*-dependence of threshold voltage of GaN MIS-HEMTs with various barrier thicknesses and with/without fluorine (F) implantation. **b** Simulated band diagram of MIS-HEMT structure in this work (*red*) at pinch-off situation, compared with MIS-HEMTs with thicker barrier layer. The band diagrams are simulated using a 1D Poisson/Schrödinger equation solver. A uniform fluorine concentration (yielding a sheet density of $1 \times 10^{13}/cm^2$) in the AlGaN barrier was used in the simulation for the purpose of simple illustration

adjustment of the output signal under different loading conditions, as shown in Fig. 12.19. Ultimately, robust sensing and protection units should be integrated to realize full-range protections against extreme operating conditions such as over-temperature, over-current, and over-voltage. It is thus desirable to develop highly integrated GaN power electronics technology with which we can implement on-chip power conditioning/protection circuits that promise to provide optimized performance, increased functionality, and enhanced reliability.

Platforms for Heterogeneous Integration

A smart power IC technology requires a platform on which both high-voltage components and low-voltage periphery devices can be integrated (Fig. 12.20) [25, 26]. The high-voltage switches prefer normally off transistors for their inherent fail-safe operation. For digital IC development, the mainstream CMOS-like architecture may have intrinsic difficulties in GaN materials due to the inferior hole mobility. The analog IC development using AlGaN/GaN has started recently.

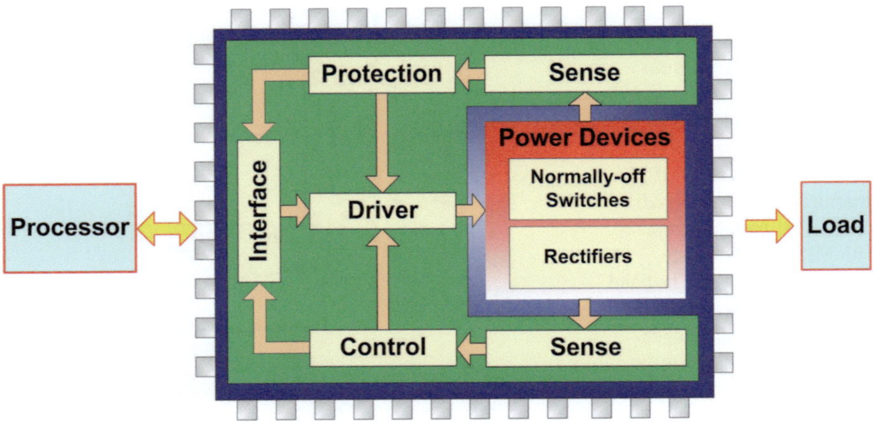

Fig. 12.19 Functional block diagram of a power converter module

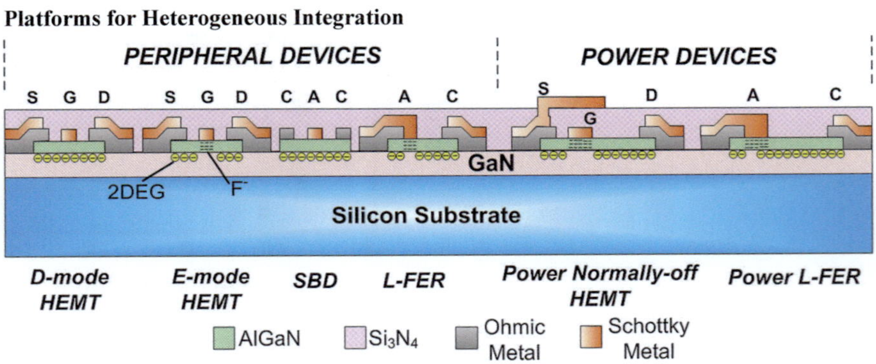

Fig. 12.20 Schematic platform of GaN smart power technology: integration of low-voltage peripheral and high-voltage power devices

Based on the robust fluorine plasma ion implantation technique, several key smart power components including high-voltage normally off transistors and lateral field-effect rectifiers [27, 28] have been developed. A smart power chip technology also requires low-voltage peripheral devices for the implementation of digital and analog blocks. Due to the lack of high-performance p-channel GaN devices, the simplest circuit configuration for GaN digital circuits is the direct-coupled FET logic (DCFL) that requires both the E-mode and D-mode n-channel HEMTs [29]. The Schottky diodes and L-FERs all exhibits graceful temperature dependences that can be explored for temperature sensing and compensation.

Lateral Field-Effect Rectifier (L-FER)

Analog circuits typically require diodes in additional to transistors. With the III-N heterostructures, diodes can be realized with three approaches including the lateral

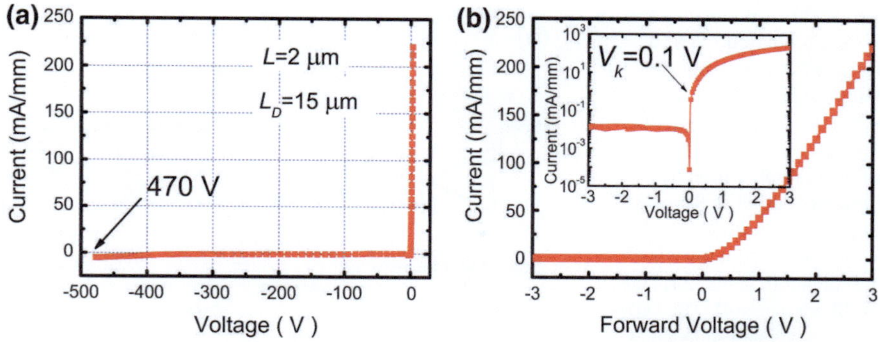

Fig. 12.21 Measured DC performances of L-FERs: **a** reverse bias, **b** forward bias

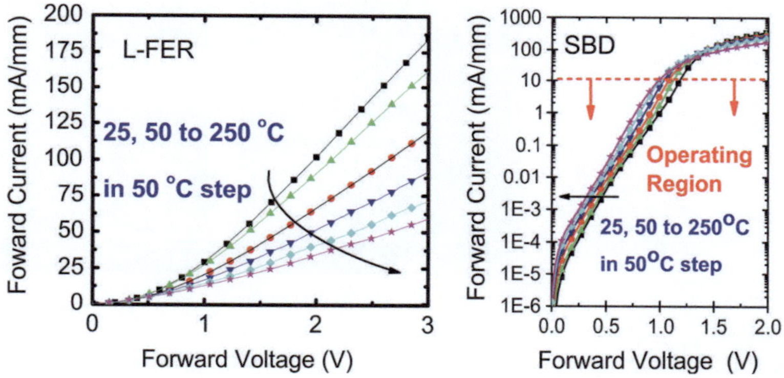

Fig. 12.22 Forward *I–V* characteristic of L-FER and SBD at varied operating temperatures

field-effect rectifier (L-FER) [27], the simple Schottky barrier diode (SBD) on heterostructures, or the lateral SBD with the Schottky junction to the 2DEG channel made from the sideway [30]. The forward and reverse characteristics of an L-FER with the anode–cathode drift region length L_D of 10 μm are plotted in Fig. 12.21. The knee voltage V_k, defined as the anode bias at a forward current of 1 mA/mm, is 0.1 V in the proposed L-FER. Figure 12.22 shows the forward *I–V* curves of L-FER and SBD at varied operating temperature. It is seen that the current of L-FER exhibits a negative temperature coefficient because its conduction current is dominated by phonon scattering. For the SBD directly fabricated on the heterostructure, the thermionic emission dominates the forward current at forward bias smaller than 1 V, and thus, its forward current has a positive temperature coefficient.

GaN Mixed-Signal ICs Based on Monolithic Integration of E/D-Mode HEMTs
Based on the fluorine implantation technique, monolithic integration of E/D-mode HEMTs has been demonstrated [29]. A 17-stage ring oscillator has been shown to operate properly, even at 375 °C [31]. Various mixed-signal GaN ICs have been

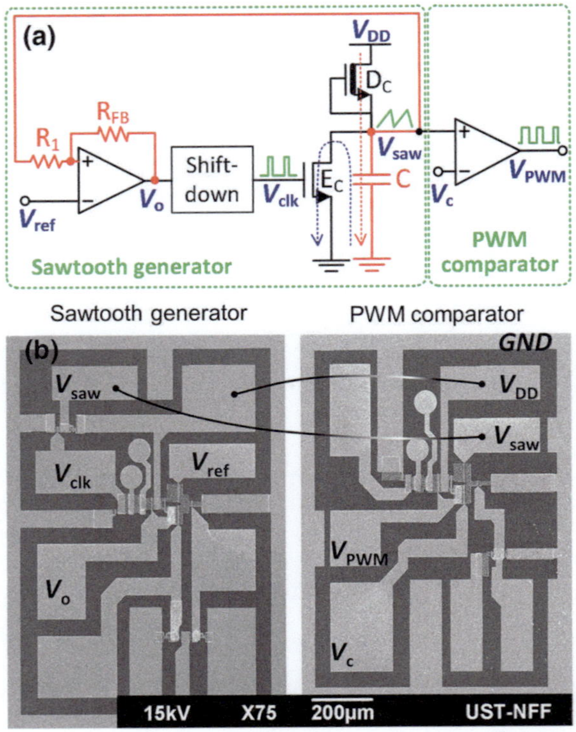

Fig. 12.23 **a** Block diagram of the GaN-based PWM IC. Three off-chip passive components are used (*marked in red line*) for the sawtooth generator. V_c is the reference voltage sampled from the converter output. **b** Scanning electron microscope (SEM) picture of the adjacent dies (*left* sawtooth generator; *right* PWM comparator). The GaN-on-Si chip is wire-bonded and mounted onto a PCB which also houses three off-chip passive components

demonstrated including a voltage reference [32], a bootstrapped comparator [33], a 2-bit quantizer and flip-flop [34], a temperature sensor [35], and a self-startup circuit [36].

The first GaN-based PWM circuit using monolithically integrated E/D-mode HEMTs and L-FERs [37] is illustrated in Fig. 12.23. The circuit is able to generate 1 MHz PWM signal with its duty cycle modulated linearly over a wide range.

The GaN-based pulse width modulator incorporates two functional blocks, i.e., a sawtooth generator and a comparator. The sawtooth generator produces a sawtooth signal (V_{saw}) and the comparator (referred to as the PWM comparator) produces the PWM signal (V_{PWM}) by comparing the reference voltage (V_c) sampled from the converter output against the sawtooth signal (V_{saw}). The pulse width of V_{PWM} is modulated by V_c, and its frequency is determined by that of V_{saw}. Circuit performances up to 1 MHz are investigated with a 5 V supply voltage. The generated sawtooth signal (V_{saw}) is injected into the PWM comparator and compared against V_c, yielding a PWM signal with its duty cycle modulated by V_c. Figure 12.24 shows the PWM waveforms when the circuit is operated at ~1.08 MHz. The pulse width of the PWM signal can be modulated over a wide range (by varying V_c from 1.3 to 2.0 V with V_{saw} oscillating between 0.89 and 2.18 V).

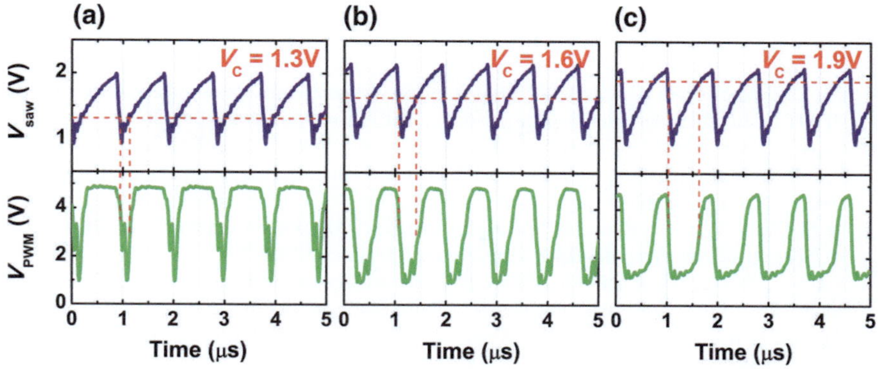

Fig. 12.24 Waveforms of GaN PWM circuit, $V_{DD} = 5$ V, $I_{DD} = 5$ mA. V_c (*red dash*) cuts V_{saw} ($f = 1.08$ MHz, $V_{saw} = 0.89$–2.18 V) and determines the duty cycle of V_{PWM}. **a** $V_c = 1.3$ V, **b** $V_c = 1.6$ V, **c** $V_c = 1.9$ V

12.4 Conclusions

Fluorine implantation, in particular the low-energy implantation using plasma tools, is shown to be able to locally incorporate negatively charged F ions into AlGaN/GaN heterojunctions. These negative fixed charges provide robust and flexible means of adjusting the threshold voltage in the 2DEG channel, enabling the realization of E-mode GaN HEMT and MIS-HEMTs on D-mode HEMT platform. Atomistic modeling based on molecular dynamics simulation reveals that F ions can be reasonably implanted into AlGaN/GaN heterojunctions assisted by the channeling effect of the (0001) C-plane AlGaN/GaN, but can be stably confined within the crystal because of the tight lattice structure of GaN and related compounds. The fluorine implantation technique offers robustness and flexibility in delivering GaN power components (transistors and rectifiers) and peripheral mixed-signal functional blocks on the same technology platform for the ultimate GaN system-on-a-chip solutions for various applications smart power amplifiers and smart power converters.

Acknowledgments The author would like to thank his former students and postdoctoral fellows for the key contributions they have made to the work described in this chapter. They are Drs. Yong Cai, Ruonan Wang, Li Yuan, Maojun Wang, Wanjun Chen, Baikui Li, King Yuen Wong, Chunhua Zhou, Alex Man Ho Kwan, Shu Yang, Cheng Liu, and Hanxing Wang. This work also greatly benefited from collaboration with Prof. Kei May Lau and Prof. Jiannong Wang at Hong Kong University of Science and Technology.

References

1. Cai Y, Zhou Y, Chen KJ, Lau KM (2005) High-performance enhancement-mode AlGaN/GaN HEMTs using fluoride-based plasma treatment. IEEE Electron Device Lett 26:435
2. Cai Y, Zhou Y, Lau KM, Chen KJ (2006) Control of threshold voltage of AlGaN/GaN HEMTs by fluoride-based plasma treatment: from depletion mode to enhancemend mode. IEEE Trans Electron Devices 53:2207
3. Jia S, Cai Y, Wang D, Zhang B, Lau KM, Chen KJ (2006) Enhancement-mode AlGaN/GaN HEMTs on silicon substrate. IEEE TransElectron Devices 53:1477
4. Yuan L, Wang MJ, Chen KJ (2008) Fluorine plasma ion implantation in AlGaN/GaN heterostructures: a molecular dynamics simulation study. Appl Phys Lett 92:102109
5. Segawa S, Kurihara M, Nakano N, Makabe T (1999) Dependence of driving frequency on capacitively coupled plasma in CF4. Jpn J Appl Phys 38:4416
6. Klein PB, Binari SC, Ikossi K, Wickenden AE, Koleske DD, Henry RL (2001) Current collapse and the role of carbon in AlGaN/GaN high electron mobility transistors grown by metalorganic vapor-phase epitaxy. Appl Phys Lett 79:21
7. Yi C, Wang R, Huang W, Tang WC-W, Lau KM, Chen KJ (2007) Reliability of enhancement-mode AlGaN/GaN HEMTs fabricated by fluorine plasma treatment. Technical digest. IEEE International Electron Device Meeting, p. 389
8. Robison RR, Law ME (2002) Fluorine diffusion: models and experiments. Technical digest. IEEE International Electron Device Meeting, p. 883
9. Hayafuji N, Yamamoto Y, Ishida T, Sato K (1996) Degradation mechanism of the AlInAs/GaInAs high electron mobility transistor due to fluorine incorporation. Appl Phys Lett 69:4075
10. Ma C, Chen H, Zhou C, Huang S, Yuan L, Roberts J, Chen KJ (2010) Reliability of enhancement-mode AlGaN/GaN HEMTs under ON-state gate overdrive. Technical digest. IEEE International Electron Device Meeting, p. 476
11. Yuan L, Wang M, Chen KJ (2008) Atomistic modeling of fluorine implantation and diffusion in III-nitride semiconductors.Technical digest. IEEE International Electron Device Meeting, p. 543
12. Wang MJ, Yuan L, Cheng CC, Beling CD, Chen KJ (2009) Defect formation and annealing behaviors of fluorine-implanted GaN layers revealed by positron annihilation spectroscopy. Appl Phys Lett 94:061910
13. Li BK, Chen KJ, Lau KM, Ge WK, Wang JN (2008) Characterization of fluorine-plasma-induced deep centers in AlGaN/GaN heterostructure by persistent photoconductivity. Phys Stat Sol C 5:1892
14. Khan MA, Hu X, Tarakji A, Simin G, Yang J, Gaska R, Shur MS (2000) AlGaN/GaN metal–oxide–semiconductor heterostructure field-effect transistors on SiC substrates. Appl Phys Lett 77:1339
15. Chen KJ, Zhou C (2011) Enhancement-mode AlGaN/GaN HEMT and MIS-HEMT technology. Stat Sol A 208:434
16. Wang R, Cai Y, Tang C-W, Lau KM, Chen KJ (2006) Enhancement-mode Si3N4/AlGaN/GaN MISHFETs. IEEE Electron Device Lett 27:793
17. Huang S, Jiang Q, Yang S, Zhou C, Chen KJ (2012) Effective passivation of AlGaN/GaN HEMTs by ALD-grown AlN thin film. IEEE Electron Device Lett 33:516
18. Tang Z, Jiang Q, Lu Y, Huang S, Yang S, Tang X, Chen KJ (2013) 600-V normally-off SiNx/AlGaN/GaN MIS-HEMT with large gate swing and low current collapse. IEEE Electron Device Lett 34:1373
19. Yang S, Tang Z, Wong K-Y, Lin Y-S, Liu C, Lu Y, Huang S, Chen KJ (2013) High-quality interface in Al2O3/GaN/AlGaN/GaN MIS structures with in situ pre-gate plasma nitridation. IEEE Electron Device Lett 34:1497
20. Lu Y, Yang S, Jiang Q, Tang Z, Li B, Chen KJ (2013) Characterization of VT-instability in enhancement-mode Al2O3-AlGaN/GaN MIS-HEMTs. Phys Stat Sol C 10:1397

21. Chu RM, Brown D, Zehnder D, Chen X, Williams A, Li R, Chen M, Newell S, Boutros K (2012) Normally-off GaN-on-Si metal-insulator-semiconductor field-effect transistor with 600-V blocking capability at 200 °C. Proceeding of 24th international symposium on power semiconductor devices and ICs, p. 237
22. Yang S, Liu S, Liu C, Chen KJ (2014) Thermally induced threshold voltage instability of III-Nitride MIS-HEMTs and MOSC-HEMTs: underlying mechanisms and optimization schemes. Technical digest. IEEE International Electron Device Meeting, p. 389
23. Liu C, Yang S, Liu S, Tang Z, Wang H, Jiang Q, Chen KJ (2015) Thermally stable enhancement-mode GaN metal-insulator-semiconductor high-electron-mobility transistor with partially recessed fluorine-implanted barrier. IEEE Electron Device Lett 36:318
24. Liu S, Yang S, Tang Z, Jiang Q, Liu C, Wang M, Chen KJ (2014) Al2O3/AlN/GaN MOS-channel-HEMTs with an AlN interfacial layer. IEEE Electron Device Lett 35:723
25. Wong K-Y, Chen W, Liu X, Zhou C, Chen KJ (2010) GaN smart power IC technology. Phys Stat Sol B 247:1732
26. Chen KJ, Kwan AMH, Jiang Q (2014) Technology for III-N heterogeneous mixed-signal electronics. Phys Stat Sol A 211:769
27. Chen W, Wong KY, Huang W, Chen KJ (2008) High-performance AlGaN/GaN lateral field-effect rectifiers compatible with high electron mobility transistors. Appl Phys Lett 92:253501
28. Chen W, Wong KY, Chen KJ (2009) Single-chip boost converter using monolithically integrated AlGaN/GaN lateral field-effect rectifier and normally-off HEMT. IEEE Electron Device Lett 30:430
29. Cai Y, Cheng Z, Yang Z, Tang WC-W, Lau KM, Chen KJ (2006) Monolithically integrated enhancement/depletion-mode AlGaN/GaN HEMT inverters and ring oscillators using CF4 plasma treatment. IEEE Trans Electron Devices 53:2223
30. Bahat-Treidel E, Hilt O, Zhytnytska R, Wentzel A, Meliani C, Würfl J, Tränkle G (2012) Fast-switching GaN-based lateral power schottky barrier diodes with low onset voltage and strong reverse blocking. IEEE Electron Device Lett 33:357
31. Cai Y, Cheng Z, Yang Z, Tang WC-W, Lau KM, Chen KJ (2007) High temperature operation of AlGaN/GaN HEMTs direct-coupled FET logic (DCFL) integrated circuits. IEEE Electron Device Lett 28:328
32. Wong K-Y, Chen WJ, Chen KJ (2010) Integrated voltage reference generator for GaN smart power chip technology. IEEE Trans Electron Devices 57:952
33. Liu X, Chen KJ (2011) GaN Single-polarity power supply bootstrapped comparator for high temperature electronics. IEEE Electron Device Lett 32:27
34. Kwan AMH, Liu X, Chen KJ (2012) Integrated gate-protected HEMTs and mixed-signal functional blocks for GaN smart power ICs. Technical digest. IEEE International Electron Device Meeting, p. 7.3.1
35. Kwan AM, Guan Y, Liu X, Chen KJ (2014) A highly linear integrated temperature sensor on a GaN smart power IC platform. IEEE Trans Electron Devices 61:2970
36. Jiang Q, Tang Z, Liu C, Lu Y, Chen KJ (2014) Substrate-coupled cross-talk effects on an AlGaN/GaN-on-Si smart power IC platform. IEEE Trans Electron Devices 61:762
37. Wang H, Kwan AM, Jiang Q, Chen KJ (2015) A GaN pulse width modulation integrated circuit for GaN power converters. IEEE Trans Electron Devices 62:1143

Chapter 13
Drift Effects in GaN High-Voltage Power Transistors

Joachim Würfl

13.1 Introduction

Drift effects in semiconductors are changing electrical device properties in dependence on their electrical, thermal and radiation treatment. In a similar manner as degradation effects, they are adversely influencing device performance. However, in contrast to degradation, drift effects are fully recoverable. This means that device performance can be brought back to its initial property by certain treatments such as device biasing at specific conditions, light exposure by heating up the unbiased device or by combining these procedures. Of course, device drift can also be triggered by other mechanisms such as moisture incorporation or by adsorption of chemicals in the gate access regions of the device (chemical sensing properties). However, this will not be considered in this section as this can be controlled by packaging and passivation techniques and is not a property of the intrinsic power device.

The knowledge on device drift effects is imperative for predicting device performance in real system environment. Additionally, a proper identification of drift effects together with an exploration of the technological background allows for technological countermeasures towards a focused device improvement.

The book chapter is structured as follows: Sect. 13.2 introduces to physical mechanisms and dependencies. Section 13.3 then discusses the most important drift phenomena observed in GaN power switching transistors and their influence on device performance. Finally, Sect. 13.4 discusses proven technological concepts to minimize device drift.

J. Würfl (✉)
Ferdinand-Braun-Institut, Leibniz-Institut für Höchstfrequenztechnik,
Gustav-Kirchhoff-Strasse 4, 12489 Berlin, Germany
e-mail: Hans-Joachim.Wuerfl@fbh-berlin.de

M. Meneghini et al. (eds.), *Power GaN Devices*,
Power Electronics and Power Systems, DOI 10.1007/978-3-319-43199-4_13

Table 13.1 Overview on important dynamic effects

Type of effect	Description	Influence on device/system performance
Dynamic increase of on-state resistance R_{on}	• Dynamic device operation increases on-state resistance compared to its static value [1–3] • Degree of effect depends on articular bias and power switching condition	• Increase losses at on-state phase of power switching. • Reduction of overall switching efficiency • Possible burnout of device due to overheating
Threshold voltage shift	• Depending on device operation condition, the threshold voltage may shift as compared to its static value [4, 5]	• Changes drive characteristics of device • May lead to spurious turn-on of device
Kink effect	• Reduction of drain current in linear device region up to knee region [6]	• Leads to additional on-state losses

13.2 Drift Effects and Their Physical Mechanisms

13.2.1 Overview

Quite a multitude of drift effects are influencing the performance of power switching devices, reduce switching efficiency and can compromise reliability. Table 13.1 summarizes the most important reversible drift mechanisms.

Often the effects described above are correlated to each other resulting in a great variety of device parameter changes.

13.2.2 Basic Physical Understanding

Most of the drift effects observed in GaN power switching transistors relate to change of charge balance in the vicinity of the active device channel region as a consequence of device operation at certain bias conditions. The charge balance can be modified by trapping/de-trapping processes [7, 8] and/or by electrostatic depletion in the presence of carriers with extremely low mobility [2, 9]. The degree of trapping depends on the availability of trap states, their energy level and capture cross section, the electric field in the particular device region and the temperature. Electric field distribution and temperature strongly correlate with the mode of device operation and even depend on the specific mission profile of the whole electronic system [10]. Thus, voltage and current levels, timing of switching events and many more parameters finally determine device properties.

Figure 13.1 provides schematic cross sections of very simple Schottky-gated GaN transistors in order to explain the physical mechanism causing recoverable drift effects. The associated basic physical principals even hold for much more complicated device structures. At full on-state, an ideal trap-free GaN transistor basically operates at its maximum 2DEG electron concentration given by the combination of epitaxial GaN/AlGaN layers in the active region [11]. Figure 13.1a depicts this "ideal" situation where the current flow through the 2DEG region is not influenced by any trapping effects. Assumed that electron trapping occurs in the buffer region between gate and drain, additional negative charges will be present there. They disturb the balance to other charges such as the 2DEG electrons or surface/interface states. In order to maintain charge neutrality in this particular device region, they need to be compensated (Fig. 13.1b). Usually, excess negative charges are balanced by a reduction of 2DEG electron concentration in the vicinity of the traps. Hence, the turn-on resistance of the 2DEG region close to the traps increases locally. An on-state resistance increase ΔR_{on} would be measureable at the device terminals. Figure 13.8b depicts this situation. Trapping can take place at various places, in the epitaxial layers, at passivation semiconductor interfaces or even at interfaces between different passivation layers. If trapping occurs close to the drain access regions, the 2DEG electron concentration reduces and leads to an increase of R_{on}. It has to be mentioned that trapping may also happen in the source access region (not shown in Fig. 13.8b). In this case, the source access region would also contribute to an additional resistance increase of R_{on} [7, 8].

If electron trapping takes place just underneath the gate, the electron concentration in the 2DEG channel reduces in this region in order to maintain charge neutrality (Fig. 13.1c). In other words, this means that when pinching of the transistor fewer electrons have to be depleted to fully close the channel. This leads to a shift of threshold voltage in positive direction. Generally, charge trapping underneath the gate, irrespective from his physical origin, leads to a shift in threshold voltage [7, 8]—in positive direction if trapping causes a surplus of negative charges (see also Fig. 13.8b) and in negative direction if the opposite is the case.

In some cases, both mechanisms are combined leading to a shift in threshold voltage together with an increase in on-state resistance. Figure 13.1d depicts this particular situation. It occurs upon off-state biasing if electrons are injected form the drain side edge of the gate into interface states (donor traps) at the AlGaN/passivation and into the AlGaN barrier layer itself [6, 12]. Also hot electrons being present in the channel region close to the drain side edge of the gate upon device operation at semi-on-state condition (close to threshold voltage) may be scattered into trap locations close to the gate [13, 14]. In both cases, a positive shift of threshold voltage and a reduction of maximum drain current would be observed.

If electrons are injected in shallow interface states along the drain access region, the excess negative charges will initially deplete the channel, resulting in a reduced drain current at low drain voltage level. However, at higher drain voltage (higher electrical fields present), the electrons are emitted from the shallow traps resulting in a recovery of the effect and an increase of drain current [6] (see also Fig. 13.8c). This effect is called "Kink effect" and will be considered more in detail in Sect. 13.3.3.

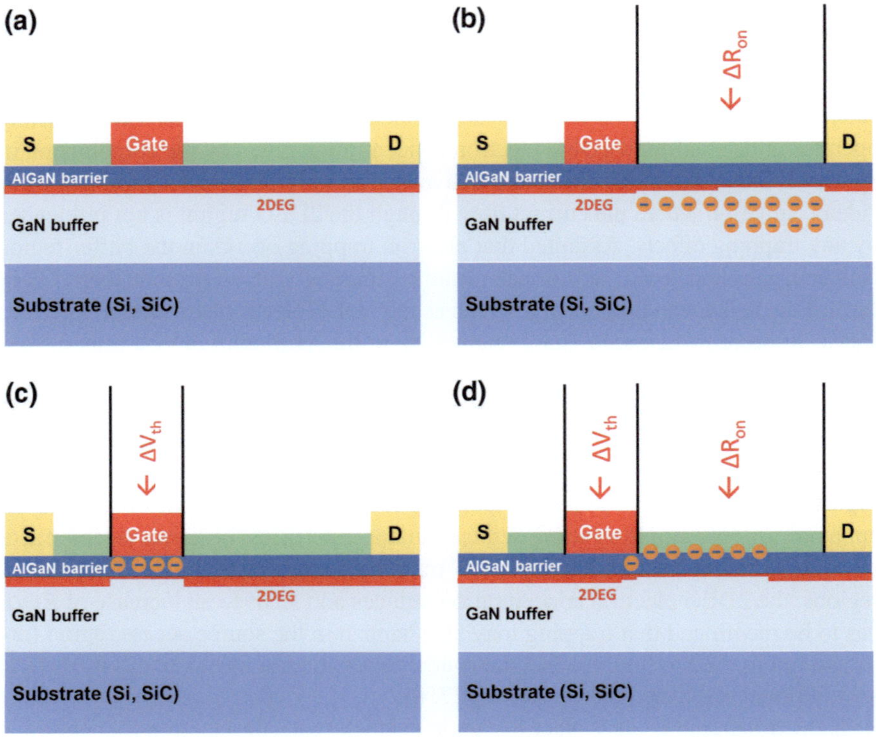

Fig. 13.1 Schematic cross section of an AlGaN/GaN HEMT at different bias and trapping conditions. For simplicity, only negative charge trapping is displayed. Trapping can be interpreted as a net change of charge equilibrium in certain device regions upon certain operation conditions. The additional charge has to be balanced immediately in order to maintain charge equilibrium of the whole system. Usually, balancing occurs by changing the electron population density of the 2DEG. This effect becomes noticeable after turning on the devices and then results in a change of on-state resistance. If negative charges are trapped in the vicinity of the source or drain access region, the 2DEG electron density will reduce resulting in an increase of on-state resistance. If trapping confines to regions underneath the gate, a positive threshold voltage shift occurs upon trapping of negative charges, and the threshold voltage can also shift in negative direction if positive charges are introduces (for example, if electrons are emitted from donor traps). **a** Ideal device without any trapping at on-state: The 2DEG is fully populated with electrons. **b** Electron trapping in the vicinity of the drain access region: This situation, for example, happens immediately after switching from off-state to on-state at high drain bias. Trapped electrons [8] (or depleted deep acceptors [9]) completely or partially deplete the 2DEG and hence impede current flow. If the source or the drain access regions are influenced by trapping, the on-state resistance changes. This gives rise to the so-called dynamic on-state resistance increase $\Delta R_{\mathrm{on\_dyn}}$. **c** Electron trapping in the vicinity of the gate (underneath the gate) results in a partial depletion of the channel just underneath the gate: As these charges aid depleting the channel, a lower gate voltage is necessary to fully turn off the device—the threshold voltage shifts to more positive values. **d** Electron trapping at the interface to passivation layer and in the barrier layer: This leads to both a threshold voltage shift and to the formation of a virtual gate at the interface oriented towards the drain $\longrightarrow$ reduction of 2DEG electron concentration, $R_{\mathrm{on\_dyn}}$ increase

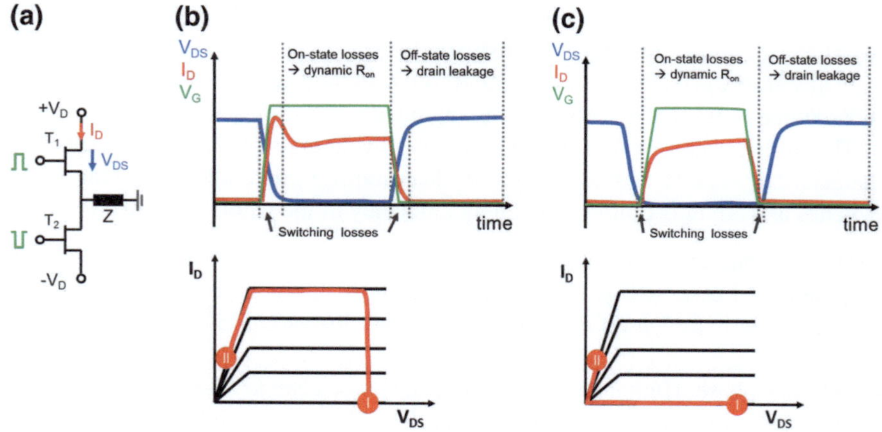

Fig. 13.2 Transient switching of power transistors in real system environment (see also [10]). **a** Example of half-bridge power switching circuit. The two transistors are switched in anti-phase against the load impedance *Z*. **b** Hard-switching device: waveforms and indication of type of losses generated during power switching (*top*) principal load (*bottom*). **c** Soft-switching device: waveforms and indication of type of losses generated during power switching (*top*) principal load (*bottom*)

13.2.3 Dependency on Device Operation Conditions

Practically all power conversion applications rely on continuous switching of power transistors from off-state to on-state and back. During each switching cycle, internal electric field distribution and device temperature change in a characteristic manner depend on device biasing and timing of switching. Figure 13.2 explains this in more detail by means of a half-bridge circuit, a quite common architecture of modern power switching systems [15]. The power transistors are connecting the load Z to the positive bias $+V_D$ and the negative bias $-V_D$ in an alternating (anti-phase) manner. During switching, the device bias point continuously alternates between conditions I and II.

In a hard-switching configuration, the devices turn on while high drain voltage is still present. Immediately after switching applying a positive gate pulse to the transistor, the internal drain/source capacitance discharges rapidly. This results in a fast increase of drain current to its maximum value. Afterwards, the drain current converges to the on-state bias point (II). Figure 13.2b shows the corresponding load line.

In contrast to hard switching, soft-switching configurations turn on only if the drain voltage of the device is negligibly low. This considerably minimizes device stressing. Soft switching can, for example, be realized in resonant configurations. Considering the circuit according to Fig. 13.2a, it can, for example, be assumed that power switches and load impedance Z form a resonant circuit. If the transistors are switched in a resonant manner such that the devices turn on only if the actual bias of the connecting point L is either $+V_D$ (T1 turns on) or $-V_D$

(T2 turns on), the devices operate according to the load line of Fig. 13.2c; switching losses are minimized considerably.

In any case, the load lines for hard and soft switching visualize the different operation device conditions which are stressing the devices differently.

The different phases of a power switching cycle are stressing the devices in a characteristic manner. Moreover, they are determining the overall losses of the power switches and are thus limiting switching efficiency of the whole system:

- On-state phase: High current density in the device, however electric fields are quite low. Losses are generated by current flow through the on-state resistance R_{on}; if this is increased dynamically, on-state losses can dominate the overall losses.
- Off-state phase: High internal electric fields, losses due to off-state drain leakage in combination with a high drain voltage. The high internal fields can give rise to charge trapping in the device which in turn results in a resistance increase at on-state.
- Switching phase: Losses depend on specific operation (hard or soft switching) and the duration of the switching event. During hard switching, both a high electric field and a high current density are present at the same time. This can be particularly stressing for the device.

In summary, drift effects depend on a quite complex interaction of device operation conditions with trapping mechanisms in the device itself.

13.3 Drift Phenomena in GaN Power Switching Transistors

13.3.1 Dynamic On-State Resistance (R_{on_dyn})

Upon power switching, the so-called dynamic on-state resistance R_{on_dyn} may be much higher than its static value R_{on}. This creates additional on-state losses and therefore compromises maximum switching efficiency [1, 2]. Figure 13.3a shows how the electric properties of switching transistors are changing if the devices are affected by an increase of on-state resistance upon dynamic operation. The particular example in Fig. 13.3b visualizes the output IV-characteristic of a GaN transistor switched to a certain bias point for a very short time after being operated at a long-term bias point for a much longer time. In the given system, the probe pulse duration has been 0.2 μs. Then, the system resides at the long-term bias point for 500 ms before starting the next probe cycle (duty cycle 1:2500). If the probing pulse time is much shorter than the trapping or de-trapping time constants, the traps are considered to be practically frozen. This means that the short probe pulses characterize the trapping state that is actually present in the long-term bias point. As switching devices are mainly switching from a high drain bias off-state bias point to an on-state point, this characterization provides a basic information on how the device might perform in system environment immediately after turn-on from off-state. According to

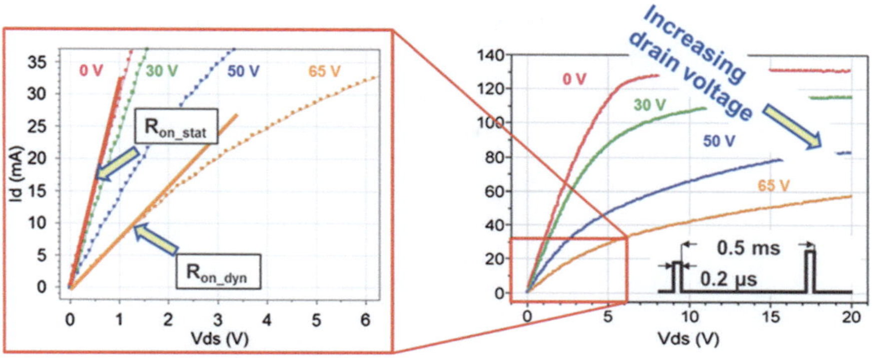

Fig. 13.3 Increase of dynamic on-state resistance upon device switching from a high drain voltage off-state bias point to any point of the output IV-characteristics, see also [16]

Fig. 13.3b, both the turn-on slope in the linear region of the IV-characteristic and the drain saturation current decrease if the device is switched from increasing drain bias levels. This effect is also referred to as “current collapse”. The turn-on slope is defined as the on-state-resistance R_{on}. It increases upon switching from a higher drain bias. In GaN power switching devices, the dynamic on-state resistance R_{on_dyn} equals to or is always higher as its static counterpart. It has to be mentioned that the particular device shown in Fig. 13.3 has been selected for demonstration purposes; it does by far not represent a modern GaN device optimized for high-voltage switching. Nowadays, the difference is practically negligible for perfectly optimized devices.

The dynamic increase of on-state resistance depends on specific device operation conditions and external influences such as:

- Duration and magnitude of off-state bias prior to switching [1–3, 8, 17, 18].
- Time delay of R_{on} measurement after switching event from off-state to on-state [8, 17, 18].
- Mode of switching in system environment (e.g. hard or soft switching) [10, 19].
- Operation temperature [7, 8, 20].
- Operation under ionizing radiation [21, 22].

13.3.1.1 Power Switching from Off-State Bias Point

As mentioned above, the dynamic on-state resistance R_{on_dyn} is caused by temporary charge trapping in device regions close to the active channel after switching from off-state, during the switching event or during operation at on-state. Continuous device switching is the standard mode of operation in an electronic power converter. Simulations according to Fig. 13.4 visualize the electric field distribution in a typical GaN HEMT device for power switching. During off-state biasing at a high drain voltage, high electric field peaks appear at the drain side edge of the gate, close to

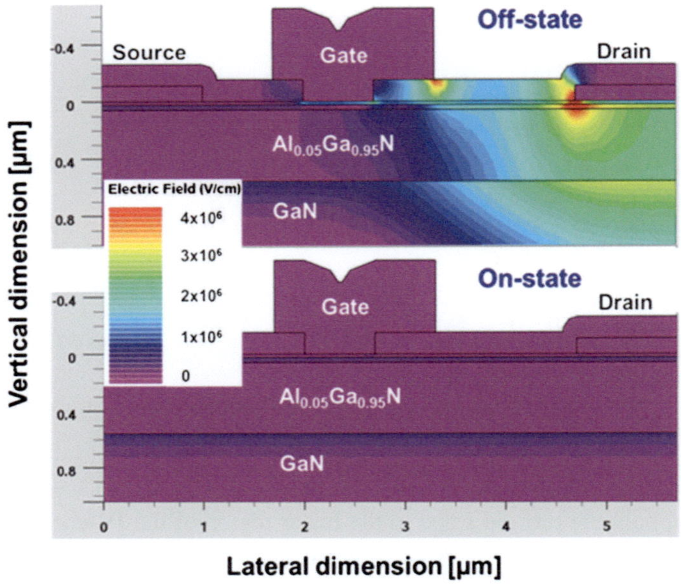

Fig. 13.4 Visualization of internal electric field distribution after switching a GaN power transistor from off-state to on-state condition (see also [2, 16]): Off-state condition: simulation of electric field distribution of a GaN HFET at reverse bias condition and at a drain voltage of 300 V. The 2DEG is fully depleted. On-state condition: Simulation of electric field distribution at steady-state on-state condition (drain voltage 0 V). The 2DEG is fully conductive in this case

the gate side edge of the drain and close to the drain side edge of the metallic gamma gate in the passivation layer. The field also deeply penetrates into the buffer. If compensating acceptors are assumed in the buffer (for example due to Fe- or C-compensation doping or due to other imperfections), the high electric field may lead to an electrostatic depletion of this region. Correspondingly, upon a high drain voltage the released holes would leave back a negatively charged space charge region in the buffer. If trap states are available, also additional electron trapping can take place. Of course, trapping is not confined to the buffer, it can also occur in the passivation layer at the drain side edge of the gate or at the interface between the AlGaN surface and the passivation layer. Finally, a trapping situation according to Fig. 13.1 may be present in the device. If the device is then switched from off-state to on-state, the electric field drops down immediately and traps are going to be de-trapped according to their own time constants. Additionally, holes electrostatically released from their acceptor states (space charge region) would slowly move back. Usually, de-trapping time constants are far larger than the turn-on time constants of the device. Therefore, as soon as the device starts to turn on, charge neutrality requires a reduced electron population of the 2DEG close to the trap-affected device regions. Hence, the 2DEG channel cannot be fully opened immediately, and electron transport is impeded until all parasitic charges are neutralized.

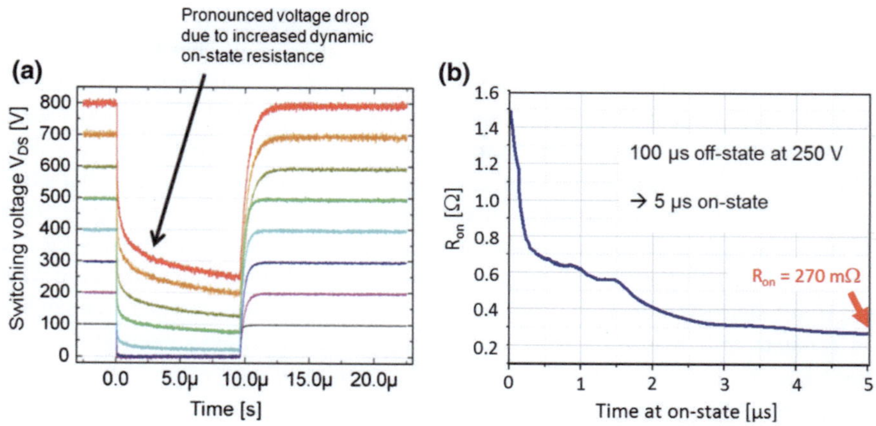

Fig. 13.5 Switching performance of GaN power devices affected by dynamic on-state resistance increase according to [2, 16]: **a** Example of dynamic switching properties of a GaN HFET showing considerable increase of dynamic on-state resistance after switching at drain bias voltages higher than 300 V. The device was fabricated on a heavily C-doped buffer structure and thus shows a very pronounced R_{on_dyn}. **b** Switching event form off-state bias at 250 V for a device with a static on-state resistance of the device is 100 mΩ: within the first 5 μs after switching, R_{on} decays to 270 mΩ in a quite complex manner with multiple time constants involved. A full recovery of R_{on} to its static value takes several seconds [2]

Figure 13.5 shows how the trapping mechanisms are influencing power switching. In an ideal case, the drain voltage should drop to a value given by the static turn-on resistance of the device immediately after turning on the transistor at the gate. According to Fig. 13.5, this is fairly the case after switching from lower drain voltages below 300 V. Switching from a higher drain voltage results in a pronounced voltage drop across the device immediately after the switching event occurs. This voltage drop is due to the dynamic increase of on-state resistance R_{on} on top of its static value ($R_{on\_dyn} = R_{on\_stat} + \Delta R_{on}$). The degree of voltage drop depends on both the drain voltage of the device just before switching and the time elapsed afterwards. R_{on_dyn} usually increases with operation voltage and can exceed the static value by more than two orders of magnitude [2, 23]. Not shown here is the fact that R_{on_dyn} also depends on the time of off-state biasing [24]. This means that for a fair benchmarking of dynamic on-state resistance data of devices from different vendors, all these boundary conditions have to be known and kept constant.

A more sophisticated study of time dependency of R_{on_dyn} unveils details of the decay as depicted in Fig. 13.5b. Immediately after switching from off-state, R_{on} rapidly decreases and then flattens out. It reaches its original value after a couple of seconds (not shown in the diagram). Interestingly, R_{on} does not really decrease exponentially. Moreover, measurements taken with higher time resolution show that R_{on_dyn} decays according to multiple exponential functions with several time constants. It is believed that this behaviour relates to different trap states contributing to the charging and discharging mechanism.

More detailed measurements in this regard have been taken by [3, 20, 24–26]. They have analysed the dependency of trapping on the duration of off-state bias stress. The drain current has been measured on a long timescale after a switching event (current transient measurements [27, 28]). Time constants between 10^{-5} s and 1000 s were covered by the measurements; therefore, it has been possible to identify fast and slow trapping centres and to attribute them to certain defects levels. For directly characterizing time dependence of dynamic on-state resistance, the devices were operated at off-state and repeatedly pulsed to the linear region of the IV-characteristics for R_{on} determination. This leads to a diagram depicting the on-state resistance increase with time of off-state stressing (Fig. 13.6). It can be clearly seen that even at a comparably low off-state drain bias of 45 V, the on-state resistance starts to increase significantly after a bias duration of about 100 s at room temperature.

Increasing the operation temperature not only leads to a higher static R_{on} due to electronic scattering in the channel but also shifts trapping time constants to shorter timescales; at 90 °C a steep increase of R_{on} after 1 s off-state stress could be observed.

This process is fully reversible: At de-trapping condition, the starting value of R_{on} is reached after more than 1000 s. The temperature dependency of trapping and de-trapping can be used for deriving capture cross sections and activation energies of the traps involved. In this case, deep acceptor traps with activation energy of 0.69 eV have been found, presumably in the buffer layer. This is supported by the fact that dynamic on-state resistance increase is practically negligible if the substrate is positively biased relative to the gate, drain and source electrodes at the front side of the wafer. Upon reverse biasing, the R_{on} increase has been even higher as compared to floating operation conditions.

The characterizations shown above provide a fingerprint on how particular device technologies might influence trapping properties. A variety of different time constants and activation energies for traps observed in GaN transistors have been compiled by [25]. The following more general statements can be given:

- The trapping time constants during off-state bias decrease with increasing drain voltage and with temperature [8, 20].
- Back-gating tests (not shown here) demonstrate that trapping responsible for dynamic R_{on} increase mainly occurs in the buffer, a negative substrate bias tends to increase trapping, whereas a positive bias has the opposite influence [24]. Especially, Fe^{+}- and C^{+}-doped buffer structures rely on deep acceptors which give rise to trapping effects during device operation [9, 26].
- The dynamic R_{on} increase often correlates with an acceptor type trap state at 0.63 eV above the valence band [26], which is related to point defects in the GaN buffer [8, 25]. Thus, material quality and pre-existing defects play a decisive role.
- Buffer trapping seems to correlate with the degree of vertical leakage [20] and gate leakage [29].

After long-term trapping experiments according to Fig. 13.6a, the on-state resistance increase saturates at certain values, see also [27]. This could be due to the fact that after long-term trapping all trap states are occupied. However, as the measurement sequence always involves a short de-trapping phase for characterizing R_{on} at

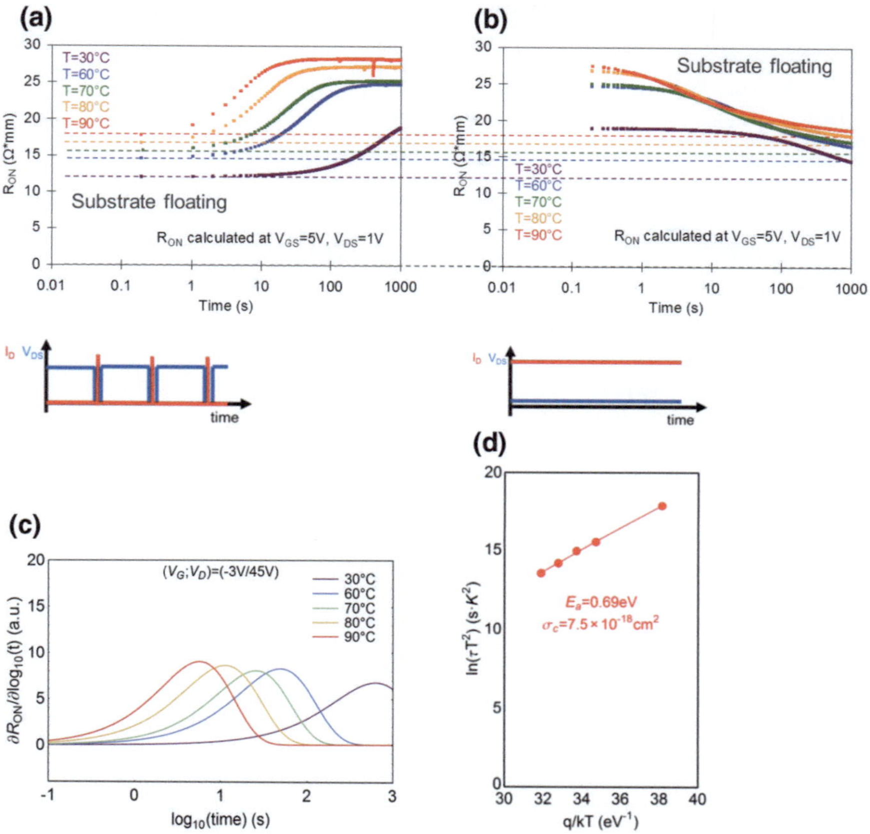

Fig. 13.6 On-state resistance evolution in dependence on different device operation conditions. The insets below the diagrams in (**a**) and (**b**) visualize the particular characterization sequence used. **a** Trapping condition: device operation at 45 V off-state drain bias, intermediate pulsed measurements of R_{on} at $V_{GS} = 5$ V; $V_{DS} = 1$ V for 50 ms. Temperature-dependent measurements from 30 to 90 °C. **b** De-trapping condition: device operation at $V_{GS} = 5$ V; $V_{DS} = 1$ V, intermediate pulsed measurements of R_{on} at $V_{GS} = 5$ V; $V_{DS} = 1$ V for 50 ms. Temperature-dependent measurements from 30 to 90 °C. **c** Derivative of trapping curve according to (**a**). **d** Determination of trap activation energy from temperature-dependent measurements in (**a**)

off-state, the saturation could also be related to a balancing effect as a result trapping at off-state and de-trapping during R_{on} measurements. In order to investigate this effect in more detail, another measurement sequence has been tested. This consisted of continuous trapping for a certain duration followed by full de-trapping and a subsequent continuation of the trapping experiment to the targeted time values (Fig. 13.7a and b). Quite a pronounced increase of R_{on} could be observed just after the initial measurement point indicating short-term trapping effects. After a limited increase or even slight reduction up to about 100 s, R_{on} continuously increases without getting into saturation up to a timescale of 1000 s. In any case, this effect is

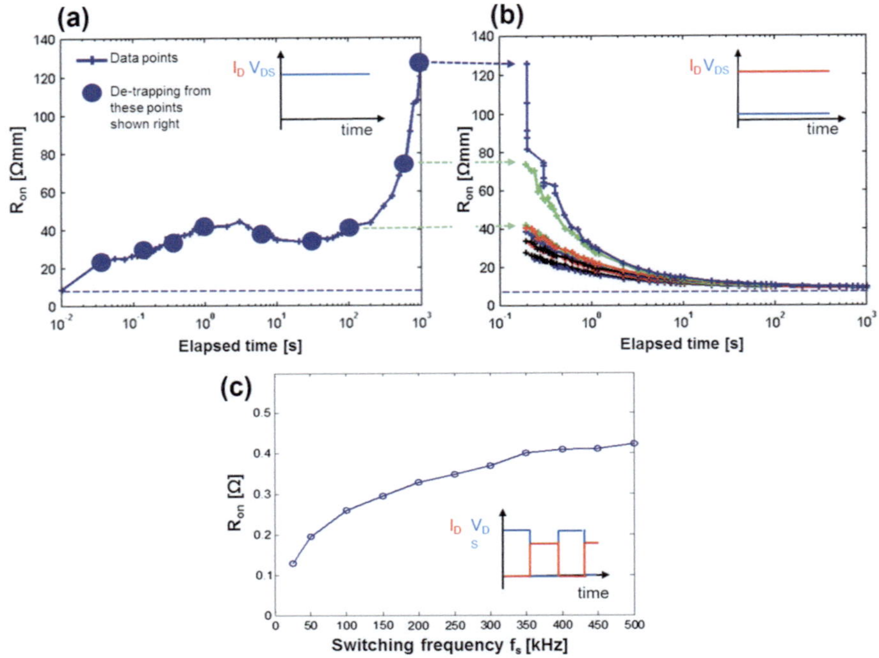

Fig. 13.7 Trapping experiment using continuous trapping at off-state and periodic switching. The insets visualize the respective measurement procedure. **a** Continuous trapping experiment at 200 V drain bias: after each measurement point, the device is fully de-trapped and then subjected to another trapping experiment with different durations. Measurement conditions: p-GaN gated normally off GaN HFET, 90 °C base-plate temperature. **b** De-trapping curves after selected off-state bias durations indicated by "*bullets*" in (**a**). De-trapping conditions: $V_{DS} = 1$ V and $V_{GS} = +1$ V, 90 °C base-plate temperature. **c** Change of R_{on} upon power switching at different frequencies up to 500 kHz and 80 V off-state drain bias. Measurement conditions: on-state drain current 4 A drain bias, duty cycle 0.5, measurements taken 250 ms after turn-on. Device under test: p-GaN gated normally off GaN HFET with 75 mΩ static on-state resistance

fully reversible, meaning that after each de-trapping cycle (Fig. 13.7b) R_{on} gets back to its initial value [30]. Similar measurements have also been taken by [18].

For transistors operating in real system environment, this behaviour means that turning off the device for a long time while a high drain voltage is present could result in a much higher on-state resistance after turning on this device again as compared to a situation where the device is turned on after a comparably short time at off-state. If R_{on} gets excessively high, this could even damage the device after first turn-on.

For power switching operation, the observed asymmetry in trapping and de-trapping time constants is of high importance and changes device performance in dependence on the specific operation conditions such as switching voltage, switching frequency and duty cycle of switching. According to Fig. 13.7c, the on-state resistance gradually increases with switching frequency. As thermal effects can be ruled out in these specific experiments, a mechanism related to balancing of trapping and de-trapping effects is proposed. Obviously, at a lower frequency de-trapping at on-state is more dominant leading to equilibrium at rather low switching frequencies. At higher frequencies, saturation of R_{on} can be observed.

13.3.1.2 Trapping Effects During On-State Operation

Negative charging may also happen at on-state conditions if electrons are captured in the channel region or are trapped in the buffer [28]. In this case, transistors face a reduction of maximum drain current (R_{on} increase). If trapping occurs underneath the gate, this effect might be accompanied by positive threshold voltage shift. Electron scattering into trap states near or underneath the gate is pronounced if devices are operated in a semi-on-state close to the threshold voltage. In this case, a high electric field at the drain side edge of the gate together with a comparably high electron density is present leading to a considerable acceleration of electrons across the high field region so that electrons are not able to thermalize there. These so-called hot electrons are quite energetic might scatter into trap states in the AlGaN barrier layer and into the buffer. It is even possible that hot electrons are creating additional trap states and thus cause permanent device degradation [13]. During hard-switching operation (see Fig. 13.2 and [10]), the transistor simultaneously faces high current and high drain voltage for a short amount of time—the ideal condition for hot electron generation. However, according to the experiments in [28] this condition not necessarily leads to significant R_{on} increase. It is believed that during the hard switching event impact ionization takes places additionally. Thus, additional holes are created which are basically balancing the trapped electrons. The investigations of [10] also show that soft switching causes much more pronounced trapping as compared to hard switching. This has been related to electron injection into the gate and drain access region during the off-state phase of soft switching.

It has to be mentioned here that not enough work on this topic has been performed and published so far. Therefore, it is not yet possible to set up a general rule on how GaN devices are reacting upon at hard- or soft-switching operation. However, above-mentioned results show that there definitely is a difference in performance which of course might be dependent on specific GaN device technology that has been tested in particular.

13.3.2 Threshold Voltage Shift

Threshold voltage shift relates to a temporary change of charge balance underneath or adjacent to the gate according to Fig. 13.1 c and d. A variety of physical effects are known leading to trapping effects in gate vicinity:

- Electron injection from Schottky gate structures under reverse gate bias [8, 28].
- Electron injection into gate isolator structures [5, 31].

- Hot electron induced trapping in the AlGaN barrier layer [13, 14].
- Interaction with ionizing irradiation [21, 22].

Any threshold voltage change also changes the driving condition of the device. Assumed that the turn-on gate voltage is kept constant, a threshold voltage shift in negative direction results in an enhanced transistor drive, whereas the opposite is true upon a positive shift of threshold voltage. In the former case, it may also happen that the transistor cannot be closed properly anymore as a higher negative gate drive voltage would be required. In system environment, this may increase sensitivity to spurious turn-on of the devices [32, 33]. In the latter case, at a given gate voltage, the turn-on resistance is reduced significantly which gives rise to higher switching losses. In any case, a threshold voltage shift needs to be avoided for reliable in system operation.

In order to better explain threshold voltage shift, Fig. 13.8 shows pulsed transfer characteristics of GaN transistors with Schottky gate. Three different conditions of pulsing are compared to each other: The first condition at both zero gate voltage and zero drain voltage ($V_{GS} = 0$ V and $V_{DS} = 0$ V) refers to the situation of the un-trapped device, the condition $V_{GS} = -7$ V and $V_{DS} = 0$ V refers to the so-called gate lag condition as the electric field is confined to the gate region only, and the situation $V_{GS} = -7$ V; $V_{DS} = 30$ V reflects the off-state condition at comparably high drain bias and provides high fields at the drain side edge of the gate or even close to the drain electrode. Pulse duration and duty cycle have been selected such that the traps are frozen at the long-term bias condition and the output IV-characteristic is only probed during the measurement pulses (see also Fig. 13.3). The transfer characteristics shown in Fig. 13.8a have been calculated from the output characteristic data at $V_{DS} = 10$ V.

The investigated devices did not show a remarkable gate lagging which means that the mechanism according to Fig. 13.8c is not very much pronounced here. In contrast, after applying drain lagging condition ($V_{GS} = -7$ V; $V_{DS} = 30$ V) a pronounced shift has been observed. As the applied field close to the gate region is much higher in this case, potentially more electrons can be trapped in the AlGaN barrier; they might even overcome the barrier and get trapped in the buffer layer region underneath the gate. These additional charges would explain the observed shift of threshold voltage.

13.3.3 Kink Effect

The Kink effect relates to a current collapse mainly in the low-voltage part of the output IV-characteristic, see Fig. 13.8c. It usually shows a strong hysteresis effect as its appearance depends on the direction how the drain voltage is scanned. Mainly the Kink effect is visible during a "forward" scan of the drain voltage [6]. If the IV-characteristic is measured in reverse orientation (e.g. ramping down from the maximum drain voltage at a given gate voltage), the Kink completely

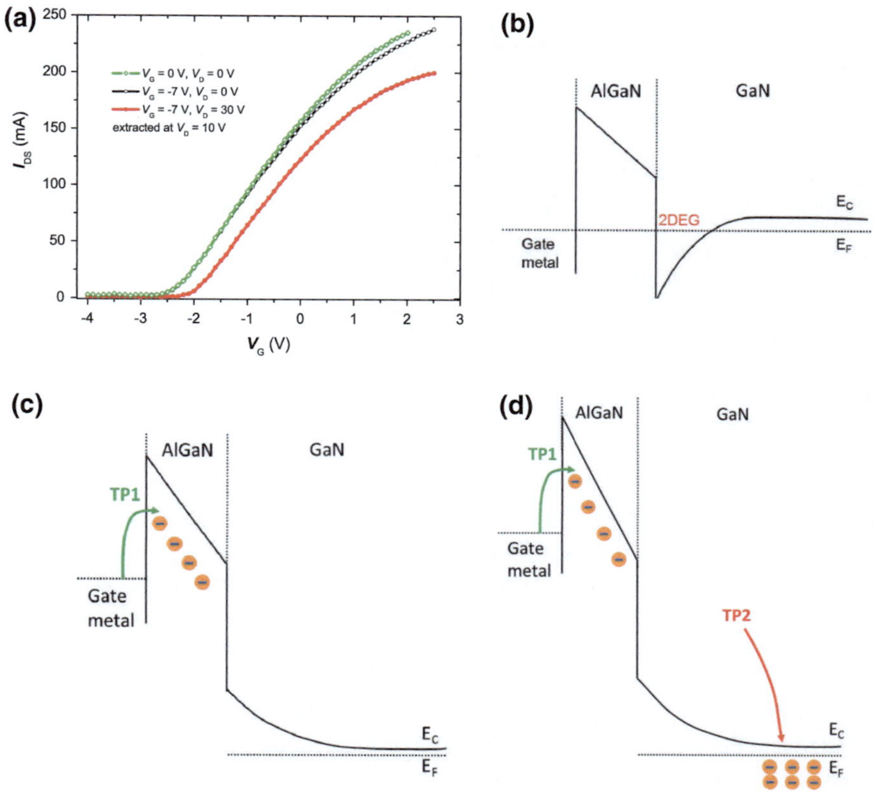

Fig. 13.8 Example and physical interpretation of threshold voltage shift and drain current degradation: **a** Pulse measurement of transfer characteristics starting on different drain and gate points. The transfer characteristics have been calculated from an output IV-characteristic measured with densely stacked branches of gate voltage levels. The *plot* represents the transfer characteristics at V_{DS} = 10 V. *Bias points* ($V_{GS} = V_{DS}$ = 0 V) ⟶ no device stressing; (V_{GS} = −7 V; V_{DS} = 0 V) ⟶ stressing of gate region; (V_{GS} = −7 V; V_{DS} = 30 V) ⟶ stressing of gate and drain region. **b** Energy band diagram of the completely de-trapped and unbiased device ($V_{GS} = V_{DS}$ = 0 V). **c** Energy band diagram at negative gate bias ($V_{GS} = V_{DS}$ = 0 V); electrons occupy states in the AlGaN layer. **d** Energy band diagram at negative gate bias and high drain bias (device turned-off by the gate) (V_{GS} = −7 V; V_{DS} = 30 V); electrons are trapped at energy states in the AlGaN layer, underneath the gate in the buffer layer and in the drain access region (not shown in the diagram)

disappears. Also the duration of I/V measurements influences the appearance of Kink effect. If the measurement of the output characteristics proceeds in a fast manner, a pronounced Kink effect has been observed; however, upon slow measurements (several seconds between measurements points) the Kink effect practically disappears [34].

The Kink effect usually relates to electron capture in trap or interface states in a region close to the drain side edge of the gate (Fig. 13.1d) [6] or in the buffer region Fig. 13.1b [35]. If electrons are trapped in a region extending from the gate to drain,

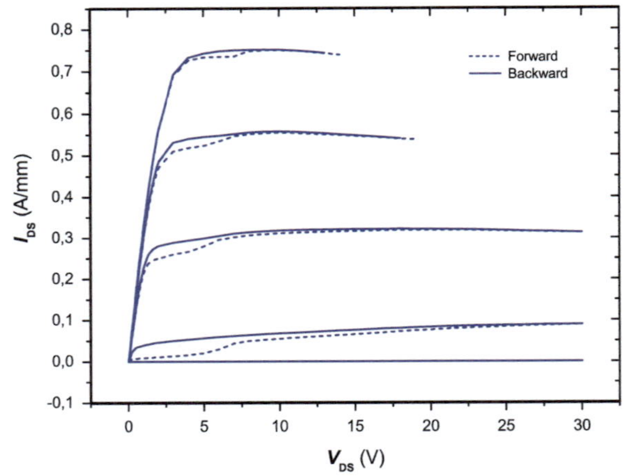

Fig. 13.9 Appearance of Kink effect in a GaN power switching device: during measurements of output characteristics, the devices show a pronounced dependence of drain current at low drain voltage from direction of IV-measurements. If drain voltage is scanned from zero to V_{DS_max}, a pronounced Kink can be observed at low drain voltage; the Kink disappears upon measurement with reversed directionality

a so-called virtual gate forms. Due to the negative charges associated with the virtual gate formation, the electron density in the 2DEG depletes and thus cuts down drain current. After emission of the trapped electrons, the virtual gate disappears and the channel conductivity fully recovers. At comparably low drain voltage, electron trapping in shallow states at surface, interface or other locations close to the channel is believed to be the dominating effect. If drain voltage or temperature increases electron emission from the trap states becomes the dominant mechanism, the traps are emptied and the drain current recovers to its original value. As possible mechanism of Kink recovery at higher drain voltage, the coexistence of holes created by weak impact ionization has been discussed by Kunihiro et al. [34]. They may compensate negatively charged trap states. Often Kink effect is also associated with threshold voltage shift [35]; in this case, regions underneath the gate electrode are additionally affected by trapping (Fig. 13.9).

13.4 Technological Countermeasures

Practically, all drift or current collapse effects are finally triggered by high electric fields in combination with pre-existing shallow- and deep-level trap states. Real degradation mechanisms are adding additional trap states to internal device regions and thus enhance drift effects.

Technological countermeasures aim at solving this root-cause problem and thus tackle the following issues:

- Understanding and improvement of epitaxial material especially in connection with built in deep trap levels. This especially relates to buffer structures containing significantly reduced deep trapping centres and by designing epitaxial layer stacks which are keeping deep trap centres off from the channel region [2, 26, 36–38].

- Reduction of critical electric field distribution in device: This can be done by optimizing epitaxial layer structure, passivation layers and field plate structures such that high field crowding, especially at the drain side edge of the gate, is reduced to a minimum [23, 39–46].
- Optimized processing especially in connection with the control of interface states and trap states in passivation or gate insulator layers [31, 47–51].

13.4.1 Optimized Epitaxial Buffer Design

Design and composition of buffer structure significantly influence dynamic properties. For high-voltage power switching devices, buffer structures have been developed that avoid punch-through effects of electrons in the channel region underneath the gate and thus increase device breakdown voltage. Furthermore, vertical breakdown through the buffer structure between the active device region and the substrate has to be prevented by optimizing thickness and composition of buffer. For avoiding punch-through effects, the buffer structure has to confine electrons to the channel region even at high drain bias by suitable potential barriers [37, 52–55]. Back-barrier buffer layers consisting of AlGaN with low Al mole fraction and compensated buffer structures relying on C+ or Fe+ doping have been applied [36, 52, 56, 57].

High values of breakdown strength of 170 V/μm have already been obtained after initial buffer optimizations towards high breakdown devices [36]. However, initially these devices were not useful for switching applications due to the high R_{on_dyn}. Therefore, different epitaxial buffer concepts were systematically compared to each other and optimized accordingly in order to find the best compromise between breakdown strength and dynamic on-state resistance increase. Figure 13.10 depicts this trade-off: The presence of deep carbon acceptors in the buffer clearly increases device (gate/drain) breakdown strength with increasing C-concentration, however, at the expense of an increasing R_{on_dyn}. According to Verzellesi et al. [58], the dynamic properties strongly depend of how carbon is incorporated in the GaN lattice. Depending on epitaxial growth technology, carbon can occupy auto-compensating donor/acceptor states on gallium and nitrogen sites, respectively, or it can be incorporated in deep acceptor levels. In the former case, nearly dispersion-free structures can be realized, whereas in the latter case dispersion effects with long-time constants have been observed. Fe-doped buffer layers and AlGaN back-barrier structures are another choice for reducing R_{on_dyn}; however, this can only be accomplished on the expense of specific breakdown strength which reduces significantly to only 40 or 50 V/μm [52, 54, 57]. A detailed overview on the impact of Fe and C buffer doping on dynamic effects in GaN transistors based on simulations has been performed in [9, 59]. A good compromise between breakdown strength and optimum dynamic properties can be obtained by combining AlGaN barrier and C-doping in such a way that the AlGaN barrier is placed close to the channel, whereas the C-doped buffer is placed in regions underneath which are not prone to large field changes during

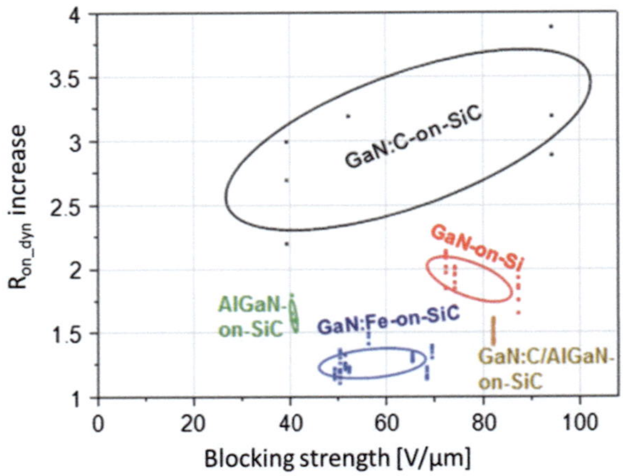

Fig. 13.10 Trade-off between dynamic on-state resistance increase and device breakdown strength in dependence on different technology tests for engineering an optimum high voltage and practically dispersion-free buffer layer [36, 57]. The measurements have been taken by pulsing from an off-state bias point at 65 V to on-state for 200 ns and then back to the original static bias point at a duty cycle of 1:2500

switching (denoted as "composite GaN:C" in Fig. 13.10). A breakdown strength of 80 V/µm and a R_{on_dyn} increase of only 10 % at 65 V switching are obtained with this combination. For 600 V operation, the increase of R_{on_dyn} has been reduced to 25 %.

13.4.2 Reduction of Electrical Field in Critical Device Regions

High electric fields are causing trapping effects and thus reduce device performance. In GaN HFETs, the highest electric fields appear in the vicinity of the gate especially at the gate edge towards the drain and, at higher drain voltage, also at the drain edge towards the gate (see also Fig. 13.4). Goal is to reduce the probability of electron trapping in the high field regions. Field plates in close vicinity to the gate or also drain field plates are solving these issues. A large variety of appropriate configurations have been discussed in the literature [39–41, 45, 60, 61]. Figure 13.11 shows a selection of possible field plate configurations which have been successfully demonstrated to spread out and more equally distribute electric fields in the vicinity of the gate and towards the drain region. They are thus reducing dispersion or current slump effects.

Field plates are well known in GaN microwave device technology, especially for S-band high-power transistors. They usually consist of an asymmetric gate with an extension of the gate wing to the drain side in combination with a single source-connected field plate overlapping the drain sided gate wing by a certain dimension. This is a proven very successful approach for drain bias levels of around 50 V. Higher drain voltages require a more sophisticated technique to spread the electric fields along the drain access region without getting field crowding in dedicated device regions. Figure 13.11 depicts a variety of technological possibilities. Often a staggered source-connected field plate with well-defined extensions towards the drain has been applied for GaN power switching

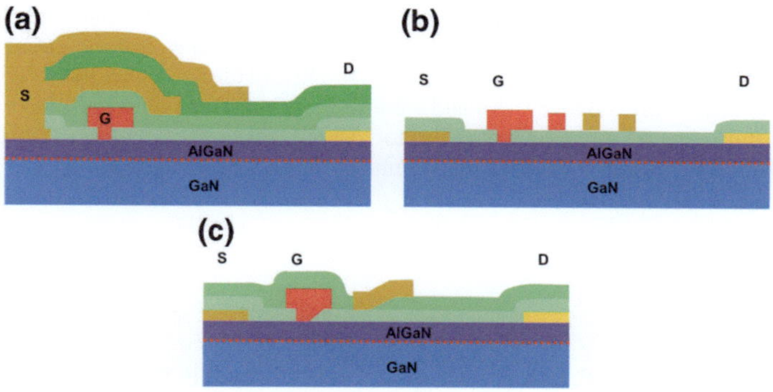

Fig. 13.11 Different field plate approaches for breakdown voltage enhancement and for reducing dispersive effects (according to [16]): **a** stacks of source-connected field plates in conjunction with an asymmetric T-gate structure; **b** multiple grading field plates which can be deliberately biased at different potentials; and **c** slanted field plate structures

devices [39] (Fig. 13.11a). The dimensions of the field plates, meaning the overlap of the drain extensions, as well as the thickness of passivation and metallization layers have to be taken into account, simulated and tested in order to realize an appropriate field distribution. In contrast to gate-connected field plates which are increasing gate to drain feedback capacitance C_{rss}, source-connected field plates add to the drain source capacitance (C_{oss}) which is less harmful in terms of device stability. Of course, source-connected field plates designed according to Fig. 13.11a are adding gate source capacitance. This can, for example, be reduced significantly by applying air-bridge-type field plate designs according to Xie et al. [60].

Multiple grating field plates (MGFPs) offer another very efficient possibility to spread out the electric field. As the drain access region of high-voltage GaN transistors has a structural dimension of typically 15–20 μm, it is comparably easy to place small metal lines in between and to connect them to either the gate or the source potential. This approach is shown in Fig. 13.11b where, for example, the MGFP close to the gate is connected to gate potential the others to source potential. A detailed analysis of MGPFs has been provided [41].

Slanted field plates avoid high field peeking at the termination of the field plate as the geometrical distance between the field plate and the channel region increases gradually. Very different structural modifications are possible. For example, as shown in Fig. 13.11c, the gate structure can be slanted at the drain side edge [42, 43, 46], and additionally, other field plate structures placed further towards the drain electrode may be slanted towards the top of an additional passivation layer to ensure a gradual fading out of the electric field [16].

Acknowledgments The author is very grateful for to the colleagues of the Ferdinand-Braun-Institut, Leibniz-Institut für Höchstfrequenztechnik, Berlin, Germany, for supporting this work. Special thanks are devoted to Dr. Olof Bengtsson, Dr. Oliver Hilt, Dr. Eldad Bahat-Treidel, Przemyslaw Kotara, Dr. Sergei Shevchenko and Rimma Zhytnytska for sharing results of their work and for fruitful scientific discussions and to Prof. Sibylle Dieckerhoff and Jan Böcker from the Technical University Berlin for special measurements on GaN power devices.

References

1. Moens P, Vanmeerbeek P, Banerjee S, Guo A, Liu C, Coppens P, Salih A, Tack M, Caesar M, Uren MJ, Kuball M, Meneghini M, Meneghesso G, Zanoni E (2015) On the impact of carbon-doping on the dynamic Ron and off-state leakage current of 650 V GaN power devices. In: 2015 Proceedings on IEEE 27th international symposium on power semiconductor devices and IC's (ISPSD), pp 37–40
2. Würfl J, Hilt O, Bahat-Treidel E, Zhytnytska R, Kotara P, Brunner F, Krueger O, Weyers M (2013) Techniques towards GaN power transistors with improved high voltage dynamic switching properties. In: 2013 Proceedings on IEEE international electron devices meeting (IEDM), pp 6.1.1–6.1.4
3. Meneghesso G, Meneghini M, Zanoni E, Vanmeerbeek P, Moens P (2015) Trapping induced parasitic effects in GaN-HEMT for power switching applications. In: 2015 Proceedings on international conference on ic design and technology (ICICDT), pp 1–4
4. DasGupta S, Sun M, Armstrong A, Kaplar RJ, Marinella MJ, Stanley JB, Atcitty S, Palacios T (2012) Slow detrapping transients due to gate and drain bias stress in high breakdown voltage AlGaN/GaN HEMTs. IEEE Trans Electron Devices 59(8):2115–2122
5. Lagger P, Ostermaier C, Pobegen G, Pogany D (2012) Towards understanding the origin of threshold voltage instability of AlGaN/GaN MIS-HEMTs. In: 2012 Proceedings on IEEE international electron devices meeting (IEDM), pp 13.1.1–13.1.4
6. Kaushik JK, Balakrishnan VR, Panwar BS, Muralidharan R (2013) On the origin of Kink effect in current-voltage characteristics of AlGaN/GaN high electron mobility transistors. IEEE Trans Electron Devices 60(10):3351–3357
7. Meneghini M, Zanoni E, Meneghesso G (2014) Gallium nitride based HEMTs for power applications: high field trapping issues. In: 2014 Proceedings on 12th IEEE international conference on solid-state and integrated circuit technology (ICSICT), pp 1–4
8. Meneghesso G, Meneghini M, Chini A, Verzellesi G, Zanoni E (2014) Trapping and high field related issues in GaN power HEMTs. In: 2014 IEEE international electron devices meeting (IEDM), 15–17 Dec 2014, pp 17.5.1, 17.5.4
9. Uren MJ, Möreke J, Kuball M (2012) Buffer design to minimize current collapse in GaN/AlGaN HFETs. IEEE Trans Electron Devices 59(12):3327–3333
10. Joh J, Tipirneni N, Pendharkar K, Krishnan S (2014) Current collapse in GaN heterojunction field effect transistors for high-voltage switching applications. In: 2014 Proceedings on IEEE international reliability physics symposium, pp 6C.5.1–6C.5.4
11. Ambacher O, Smart J, Shealy JR, Weimann NG, Chu K, Murphy M, Schaff WJ, Eastman LF, Dimitrov R, Wittmer L, Stutzmann M, Rieger W, Hilsenbeck J (1999) Two-dimensional electron gases induced by spontaneous and piezoelectric polarization charges in N- and Ga-face AlGaN/GaN heterostructures. J Appl Phys 85(6):3222–3233
12. Zanoni E, Meneghini M, Chini A, Marcon D, Meneghesso G (2013) AlGaN/GaN-based HEMTs failure physics and reliability: mechanisms affecting gate edge and Schottky junction. IEEE Trans Electron Devices 60(10):3119–3131
13. Meneghini M, Stocco A, Silvestri R, Ronchi N, Meneghesso G, Zanoni E (2012) Impact of hot electrons on the reliability of AlGaN/GaN high electron mobility transistors. In: 2012 IEEE international reliability physics symposium (IRPS), pp 2C.2.1–2C.2.5

14. Zanoni E, Meneghini M, Meneghesso G (2012) Hot electrons and time-to-breakdown induced degradation in AlGaN/GaN HEMTs. In: 2012 19th international conference on microwave radar and wireless communications (MIKON), vol 2, pp 593–598
15. Nomura T, Ishii S, Masuda M, Yoshida S, Yamate T, Sudo Y, Takeda J (2006) High temperature operation at 225 °C of a half bridge module using GaN HFETs. In: 2006 Proceeding on IEEE vehicle power and propulsion conference, pp 1–5
16. Würfl J (2016) GaN high-voltage power devices. In: Medjdoub F (ed) Gallium nitride (GaN): physics, devices, and technology. CRC Press, Boca Raton, pp 1–44
17. Würfl J, Hilt O, Bahat-Treidel E, Zhytnytska R, Kotara P, Krüger O, Brunner F, Weyers M (2013) Breakdown and dynamic effects in GaN power switching devices. Phys Stat Sol C 10(11):1393–1396
18. Wespel M, Dammann M, Polyakov V, Reiner R, Waltereit P, Weiss B, Quay R, Mikulla M, Ambacher O (2015) High-voltage stress time-dependent dispersion effects in AlGaN/GaN HEMTs. In: 2015 Proceedings on IEEE international reliability physics symposium (IRPS), pp CD.2.1–CD.2.5
19. Lu B, Palacios T, Risbud D, Bahl S, Anderson DI (2011) Extraction of dynamic on-resistance in GaN transistors: under soft- and hard-switching conditions. In: 2011 Proceedings IEEE compound semiconductor integrated circuit symposium (CSICS), pp 1–4
20. Meneghini M, Silvestri R, Dalcanale S, Bisi D, Zanoni E, Meneghesso G, Vanmeerbeek P, Banerjee A, Moens P (2015) Evidence for temperature-dependent buffer-induced trapping in GaN-on-silicon power transistors. In: 2015 Proceedings on IEEE international reliability physics symposium (IRPS), pp 2E.2.1–2E.2.6
21. Koehler AD, Anderson TJ, Weaver BD, Tadjer MJ, Hobart KD, Kub FJ (2013) Degradation of dynamic ON-resistance of AlGaN/GaN HEMTs under proton irradiation. In: 2013 Proceedings on IEEE workshop on wide bandgap power devices and applications (WiPDA), pp 112–114
22. Sasikumar A, Zhang Z, Kumar P, Zhang EX, Fleetwood DM, Schrimpf RD, Saunier P, Lee C, Ringel SA, Arehart AR (2015) Proton irradiation-induced traps causing VT instabilities and RF degradation in GaN HEMTs. In: 2015 Proceedings on IEEE international reliability physics symposium (IRPS), pp 2E.3.1–2E.3.6
23. Saito W, Nitta T, Kakiuchi Y, Saito Y, Tsuda K, Omura I, Yamaguchi M (2007) Suppression of dynamic on-resistance increase and gate charge measurements in high-voltage GaN-HEMTs with optimized field-plate structure. IEEE Trans Electron Devices 54(8):1825–1830
24. Bisi D, Meneghini M, Marino FA, Marcon D, Stoffels S, Van Hove M, Decoutere S, Meneghesso G, Zanoni E (2014) Kinetics of buffer-related RON-increase in GaN-on-silicon MIS-HEMTs. IEEE Electron Device Lett 35(10):1004–1006
25. Bisi D, Meneghini M, de Santi C, Chini A, Damman M, Brueckner P, Mikulla M, Meneghesso G, Zanoni E (2013) Deep-level characterization in GaN HEMTs-part I: advantages and limitations of drain current transient measurements. IEEE Trans Electron Devices 60(10):3166–3175
26. Meneghini M, Rossetto I, Bisi D, Stocco A, Cester A, Meneghesso G, Zanoni E, Chini A, Pantellini A, Lanzieri C (2014) Role of buffer doping and pre-existing trap states in the current collapse and degradation of AlGaN/GaN HEMTs. In: 2014 Proceedings on IEEE international reliability physics symposium, pp 6C.6.1–6C.6.7
27. Meneghini M, Rossetto I, Bisi D, Stocco A, Chini A, Pantellini A, Lanzieri C, Nanni A, Meneghesso G, Zanoni E (2014) Buffer traps in Fe-doped AlGaN/GaN HEMTs: investigation of the physical properties based on pulsed and transient measurements. IEEE Trans Electron Devices 61(12):4070–4077
28. Joh J, del Alamo JA (2011) A current-transient methodology for trap analysis for GaN high electron mobility transistors. IEEE Trans Electron Devices 58(1):132–140
29. Bisi D, Stocco A, Meneghini M, Rampazzo F, Cester A, Meneghesso G, Zanoni E (2014) Characterization of high-voltage charge-trapping effects in GaN-based power HEMTs. In: 2014 Proceedings on 44th European solid state device research conference (ESSDERC), pp 389–392

30. Würfl J, Troppenz M, Hilt O, Bahat-Treidel E, Badawi N, Böcker J, Dieckerhoff S (2015) Dynamics of drift effects in GaN power switching transistors. In: Proceedings WOCSDICE 2015, pp 61–64
31. Lagger P, Reiner M, Pogany D, Ostermaier C (2014) Comprehensive study of the complex dynamics of forward bias-induced threshold voltage drifts in GaN based MIS-HEMTs by stress/recovery experiments. IEEE Trans Electron Devices 61(4):1022–1030
32. Wang J, Chung HS-H (2013) New insight into the mechanism of the spurious triggering pulse in the bridge-leg configuration. In: 2013 Proceedings on 5th international conference on power electronics systems and applications (PESA), pp 1–6
33. Wang J, Chung HS-H (2015) A novel RCD level shifter for elimination of spurious turn-on in the bridge-leg configuration. IEEE Trans Power Electron 30(2):976–984
34. Kunihiro K, Kasahara K, Takahashi Y, Ohno Y (1999) Experimental evaluation of impact ionization coefficients in GaN. IEEE Electron Device Lett 20(12):608–610
35. Meneghesso G, Zanon F, Uren MJ, Zanoni E (2009) Anomalous Kink effect in GaN high electron mobility transistors. IEEE Electron Device Lett 30(2):100–102
36. Hilt O, Bahat-Treidel E, Cho E, Singwald S, Würfl J (2012) Impact of buffer composition on the dynamic on-state resistance of high-voltage AlGaN/GaN HFETs. In: 2012 Proceedings on 24th international symposium on power semiconductor devices and ICs (ISPSD), pp 345–348
37. Bahat-Treidel E (2012) GaN based HEMTs for high voltage operation; design, technology and characterization. Ph.D., Technical University Berlin, Germany
38. Ando Y, Takenaka I, Takahashi H, Sasaoka C (2015) Correlation between epitaxial layer quality and drain current stability of GaN/AlGaN/GaN heterostructure field-effect transistors. IEEE Trans Electron Devices 62(5):1440–1447
39. Huili X, Dora Y, Chini A, Heikman S, Keller S, Mishra U (2004) High breakdown voltage AlGaN-GaN HEMTs achieved by multiple field plates. IEEE Electron Device Lett 25(4): 161–163
40. Dora Y, Chakraborty A, McCarthy L, Keller S, DenBaars SP, Mishra U (2006) High breakdown voltage achieved on AlGaN/GaN HEMTs with integrated slant field plates. IEEE Electron Device Lett 27(9):713–715
41. Bahat-Treidel E, Hilt O, Brunner F, Sidorov V, Würfl J, Tränkle G (2010) AlGaN/GaN/AlGaN DH-HEMTs breakdown voltage enhancement using multiple grating field plates. IEEE Trans Electron Devices 57(6):1208–1216
42. Li Z, Chu R, Zehnder D, Khalil S, Chen M, Chen X, Boutros K (2014) Improvement of the dynamic on-resistance characteristics of GaN-on-Si power transistors with a sloped field-plate. In: 2014 Proceedings on 72nd annual device research conference (DRC), pp 257–258
43. Chiu H-C, Yang C-W, Wang H-C, Huang F-H, Kao H-L, Chien F-T (2013) Characteristics of AlGaN/GaN HEMTs with various field-plate and gate-to-drain extensions. IEEE Trans Electron Devices 60(11):3877–3882
44. Xie G, Xu E, Lee J, Hashemi N, Ng WT, Zhang B, Fu FY (2012) Breakdown voltage enhancement for power AlGaN/GaN HEMTs with air-bridge field plate. In: 2012 Proceedings on 24th international symposium on power semiconductor devices and ICs (ISPSD), pp 337–340
45. Li Z, Chu R, Zehnder D, Khalil S, Chen M, Chen X, Boutros K (2014) Improvement of the dynamic on-resistance characteristics of GaN-on-Si power transistors with a sloped field-plate. In: 2014 Proceedings of 72nd annual device research conference (DRC), pp 257–258
46. Kuzuhara M, Tokuda H (2015) Low-loss and high-voltage III-nitride transistors for power switching applications. IEEE Trans Electron Devices 62(2):405–413
47. Lin S, Wang M, Xie B, Wen C, Min P, Yu Y, Wang J, Hao Y, Wu W, Huang S, Chen KJ, Shen B (2015) Reduction of current collapse in GaN high-electron mobility transistors using a repeated ozone oxidation and wet surface treatment. IEEE Electron Device Lett 36(8):757–759
48. Chevtchenko SA, Kurpas P, Chaturvedi N, Lossy R, Würfl J (2011) Investigation and reduction of leakage current associated with gate encapsulation by SiN_x in AlGaN/GaN HFETs. In: 2011 Proceedings international conference on compound semiconductor manufacturing technology (CS ManTech 2011), pp 237–240

49. Choi W, Ryu H, Jeon N, Lee M, Cha H-Y, Seo K-S (2014) Improvement of *V*th instability in normally-off GaN MIS-HEMTs employing PEALD-SiNx as an interfacial layer. IEEE Electron Device Lett 35(1):30–32
50. Liu ZH, Ng GI, Zhou H, Arulkumaran S, Maung YKT (2011) Reduced surface leakage current and trapping effects in AlGaN/GaN high electron mobility transistors on silicon with SiN/ Al_2O_3 passivation. Appl Phys Lett 98(11):113506
51. Lee Y-C, Kao T-T, Merola JJ, Shen S-C (2014) A remote-oxygen-plasma surface treatment technique for III-nitride heterojunction field-effect transistors. IEEE Trans Electron Devices 61(2):493–497
52. Würfl J, Bahat-Treidel E, Brunner F, Cho M, Hilt O, Knauer A, Kotara P, Weyers M, Zhytnytska R (2012) Device breakdown and dynamic effects in GaN power switching devices: dependencies on material properties and device design. ECS Trans 50(3):211–222
53. Würfl J, Hilt O, Bahat-Treidel E, Zhytnytska R, Klein K, Kotara P, Brunner F, Knauer A, Krüger O, Weyers M, Tränkle G (2013) Technological approaches towards high voltage, fast switching GaN power transistors. ECS Trans 52(1):979–989
54. Bahat-Treidel E, Brunner F, Hilt O, Cho M, Würfl J, Tränkle G (2010) AlGaN/GaN/GaN: C back-barrier HFETs with breakdown voltage of over 1 kV and low RON × A. IEEE Trans Electron Devices 57(11):3050–3057
55. Bahat-Treidel E, Hilt O, Brunner F, Würfl J, Tränkle G (2010) Punchthrough-voltage enhancement of AlGaN/GaN HEMTs using AlGaN double-heterojunction confinement. IEEE Trans Electron Devices 55(12):3354–3359
56. Uren MJ, Nash KJ, Balmer RS, Martin T, Morvan E, Caillas N, Delage SL, Ducatteau D, Grimbert B, De Jaeger JC (2006) Punch-through in short-channel AlGaN/GaN HFETs. IEEE Trans Electron Devices 53(2):395–398
57. Würfl J, Bahat-Treidel E, Brunner F, Cho M, Hilt O, Knauer A, Kotara P, Weyers M, Zhytnytska R (2012) Device breakdown and dynamic effects in GaN power switching devices: dependencies on material properties and device design. ECS Trans 41(8):127–138
58. Verzellesi G, Morassi L, Meneghesso G, Meneghini M, Zanoni E, Pozzovivo G, Lavanga S, Detzel T, Haberlen O, Curatola G (2014) Influence of buffer carbon doping on pulse and AC behavior of insulated-gate field-plated power AlGaN/GaN HEMTs. IEEE Electron Device Lett 35(4):443–445
59. Uren MJ, Silvestri M, Casar M, Hurkx GAM, Croon JA, Sonsky J, Kuball M (2014) Intentionally carbon-doped AlGaN/GaN HEMTs: necessity for vertical leakage paths. IEEE Electron Device Lett 35(3):327–329
60. Xie G, Xu E, Lee J, Hashemi N, Ng WT, Zhang B, Fu FY (2012) Breakdown voltage enhancement for power AlGaN/GaN HEMTs with air-bridge field plate. In: 2012 24th international symposium on power semiconductor devices and ICs (ISPSD), pp 337–340
61. Petru A (2010) Breakdown voltage enhancement in lateral AlGaN/GaN heterojunction FETs with multiple field plates. In: 2010 10th IEEE international conference on solid-state and integrated circuit technology (ICSICT), pp 1344, 1346

Chapter 14
Reliability Aspects of 650-V-Rated GaN Power Devices

Peter Moens, Aurore Constant and Abhishek Banerjee

14.1 Introduction

GaN devices are promising candidates for the next-generation power devices for energy-efficient applications. Although astounding performance is already proven by many research papers, the widespread adoption of GaN power devices in the market is still hampered by the following: (1) yield and reproducibility; (2) cost; and (3) reliability. All three factors are to be considered, but to convince customers to adopt GaN power devices in their next-generation products, proven device and product reliability is a must.

This chapter will focus on the main intrinsic reliability mechanisms for GaN power devices. It will cover gate ohmic contact reliability, gate dielectric reliability, and buffer stack reliability. The need to do reliability investigations based on statistical data on large-area power transistors (100^+ mm gate width) instead of small test structures will be emphasized. Acceleration models and statistical distribution models (Weibull) are discussed.

14.2 Reliability of Au-Free Ohmic Contacts

14.2.1 Introduction to Ohmic Contact Reliability

Due to the high electric field and self-heating generated in the drain area, degradation of the ohmic contacts can affect device lifetime [1]. Au-based contacts have proven to be a significant reliability issue at elevated temperatures in GaAs- and InGaAs-based

P. Moens (✉) · A. Constant · A. Banerjee
ON Semiconductor, Oudenaarde, Belgium
e-mail: Peter.Moens@onsemi.com

M. Meneghini et al. (eds.), *Power GaN Devices*,
Power Electronics and Power Systems, DOI 10.1007/978-3-319-43199-4_14

HEMTs [2]. In AlGaN-/GaN-based HEMTs, ohmic contacts with standard Au-based metallization seem to warrant sufficient stability under elevated temperature life testing to a certain limit. Ti/Al/Pt/Au ohmic contacts subjected to step stress of 48 h have shown marginal degradation up to 300 °C (estimated) junction temperature. Beyond 300 °C, self-heating results in poor device performance through ohmic contact degradation [3].

Similar degradation has been observed after thermal storage tests that resulted in junction temperatures equivalent to the electrical stress tests [4]. Instabilities in Ti/Al/Ni/Au-based ohmic contacts have been reported after aging tests over 2000 h at temperatures up to 340 °C. An increase of the contact resistance has been observed at 340 °C as well as surface roughness due to growth of Au-rich grains creating cracks in the passivation layer. Two degradation mechanisms have been identified: Au interdiffusion within the metal layers and Ga out-diffusion from AlGaN into the metal layer at elevated temperatures. The use of refractory metals and barriers in the metallization has demonstrated to be beneficial for reliability by preventing interdiffusion. Indeed, Ti/WSiN/Au ohmic contacts have shown to be fairly stable up to 800 °C during temperature stress with 20-min step duration [5]. Also Ti/Al/Ti/Au ohmic contacts with a WSiN barrier layer have shown marginal degradation after 120 h thermal stress up to 500 °C [6–8].

Although large-scale production of AlGaN/GaN HEMT devices in silicon CMOS fabs requires Au-free processes, Au-based metallization is the most widely used in reliability studies on ohmic contacts. The stability in Au-free ohmic contacts subject to electrical stress has not been thoroughly studied and the failure mechanism inherent to the metallization scheme is still not well known. Very few reports have been published on the robustness and reliability of Au-free ohmic contacts [9, 10].

14.2.2 Au-Free Ohmic Contacts Processing

Au-free ohmic contacts were processed on undoped AlGaN/GaN heterostructures grown on 6-in. (111) Si wafers, using a CMOS production line [10]. A high-quality in situ grown SiN is used as a capping layer. The contact width W and length L_c are 100 μm, and the contact spacing L is ranging from 2 to 18 μm. The sheet resistance of the 2DEG (R_{DEG}) as measured from TLM and Van der Pauw structures is ~450 Ω/sq. The mobility of electrons in the channel and the 2DEG density are ~1750 cm^2/V s and ~9×10^{12} cm^{-2}, respectively. No AlGaN barrier recess has been performed before metallization of the ohmic contacts. The contact formation has been done by sputter deposition of Ti, $AlCu_{0.5\,\%}$ and TiN layers, annealed under N_2 ambient in a rapid thermal processing furnace. The typical contact resistance R_c values are below 0.7 Ω mm at room temperature.

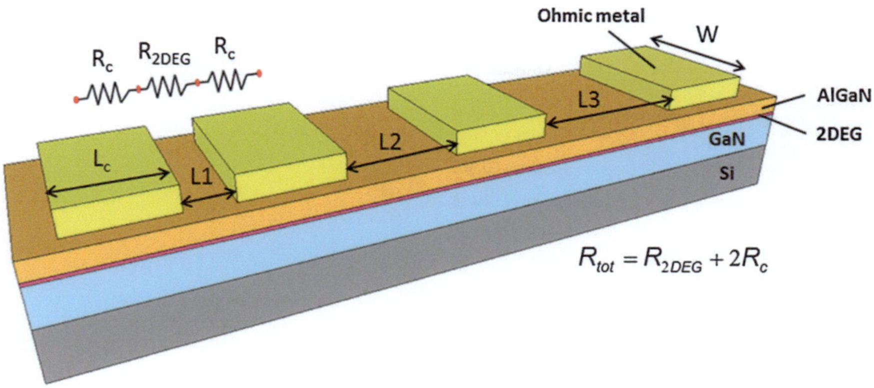

Fig. 14.1 Schematic of the TLM structure on AlGaN/GaN with various contact spacings (L_1, L_2, L_3 …). Resistances and other dimensions are as defined in the text

14.2.3 Stressing and Measurement Procedure

The reliability of Au-free ohmic contacts under electrical and temperature stress is evaluated using transmission line method (TLM) structures, see Fig. 14.1. A typical *I-V* characteristic measured at room temperature is shown in Fig. 14.2. As can be noted from Fig. 14.2, the electrical characteristics of the TLM structures vary with the contact spacing, and the saturation current increases with increasing contact spacing. In order to have a measurable degradation in a reasonable amount of stress time (typically 10^4 s), the TLM structures have to be stressed in saturation. Since the saturation current is different for each TLM spacing, it is a good practice to stress the structures at constant power ($V_{stress} \times I_{stress}$ kept constant) [10]. The saturation current is monitored and the stress voltage is continuously adapted over time to maintain a constant power during stress measurement. The total resistance is extracted in linear regime, the measurement of which induces minimal degradation due to the much lower voltage/power compared to the stress conditions. The total resistance is the 2DEG resistance plus the resistances from the two contacts $R_{2DEG} + 2\,R_c$ (Fig. 14.1). The 2DEG resistance is dependent on the contact spacing, whereas the contact resistance is not.

An intrinsic problem when evaluating the reliability of ohmic contacts is that under DC stress, the degradation exhibits both a recoverable as well as a permanent degradation component [10, 11]. The recoverable degradation component is attributed to surface depletion between the contacts due to charge trapping at the SiN/AlGaN interface [12, 13], whereas the permanent degradation is believed to be located at the edge of the metal/AlGaN interface and due to the self-heating of the structure [14]. Both effects are thermally accelerated [11]. To discriminate between permanent and recoverable degradation, a recovery step is performed after each stress step, resulting in nearly complete detrapping of

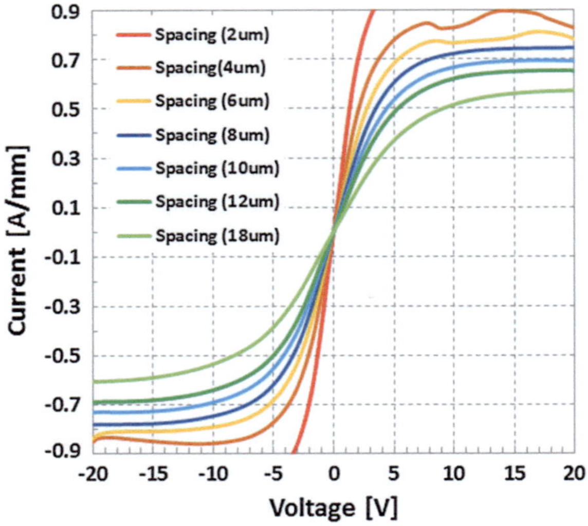

Fig. 14.2 Electrical characteristics of the TLM structures measured at room temperature for different spacing ranging from 2 to 18 μm

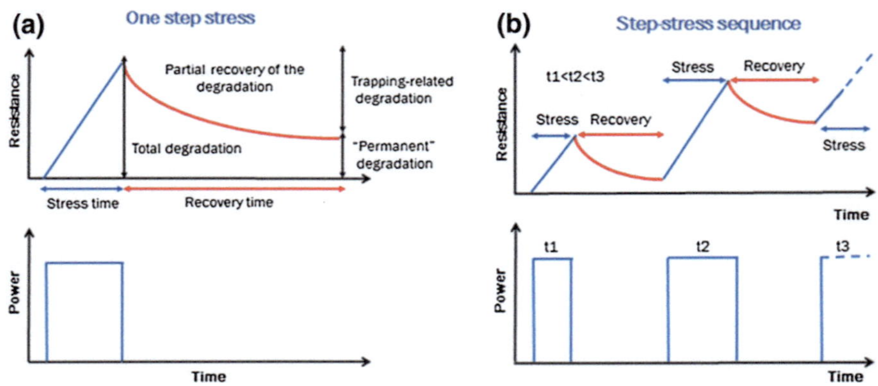

Fig. 14.3 Schematic of resistance and power evolution to access the "permanent" degradation. **a** One step stress and recovery sequence; **b** Multiple sequences step stress and recovery sequences

surface interface-related trapping. See Fig. 14.3 for the stress-recovery sequence. Recovery time varies from 10^4 s at 30 °C up to 200 s at 200 °C independently of the stress power and contact spacing. The final resistance change is defined as "permanent" degradation, free of charge trapping influence. Typical results are shown in Fig. 14.4.

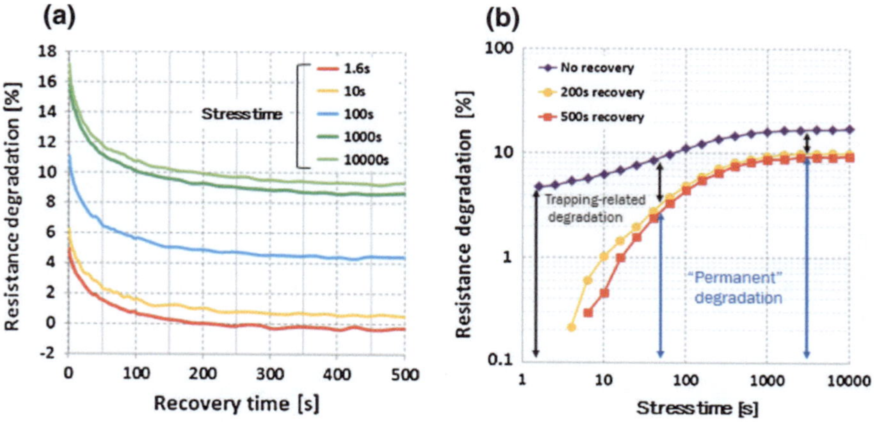

Fig. 14.4 **a** Time evolution of the total resistance during recovery time as a function of stress step time during 1.25 W constant power stress. **b** Corresponding time evolution of the total resistance during stress time as a function of recovery time (at 125 °C for 6 μm contact spacing)

14.2.4 Reliability Evaluation of Au-Free Ohmic Contacts

14.2.4.1 Degradation as a Function of Contact Spacing

Figure 14.5a shows the time evolution of the permanent total resistance measured during a 1.55-W constant power stress at 125 °C on TLMs with 6, 10, and 18 μm spacings. The permanent resistance degradation features a fast degradation with stress time, followed by a hard saturation around 10 %. TLMs with larger spacings just shift the degradation curve to longer stress times, but the overall shape of the curve is unaltered. Figure 14.5b shows that the log(t_{fail}) (arbitrarily chosen as the time to reach 2 % of permanent total resistance change) decreases with the contact spacing, for given power stress.

14.2.4.2 Degradation as a Function of Stress Power

Figure 14.6a shows the time evolution of the permanent total resistance degradation measured for different constant power stresses at 125 °C on TLMs with 18 μm spacing. The time evolution of the degradation is identical for all stress powers, and higher stress powers shift the curves to lower time. Again all the degradation curves saturate around 10 %. Although the saturation current is similar for the different powers, TLMs operate at higher voltage for higher power stress (Fig. 14.2). This is a strong evidence for voltage and not saturation current as the dominant degradation driver during stress of TLMs with a given spacing. Figure 14.6b shows the log(t_{fail}) as a function of stress power. The data are fitted with two models: an exponential model (Eq. 14.1) and a power law model (Eq. 14.2).

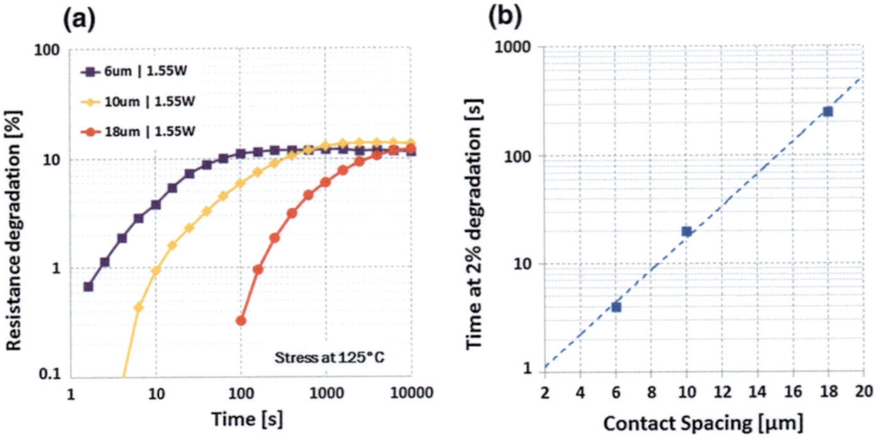

Fig. 14.5 **a** Time evolution of the permanent total resistance degradation during 1.55 W constant power stress at 125 °C for 6, 10, and 18 µm contact spacings. **b** Time extracted at 2 % degradation as a function of contact spacing

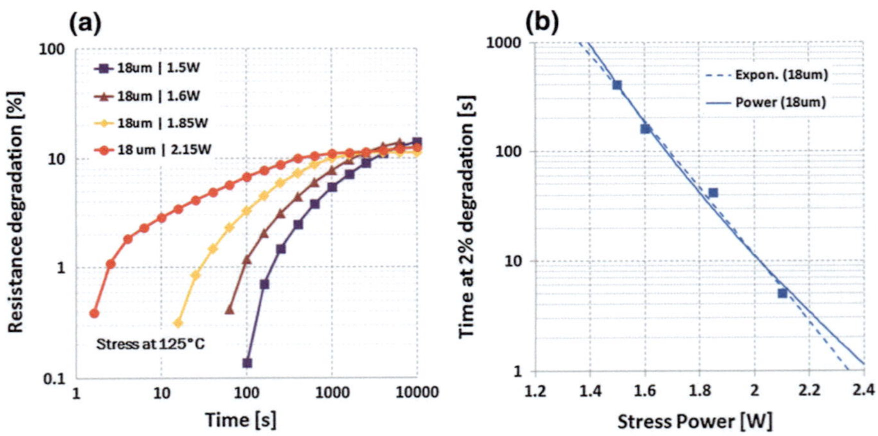

Fig. 14.6 **a** Time evolution of the permanent total resistance degradation as a function of constant power stress measured at 125 °C on 18 µm contact spacing. **b** Time extracted at 2 % degradation as a function of constant stress power. Data are fitted by means of power law (*continuous line*) and of exponential law (*dashed line*)

$$t_{\mathrm{fail}}(P) = t_0 \mathrm{e}^{-\gamma .P} \tag{14.1}$$

$$t_{\mathrm{fail}}(P) = k_0 P^{-n} \tag{14.2}$$

where t_{fail} is the time-to-failure; t_0, k_0, γ; and n are constants; and P is the stress power.

To make a lifetime prediction of the permanent resistance degradation as a function of contact spacing and stress power, time-to-failure has been extracted at 10 %

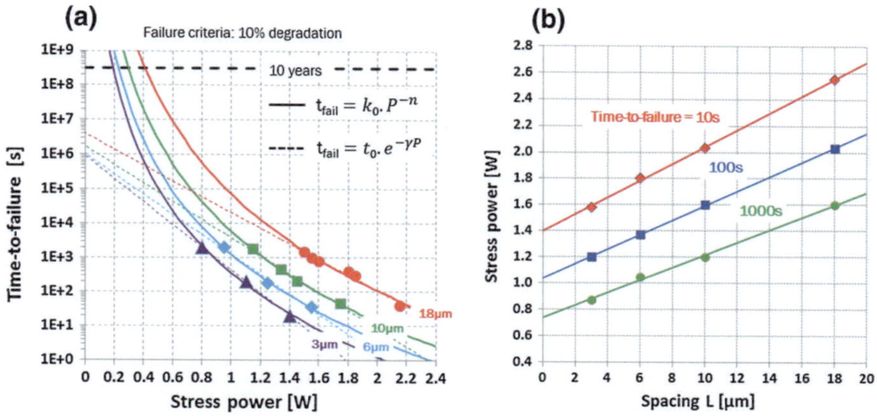

Fig. 14.7 **a** Evolution of time-to-failure extracted at 10 % degradation as a function of constant stress power at 125 °C for different contact spacing. Data are fitted by means of power law (*continuous line*) and of exponential law (*dashed line*). **b** Stress power as a function of contact spacing for extrapolated time-to-failures 10, 100, and 1000 s

degradation as failure criteria for all stress conditions. The time to reach a predefined failure criterion such as a 10 % change in permanent resistance is plotted in Fig. 14.7. The data yields an estimate of the mean time-to-failure for each stress power and contact spacing. Lifetime extrapolation is based on the fitting of both exponential and power law models to the experimental data. Both models fit the accelerated data well. However, the exponential model predicts a finite time-to-failure with zero power-applied stress condition, which is unrealistic. Therefore, the power law model is estimated to be closer to reality. As such it is possible to extrapolate time-to-failure toward lower power, and an operating power of 0.4 W at 125 °C guarantees 10 years of lifetime at the 10 % level considering 18 μm spacing.

14.2.4.3 Temperature Dependence and Activation Energy

The interacting effects of temperature and power during stress have been studied for 6 μm spacing stressed at 1.25 W under dark conditions at various temperatures. Figure 14.8a shows the time evolution of the permanent total resistance degradation as a function of the chuck temperature ranging from 30 to 200 °C. Although all the curves saturate ~10 %, time-to-failure at which irreversible degradation occurs strongly depends on temperature, and all degradation curves are shifted along the stress time axis, the shift being determined by temperature. To a first order, the slope of the degradation curves is the same. On the other hand, the resistance degradation curves taken immediately after the stress (hence without recovery) show a different trend, see Fig. 14.8b. All curves seem to reach a maximum degradation of ~20 %, but the degradation speed (i.e., the slope of the degradation curve before reaching saturation) is increasing with temperature. Furthermore, the resistance degradation

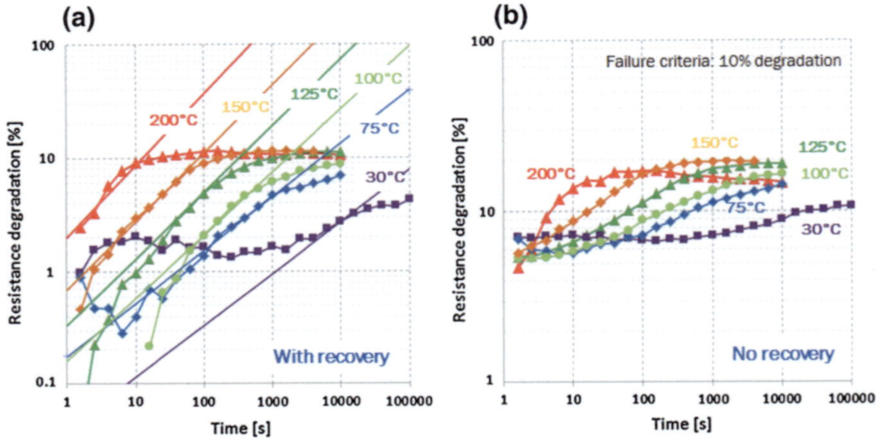

Fig. 14.8 Time evolution of the total resistance degradation as a function of chuck temperature for 6 μm contact spacing stressed at a constant power of 1.25 W **a** permanent degradation, i.e., after recovery; **b** immediately after stress, i.e., without recovery

decreases after reaching a maximum, as can be clearly noted from the data for $T = 200$ °C. It is clear that the dynamics of surface trapping effects are more complex than the permanent degradation.

Assuming a power law dependency for the time-to-fail as a function of stress power, the time-to-fail of the total permanent resistance degradation can be written as (Eq. 14.3):

$$t_{\mathrm{fail}}(P,T) = k_0(T)P^{-n}\mathrm{e}^{E_a/kT_{\mathrm{eff}}} \tag{14.3}$$

where t_{fail} is the time-to-failure, k_0 and n are constants, P is the stress power, E_a is the activation energy, T_{eff} is the effective temperature, and k is Boltzmann's constant. In order to assess the activation energy, the effective temperature under the contacts can be estimated as $T_{\mathrm{eff}} = R_{\mathrm{th}}P + T_{\mathrm{chuck}}$ [15]. This effective temperature depends on the applied power P, the thermal resistance R_{th}, and the chuck temperature T_{chuck}. The thermal resistance can be roughly approximated by $R_{\mathrm{th}} = R_{\mathrm{L,th}} + 2R_{\mathrm{c,th}} \approx 1/\left(2k\sqrt{LW}\right) + 2R_{\mathrm{c,th}}$ where k is the thermal conductivity, $R_{\mathrm{L,th}}$ is the thermal resistance along the channel, and $R_{\mathrm{c,th}}$ is the thermal resistance through the ohmic contacts [15]. The chuck temperature has been varied from 30 to 200 °C resulting in variation of the effective temperature from 225 to 450 °C depending on power dissipation in the TLM structure.

The time to reach a predefined failure criterion such as a 10 % change in resistance for different effective temperature is extrapolated in Fig. 14.9a. A fit to the mean time-to-failure data plotted versus the inverse of the effective temperature yields an activation energy of 1.04 eV for "permanent" degradation free of charge trapping influence. An activation energy of 0.90 eV is found when taking into account

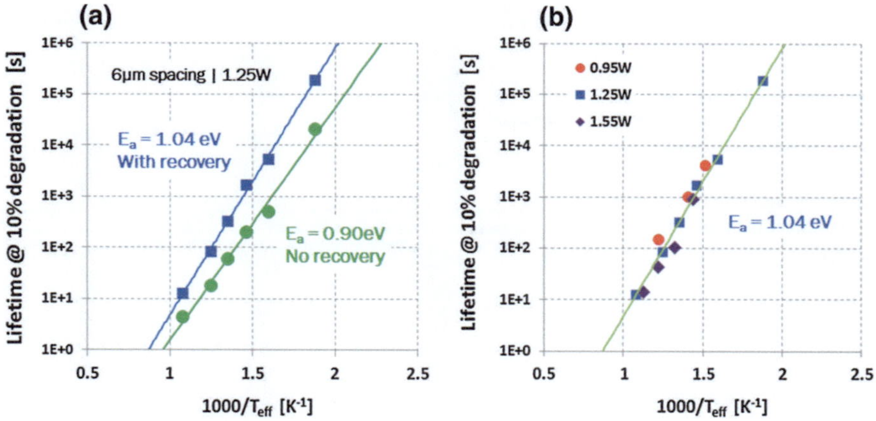

Fig. 14.9 **a** Arrhenius plot: mean time-to-failure versus the inverse of the effective temperature extracted for step stress with and without recovery step after stress at 1.25 W on 6 μm contact spacing; **b** Arrhenius plot: mean time-to-failure versus the inverse of the junction temperature extracted for step stress with recovery step after stress 0.95, 1.25, and 1.55 W on 6 μm contact spacing

trapping-related degradation. The relatively large difference between the two activation energies is mainly due to the complexity introduced by trapping-related degradation. Since traps exist in different energy levels and positions inside the SiN/AlGaN/GaN stack, it is difficult to interpret the activation energies obtained from data that mixes trapping-related effect. Nevertheless, the activation energy for permanent resistance degradation closely matches values reported in the literature 0.84–0.91 eV for experiments performed on similar technology [11]. Figure 14.9b plots the mean time-to-failure versus the inverse of the effective temperature for different stress powers. The same activation energy is extracted for the permanent degradation, supporting the validity of our approach.

14.2.4.4 Failure Mechanisms

As discussed above, the TLM structures show an increase of the total resistance when subjected to constant power stress as a consequence of the permanent degradation of the current. Emission microscopy (EMMI) has been performed on TLM structures with different contact spacings. As shown in Fig. 14.10a, by increasing the power from 2.4 up to 2.6 W on 18 μm contact spacing, the electroluminescence (EL) shows hot spots located at the edge of the metal contact on which the voltage is applied. The emission irregularity along the metal is most likely due to defects located at the edge of the contact, causing a non-uniform distribution of the electric field. On the contrary, for smaller contact spacing EL exhibits hot spots at the edge of both metal contacts at a power of ~2.3–2.5 W (Fig. 14.10b–e). In addition, the size and the amount of hot spots increase with decreasing the contact spacing from 18 to 6 μm.

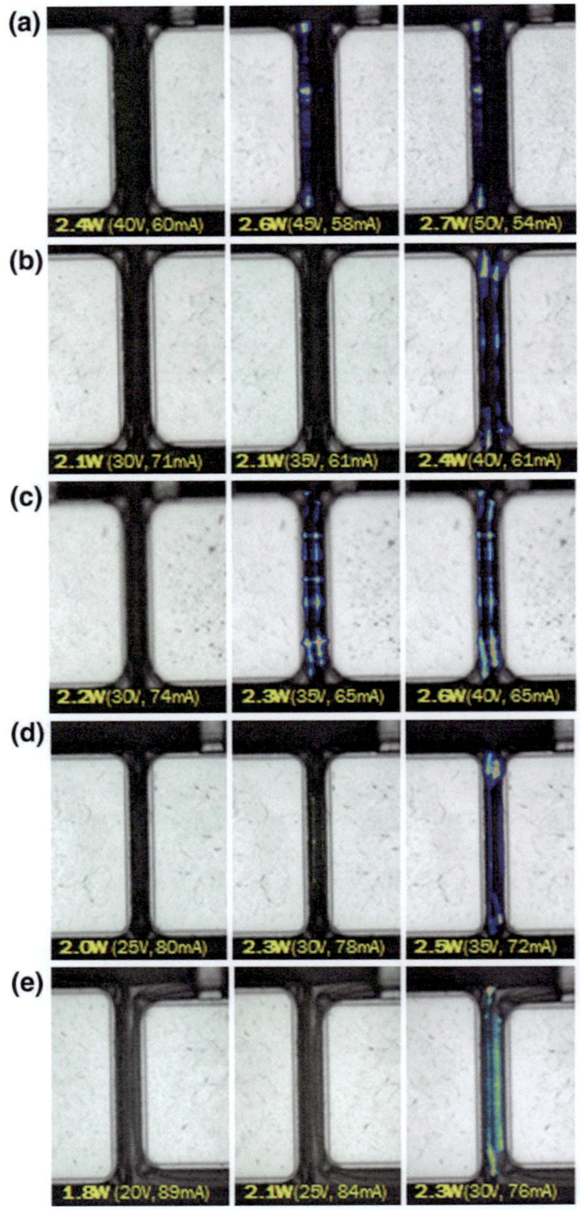

Fig. 14.10 Electroluminescence signature in TLM structures with 18, 12, 10, 8, and 6 μm contact spacing stressed at room temperature for 150–300 s at different stress powers. Voltage is applied to the left contact; the right contact is at ground potential

High-field current flow generates resistive heating along the AlGaN/ohmic contacts region, in addition to the self-heating in the channel. This localized current density and associated heating is believed to result in the formation of defects at the edge of the contacts, leading to the degradation of the total resistance.

14.2.5 Conclusions

The reliability of Au-free ohmic contacts on GaN is assessed by constant power stress in saturation. The degradation of the total resistance (contact resistance + 2DEG sheet resistance) contains both a recoverable and a permanent component, attributed to degradation at the SiN/AlGaN and metal/AlGaN interfaces, respectively. The permanent degradation exhibits a power law dependency as a function of stress time, with complete saturation at around 10 % degradation. The permanent degradation has an activation energy of 1.04 eV, independent of stress power.

14.3 Intrinsic Reliability of MISHEMT Gate Dielectrics

14.3.1 Introduction

Gallium nitride (GaN)-based power should exhibit low gate leakage, especially at elevated ambient/operating temperatures [16, 17]. Standard gate design for these devices is based on metal–insulator–semiconductor (MIS) architecture in order to suppress gate leakage current during on (forward gate bias)- and off (reverse gate bias)-state condition. Hence, the right choice of gate dielectric enhances the device performance and reliability. MOCVD grown in situ SiN has been the most ideal choice so far due to its lattice matching qualities with the (Al)GaN crystal. Furthermore, GaN devices are by nature normally on devices and a large negative gate bias is required to switch them off. The magnitude of the negative bias (threshold voltage) is directly dependent on the thickness of the gate dielectric. A thicker dielectric layer will render the HEMTs with a very negative threshold voltage V_{TH} which is generally not desirable, especially in cascode configuration as it might drive the low-voltage silicon MOSFET into avalanche conditions [18]. Furthermore, SiN is known to suffer from memory effect, so the amount of charge trapping in the bulk of the dielectric layer will be directly proportional to the thickness of the layer [19]. On the other hand, in case of a thinner dielectric, the electric field across the dielectric layer is too high, resulting in carrier tunneling through the dielectric layer, leading to higher gate leakage current. So, in order to achieve optimum performance, both the choice and the architecture of the gate dielectric are equally important.

In the first section, the conduction mechanisms under forward and reverse bias condition through the metal/SiN/AlGaN gate dielectric stack are discussed, using gate current characteristics over a wide range of bias and temperature. Also, the implementation of several conduction models such as Poole–Frenkel (P–F), Fowler–Nordheim (F–N), etc., to fit the leakage current is demonstrated. The second section focuses on the TDDB measurements of the above-mentioned gate dielectric stack under forward bias condition and extraction of operating voltage for specific operational lifetime.

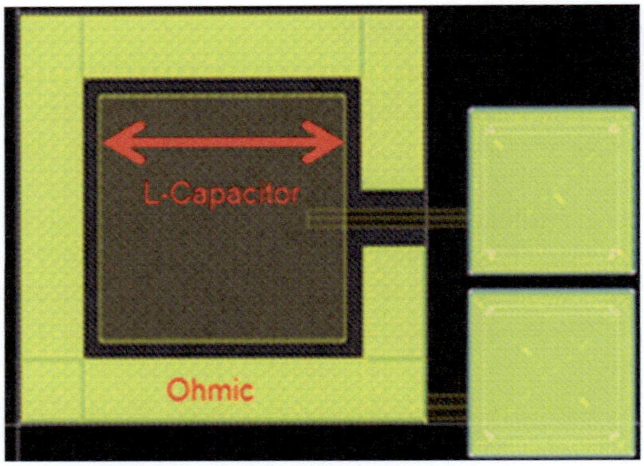

Fig. 14.11 Schematic of the MIS gate capacitor used for the gate leakage analysis

14.3.2 Experiments

Gate capacitors, 150 μm × 150 μm in size (i.e., 0.0225 mm^2) with in situ SiN and TiN/Al/TiN gate metal (shown in Fig. 14.11), were used for the analysis and the 2DEG is contacted by Ti/Al/TiN-based ohmic contacts. Details of the epitaxial structure and processing are outlined in [20].

Typical gate current density, measured on capacitor structures as a function of gate field for temperatures ranging from 25 to 200 °C, is shown in Fig. 14.12. Both in forward and reverse bias condition, the gate current density increases with temperature. Note that the *x*-axis is labeled "field," since the exact gate field is not a priori known. The reason is shown in Fig. 14.13a, b, which is the result of a TCAD simulation of the gate capacitor structure under forward and reverse bias conditions, respectively. Under forward bias (Fig. 14.13a), there is a fixed field across the AlGaN barrier (polarization field), and the complete potential drop is across the SiN dielectric. Under reverse bias conditions, however, the voltage drop is distributed between the SiN, AlGaN barrier, and the (depleted) GaN layer underneath. The exact field distribution will depend on the exact donor-state ionization and on the gate bias V_G condition. As from $V_G \leq V_{TH}$ (near channel pinch-off), the field across the SiN and AlGaN barrier remains constant, and the potential is dropped across the GaN layer. Since the many unknowns, the analysis and conclusions out of the reverse bias data will not make use of the electric field. Nevertheless, important conclusions can be drawn:

- In forward bias condition, electrons out of the 2DEG will be injected in the AlGaN barrier and in the SiN. The energy barriers between the 2DEG/AlGaN and AlGaN/SiN are small. Hence, any current measured in the gate metal will be due to the leakage current of the SiN dielectric itself.
- In reverse bias condition, electrons are injected from the gate metal in the SiN dielectric. Since a substantial energy barrier is present, electrons either tunnel

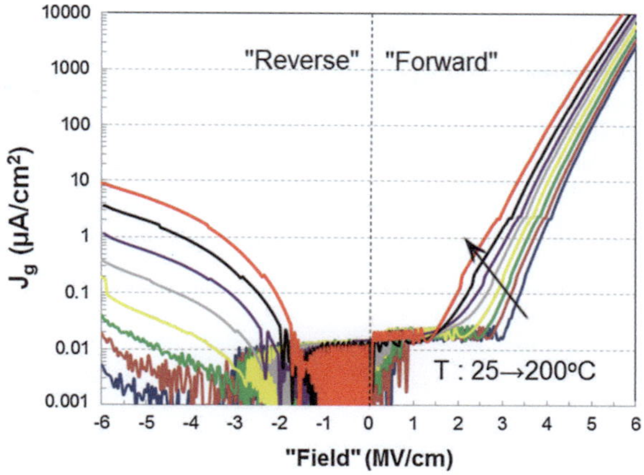

Fig. 14.12 J_g–E_g curves measured on the gate capacitor structures as a function of ambient temperature

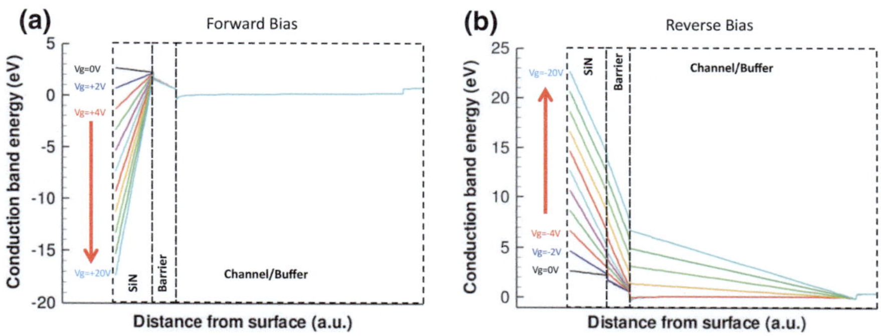

Fig. 14.13 Band conduction (E_C) band diagram under **a** forward bias condition; **b** reverse bias condition

through the barrier (Fowler-Nordheim tunneling) or are injected over the barrier (thermionic emission). After that, electrons will follow certain leakage paths through the SiN, AlGaN, and GaN before they are collected at the ohmic contact.

14.3.3 Analysis of Leakage Current Under Forward Bias Condition

Under forward bias, the voltage drop is completely over the SiN; hence, the field can be easily estimated. Figure 14.14 plots the forward bias current density in a Poole–Frenkel (P–F) plot, i.e., Ln(J/E) versus sqrt(E), as shown in Eq. 14.4 [20, 21].

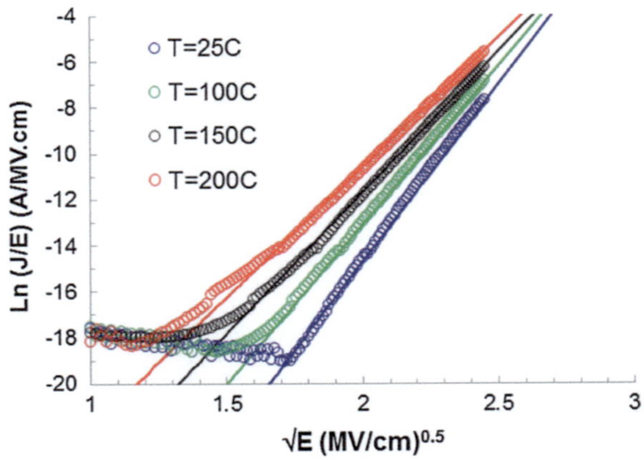

Fig. 14.14 Poole–Frenkel (P–F) plot under forward bias condition for different ambient temperatures. *Symbols* denote experimental data, full lines are the fit of the PF model. Data obtained on capacitor structures (see Fig. 14.11)

$$J_{PF} = AE \exp\left[-\frac{q(\phi_B - \sqrt{qE/\pi\varepsilon_S})}{kT}\right] \tag{14.4}$$

where *A* is a fitting constant, ϕ_B is the triangular barrier height/conduction band offset between the AlGaN and the SiN and ε_S is effective permittivity of the SiN/AlGaN-combined dielectric stack. Equation 14.4 can be rearranged as follows:

$$\ln\left(\frac{J_{PF}}{E}\right) = B\sqrt{E} + C \tag{14.5}$$

where

$$B = \frac{q}{kT}\sqrt{\frac{q}{\pi\varepsilon_S}} \tag{14.6}$$

and

$$C = -\frac{q\phi_B}{kT} + Ln(A) \tag{14.7}$$

Using Eq. 14.5, a straight line is expected with a slope B decreasing with increase in temperature. Fitting the experimental data self-consistently (i.e., the slope is calculated using Eq. 14.6 using one value for ε_s, and *T* as from the experiment), a very good agreement between the model and the experimental data is achieved, confirming the fact that under forward bias condition the conduction mechanism is P–F emission. Figure 14.15 shows the activation energy extraction at selected fields by plotting Ln(*J*/*E*) versus *q*/*kT*. As can be noted, the extracted activation energy E_a reduces with the electric field, in line with the field-dependent barrier lowering proposed by the P–F model. Figure 14.16 plots the activation energy E_a versus the

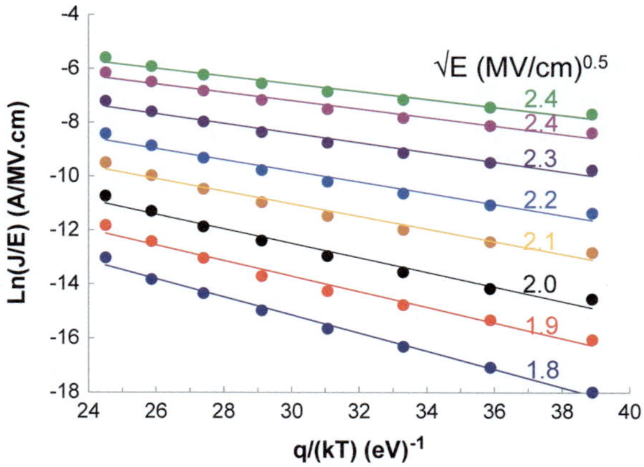

Fig. 14.15 Ln(J/E) versus temperature for different gate fields. Forward bias condition. Data obtained on capacitor structures

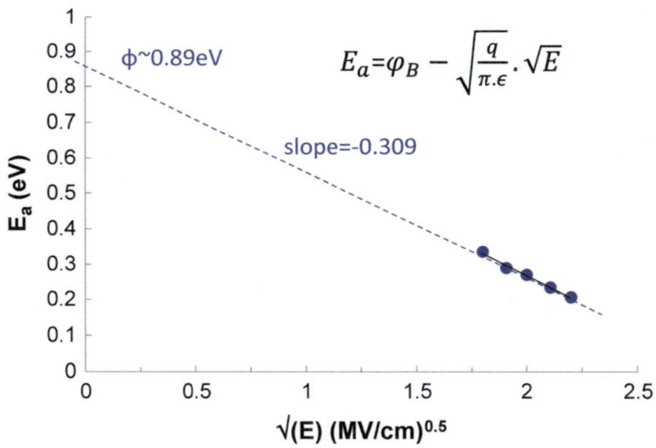

Fig. 14.16 Extracted activation energy E_a (see Fig. 14.15) versus $\sqrt{E}$. Forward bias condition. Data obtained on capacitor structures

sqrt(E), showing the expected linear dependence (Eq. 14.8). The field-dependent activation energy $E_a(E)$ can be calculated using the following equation:

$$E_a = \phi_B - \sqrt{\frac{q}{\pi \varepsilon_S}}\sqrt{E} \tag{14.8}$$

Extrapolating Eq. 14.8 to zero field yields the barrier height of the trap, or trap energy level. From Fig. 14.16, a trap energy of 0.89 eV is extracted. As a consistency check, the slope of E_a versus sqrt(E), i.e., −0.309, can be compared to the theoretical expected slope −sqrt($q/\pi\varepsilon$) = −0.27, again in fair agreement.

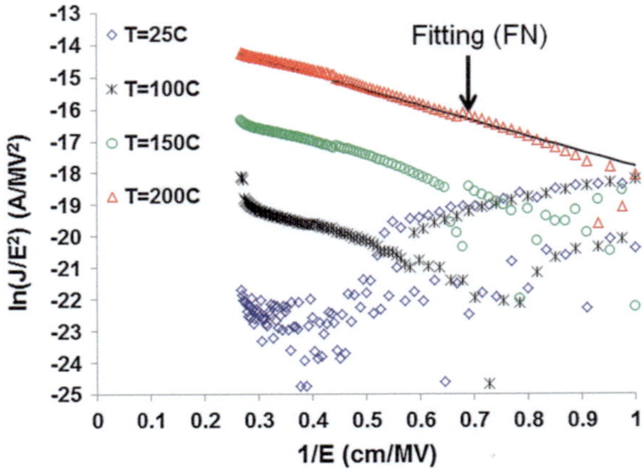

Fig. 14.17 F–N plot under reverse bias condition for different ambient temperatures. *Symbols* denote experimental data and full line is the fit of the FN model

14.3.4 Analysis of Leakage Current Under Reverse Bias Condition

Under reverse gate bias, the field is distributed across several dielectric layers (SiN, AlGaN, GaN). The *J*–*V* characteristics under the reverse bias (−15 → 0 V) show strong electric field and temperature dependency, indicating Fowler-Nordheim (FN) conduction along with a significant temperature-assisted leakage component. The J_{FN}–E relation is given by the following [21, 22]:

$$J_{FN} = AE^2 exp\left[-\frac{8\pi\sqrt{2m_e^*(q\phi_R)^3}}{3qhE}\right] \tag{14.9}$$

where A is a F–N constant, m_e^* is the effective electron mass in the dielectric, h is the Planck's constant, and ϕ_R is the effective barrier height. For F–N conduction, a plot of $\mathrm{Ln}(J_{FN}/E^2)$ versus $1/E$ must yield a straight line. Figure 14.17 shows that the experimental data is in line with the theoretical model of F–N conduction.

Plotting the gate current density J versus temperature for some selected gate bias conditions, i.e., varying gate fields, one clearly observes that the extracted activation energy E_a is independent of the field, shown in Fig. 14.18. This strongly suggests a thermionic emission process, with activation energy $E_a = 0.7$ eV. The extracted E_a of 0.7 eV is much lower than what is estimated from the electron affinities of the gate metal and the SiN film. *Several literatures report different electron affinity (e.a.) values for SiN films ranging between* 2 *and* 3 eV. *For this calculation, an e.a. value of* 2 eV *is used.* Based on the work function and electron affinity of TiN and SiN to be ~4 and ~2 eV, respectively, the expected barrier difference under reverse bias condition was estimated to be ~2 eV, indicating toward a defect/trap state in the bulk of the SiN film.

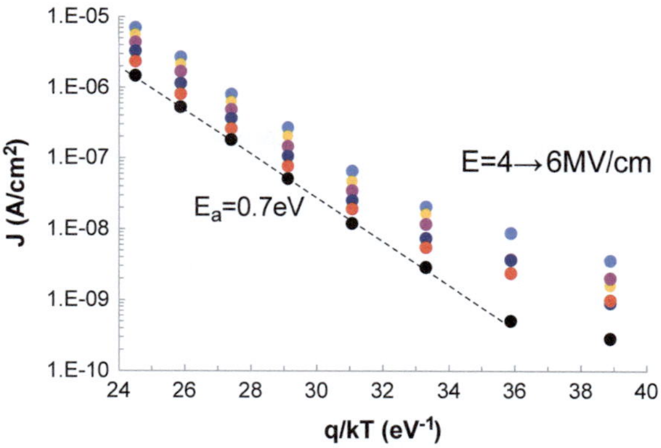

Fig. 14.18 Gate current density J versus ambient temperature, measured under reverse bias condition on a gate capacitor, for different gate fields

14.3.5 Analysis of Defect States in Bulk SiN

Several MIS capacitors were fabricated of varying SiN film thicknesses and the J-(V) characteristics were compared. Comparison data shown in Fig. 14.19a are for 25 and 200 °C only. Reduction (~two orders of magnitude) in leakage was observed at 200 °C under forward bias condition for thicker SiN layer. For a thicker SiN, the tunneling of electrons through the trapezoidal barrier at the SiN/AlGaN interface becomes dominant at low fields; hence, the onset of P–F conduction is delayed (i.e., occurs only at higher voltages, ~4 V higher for the cases considered in Fig. 14.19). On the contrary, only a very small reduction was observed under reverse bias condition even at 200 °C, indicating a limitation in the intrinsic property of MOCVD SiN films. In this case, the leakage current due to injection of electrons from the metal into the SiN is independent of the thickness of the SiN. Several literature studies have been reported regarding the defect states in the bulk SiN films indicate defect energies between 1.6 and 1.9 eV (Si^0 donor and Si^- acceptor energy, respectively) from the conduction band. Reverse calculations based on the intrinsic properties of the gate metal (work function of TiN ~4 eV) and the SiN (electron affinity ~2 eV) reveal a defect energy of ~1.3 eV in the SiN bulk from the conduction band E_C. The defect energy is fairly in line with what has been reported previously [23, 24]. Figure 14.19b shows the estimated defect energy state in the SiN film.

14.3.6 TDDB Study

Having established the current conduction mechanisms in the SiN dielectric, one can proceed to the dielectric stressing experiments: constant field stress (TDDB). Measurements are performed on large power devices, with $W > 100$ mm. Unless specified otherwise, measurement temperature is kept at 150 °C. Figure 14.20 depicts the gate

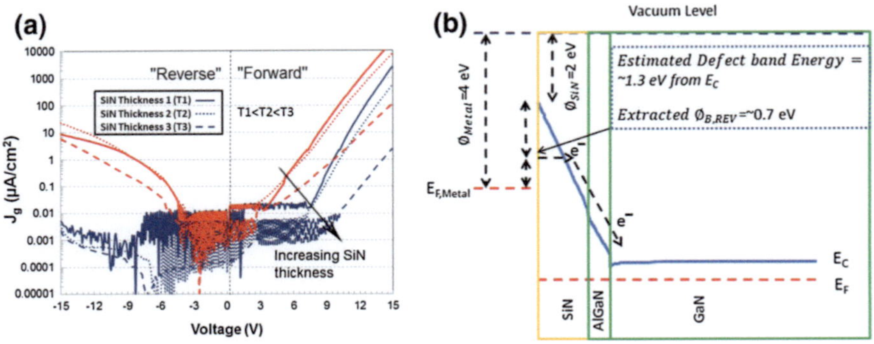

Fig. 14.19 **a** J_g–E curves measured on the gate capacitor as a function of temperature for three different SiN thicknesses. The *blue-colored lines* represent the data measured at 25 °C and the *red-colored lines* represent the data measured at 200 °C; **b** band diagram showing the estimated intrinsic defect energy within the SiN film (*Electron affinity for the MOCVD SiN film was assumed to be* ~2 eV *for this calculation*)

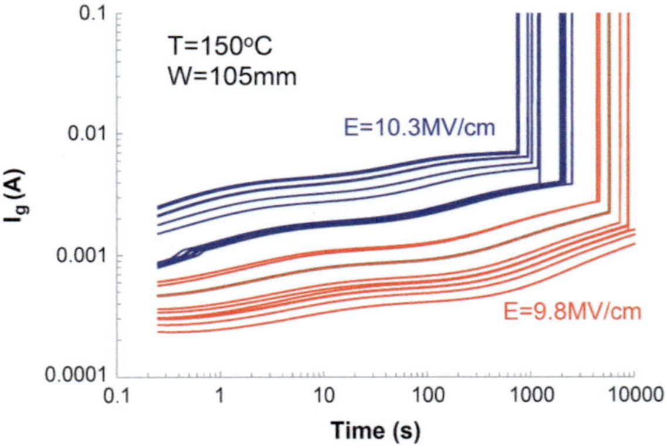

Fig. 14.20 I_g-time curve characteristics measured on $W = 105$ power transistors. Constant forward gate field stress (TDDB), $T = 150$ °C

current versus stress time, for two different gate stress fields (forward gate stress). Note that contrary to expectation, the current increases with stress time. This is believed to be due to the creation of extra defects in the SiN dielectric or the SiN/AlGaN interface during stress, increasing the P-F mobility (more defects = faster hopping). Figure 14.21 shows the Weibull distribution at 5 different stress fields. The full lines represent the Weibull model fit, using the P–F model for field acceleration. The field acceleration factor is γ. The field acceleration model used is as follows:

$$t_{BD} \propto E \exp\left(-\gamma\sqrt{E}\right) \tag{14.10}$$

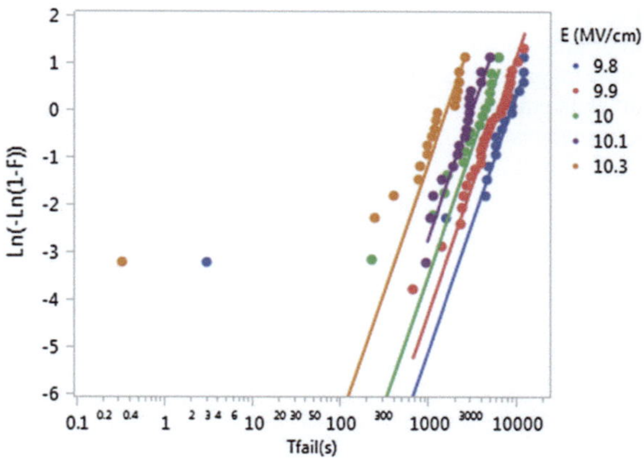

Fig. 14.21 Weibull distribution and model fit at $T = 150$ °C, for five different gate stress fields (forward conduction). The field extrapolation model used is P–F

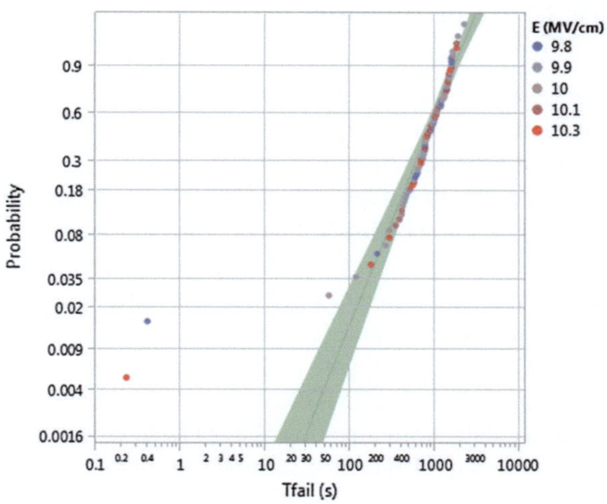

Fig. 14.22 Normalization of the TDDB data of Fig. 14.21 to one stress field, using the P–F field acceleration model. Weibull slope $\beta = 2.35$

Lifetime extrapolation calculations using P–F model revealed an operational voltage of ~5 V for 10 years of operation at $T = 150$ °C for the SiN only dielectric calculated at 100 ppm. Normalization of all the data to one stress field results in Fig. 14.22. Clearly all the data are one single Weibull curve with slope $\beta = 2.35$, data from different fields are nicely distributed, supporting the validity of our model and assumptions. Forward voltage stress is also carried out at different temperatures. Figure 14.23 shows the TDDB data at three different fields and at three different temperatures. Little temperature acceleration is observed.

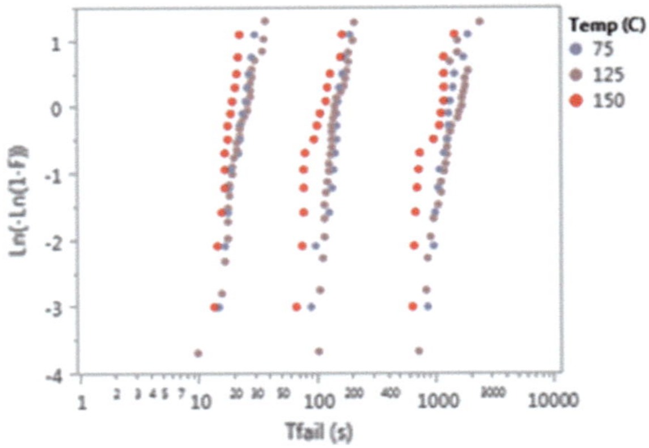

Fig. 14.23 TDDB data for three different stress fields (forward gate stress) at three different ambient temperatures

14.3.7 Conclusions

Leakage current through gate dielectric/AlGaN barrier is measured and possible conduction mechanisms are proposed using TCAD-based band diagrams. Under forward gate bias condition, current conduction through the dielectric follows P–F emission, through traps in the SiN with zero field activation energy of ~0.89 eV. On the other hand, Fowler-Nordheim conduction was observed to be main conduction mechanism under reverse bias condition along with a thermally assisted component. The extracted barrier of ~0.7 eV was attributed to a defect state in the SiN film, pointing toward an intrinsic issue.

TDDB of the SiN gate dielectric under forward bias conditions shows Weibull distributed data. The TDDB data are field accelerated, but the temperature acceleration is small. Normalization of the TDDB data using a P–F field acceleration model to one stress field indicated convergence of all the data points in one curve, validating the P–F-based acceleration model. The extracted Weibull slope was found to be ~2.35.

14.4 Buffer Stack Reliability—Off-State High-Voltage Drain Stress

14.4.1 Introduction

AlGaN-/GaN-based HEMTs are actively being researched as next-generation power devices in the 100–650 V range. A key concern inhibiting the widespread adoption in the market is their reliability, especially during long-term off-state stress at high temperature (HTRB). Under HTRB conditions, the 2DEG is depleted and the GaN stack behaves as a dielectric, with the threading dislocations serving as leakage paths. Voltage acceleration between 100 and 130 V under HTRB condition has been observed for 100 V GaN-on-Si devices [25]. The importance of the buffer epi stack for off-state stress of 600-V GaN-on-Si devices is shown in [26], but without reporting any voltage acceleration data or model. Hence, it is important to study the voltage acceleration mechanisms in high-voltage AlGaN/GaN-based power devices.

In the sections below, it will be shown that due to the particular nature of the current conduction mechanism in undoped GaN stacks, devices show no voltage-accelerated degradation under HTRB stress between 420 and 850 V. This is explained by the GaN buffer stack becoming resistive above a certain critical voltage (trap filling level V_{TFL}), which allows the trapped charge to leak away. The important role of buffer traps (more specifically C_N acceptor traps) is highlighted.

14.4.2 Current Conduction Mechanism

Experiments are performed on AlGaN/GaN-on-Si power devices, processed on 6-inch wafers. The devices are 100 mΩ power transistors with a ~20A current rating [27]. Growing GaN-on-Si results in ~10^9 cm^{-2} threading dislocations which serve as leakage paths through the buffer stack. This means that every power device of ~ few mm^2 has >10^6 dislocations, or potential paths for vertical current conduction, i.e., current between the top ohmic contacts and the Si substrate. Figure 14.24 shows the vertical leakage current through the GaN stack as a function of temperature, under forward conduction, meaning that the silicon substrate is grounded, and the top ohmic contact is biased positive. The J–V characteristic is typical for space-charge-limited (SCL) current, or the conduction in a dielectric through spillover from a metal [28]. V_{TFL} is the voltage at which the traps in the buffer are ionized, so the quasi-Fermi levels are depinned and move up to the band-edge; hence, the steep increase in current with voltage ($J \sim V^n$ behavior). The traps are identified using current DLTS, as C-atoms on a N-site at $E_V + 0.85$ eV [27]. Above V_{TLF}, the vertical field becomes large enough to stimulate field-enhanced Poole-Frenkel current conduction which allows the stored charge in the C_N acceptors to leak away. As a result, the GaN buffer becomes resistive instead of capacitive [29], and no charge is stored in the buffer. A good estimate of the amount of traps can be obtained from V_{TLF} [28]. In our particular case, the amount of C_N traps is estimated to be ~10^{18} cm^{-3}.

14.4.3 High-Temperature Reverse Bias

The model stating that above V_{TFL}, no charge is stored in the GaN layer and the device becomes insensitive to the trap dynamics and has important consequences for any reliability test that relies on voltage acceleration. It basically means that above V_{TFL}, charge is emitted at the same rate as it is trapped, and if no additional traps are created,

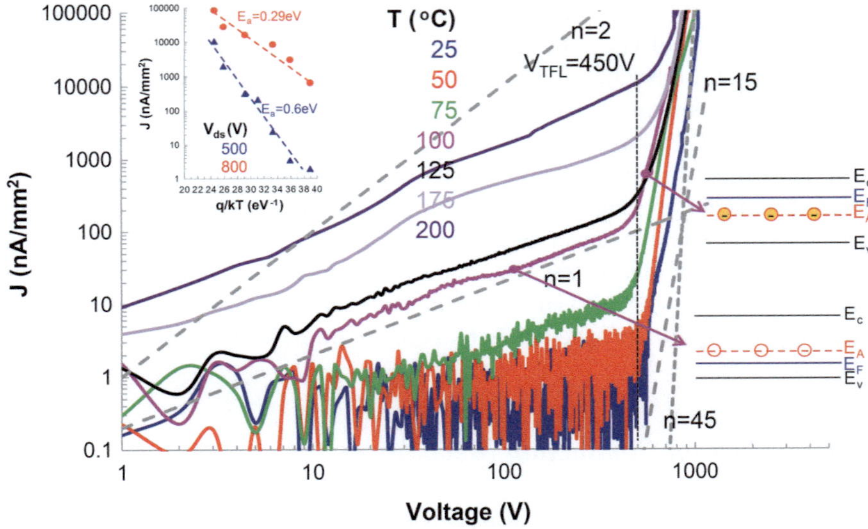

Fig. 14.24 ln(J)–ln(V) characteristic of the vertical leakage current as a function of temperature. The trap filling voltage V_{TFL} is the voltage at which all acceptor traps are ionized (see [28]), resulting in depinning of the Fermi level. Note the Ohmic conduction ($n = 1$) till $V = V_{\text{TFL}}$. Above V_{TFL}, the current through the buffer increases rapidly. Electrons are injected from the Si substrate by thermionic emission with $E_{\text{a}} = 0.6$ eV

the structure acts as a resistor. We will focus on degradation of "dynamic" R_{on} (measured 4 ms after releasing the stress voltage) and "static" R_{on} (measured 30 s after releasing the stress voltage). Figure 14.25a shows the degradation of dynamic R_{on} at $T = 150$ °C, $V_{\text{ds}} = 520$ V, as a function of stress time for different L_{gd}. The slight increase in dynamic R_{on} (following a ln(t) behavior, see also [25]) indicates that slightly more charge is trapped following the stress. Shorter L_{gd} gives better performance (total amount of trapped charge in the access region is smaller). However, for too short L_{gd}, degradation becomes larger due to too high lateral field, indicating the subtle balance between device design and buffer epi design. Figure 14.25b shows the data at $T = 150$ °C, for one L_{gd}, with different stress voltages. The most striking feature is that between 420 and 600 V no voltage acceleration is observed, in line with the model that above V_{TLF} no charge is stored in the buffer stack. Figure 14.25c shows the extrapolated time-to-fail at 10 % degradation in dynamic R_{on} using an ln(t) extrapolation. These data suggest that for half of the population, the devices can be stressed at $T = 150$ °C for 1 year at $V_{\text{ds}} = 520$ V, with a shift in dynamic R_{on} of less than 10 %. The devices show full recovery after 10^2 s at room temperature. Next, the devices are stressed at $T = 150$ °C, from 500 V up to 950 V, see Fig. 14.26, showing R_{on} as a function of stress time. Up to 800 V, R_{on} is stable (within 10 %), but as from $V_{\text{ds}} = 900$ V, R_{on} starts to increase and other mechanisms start to occur. However, removing the stress and letting the device relax for 16 h results in almost full recovery (within less than 10 % of the original value).

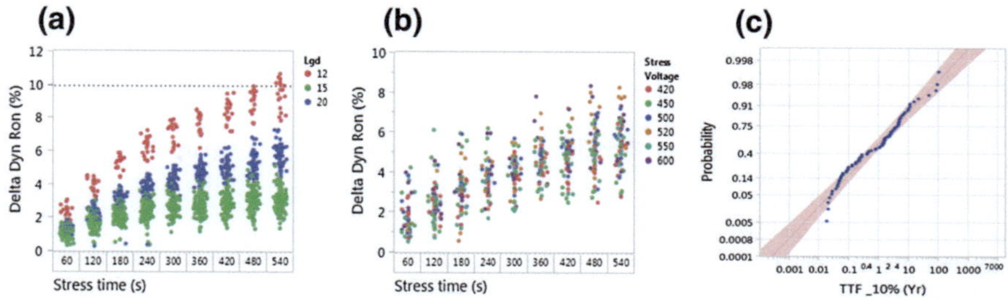

Fig. 14.25 Measurements on 100 mΩ power transistors at $T = 150$ °C. **a** Dynamic R_{on} increase measured as a function of HTRB stress time at $V_{ds} = 520$ V, for different L_{gd}. **b** Dynamic R_{on} increase as a function of HTRB stress time for $L_{gd} = 20$ µm and for stress voltages ranging from 420 to 600 V. Dynamic R_{on} is measured at the drain stress condition. **c** Time-to-failure for dynamic R_{on} at $T = 150$ °C, $V_{ds} = 520$ V failure criterion is 10 % degradation in dynamic R_{on}. Data are obtained by extrapolating the results of each device of Fig. 14.25a ($L_{gd} = 15$ µm only) using a ln(t) behavior. These data suggest that for half of the population, the devices can be stressed at $T = 150$ °C for 1 year at $V_{ds} = 520$ V, with a shift in dynamic R_{on} of less than 10 %

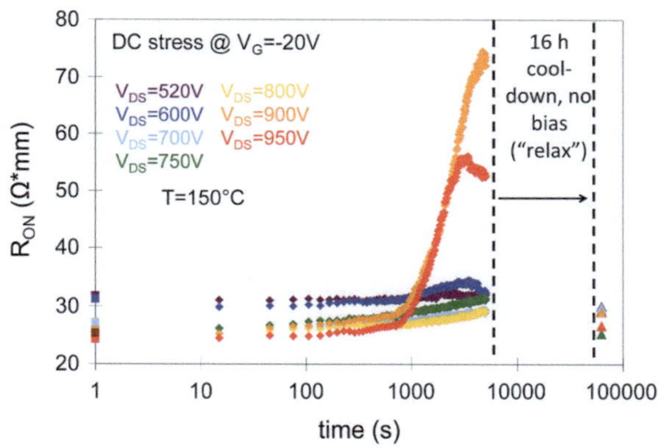

Fig. 14.26 R_{on} as a function of stress time for different V_{ds_stress} conditions from 520 V up to 950 V and $T = 150$ °C. Stress is stopped after 6000 s, after which the devices are cooled down ("relax"). Devices fully recover after 16 h of relax. Note the dynamics in the static R_{on} at $V_{ds} = 900$ and 950 V, showing a partial recovery even during the off-state stress

14.4.4 High-Voltage off-State Drain Stress

Since the GaN stack behaves as a dielectric, one can apply high-voltage TDDB during which the GaN buffer stack is stressed in off-state at high voltage until failure. High-voltage off-state stress on high-voltage power transistors is reported in [30, 31]. In [30], large-area devices were stressed till failure at RT, at $V_{ds} = 700$ V and $V_{ds} = 750$ V. In [31], large-area power transistors were stressed at $T = 80$ °C, at $V_{ds} = 1100$ and 1150 V. An inverse power law was used for field acceleration.

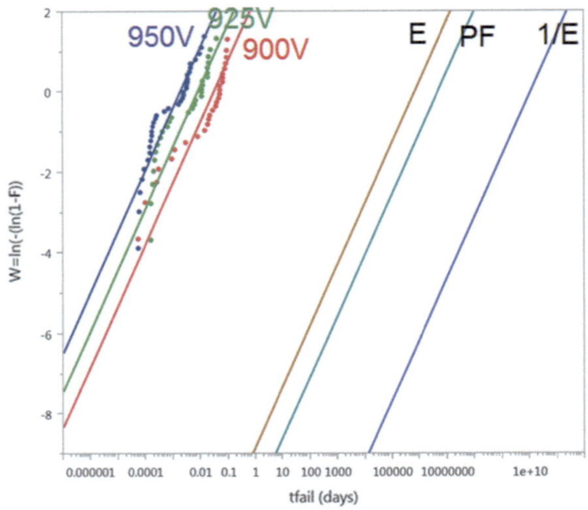

Fig. 14.27 High-voltage off-state stress at $T = 200\ °C$, on 100 mΩ power transistors, at $V_{ds} = 900$, 925, and 950 V. Data are plotted on a Weibull plot; three different extrapolation models are used: *E*, 1/*E*, and Poole-Frenkel. Extrapolation is done at $V_{ds} = 600$ V

Large-area power transistors ($W > 100$ mm) are stressed with source and substrate grounded, gate below pinch-off, and drain put at a high voltage. Time-to-failure distributions at $V_{ds} = 900$, 925, and 950 V are plotted in Fig. 14.27. To induce failure of the GaN stack within a reasonable measurement time, the temperature had to be increased up to 200 °C. The data fit a Weibull distribution (as expected for a dielectric), albeit that apparently a bi-model distribution is obtained. This is attributed to within-wafer and wafer-to-wafer non-uniformities in the buffer stack. Figure 14.27 also plots the extrapolation of the data to models are used: *E*, 1/*E* and Poole-Frenkel, to $V_{ds} = 600$ V. Based on the discussion in Sect. 14.4.2, the Poole-Frenkel model should be applied.

14.4.5 Conclusions

For voltages above V_{TFL} (450 V), dynamic R_{on}, as well as the degradation of static R_{on} under HTRB stress, is independent of the stress voltage, a consequence of current transport by a SCL current. Only at extreme voltages (>900 V), a shift in static R_{on} is observed, which is recoverable. From the data, it follows that above V_{TFL}, the buffer structure becomes resistive, suppressing charge storage in C_N acceptor traps. Stressing the power devices in off-state till failure requires high voltage ($V_{ds} > 900$ V) and high temperature ($T > 150$ °C). A bi-model Weibull distribution is obtained.

References

1. Meneghesso G, Meneghini M, Tazzoli A, Ronchi N, Stocco A, Chini A, Zanoni E (2010) Reliability issues of gallium nitride high electron mobility transistors. Int. J. Microw Wireless Technol 2:39–50
2. Meneghesso G, Magistrali F, Sala D, Vanzi M, Canali C, Zanoni E (1998) Failure mechanisms due to metallurgical interactions in commercially available AlGaAs/GaAs and AlGaAs/InGaAs HEMTs. Microelectron Rel 38:497–506
3. Chou YC, Leung D, Smorchkova I, Wojtowicz M, Grundbacher R, Callejo L, Kan Q, Lai R, Liu PH, Eng D, Oki A (2004) Degradation of AlGaN/GaN HEMTs under elevated temperature lifetesting. Microelectron Rel 44:1033–1038
4. Piazza M, Dua C, Oualli M, Morvan E, Carisetti D, Wyczisk F (2009) Degradation of TiAlNiAu as ohmic contact metal for GaN HEMTs. Microelectron Rel 49:1222–1225
5. Daumiller L, Kirchner C, Kamp M, Ebeling KJ, Kohn E (1999) Evaluation of the temperature stability of AlGaN/GaN heterostructure FETs. IEEE Elec Dev Lett 20:448–450
6. Wurfl J, Hilsenbeck J, Nebauer E, Trankle G, Obloh H (1999) Technology and thermal stability of AlGaN/GaN HFETs", proc. GaAs Appl Symp (GAAS), pp 430–434
7. Hilsenbeck J, Nebauer E, Wurfl J, Trankle G, Obloh H (2000) Aging behaviors of AlGaN/GaN HFETs with advanced ohmic and schottky contacts. Elec Lett 36:980–981
8. Wurfl J, Hilsenbeck J, Nebauer E, Trankle G, Obloh H, Osterle W (2000) Reliability of AlGaN/GaN HFETs comprising refractory ohmic and schottky contacts. Microelectron Rel 40:1689–1693
9. Wu TL, Marcon D, Stoffels S, You S, De Jaeger B, Van Hove M, Groeseneken G, Decoutere S (2009) Stability evaluation of Au-free Ohmic contacts on AlGaN/GaN HEMTs under a constant current stress. Microelectron Rel 49:1222
10. Constant A, Cano J-F, Banerjee A, Coppens P, Moens P (2015) Reliability of Au-free ohmic contacts for high-voltage AlGaN/GaN power devices. In: proceedings of the reliability of compound semiconductors (ROCS) (to be published)
11. Wu Y, Chen C-Y, del Alamo JA (2014) Activation energy of drain-current degradation in GaN HEMTs under high-power DC stress. Microelectron Rel 54:2668–2674
12. Kuzmik J, Bychikhin S, Pogany D, Gaquiere C, Morvan E (2006) Current conduction and saturation mechanism in AlGaN/GaN ungated strcutures. J Appl Phys, 99, pp 123720-1-7
13. Hu C-Y, Hashizume T, Ohi K, Tajima M (2010) Trapping effect evaluation of gateless AlGaN/GaN heterojunction field-effect transistors using transmission-line-model method. Appl Phys Lett 97:222103-1-3
14. Benbakhti B, Rousseau M, Soltani A, De Jaeger J-C (2006) Analysis of thermal effect influence in gallium-nitride-based TLM structures by means of a transport–thermal modeling. IEEE Trans Elect Dev 53:2237–2242
15. Cahill DG, Braun PV, Chen G, Clarke DR, Fan S, Goodson KE, Keblinski P, King WP, Mahan GD, Majumdar A, Maris HJ, Phillpot SR, Pop E, Shi L (2014) Nanoscale thermal transport. II 2003–2012. Appl Phys Rev, 1, pp 011305-1-45
16. Treu M, Vecino E, Pippan M, Haberlen O, Curatola G, Deboy G, Kutschak M, Kirchner U (2012) The role of silicon, silicon carbide and gallium nitride in power electronics. In: proceedings of the IEEE international electron devices meeting (IEDM), pp 711–714
17. del Alamo JA, Joe J (2009) GaN HEMT reliability. Microelectron Rel 49:1200–1206
18. Huang X, Du W, Liu Z, Lee FC, Li Q (2014) Avoiding Si MOSFET avalanche and achieving true zero-voltage-switching for cascode devices. In: proceedings of the IEEE energy conversion congress and exposition (ECCE), pp 106–112
19. Vianello E, Driussi F, Blaise P, Palestri P, Esseni D, Perniola L, Molas G, De Salvo B, Selmi L (2011) Explanation of the charge trapping properties of silicon nitride storage layers for NVMs—Part II: atomistic and electrical modeling. IEEE Trans Elec Dev 58:2490–2499

20. Moens P, Liu C, Banerjee A, Vanmeerbeek P, Coppens P, Ziad H, Constant A, Li Z, De Vleeschouwer H, Roig-Guitart J, Gassot P, Bauwens F, De Backer F, Padmanabhan B, Salih A, Parsey J, Tack M (2014) An industrial process for 650 V rated GaN-on-Si power devices using in-situ SiN as a gate dielectric, ISPSD'14: Proceedings of the 26th IEEE international symposium on power semiconductor devices and ICs, pp 374–377
21. Turuvekere S, Karumuri N, Rahman AA, Bhattacharya A, Dasgupta A, DasGupta N (2013) Gate leakage mechanisms in AlGaN/GaN and AlInN/GaN HEMTs: comparison and modelling. IEEE Trans Elec Dev 60:3157–3165
22. Sze SM (2001) "MIS diode and charge-coupled devices", physics of semiconductor devices, 2nd edn. Wiley, New York, pp 402–407
23. Ko C, Joo J, Han M (2006) Annealing effects on the Photoluminescence of Amorphous Silicon Nitride films. J Korean Phys Soc 48:1277–1280
24. Li D, Huang J, Deren Y (2008) Electroluminescence of silicon–rich silicon nitride light-emitting devices. In: Proceedings of the 5th IEEE international conference on group IV photonics, pp 119–121
25. Lidow A, Strittmatter R, Chunhua Z, Yanping M (2015) Enhancement mode gallium nitride transistor reliability. In: Proceedings of the IEEE international reliability physics symposium (IRPS), pp 2E11–2E15
26. Kwan MH, Wong KY, Lin YS, Yao FW, Tsai MW, Chang YC, Chen PC, Su RY, Wu CH, Yu JL, Yang FJ, Lansbergen GP, Wu H–Y, Lin M–C, Wu CB, Lai Y-A, Hsiung C-W, Liu P-C, Chiu H-C, Chen C-M, Yu CY, Lin HS, Chang MH, Wang S-P, Chen LC, Tsai JL, Tuan HC, Kalnitsky A (2014) CMOS–compatible GaN–on–Si field–effect transistors for high voltage power applications. In: proceedings of the IEEE international electron devices meeting (IEDM), pp 1761–1764
27. Moens P, Vanmeerbeek P, Banerjee A, Guo J, Liu C, Coppens P, Salih A, Tack M, Caesar M, Uren MJ, Kuball M, Meneghini M, Meneghesso G, Zanoni E (2015) On the impact of carbon-doping on the dynamic R_{on} and off-state leakage current of 650 V GaN power devices, ISPSD'15: Proceedings of the 27th IEEE international symposium on power semiconductor devices and ICs, pp 37–40
28. Lampert MA (1956) Simplified theory of space-charge-limited currents in an insulator with traps. Phys Rev 103:1648–1656
29. Uren M, Silvestri M, Casar M, Hurkx GAM, Croon JA, Sonsky J, Kuball M (2014) Intentionally carbon-doped AlGaN/GaN HEMTs: necessity for vertical leakage paths. IEEE Elec Dev Lett 35:327–329
30. Briere MA (2014) Commercially viable GaN-based power devices. In: Proceedings of the applied power electronics conference and exposition (APEC)
31. Kikkawa T, Hosoda T, Shono K, Imanishi K, Asai Y, Wu Y, Shen L, Smith K, Dunn D, Chowdhury S, Smith P, Dunn D, Chowdhury S, Smith P, Gritters J, McCarthy L, Barr R, Lal R, Mishra U, Parikh P (2015) Commercialization and reliability of 600 V GaN power switches. In: Proceedings of the IEEE international reliability physics symposium (IRPS), pp 6C11–6C16

Chapter 15
Switching Characteristics of Gallium Nitride Transistors: System-Level Issues

Fred Lee, Qiang Li, Xiucheng Huang and Zhengyang Liu

The use of gallium nitride devices is gathering momentum, with a number of recent market introductions for a wide range of applications such as point-of-load (POL) converters, off-line switching power supplies, battery chargers, and motor drives. GaN devices have a much lower gate charge and lower output capacitance than silicon MOSFETs and, therefore, are capable of operating at a switching frequency 10 times greater. This can significantly impact the power density of power converters, their form factor, and even current design and manufacturing practices. To realize the benefits of GaN devices resulting from significantly higher operating frequencies, a number of issues have to be addressed, such as converter topology, magnetics, control, packaging, and thermal management. This chapter studies the switching characteristics of high-voltage GaN devices including some specific issues related to the cascode GaN. An evaluation is presented of the cascode GaN based on a buck converter in hard-switching and soft-switching modes, which shows the necessity of soft switching for cascode GaN devices at high frequencies. High d*v*/d*t*- and d*i*/d*t*-related gate drive issues associated with the higher switching speed of GaN devices are addressed, and many important design considerations are presented. Additionally, this chapter illustrates the utilization of GaN in a wide range of emerging applications.

F. Lee · Q. Li (✉) · X. Huang · Z. Liu
Virginia Tech, Blacksburg, USA
e-mail: lqvt@vt.edu

M. Meneghini et al. (eds.), *Power GaN Devices*,
Power Electronics and Power Systems, DOI 10.1007/978-3-319-43199-4_15

15.1 Switching Characteristics of E-mode and Cascode GaN

Understanding the switching characteristics of GaN switches is essential to use GaN devices in circuit design correctly and efficiently. In the following section, the switching loss mechanism of both e-mode and cascode GaN will first be explained and compared; then, the package impact on switching performance of GaN device will be illustrated in details; a stack-die package is also proposed for cascode GaN devices to minimize the impact of package parasitic inductance on switching performance; finally, the soft-switching operation is compared with hard-switching operation for GaN devices, which shows zero voltage switching (ZVS) is strongly desired to fully exploit the potential of the GaN devices.

15.1.1 Switching Loss Mechanism

In hard-switching converters, the overlap of voltage and current across the drain and source of the device leads to significant power losses during switching events. To better illustrate the switching characteristics of high-voltage GaN switches, typical switching waveforms of the top switch in a buck converter are shown in Fig. 15.1. During the turn-on transition, a large current overshoot is induced by the junction capacitor charge of the bottom switch. The cascode GaN has additional charge due to the reverse recovery of the low-voltage silicon MOSFET in the cascode configuration. The integral of voltage and current during the turn-on transition generates significant power dissipation, which is in the magnitude of tens of μJ. However, the turn-on switching loss is much lower than with the Si MOSFET, which has the same breakdown voltage and same on-resistance.

On the other hand, the cross-time of the drain–source voltage and current during the turnoff transition is quite short and the energy dissipation is less than a few μJ. The major reason for such a small loss is due to the high transconductance of GaN devices. In addition, the cascode GaN has a smaller turnoff loss at higher current conditions due to the intrinsic current–source driving mechanism [1, 2]. The cascode structure minimizes the Miller effect during the turnoff transition; therefore, the loss is not sensitive to the turnoff current.

The switching energy of the 600 V e-mode GaN and the cascode GaN, which has the same on-resistance, is shown in Fig. 15.1c, d. Generally speaking, the turn-on loss is much higher than the turnoff loss. The e-mode GaN has a smaller turn-on switching loss, which takes advantage of the lack of reverse recovery charge. However, the cascode GaN has a smaller turnoff switching loss at a higher turnoff current due to its structural benefits.

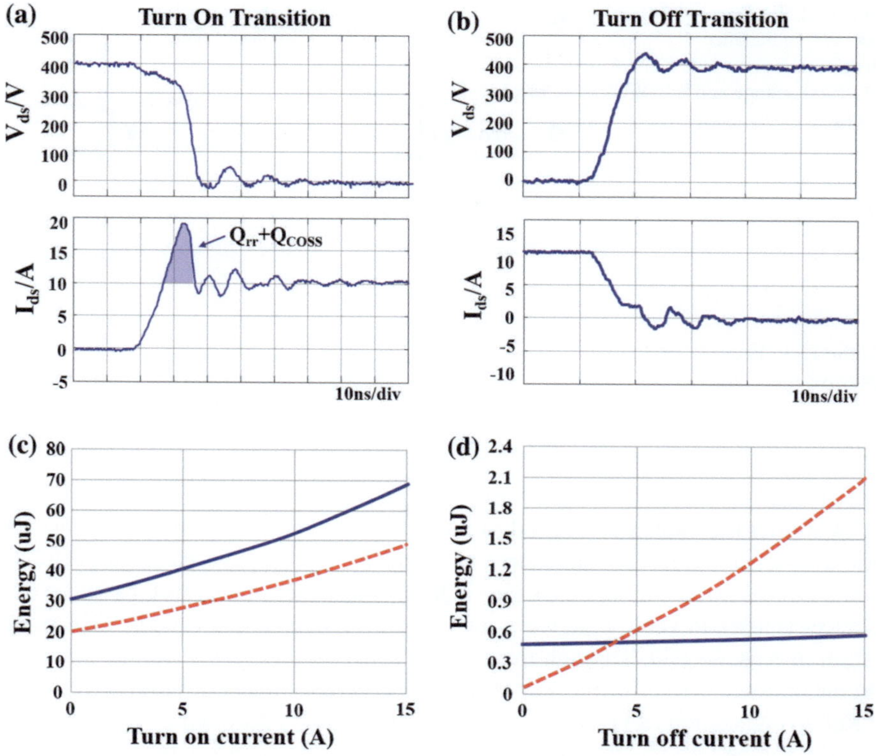

Fig. 15.1 Switching characteristics of high-voltage GaN switches: **a** Turn-on transition waveforms, **b** turnoff transition waveforms, **c** turn-on switching energy (*solid line* Cascode GaN; *dashed line* e-mode GaN), and **d** turnoff switching energy (*solid line* Cascode GaN; *dashed line* e-mode GaN)

15.1.2 Packaging Influence

Both e-mode GaN devices and cascode GaN devices are able to switch very fast. However, parasitic inductance introduced by the bulky package becomes a limit which results in large switching loss and severe oscillations during switching transitions.

Regarding packaging influence, much effort has been spent on studying the Si MOSFET with monolithic structure [3–5]. It is well-documented that the common source inductance (CSI), which is defined as the inductance shared by the power loop and driving loop, is the most critical parasitic source. The CSI acts as negative feedback to slow down the driver during the turn-on and turnoff transitions and thus prolongs the voltage and current crossover time and significantly increases the switching loss.

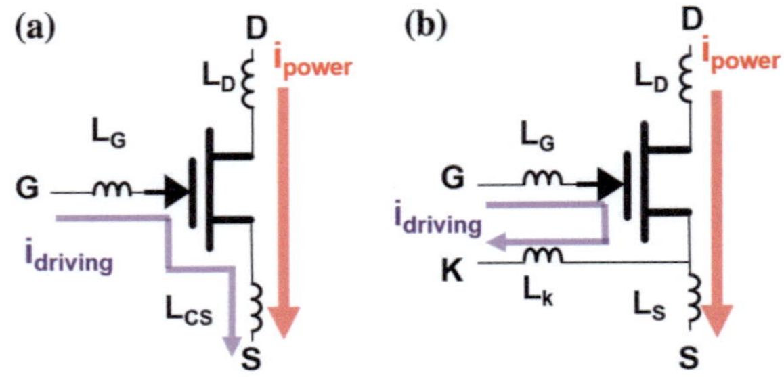

Fig. 15.2 Package-related parasitic inductance distribution of e-mode GaN: **a** TO 220 package and **b** PQFN package

Table 15.1 Package-related parasitic inductance values

	L_G (nH)	L_D (nH)	L_S (nH)	L_K
TO-220	3.6	2.3	3.9	N/A
PQFN	2.4	1.3	0.9	1.3 nH

A similar theory can be applied to e-mode GaN devices. Figure 15.2 shows the package-related parasitic inductance of a typical through-hole package (TO-220) and a typical surface mount package (PQFN), while their inductance based on real devices is listed in Table 15.1. Here, the inductance value is extracted by FEA simulation in Ansoft Q3D. Through this comparison, it can be observed that a surface mount package is an effective way to reduce the value of parasitic inductance, while a Kelvin connection (K) is able to decouple the power loop from the driving loop so that the source inductance is no longer the common source inductance.

Unlike e-mode GaN devices, the cascode GaN device has very complex parasitic inductance distribution and the identification of the CSI is not straightforward [6, 7]. As proposed by [6, 7], the first step is to analyze the CSI of the low-voltage Si MOSFET and of the high-voltage GaN HEMT separately. In terms of the Si MOSFET (Fig. 15.3a), L_{int3} and L_S are the CSIs of the Si MOSFET; while in terms of the GaN HEMT (Fig. 15.3b), L_{int3} and L_{int1} are the CSIs of the GaN HEMT. Therefore, in terms of the cascode GaN device (Fig. 15.3c), since L_{int3} is the CSI for both the GaN HEMT and the Si MOSFET, it is the most critical parasitic inductance. L_{int1} is the second most critical inductance, since it is the CSI of the high-voltage GaN HEMT, which has the most significant switching loss. Finally, L_S is the third most critical inductance.

A simulation model is used to verify this analysis. A circuit-level simulation is conducted based on a buck converter design working in continuous-current-mode hard-switching conditions. The switching loss in the hard-switching condition is the primary concern. Since the turn-on switching loss is the dominant loss, the turn-on energy at 400 V/10 A conditions is chosen for comparison. The simulated turn-on switching loss is shown in Fig. 15.4. Based on the TO-220 package, the total turn-on energy is 23.6 μJ at 400 V/10 A, which is set as the benchmark. By eliminating the impact of L_D, L_{int2}, or L_G, the reduction of the turn-on switching loss is negligible (around 1% of 23.6 μJ). On the other hand, when the critical parasitic inductance such as L_S, L_{int1}, or L_{int3} is removed, the turn-on switching loss is reduced significantly

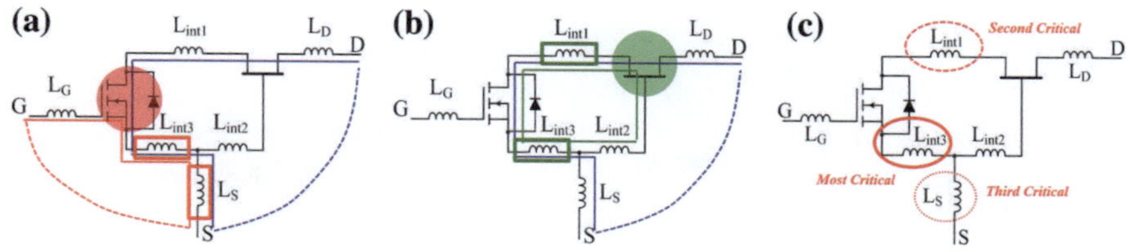

Fig. 15.3 Common-source inductance distribution in cascode GaN device: **a** in a Si device, **b** in a GaN device, and **c** for cascode structure

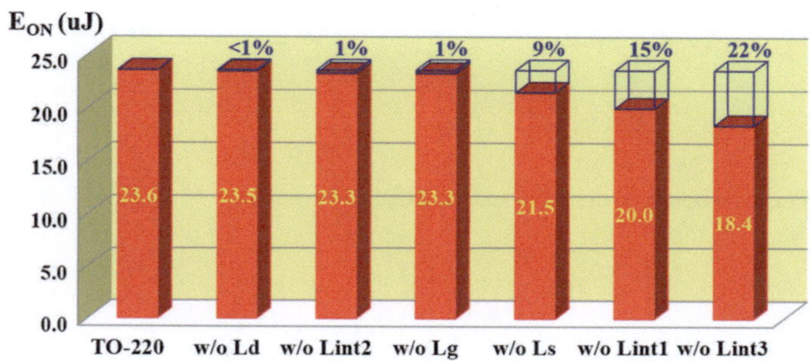

Fig. 15.4 Device turn-on loss comparison at 400 V/10 A

(9, 15, and 22 % of 23.6 μJ, respectively). So the simulation results quantitatively verify the theoretical analysis that the L_{int3}, L_{int1}, and L_S are the most critical package parasitic inductances.

According to the analysis, the traditional package has significant effects on the device switching performance. The stack-die package is proposed to solve the package-related issues [5, 8–10].

For the TO-220 package, there are three common source inductances: L_{int1}, L_{int3}, and L_S. However, the stack-die package is able to eliminate all common source inductances. As shown in Fig. 15.5b, in the stack-die package of the cascode GaN device, the Si MOSFET (drain pad) is mounted directly on the top of the GaN HEMT (source pad). The interconnection between the two dies is minimized in this way; thus, the stack-die package is considered to be the optimized package for the cascode GaN device.

The stack-die package is built and tested in the same setup as the TO-220 package. Experimental results show that the stack-die package makes significant improvement in hard switching.

Figure 15.6 shows the turn-on waveforms under 400 V/5 A conditions. In accordance with the previous analysis, the stack-die package has a faster switching transition speed compared to the TO-220 package. The d*v*/d*t* is increased from 90 to 130 V/ns, while the main transition time and turn-on switching energy are

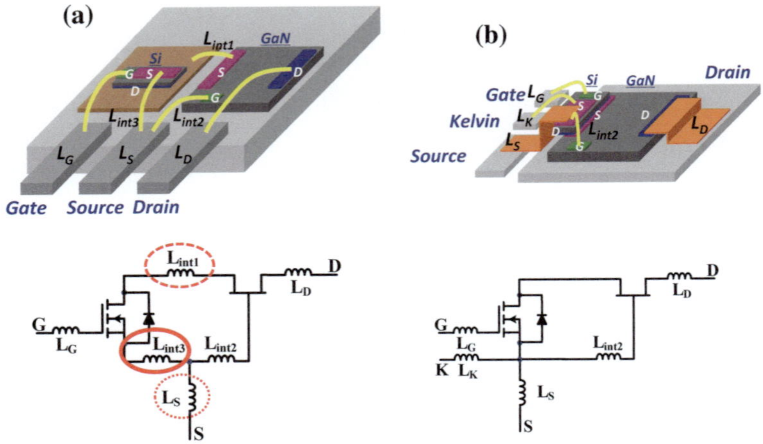

Fig. 15.5 Package bonding diagram and schematic for cascode GaN HEMT: **a** TO-220 and **b** stack-die

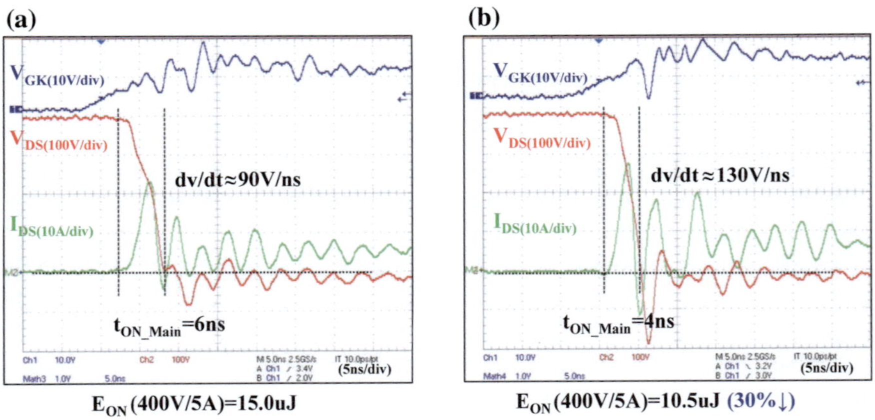

Fig. 15.6 Turn-on waveform comparison between **a** TO-220 package and **b** stack-die package

reduced from 6 ns and 15.0 μJ to 4 ns and 10.5 μJ, respectively. Thirty percent of the turn-on energy is saved at this case.

Figure 15.7 shows the measured switching energy under different load current conditions. In all load ranges, the turn-on energy of the stack-die package is reduced significantly.

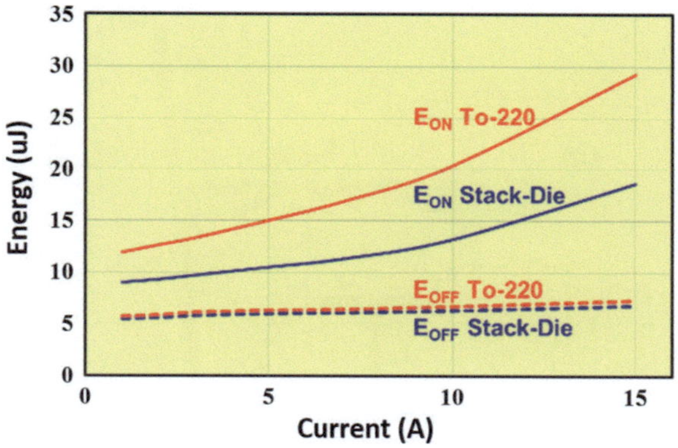

Fig. 15.7 Switching energy for different packages

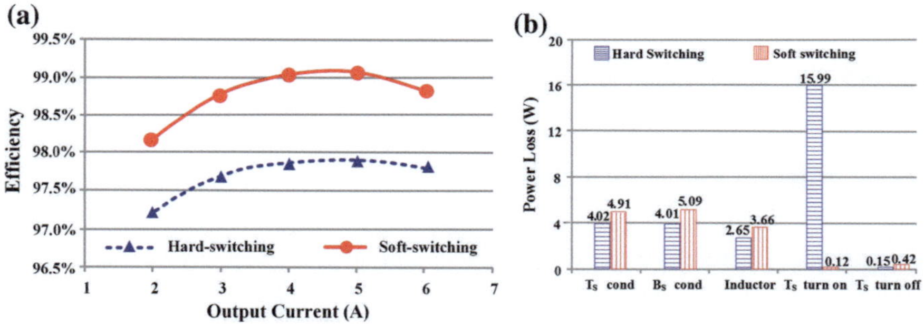

Fig. 15.8 Soft switching versus hard switching **a** efficiency comparison and **b** loss breakdown at 6 A

15.1.3 Comparison Between Hard Switching and Soft Switching

Even with a better package, the turn-on switching loss of a high-voltage GaN switch under hard-switching conditions is significant and dominant in high-frequency applications where the loss may be 10–20 W loss at 500 kHz operation, for example. ZVS turn-on is strongly desired to fully exploit the potential of the GaN switch. Critical conduction mode (CRM) operation is the most simple and effective way to achieve ZVS turn-on and is widely used in medium-power application and low-power application.

Figure 15.8a shows a comparison of the efficiency of soft-switching and hard-switching buck converters with GaN devices. The switching frequency at 6 A output is designed to be 500 kHz. Figure 15.8b shows the loss breakdown with 6 A output. The chart clearly shows that the turn-on loss is minimized with ZVS and introduces only a little more conduction loss. The increase of conduction loss is due to the circulating energy used to achieve ZVS.

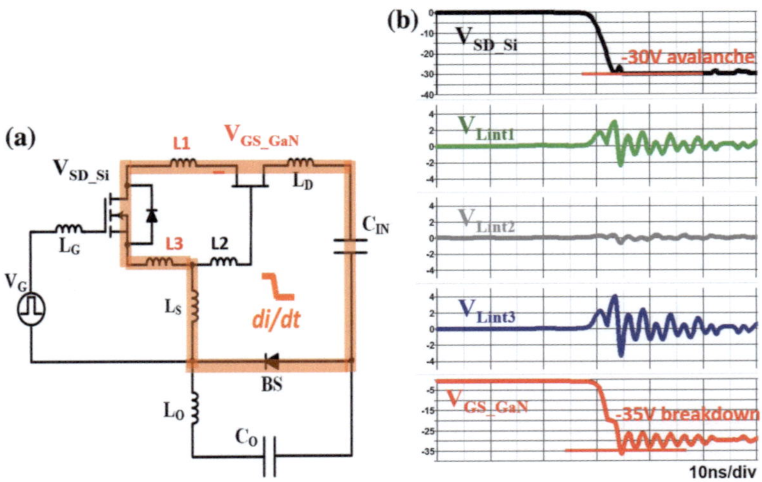

Fig. 15.9 Gate breakdown mechanism: **a** d*i*/d*t* loop on circuit and **b** simulation waveforms

15.2 Special Issues of Cascode GaN

The cascode structure is usually applied to a normally-on GaN device and makes it a normally-off device. However, the interaction between the high-voltage normally-on GaN and the low-voltage Si MOSFET may induce undesired features. The interconnection-related parasitic inductance and junction capacitor ratio between the MOSFET and the normally-on GaN switch should be designed properly to avoid reliability issues and improve switching performance.

15.2.1 Impact of Packaging on Gate Breakdown

As the switching transition of GaN devices is extremely fast, very high d*v*/d*t* (200 V/ns) and d*i*/d*t* (10 A/ns) are observed. One device failure mode is discovered when the cascode GaN device switches under high d*i*/d*t* conditions.

The device failure mechanism is shown in Fig. 15.9. For example, if the drain–source breakdown voltage of a Si MOSFET is 30 V and the gate–source breakdown voltage of a GaN HEMT is −35 V, then the avalanche of the Si MOSFET, which clamps the V_{DS_Si} at 30 V, acts like a last defense to protect against gate breakdown of the GaN HEMT.

The insurance offered by the Si MOSFET fails due to high d*i*/d*t* and package-related parasitic inductance. During the turnoff transition, voltage spikes and parasitic ringing are observed due to the *L*d*i*/d*t* mechanism, which leads to lower than −35 V undershoot on V_{GS_GaN} and potential device degradation and failure.

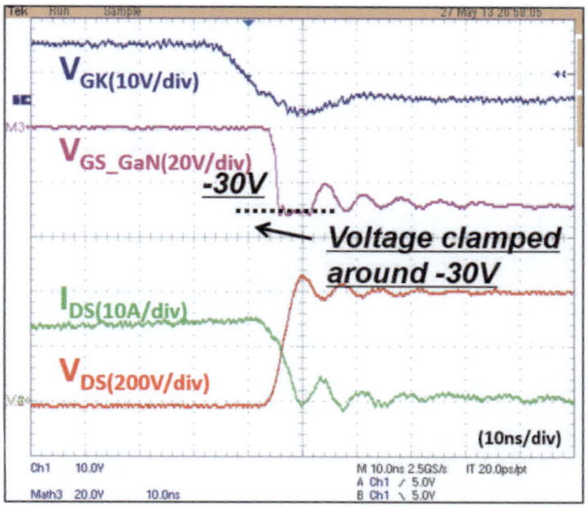

Fig. 15.10 Turnoff waveform of proposed cascode GaN stack-die package

The proposed stack-die package is able to solve this issue [7]. As L_1 and L_3 are totally eliminated, there is almost no d*i*-/d*t*-induced voltage spike in the internal GaN driving loop; thus, the V_{GS_GaN} follows—V_{DS_Si} closely, or in other words, the Si avalanche protection becomes very effective. Figure 15.10 shows the measured turnoff waveforms of a stack-die package under 400 V/15 A conditions. The purple curve shows that V_{GS_GaN} is clamped at −30 V, while no voltage undershoot is observed.

15.2.2 Impact of Capacitor Mismatch

It quite often occurs that the junction capacitor charge of the high-voltage GaN is higher than that of the low-voltage Si MOSFET in the cascode GaN devices. A series of consequences may occur when the charge of the GaN and Si MOSFET is mismatched. The Si MOSFET may reach avalanche at the turnoff transition, which would cause additional loss and reliability concerns. The GaN switch cannot achieve ZVS turn-on even when the external waveform of the cascode GaN looks like ZVS [11, 12]. The worst occasion is that the cascode GaN may be subject to divergent oscillation under high-current turnoff conditions.

15.2.2.1 Si Avalanche

The voltage distribution between the GaN switch and the Si MOSFET during the turnoff transition is largely determined by the junction capacitor charge. The key waveforms during the turnoff transition and the equivalent circuits are shown in Fig. 15.11.

The Si MOSFET is completely turned off at t_1. The drain–source voltage of the Si MOSFET, which is also the gate–source voltage of the GaN with inverse polarity,

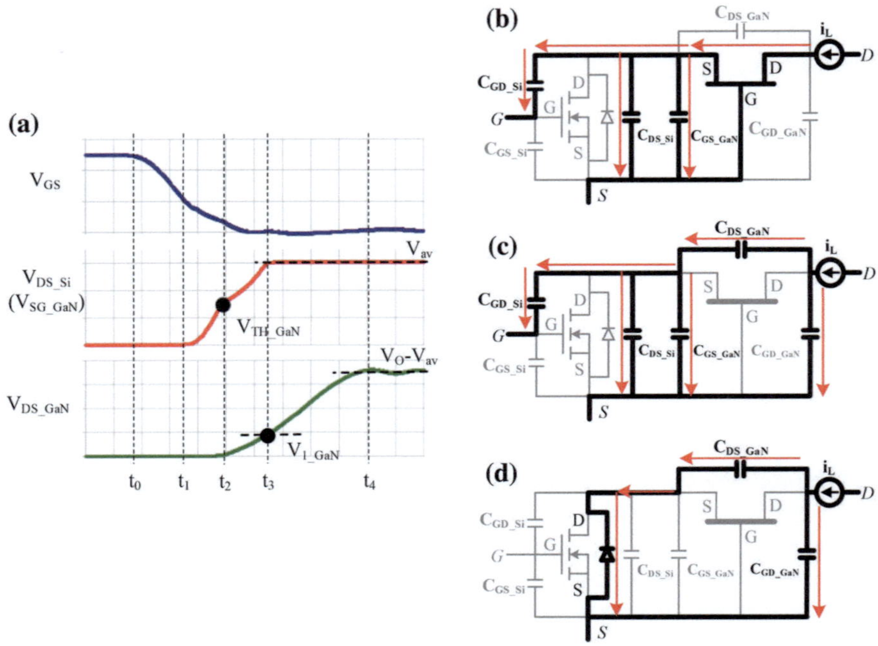

Fig. 15.11 Voltage distribution of cascode GaN during turnoff transition: **a** Si MOSFET reach avalanche, **b** stage I: t_1–t_2, **c** stage II: t_2–t_3, and **d** stage III: t_3–t_4

is charged by an external current source through the GaN channel. The GaN channel is pinched off at t_2 when V_{DS_Si} ($-V_{GS\_GaN}$) reaches V_{TH_GaN}. Then, V_{DS_Si} increases simultaneously with V_{DS_GaN}. V_{DS_Si} is driven to avalanche at t_3, while V_{DS_GaN} only rises to V_{1_GaN}, which is quite a bit lower than the steady-state value. During Stage II, the total amount of charge stored in C_{OSS_Si} and C_{GS_GaN} is defined as Q_{II}. There is the same amount of charge stored in C_{DS_GaN}, since they are in series on the current path. During Stage III, C_{DS_GaN} is charged independently through the avalanche path of the Si MOSFET, as shown in Fig. 15.11c, and the V_{DS_GaN} rises from V_{1_GaN} to the steady-state value. The total amount of charge stored in C_{DS_GaN} during Stage III is defined as Q_{III}. This amount of charge flows through the avalanche path, and it causes additional loss in every switching cycle, which can be calculated as follows:

$$P_{av} = V_{av} \times Q_{III} \times f_s \tag{15.1}$$

According to (15.1), P_{av} is proportional to the switching frequency. This is undesirable, especially in high-frequency applications.

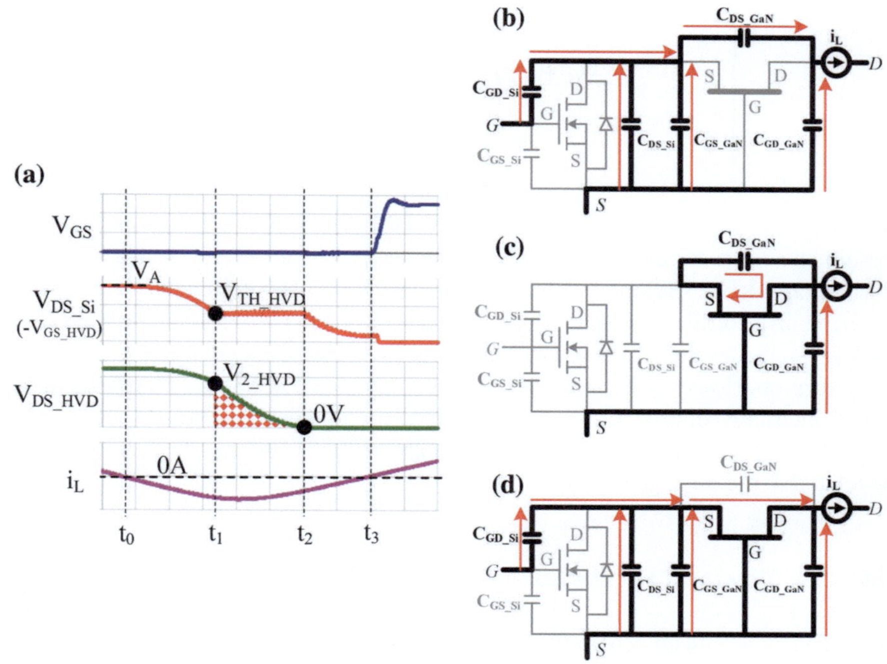

Fig. 15.12 Voltage distribution of cascode GaN during ZVS turn-on transition: **a** GaN internal turn-on, **b** stage I: t_0–t_1, **c** stage II: t_1–t_2, and **d** stage III: t_2–t_3

15.2.2.2 Failure to Achieve ZVS

Figure 15.12 shows the key waveforms during the ZVS turn-on transition and the equivalent circuits. The negative inductor current is used to discharge the junction capacitors to achieve ZVS.

At t_0, C_{DS_GaN} together with C_{OSS_Si} and C_{GS_GaN} is discharged by negative inductor current. Since the charge stored in the GaN switch is much larger than the charge stored in the Si MOSFET, V_{DS_GaN} only decreases to V_{2_GaN} when V_{DS_Si} decreases to V_{TH_GaN} at t_1. The channel of the GaN switch is conductive during Stage II, as shown in Fig. 15.12b. The remaining charge of C_{DS_GaN}, which is Q_{III}, is dissipated through the channel directly, and this induces additional turn-on loss that is proportional to the switching frequency. During Stage II, the voltage decrease slopes of C_{DS_GaN} and C_{GD_GaN} are consistent. The majority of the inductor current flows through C_{GD_GaN}, and the circuit satisfies the following equations:

$$\begin{cases} v_{DS\_GaN} + v_{DS\_Si} = v_{GD\_GaN} \\ C_{GD\_GaN} \dfrac{dv_{GD\_GaN}}{dt} = i_L \\ C_{DS\_GaN} \dfrac{dv_{DS\_GaN}}{dt} = g_{f\_GaN}\left(-v_{DS\_Si} - V_{TH\_GaN}\right) \end{cases} \quad (15.2)$$

where g_{f_GaN} is the transconductance of the GaN switch. A small decrease in V_{DS_Si} results in a large increase of the GaN switch displacement current, which leads to a fast voltage decrease slope. Therefore, V_{DS_Si} stays almost constant during this stage to maintain a consistent voltage slope, which is shown in Fig. 15.12a. The waveform of V_{DS_GaN} makes the terminal waveform of the cascode GaN device appear to have ZVS turn-on. However, the majority of the energy stored in C_{DS_GaN} is actually dissipated internally due to a mismatch in charge. This phenomenon always occurs, regardless of what kind of ZVS techniques are applied. The internal switching loss is related to the mismatch charge.

15.2.2.3 Divergent Oscillation

The capacitance mismatch may cause the cascode GaN to have divergent oscillation issue under high-current turnoff conditions. The voltage ringing formed by the loop inductance and junction capacitors may exceed the threshold to trigger the GaN's internal turn-on mechanism and therefore leads to divergent oscillation.

The turnoff process is the same as that shown in Fig. 15.10. V_{DS_Si} reaches avalanche, while V_{DS_GaN} only increases to V_{1_GaN}. Then, the Si MOSFET stays in the avalanche region, and V_{DS_GaN} is charged up to the peak value V_{peak} through the avalanche path.

After V_{DS_GaN} reaches V_{peak}, the turnoff transition period is over, and the junction capacitance of the cascode GaN device oscillates with loop inductance L_P. The initial voltages of V_{DS_GaN} and V_{DS_Si} are V_{peak} and V_{av}, respectively. The initial oscillation current of L_P is 0 A. The ideal oscillation waveforms of V_{DS_GaN} and i_{Lp} without considering the damping effect are the dashed curves shown in Fig. 15.13. The ideal oscillation amplitude is V_{peak}. The detailed operation can be described as follows:

[t_0–t_1]: L_P resonates with two capacitance branches. One is C_{DS_GaN}, which is in series with C_{OSS_Si} and C_{GS_GaN}. The other capacitance branch is C_{GD_GaN}. The equivalent circuit is shown in Fig. 15.14a. C_{DS_GaN} is discharged in series with C_{OSS_Si} and C_{GS_GaN} by part of the loop inductance current i_{LP} during this stage. At t_1, V_{GS_GaN} reaches V_{TH_GaN}, while V_{DS_GaN} drops ΔV, which is smaller than the ideal resonant peak–peak amplitude. The charge removed by i_{Lp} during this period is defined as ΔQ.

[t_1–t_2]: After t_1, GaN is internally turned on and C_{DS_GaN} is bypassed by the GaN channel directly. Therefore, the loop inductance L_p only resonates with C_{GD_GaN}, and the equivalent circuit is shown in Fig. 15.14b. The resonant period is reduced due to the smaller capacitance. Moreover, the voltage drop of V_{GD_GaN} is greater than it is

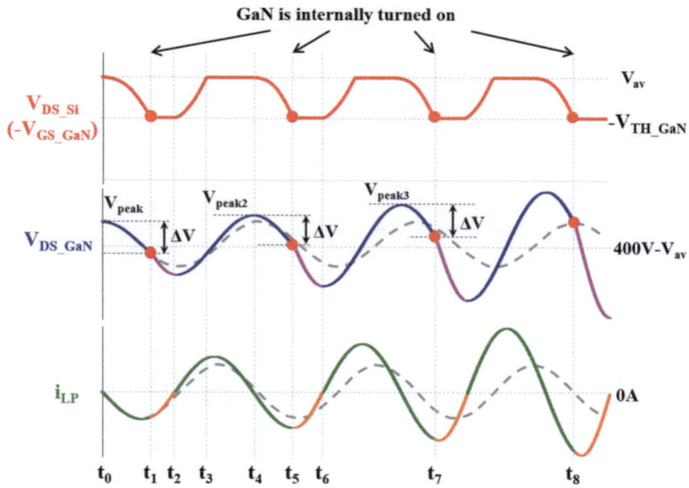

Fig. 15.13 Illustration of divergent oscillation

with ideal oscillation, since C_{GD_GaN} is discharged by i_{Lp}. V_{GS_GaN} remains almost constant during this stage for the reason described in Sect. 15.2.2.2. Although C_{DS_GaN} does not participate in oscillation during this stage, V_{DS_GaN} and V_{GD_GaN} still satisfy the KVL law. $V_{DS\_GaN} = V_{GD\_GaN} - V_{TH\_GaN}$. Therefore, the voltage drop of V_{DS_GaN} is also larger than the ideal case. At t_2, i_{Lp} reaches 0A, and V_{DS_GaN} reaches the valley point.

[t_2–t_3]: After t_2, the next oscillation period starts. The values of V_{DS_Si}, $V_{DS\_GaN,}$ and i_{Lp} at t_2 become the initial conditions of the next oscillation cycle. It should be noted that the initial condition of V_{DS_GaN} is lower than the ideal case. V_{DS_Si} reaches avalanche at t_3, and the charge stored in C_{DS_GaN} during this period is ΔQ, which is the same as the charge removed during t_0–t_1.

[t_3–t_4]: V_{ds_Si} stays in the avalanche region, and V_{DS_GaN} continues to increase until it reaches the peak value V_{peak2}. V_{peak2} should be larger than V_{peak} due to the initial conditions. Similarly, the resonant current is also greater than in the ideal case.

[t_4–t_5]: This stage is similar to the stage [t_0–t_1]. V_{DS_Si} and V_{DS_GaN} decrease simultaneously, and V_{DS_Si} reaches V_{TH_GaN} when ΔQ is removed by resonant current at t_5. GaN is internally turned on earlier than in stage [t_0–t_1] since V_{peak2} is higher than V_{peak}. This allows more resonant current to discharge C_{GD_GaN} in the next stage, and therefore, V_{DS_GaN} drops to an even lower value. As a result, in the following oscillation periods, the GaN is internally turned on in every cycle and each turn-on instant is earlier than in the previous cycle. The ringing amplitudes of V_{DS_GaN} and i_{Lp} increase with each cycle, and the oscillation eventually becomes divergent.

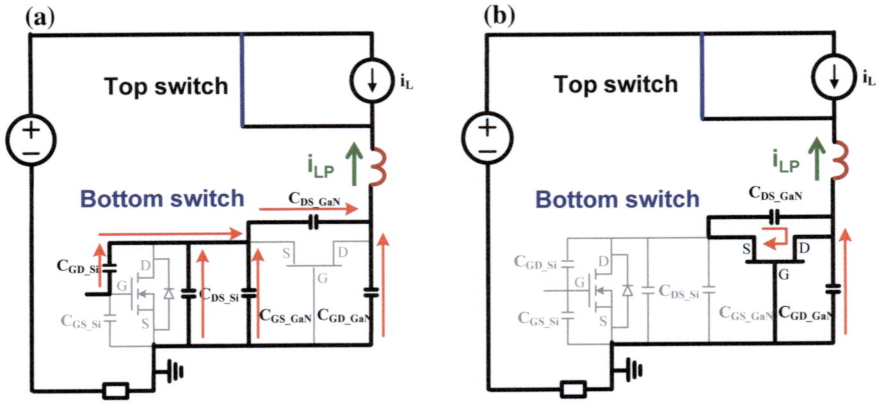

Fig. 15.14 Equivalent circuits **a** t_0–t_1 equivalent circuit and **b** t_1–t_2 equivalent circuit

15.2.2.4 Solution to Solve Capacitor Mismatch Issue

The largest contributing factor to the three issues detailed above is the capacitance mismatch between the Si MOSFET and GaN in the cascode configuration. A straightforward method to match the capacitance is to select an appropriate Si MOSFET with a larger junction capacitance according to the junction capacitance of the GaN switch. However, doing this means the total gate charge of the Si MOSFET will also increase, which will significantly increase the driving loss at high frequencies. Moreover, the increase of C_{GD_Si} will elongate the Si MOSFET turnoff transition and increase switching loss due to a strong Miller effect.

Based on the analysis in Section II, the total mismatch charge in the cascode device is Q_{III}. Therefore, an additional capacitance C_X is added in parallel with the drain–source terminals of the Si MOSFET to compensate the charge mismatch, as shown in Fig. 15.15. The required minimum value of C_X should guarantee that C_{DS_GaN} achieves its steady-state voltage before the Si MOSFET reaches avalanche. Therefore, the expression of C_X is as follows:

$$C_X \geq \frac{Q_{III}}{V_{av} - V_{TH\_GaN}} \tag{15.3}$$

Paralleling C_X between the drain–source terminals of the Si MOSFET will not increase its driving loss, and the turnoff loss is still very small due to the merits of the cascode structure, as mentioned in [1, 2].

The issues discussed in Sect. 15.2.2 may not occur under low-voltage conditions, such as applications that operate below 200 V. The proposed solution may slow down the device switching speed, but with very limited increase of the turnoff switching loss. Generally speaking, the proposed solution is targeted to solving the

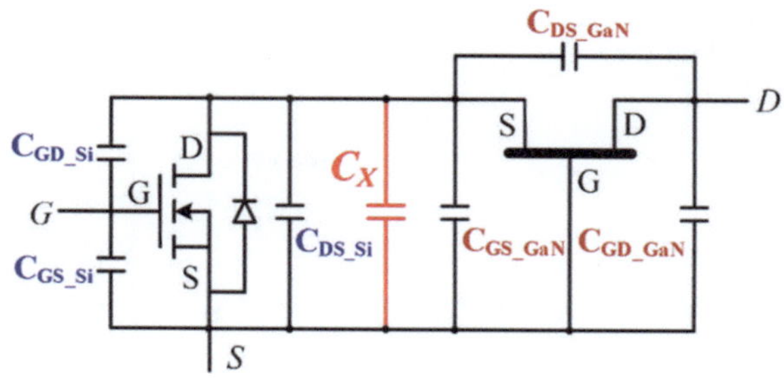

Fig. 15.15 Adding an extra capacitor in cascode device

issues for 600–1200 V-rated cascode devices, which are generally used in 400–800 V applications.

15.3 Gate Driver Design for GaN Device

GaN devices have a much lower gate charge and lower output capacitance than Si MOSFETs and, therefore, are capable of operating at a switching frequency ten times greater. In the meantime, the switching speed, in terms of d*v*/d*t* and d*i*/d*t*, is 3–5 times higher than that of Si MOSFETs [13]. These conditions need to be understood well and tackled properly in order to design circuits to fully utilize GaN devices.

15.3.1 The di/dt *Issue*

By using a GaN device in MHz applications, especially when CRM is used to achieve soft-switching, high-current turnoff can induce high d*v*/d*t* and d*i*/d*t* issues.

When the GaN switch is turned off, the falling d*i*/d*t* slope induces negative voltage on the CSI. This negative voltage will induce an opposing voltage across the gate–source of the GaN, which is intended to turn-on the device.

Figure 15.16 shows an example of a buck converter with cascode GaN switches. The total common source inductance L_S consists of the common source inductance in the device package and the parasitic inductance of the PCB trace. When the high-side switch is turned off, the high d*i*/d*t* will induce inverse voltage on L_S. A 7 V spike appears on the terminal of the top switch gate signal, as shown in Fig. 15.17. This spike may spur a false turn-on of the device if the internal gate signal reaches its threshold voltage and causes shoot through.

The best way to improve the d*i*/d*t* immunity is to minimize CSI by improving packaging and the PCB layout. Separating the gate and power loops with a Kelvin connection is helpful to reduce CSI. The internal source inductance of the GaN device also should be minimized. Figure 15.18 shows the experimental waveforms

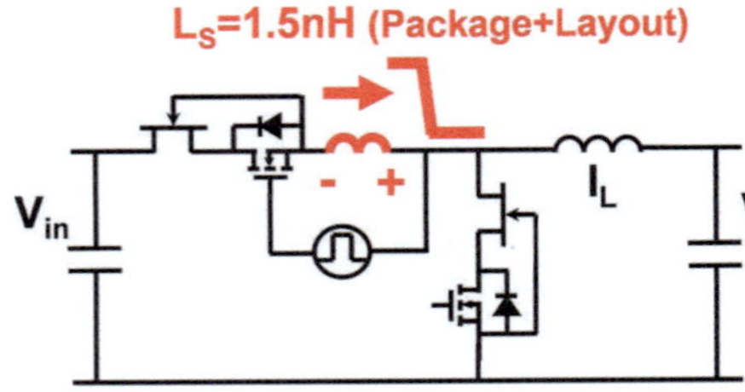

Fig. 15.16 d*i*/d*t* impact on gate drive loop

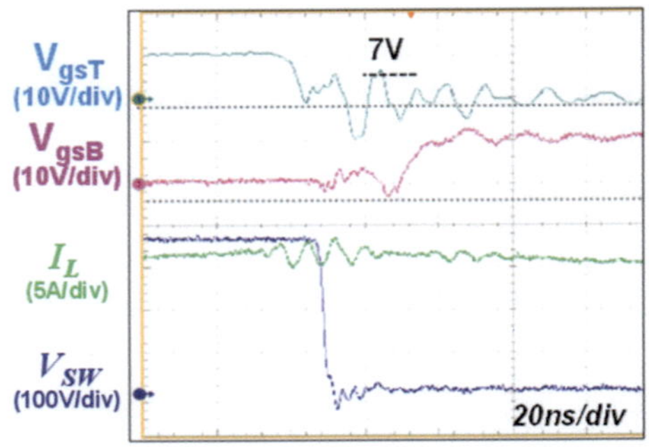

Fig. 15.17 High d*i*/d*t* induces voltage spike across CSI

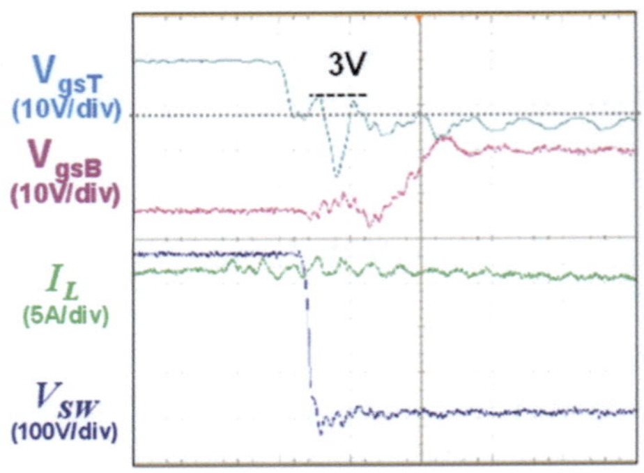

Fig. 15.18 Smaller voltage spike with less CSI

with better packaging and layout. The voltage spike is much more significantly suppressed than in Fig. 15.17.

15.3.2 The d*v*/d*t* *Issue*

The d*v*/d*t*-related driving issue is more complicated and difficult than the d*i*/d*t* problem. In most cases, a level shifter or isolator is used for the high-side driver. Parasitic capacitance of this part is a high-frequency noise path for the common-mode current generated by the switch node d*v*/d*t*. For a positive d*v*/d*t* event, the

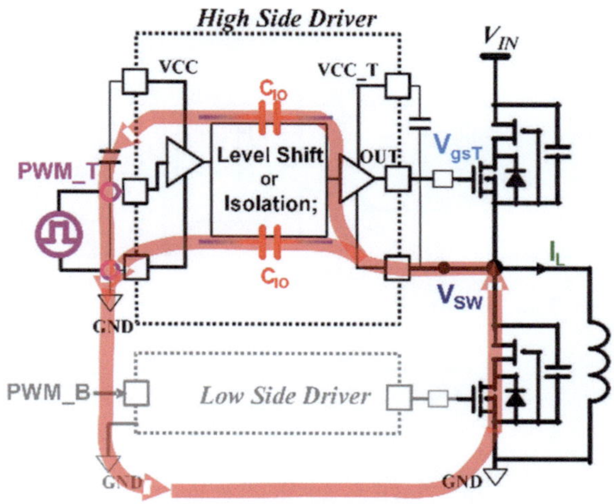

Fig. 15.19 High d*v*/d*t* causes common-mode current across parasitic capacitance of level shifter or isolator (cascode GaN is used as an example)

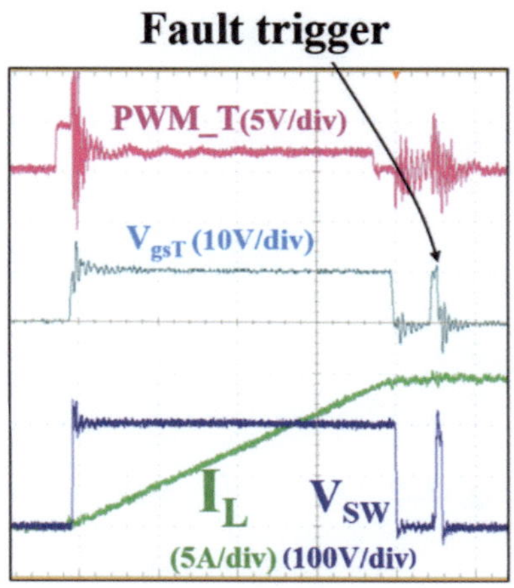

Fig. 15.20 High d*v*/d*t* deteriorates the PWM signal

high-voltage slew rate across capacitor C_{IO} generates the common-mode current which flows in the loops, as shown in Fig. 15.19. This common-mode current causes ground bounce on the PWM input side and can cause changes in the logic state. For a negative d*v*/d*t* event, the common-mode current flows clockwise and can deteriorate the PWM signal, as shown in Fig. 15.20.

Generally speaking, a bootstrap IC, optocoupler, driving transformer, and digital isolator are typical candidates for the high-side driver. Among these four candidates, the digital isolator has the lowest parasitic capacitance and smallest propagation delay. These features make it suitable for GaN application. The power for the high-side digital isolator and high-speed driver usually comes from an isolated power supply module. However, the isolated power supply is bulky and more importantly susceptible to noise due to relative large capacitance. A bootstrap diode can be used

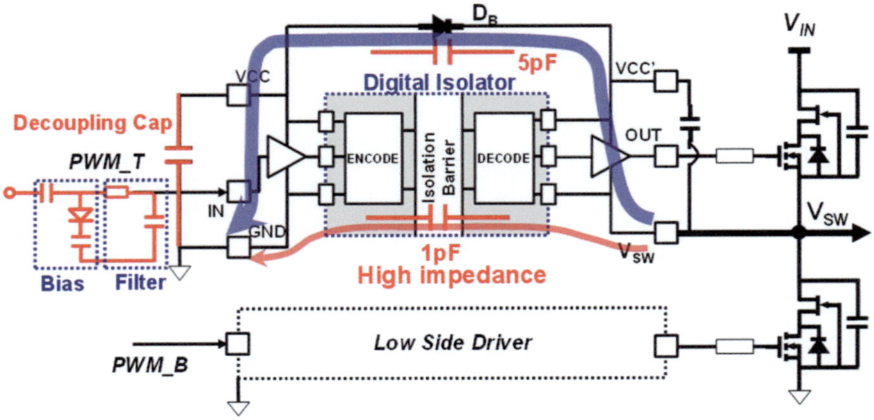

Fig. 15.21 Simplified high-side driving circuit

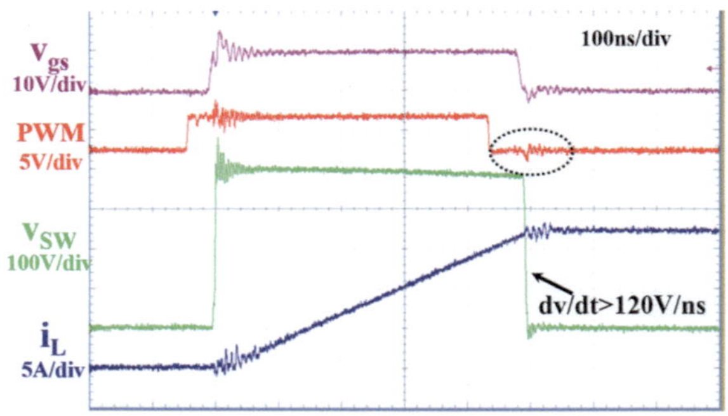

Fig. 15.22 Experimental waveform of improved d*v*/d*t* immunity for high-side driver

to simplify the power supply for the high side of the digital isolator and driver, as shown in Fig. 15.21. The decoupling capacitor should be put as close to bootstrap diode and d*v*/d*t* noise source ground as possible to minimize the common-mode current loop. To further improve the d*v*/d*t* immunity, a negative bias circuit and RC filter can be used at the input PWM terminal.

The PCB layout also plays an important role in improving the d*v*/d*t* immunity for the high-side driver. Avoiding PCB layout overlap between the ground and the high side can effectively reduce the parasitic capacitance. The experimental waveforms resulting from the improved driving circuit and PCB layout are shown in Fig. 15.22. The voltage ringing on the PWM signal is minimized, and the d*v*/d*t* immunity is over 120 V/ns.

15.4 System-Level Impact

Performance improvement and size reduction are the key drivers and metrics for the advancement of power conversion technology. Improving the semiconductor device is the first step to meeting the requirements of modern systems. The emerging GaN device, with much improved figures of merit, opens the door for operating frequencies well into the MHz range. A number of design examples are illustrated in this chapter and other publications that show impressive efficiency and power density improvements. Moreover, the potential impact of GaN goes beyond the simple measures of efficiency and power density. It is feasible to design a system with a more integrated approach at higher frequencies, and therefore, it is easier for automated manufacturing. This will bring significant cost reductions in power electronics equipment and unearth numerous new applications which have been previously precluded due to high cost.

15.4.1 3D Integrated Point-of-Load Converter

Non-isolated DC/DC converters are widely used in many different applications, such as telecommunications, computing, portable electronics, and automobiles. These non-isolated DC/DC converters usually are located right next to the load, so they are also called POL converters. These POL converters are normally constructed with discrete components and operate at relatively low frequencies. As a result, passive components like inductors are bulky, and they occupy considerable space on the motherboard. Because of the power demands for POL converters and the limited real estate of the motherboard, POL converters must be made significantly smaller than in the past. To achieve these goals, two things have to happen simultaneously: One is a significant increase in the switching frequency to reduce the size and weight of the inductors and capacitors, and the second is to integrate passive components, especially magnetic components, with the active components to realize the needed power density. The POL converters normally have 5 or 12 V input voltage. Therefore, low-voltage (12–30 V) Si power MOSFETs are usually used to build these POL converters. Due to the limitations of Si power MOSFETs, industry only can achieve high-frequency (>2 MHz) operation for very low-power POL converters, i.e., those in the 1–5 W level. In order to address the intense thirst from the computing and telecommunication industries for high-power-density POL converters, we also need to increase the switching frequency of high-power (>20 W) POL converters to above 1 MHz. However, it is very difficult to push the switching frequency above 1 MHz for a high-power POL converter with today's Si power MOSFET due to excessive switching loss.

Among the different methods for integrating power converters, three-dimensional (3D) integration, which uses magnetic component as a substrate to achieve a very high-power density, is one of the more promising methods [14, 15]. A conceptual

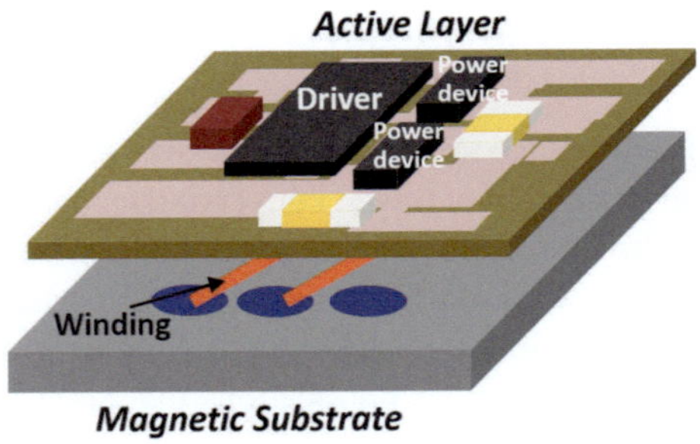

Fig. 15.23 Three-dimensional integration concept

diagram of this 3D integration is shown in Fig. 15.23. The basic concept of 3D integration is to integrate the whole converter in a vertical way. First, very low-profile passive layers are built as a substrate for the whole converter, and the active layers are built above the passive layers to realize footprint saving and full space utilization.

Recently, by using a GaN HEMT, several 3D integrated POL converters have been successfully demonstrated by the Center for Power Electronics Systems (CPES) with power densities as high as 1000 W/in.^{3} [16–18].

Figure 15.24a shows a prototype of a 2 MHz 3D integrated POL converter with EPC's GaN devices. Figure 15.24b shows the testing results for efficiency. The third-generation prototype has a design with a lateral high-frequency power loop, while the fourth-generation prototype has an improved design with a vertical high-frequency power loop. It can be seen that the active layer design is very critical to improving converter efficiency. By using a shielding layer to reduce parasitic inductance, the third-generation prototype is able to become 1–2 % more efficient. The fourth-generation prototype has a vertical high-frequency power loop, which has very small parasitic inductance. Therefore, the fourth-generation prototype without a shielding layer can have similar full-load efficiency, but even better light-load efficiency than the third-generation prototype with a shielding layer.

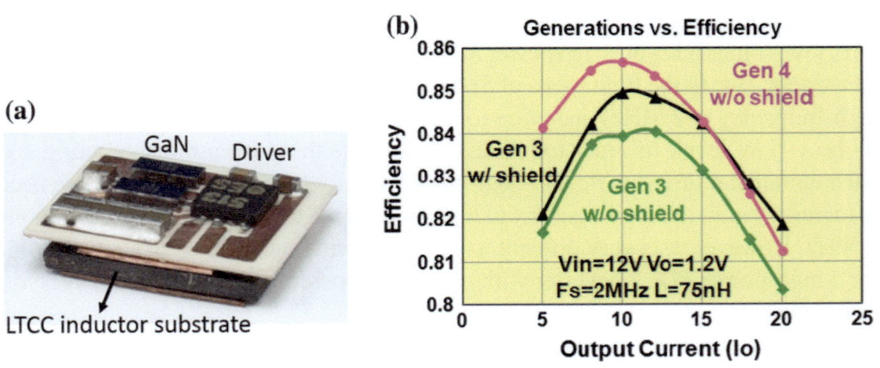

Fig. 15.24 3D integrated POL converters with EPC GaN devices: **a** prototype of 3D integrated POL converter and **b** test results of converter efficiency

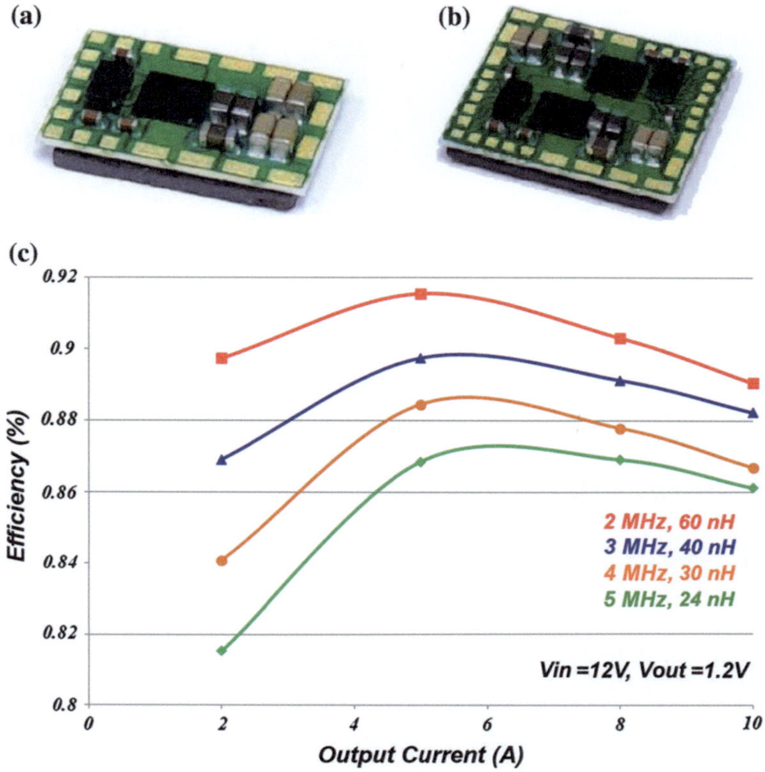

Fig. 15.25 Three-dimensional integrated POL with IR GaN devices: **a** single-phase version; **b** two-phase version; and **c** efficiency of single-phase version

Figure 15.25 shows two prototypes of 5 MHz 3D integrated POL converters with IR's GaN devices. Figure 15.25a is single-phase version with a 10 A output current; Fig. 15.25b shows a two-phase version with a 20 A output current. Figure 15.25c shows the efficiency of the single-phase POL converter with different frequencies. At 2 MHz, this integrated POL converter has over 91 % peak efficiency; at 5 MHz, its peak efficiency is still as high as 87 %. The total converter footprint of the single-phase version is only 85 mm^2. For the single-phase converter, a series of LTCC inductor substrates are designed for 1–5 MHz operation frequencies. These inductors have the same footprint, but have different core thicknesses to achieve the designed inductance. At a 5 MHz operation frequency, the inductor core thickness is only 0.9 mm. With this ultra-thin inductor substrate, the power density of the 3D integrated POL converter is as high as 800 W/in.^{3} at 5 MHz. The two-phase module is built with an inverse-coupled inductor.

The inverse-coupled inductor can have a higher inductance density than a non-coupled inductor due to the DC flux canceling effect. Thus, the core thickness of an inverse-coupled inductor is thinner than that of a non-coupled inductor. At 5 MHz,

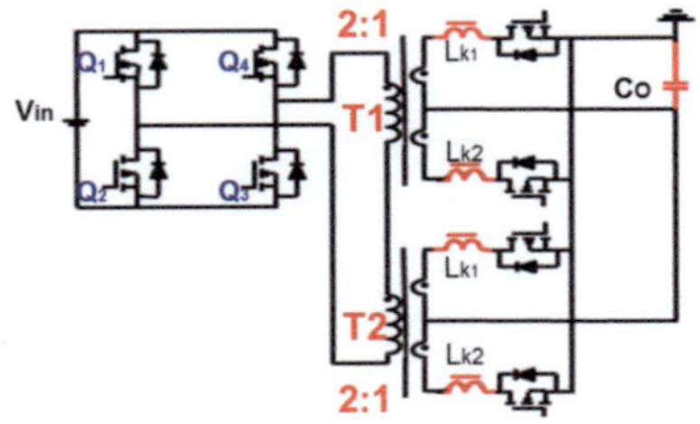

Fig. 15.26 High-frequency quasi-resonant converter

the inverse-coupled inductor has a core thickness of only 0.4 mm. As a result, the two-phase POL converter has higher power density than the single-phase POL converter. At 5 MHz, It achieves power density as high as 1000 W/in.^{3}.

15.4.2 Isolated DC/DC Converter

15.4.2.1 48–12 V DCX

Intermediate bus architecture (IBA) is commonly used in CPU and telecom applications. In IBA, the first stage takes a nominal 48 V input and steps it down to a range of 8–12 V. This allows for small, high-efficiency POL converters to regulate the second-stage voltage to the final output.

The drawbacks of the state-of-the-art IBA design include being limited to use with low frequencies (a couple hundred kilohertz); bulky passive components; and a large leakage inductance of the transformer, which results in duty cycle loss, body diode loss, and large transient voltage spikes.

An improved high-frequency topology is proposed in [19], as shown in Fig. 15.26. This topology can reduce switching loss and resolve ringing issues by using a resonant technique that utilizes the transformer's magnetizing inductance and resonance of the leakage inductance with a small capacitance to achieve ZVS, limit the turnoff current, and eliminate body diode conduction. The improved topology allows for high efficiency at high switching frequencies.

Gallium nitride transistors further facilitate high-frequency operation of the bus converter. Thus, the high-frequency transformer design becomes the bottleneck to achieve higher efficiency. Traditional transformer design suffers from high leakage inductance and high conduction loss due to eddy currents and the proximity effect and is not suitable for MHz-level high-frequency operation.

The proposed high-frequency transformer design uses distributed transformers, which reduces resistance and leakage inductance. A magnetic core integration method utilizing flux cancelation is also proposed to further reduce the core loss introduced by distributed transformers.

A 1.6 MHz GaN converter with $V_{in} = 48$ V, $V_o = 12$ V, $I_o = 30$ A, and using the proposed integrated transformer, is shown in Fig. 15.27 [20]. A power density of

Fig. 15.27 48–12 V, 1.6 MHz isolated buck converter with GaN transistors

900 W/in.[3] is achieved, which is double that of the state-of-the-art Si-based design. The corresponding efficiency is shown in Fig. 15.28.

15.4.2.2 400–12 V DCX

A 400–12 V LLC resonant converter is built to demonstrate the high-frequency capability of the GaN device and its impact on circuit design, as shown in Fig. 15.29. A 600 V GaN device is used on the primary side, and the frequency is pushed to 1 MHz. Therefore, it is possible to use a matrix transformer and PCB winding [21].

Instead of a traditional single-core structure with litz wire winding, the proposed design has four transformers integrated into two cores and PCB winding. The turns ratio of each transformer is only 4:1, which makes the winding structure much simpler than in the traditional structure. Furthermore, the current sharing is much better because, instead of parallel SRs, the proposed design parallels the secondary-side windings.

The matrix transformer can help to reduce leakage inductance and the ac resistance of the windings so that the flux cancelation method can then be utilized to reduce core size and loss. Synchronous rectifier (SR) devices and output capacitors are

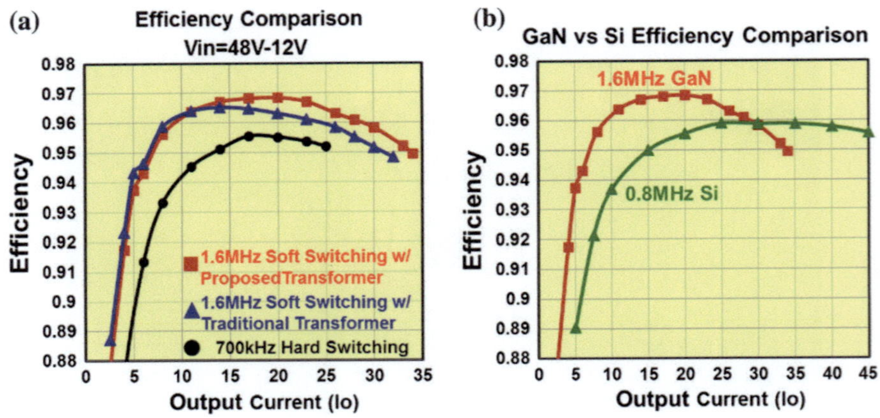

Fig. 15.28 Efficiency curves showing **a** hard-switching/soft-switching comparison and **b** GaN transistors versus Si MOSFETs

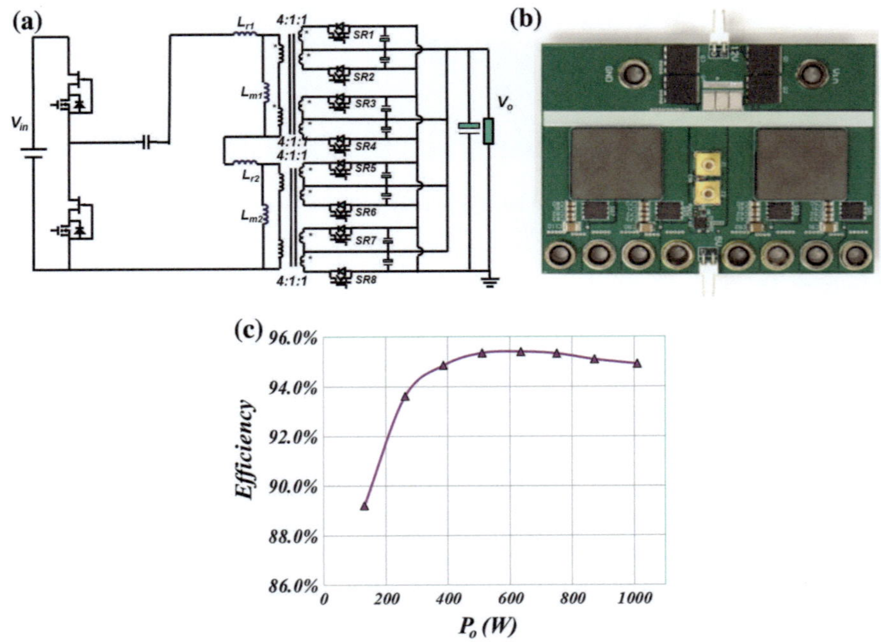

Fig. 15.29 Topology, prototype, and efficiency of 1 kW 400/12 V LLC resonant converter with matrix transformer: **a** circuit diagram, **b** prototype, and **c** measured efficiency

integrated into the secondary windings to eliminate termination-related winding losses and via loss and to reduce leakage inductance. All transformer windings use only a four-layer PCB board. No litz wire is applied.

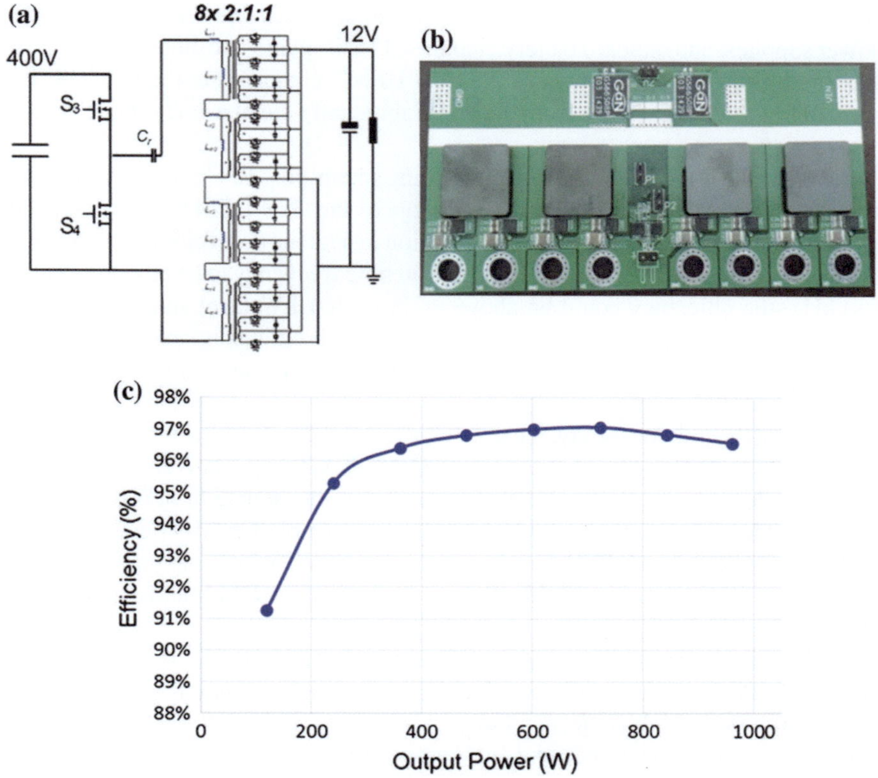

Fig. 15.30 A further improved design of 1 kW LLC with matrix transformer: **a** circuit diagram, **b** prototype, and **c** measured efficiency

Combining all of these benefits, the peak efficiency of the converter using the proposed design is 95.5 % and the power density achieved is 700 W/in.^{3}, which is 5–10 times higher than state-of-the-art industry practice. At the same time, it has a very low-profile and automatic manufacturing capability.

Figure 15.30 shows a further improved design of this 1 kW LLC converter. With eight transformers integrated in four cores, an e-mode GaN device on the primary side, and improved magnetic design, 97.1 % peak efficiency is achieved with the same 700 W/in.^{3} power density.

15.4.3 MHz Totem-Pole PFC Rectifier

With the advent of 600 V gallium nitride (GaN) power semiconductor devices, the totem-pole bridgeless power factor correction (PFC) rectifier [22, 23], which was a nearly abandoned topology, has suddenly become a popular front-end candidate for applications such as two-stage high-end adaptors, server and telecommunication power supplies, and onboard battery chargers. This is mostly attributed to the significant performance improvement of the GaN HEMT compared to the Si MOSFET, particularly the better figure-of-merit and significantly smaller body diode reverse recovery effect.

The design of a GaN-based hard-switching totem-pole PFC rectifier is demonstrated in [24]. As the reverse recovery charge of the GaN HEMT is much smaller than the Si MOSFET, hard-switching operation in a totem-pole bridge configuration turned out to be practical. By limiting the switching frequency to be around or below 100 kHz, the efficiency could be above 98 % for a 1 kW-level single-phase PFC rectifier. Even though the simple topology and high efficiency are attractive, the system-level benefit is limited because the switching frequency is still similar to that of a Si-based PFC rectifier.

Based on the previous study, it can be concluded that soft switching truly benefits the cascode GaN HEMT. As the cascode GaN HEMT has relatively high turn-on loss and extremely small turnoff loss due to the current–source turnoff mechanism, critical-mode (CRM) operation with ZVS is very suitable for GaN device. A MHz GaN-based CRM boost PFC rectifier is demonstrated which has the significant system benefits as the volume of the boost inductor and the DM filter is dramatically reduced [25, 26].

With a similar system-level vision, the cascode GaN HEMT is applied in the totem-pole PFC rectifier while pushing the operating frequency to above 1 MHz. Several important high-frequency issues, which used to be less significant at low frequencies, are emphasized, and the corresponding solutions are proposed and experimentally verified [27]. They include ZVS extension in order to eliminate the switching loss caused by non-ZVS valley switching, variable on-time control to

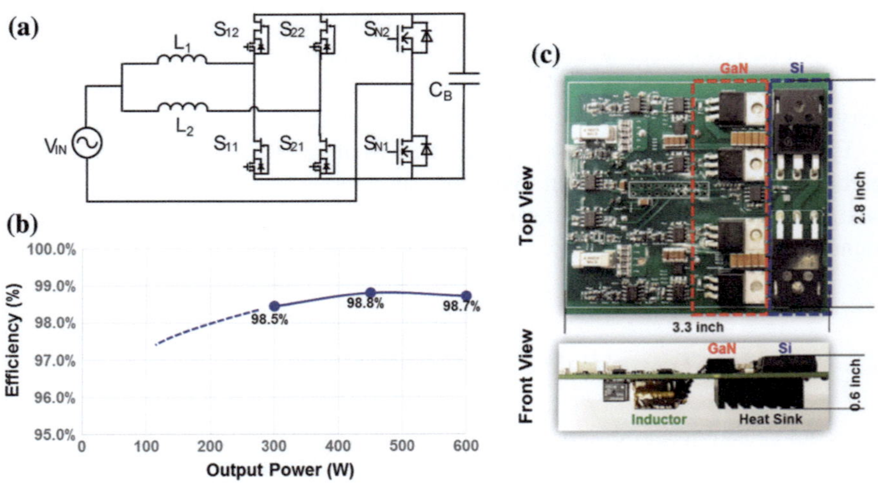

Fig. 15.31 GaN-based MHz totem-pole PFC **a** topology, **b** measured efficiency, and **c** prototype

improve the power factor, particularly the zero-crossing distortion caused by traditional constant on-time control, and interleaving control for input current ripple cancelation.

A 1.2 kW dual-phase interleaved MHz totem-pole PFC rectifier is built with 99 % peak efficiency and 200 W/in.^{3} power density, as shown in Fig. 15.31 [27]. In this rectifier, the inductor is 80 % smaller than used in the state-of-the-art industrial practice [28].

The volume of the DM filter is reduced significantly by pushing the frequency to several MHz and using a two-phase interleaving configuration [29]. At operating frequencies from 100 kHz to 1 MHz, a simple one-stage filter is sufficient to suppress the noise so that the DM filter size is reduced by 50 %. Then, moving

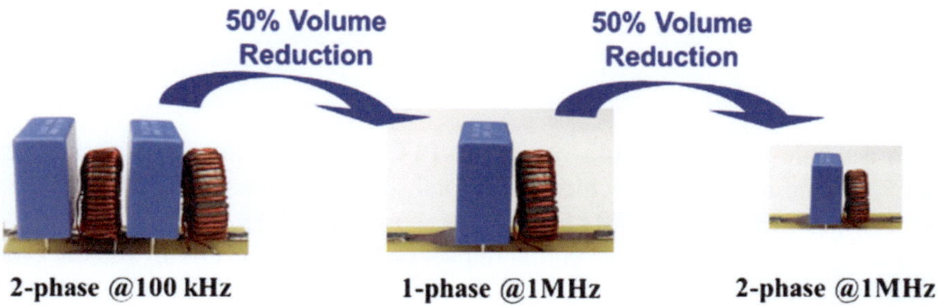

Fig. 15.32 DM filter size comparison

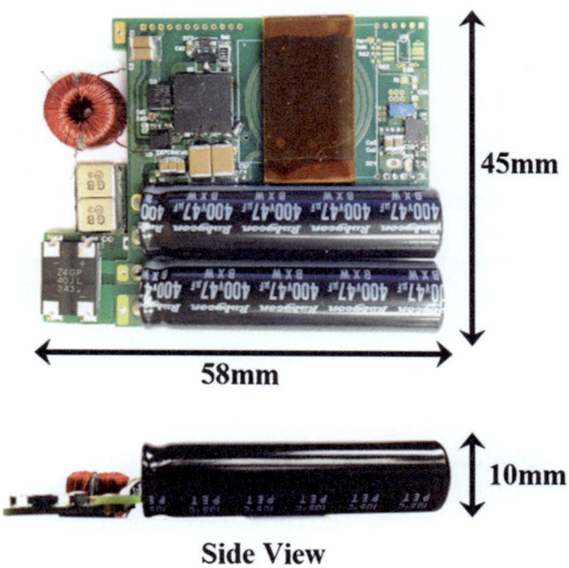

Fig. 15.33 Prototype of MHz flyback converter

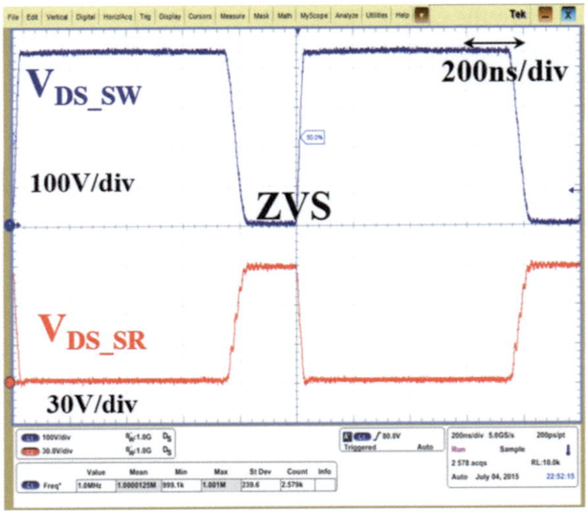

Fig. 15.34 ZVS achievement of active clamp flyback converter

from a single-phase PFC to a two-phase PFC with good interleaving, another 50 % volume reduction is achieved. In total, the DM filter of a 1 MHz PFC is only one-quarter that of a 100 kHz PFC, as shown in Fig. 15.32. Hence, the high-frequency operation has a substantial impact on both the PFC circuit and the EMI filter.

15.4.4 High-Density Wall Adapter

The adapter is strongly driven by efficiency and power density for all forms of portable electronics. Most adapters only operate at relatively low frequencies (<100 kHz), with state-of-the-art efficiency up to 91.5 %. However, low-frequency operation limits the adapter power density to 6–9 W/in.^{3}. The emerging GaN device is deemed a game-changing device in this particular application, with improved efficiency and significant size reduction.

Flyback converters are the dominant topology for low-power adapter application due to their simplicity and low cost. With an active clamp circuit, the leakage energy can be recycled, and the voltage ringing is minimized. Moreover, ZVS for both the main switch and clamping switch can be realized by proper design as well. The traditional flyback transformer is handmade, which is an intensive, labor-involved manufacturing process. The manufacturing cost is a concern, and the parameter variation is another circuit design issue. A PCB winding-based transformer is only feasible when the switching frequency is over several hundred kHz due to its capacity for fewer turns and a smaller core size. The leakage inductance and parasitic capacitance of the transformer can be well controlled by PCB manufacturers. Moreover, shielding can be easily integrated in the PCB winding to reduce the CM noise [30].

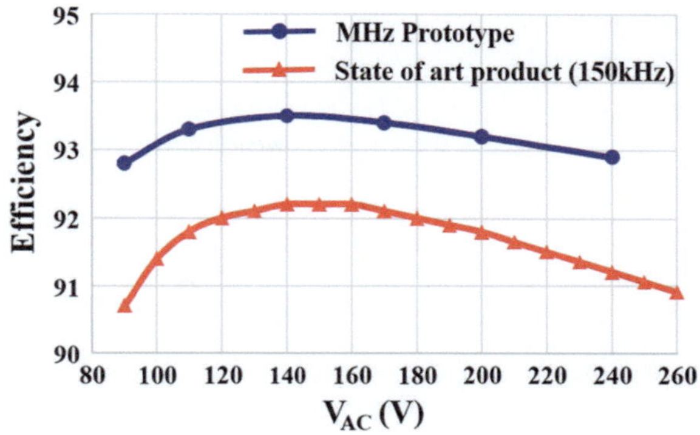

Fig. 15.35 Efficiency comparison of MHz prototype and state-of-the-art product

Figure 15.33 shows the prototype of a MHz active clamp flyback front-end converter. Figures 15.34 and 15.35 show the key experimental data. The size of the flyback transformer, EMI filter, and output filter is significantly reduced, with 10 times higher switching frequency than industry practice. The power density excluding the case is over 40 W/in.^{3}, which is two times higher than the state-of-the-art product. ZVS is achieved for a wide input voltage range, and there is no voltage spike during the main switch-off period, which is beneficial for EMI noise. The measured full-load efficiency over a wide input range is 1–2 % higher than the state-of-the-art product.

15.5 Summary

GaN devices offer superior performance to Si MOSFETs, but several issues still have to be address in order to better understand and utilize GaN devices in circuit design. The switching characteristics of high-voltage GaN devices are evaluated, and the analysis indicates soft switching is desired for high-frequency applications. The cascode GaN has special issues due to its two-device structure. Better packaging and balanced capacitance can solve the issues and fully explore the potential of the cascode GaN. Some design considerations for high-speed GaN devices are addressed to tackle high d*v*/d*t* and d*i*/d*t* issues. Several system design examples are presented, including GaN-based 5 MHz POL, 1.6 MHz 48–12 V DCX, 1 MHz 400–12 V LLC DCX, 1 MHz two-phase PFC, and 1 MHz flyback wall adapter.

All of these design examples show the impact of GaN devices on power electronics goes far beyond efficiency and power density improvement. It changes the method of system design with much higher switching frequencies. Even though GaN is still in an early stage of development, it is presumably a game-changing device with a scale of impact yet to be defined.

References

1. Huang X, Liu Z, Li Q, Lee FC (2014) Evaluation and application of 600 V GaN HEMT in cascode structure. IEEE Trans Power Electron 29(5):2453–2461
2. Huang X, Li Q, Liu Z, Lee FC (2014) Analytical loss model of high voltage GaN HEMT in cascode configuration. IEEE Trans Power Electron 29(5):2208–2219
3. Ren Y, Xu M, Zhou J, Lee FC (2006) Analytical loss model of power MOSFET. IEEE Trans Power Electron 21(2):310–319
4. Yang B, Zhang J (2005) Effect and utilization of common source inductance in synchronous rectification. In Proceedings of IEEE APEC, pp 1407–1411
5. Jauregui D, Wang B, Chen R (2011) Power loss calculation with common source inductance consideration for synchronous buck converters. TI application note. Available online www.t i.com
6. Liu Z, Huang X, Lee FC, Li Q (2014) Package parasitic inductance extraction and simulation model development for the high-voltage cascode GaN HEMT. IEEE Trans Power Electron 29(4):1977–1985
7. Liu Z, Huang X, Zhang W, Lee FC, Li Q (2014) Evaluation of high-voltage cascode GaN HEMT in different packages. In: Proceedings of IEEE APEC, pp 168–173
8. Zhang W, Huang X, Liu Z, Lee FC, She S, Du W, Li Q. A new package of high-voltage cascode gallium nitride device for megahertz operation. IEEE Trans Power Electron (early access)
9. She S, Zhang W, Huang X, Du W, Liu Z, Lee FC, Li Q. Thermal analysis and improvement of cascode GaN HEMT in stack-die structure. In: Proceedings of IEEE ECCE, pp 5709–5715
10. She S, Zhang W, Liu Z, Lee FC, Huang X, Du W, Li Q (2015) Thermal analysis and improvement of cascode GaN device package for totem-pole bridgeless PFC rectifier. Appl Therm Eng
11. Huang X, Liu Z, Lee FC, Li Q (2015) Characterization and enhancement of high-voltage cascode GaN devices. IEEE Trans Electron Devices 62(2):270–277
12. Huang X, Du W, Lee FC, Li Q, Liu Z. Avoiding Si MOSFET avalanche and achieving zero-voltage-switching for cascode GaN devices. IEEE Trans Power Electron (early access)
13. Zhang W, Huang X, Lee FC, Li Q (2014) Gate drive design considerations for high voltage cascode GaN HEMT. In: Proceedings of IEEE APEC, pp 1484–1489
14. Li Q, Lee FC (2009) High inductance density low-profile inductor structure for integrated point-of-load converter. In: 2009 IEEE applied power electronics conference and exposition (APEC), Washington, District of Columbia, 15–19 Feb 2009, pp 1011–1017
15. Li Q, Dong Y, Lee FC, Gilham D (2013) High-density low profile coupled inductor design for integrated point-of-load converters. IEEE Trans Power Electron 28(1):547–554
16. Reusch D, Lee FC, Gilham D, Su Y (2012) Optimization of a high density gallium nitride based non-isolated point of load module. In: Proceedings of IEEE ECCE, pp 2914–2920
17. Ji S, Reusch D, Lee FC (2013) High frequency high power density 3D integrated gallium-nitride-based point of load module design. IEEE Trans Power Electron 28(9):4216–4226
18. Su Y, Li Q, Lee FC (2013) Design and evaluation of a high-frequency LTCC inductor substrate for a three-dimensional integrated DC/DC converter. Special issue: "power supply on chip. IEEE Trans Power Electron 28(9):4354–4364
19. Ren Y, Xu M, Sun J, Lee FC (2005) A family of high power density unregulated bus converters. IEEE Trans Power Electron 20(5):1045–1054
20. Reusch D, Lee FC (2012) High frequency isolated bus converter with gallium nitride transistors and integrated transformer. In: 2012 IEEE energy conversion congress and exposition (ECCE), pp 3895, 3902
21. Huang D, Ji A, Lee FC (2014) LLC resonant converter with matrix transformer. IEEE Trans Power Electron 29(8):4339–4347
22. Su B, Zhang J, Lu Z (2011) Totem-pole boost bridgeless PFC rectifier with simple zero-current detection and full-range ZVS operating at the boundary of DCM/CCM. IEEE Trans Power Electron 26(2):427–435

23. Marxgut C, Krismer F, Bortis D, Kolar JW (2014) Ultraflat interleaved triangular current mode (TCM) single-phase PFC rectifier. IEEE Trans Power Electron 29(2):873–882
24. Zhou L, Wu Y-F, Mishra U (2013) True bridgeless totem-pole PFC based on GaN HEMTs. PCIM Europe 2013, pp 1017–1022
25. Liu Z, Huang X, Mu M, Yang Y, Lee FC, Li Q (2014) Design and evaluation of GaN-based dual-phase interleaved MHz critical mode PFC converter. In Proceedings of IEEE ECCE, pp 611–616
26. Yang Y, Liu Z, Lee FC, Li Q (2014) Analysis and filter design of differential mode EMI noise for GaN-based interleaved MHz critical mode PFC converter. In: Proceedings of IEEE ECCE, pp 4784–4789
27. Liu Z, Lee FC, Li Q, Yang Y (2015) Design of GaN-based MHz totem-pole PFC rectifier. In: Proceedings of IEEE ECCE
28. Mu M, Lee FC (2014) Comparison and optimization of high frequency inductors for critical model GaN converter operating at 1 MHz. In: 2014 international power electronics and application conference and exposition (PEAC), pp 1363–1368
29. Yang Y, Mu M, Liu Z, Lee FC, Li Q (2015) Common mode EMI reduction technique for interleaved MHz critical mode PFC converter with coupled inductor. In: Proceedings of IEEE ECCE
30. Yang Y, Huang D, Lee FC, Li Q (2013) Transformer shielding technique for common mode noise reduction in isolated converters. In: Proceedings of IEEE ECCE, pp 4149–4153

Author index

M. Meneghini et al. (eds.), *Power GaN Devices*,
Power Electronics and Power Systems, DOI 10.1007/978-3-319-43199-4

Subject index

M. Meneghini et al. (eds.), *Power GaN Devices*,
Power Electronics and Power Systems, DOI 10.1007/978-3-319-43199-4

MIX
Papier aus verantwortungsvollen Quellen
Paper from responsible sources
FSC® C105338

If you have any concerns about our products,
you can contact us on
ProductSafety@springernature.com

In case Publisher is established outside the EU,
the EU authorized representative is:
Springer Nature Customer Service Center GmbH
Europaplatz 3, 69115 Heidelberg, Germany

Printed by Libri Plureos GmbH
in Hamburg, Germany